Infrared and Raman Characteristic Group Frequencies

Infrared and Raman Characteristic Group Frequencies

Tables and Charts

Third Edition

GEORGE SOCRATES

Formerly of Brunel, The University of West London, Middlesex, UK

JOHN WILEY & SONS, LTD

Chichester • New York • Weinheim • Toronto • Brisbane • Singapore

Other Wiley Editorial Offices

John Wiley & Sons Inc., 111 River Street, Hoboken, NJ 07030, USA

Jossey-Bass, 989 Market Street, San Francisco, CA 94103-1741, USA

Wiley-VCH Verlag GmbH, Boschstr. 12, D-69469 Weinheim, Germany

John Wiley & Sons Australia Ltd, 33 Park Road, Milton, Queensland 4064, Australia

John Wiley & Sons (Asia) Pte Ltd, 2 Clementi Loop #02-01, Jin Xing Distripark, Singapore 129809

John Wiley & Sons Canada Ltd, 22 Worcester Road, Etobicoke, Ontario, Canada M9W 1L1

Wiley also publishes its books in a variety of electronic formats. Some content that appears in print may not be available in electronic books.

Library of Congress Cataloguing in Publication Data

Socrates, G. (George)
    Infrared and Raman characteristic group frequencies: tables and charts / George.
    Socrates. – 3rd ed.
        p. cm.
    Rev. ed. of: Infrared and Raman characteristic group frequencies. 2nd ed. c1994.
    Includes bibliographical references and index.
    ISBN 0-471-85298-8
    1. Infrared spectroscopy. 2. Raman spectroscopy .I. Socrates, G. (George). Infrared characteristic group frequencies. II. Title.

    QC457 .S69 2000
    543′.08583 – dc21   00-032096

British Library Cataloguing in Publication Data

A catalogue record for this book is available from the British Library

ISBN 978-0-470-09307-8 (P/B)

Typeset in 10/12pt Times by Laser Words, Madras, India
Printed and bound in Great Britain by CPI Group (UK) Ltd, Croydon, CR0 4YY

C9780470093078_120724

Contents

Contents _____ vii

Contents _____ ix

List of Charts and Figures

List of Tables

Symbols Used

Ar	aromatic	Ph	phenyl
asym	asymmetric	R	alkyl
br	broad	s	strong
comp	compound	sat	saturated
CPDE	cyclopentadienyl	sh	sharp
def	deformation	skel	skeletal
dp	depolarised	str	stretching
EDTA	ethylene diamine tetraacetic acid	sym	symmetric
Et	ethyl	unsat	unsaturated
G	aliphatic or aromatic	v	variable
m	medium	vib	vibration
M	metal atom	vs	very strong
Me	methyl	vw	very weak
oop	out-of-plane	w	weak
p	polarised		

Preface

The purpose of this book is to provide a simple introduction to characteristic group frequencies so as to assist all who may need to interpret or examine infrared and Raman spectra. The characteristic absorptions of functional groups over the entire infrared region, including the far and near regions, are given in tables as well as being discussed and amplified in the text.

A section dealing with spurious bands that may appear in both infrared and Raman spectra has been included in the hope that confusion may be avoid by prior knowledge of the reasons for such bands and the positions at which they may occur.

In order to assist the analyst, three basic infrared correlation charts are provided. Chart 1.4 may be used to deduce the absence of one or more classes of chemical compound by the absence of an absorption band in a given region. Chart 1.5 may be used to determine which groups may possibly be responsible for a band at a given position. Chart 1.6 may be used if the class of chemical is known (and hence the functional groups it contains) in order to determine at a glance the important absorption regions. Chart 1.7 gives the band positions and intensities of functional groups observed when Raman spectroscopy is used. Having identified a functional group as possibly being responsible for an absorption band, by making use of the charts provided, the information in the relevant chapter (or section) and table should both be used to confirm or reject this assumption. If the class of chemical is known then the relevant chapter may be turned to immediately. It may well be that information contained in more than one chapter is required, as, for example, in the case of aromatic amines, for which the chapters on aromatics and on amines should both be referred to. In order to assist the reader, absorptions of related groups may also be dealt with in a given chapter.

Unless otherwise stated, in the text and tables, the comments in the main refer to infrared rather than Raman. Comments specifically aimed at Raman state that this is the case. The reason for this, is that infrared is by far the more commonly used technique.

Throughout the text, tables, and charts, an indication of the absorption intensities is given. Strictly speaking, absorptivity should be quoted. However, there are insufficient data in the literature on the subject and, in any case, the intensity of an absorption of a given functional group may be affected by neighboring atoms or groups as well as by the chemical environment (e.g. solvent, etc.). The values of the characteristic group frequencies are given to the nearest $5\,\mathrm{cm}^{-1}$.

Normally, the figures quoted for the absorption range of a functional group refer to the region over which the maximum of the particular absorption band may be found. In the main, the absorption ranges of functional groups are quoted for the spectra of dilute solutions using an inert solvent. Therefore, if the sample is not in this state, e.g. is examined as a solid, then depending on its nature some allowance in the band position(s) may need to be made.

It is important to realise that the absence of information in a column of a table does not indicate the absence of a band – rather, it suggests the absence of definitive data in the literature.

The near infrared region is discussed briefly in a separate chapter as are the absorptions of inorganic compounds.

The references given at the end of each chapter and in the appendix provide a source of additional information.

The chapter dealing with polymers contains the minimum theory required for the interpretation and understanding of polymer spectra. It deals with the most common types of polymer and also contains a section dealing with plasticisers. A flowchart is also provided to assist those interested in the identification of polymers. The chapter on biological samples molecules covers the most commonly occurring types of biological molecule. The inorganic chapter is reasonably extensive and contains many useful charts.

I wish to thank Dr. K. P. Kyriakou for his encouragement and Isaac Lequedem for his continued presence in my life. There are no words which can adequately express my thanks to my wife, Jeanne, for her assistance throughout the preparation of this book.

G. S.

1 Introduction

Both infrared and Raman spectroscopy are extremely powerful analytical techniques for both qualitative and quantitative analysis. However, neither technique should be used in isolation, since other analytical methods may yield important complementary and/or confirmatory information regarding the sample. Even simple chemical tests and elemental analysis should not be overlooked and techniques such as chromatography, thermal analysis, nuclear magnetic resonance, atomic absorption spectroscopy, mass spectroscopy, ultraviolet and visible spectroscopy, etc., may all result in useful, corroborative, additional information being obtained.

The aim of this book is both to assist those who wish to interpret infrared and/or Raman spectra and to act as a reference source. It is not the intention of this book to deal with the theoretical aspects of vibrational spectroscopy, infrared or Raman, nor to deal with the instrumental aspects or sampling methods for the two techniques. There are already many good books which discuss these aspects in detail. However, it is not possible to deal with the subject of characterisation without some mention of these topics but this will be kept to the minimum possible, consistent with clarity.

Although the technique chosen by an analyst, infrared or Raman, often depends on the task in hand, it should be borne in mind that the two techniques do often complement each other. The use of both techniques may provide confirmation of the presence of particular functional groups or provide additional information.

In recent years, despite the great improvements that have been made in laser Raman spectroscopy, some analysts still consider (wrongly, in my view) that the technique should be reserved for specialist problems, some of their reasons for this view being as follows:

1. Infrared spectrometers are generally available for routine analysis and the technique is very versatile.
2. Raman spectrometers tend to be more expensive than infrared spectrometers and so less commonly available.
3. Until recently, infrared spectrometers, techniques and accessories had improved much faster than those of Raman.

4. There are vast numbers of infrared reference spectra in collections, databases (digital format) and the literature, which can easily be referred to, whereas this is not the case for Raman. Although much better now, the quantity of reference spectra available for Raman simply does not compare with that for infrared.
5. Often, in order to obtain good Raman spectra, a little more skill is required by the instrument operator than is usually the case in infrared. Over the years, both techniques have become more automated and require less operator involvement.
6. Until recently, the acquisition of Raman spectral data has been a relatively slow process.
7. Fluorescence has, in the past, been a major source of difficulty for those using Raman spectroscopy although modern techniques can minimise the effects of this problem.
8. Localised heating, due to the absorption of the radiation used for excitation, may result in numerous problems in Raman spectroscopy – decomposition, phase changes, etc.
9. Quantitative measurements are a little more involved in Raman spectroscopy.
10. With older instruments and certain types of samples, liquids and solids should be free of dust particles to avoid the Raman spectrum being masked by the Tyndall effect.

On the other hand, it should be noted that:

1. In many cases, sample preparation is often simpler for Raman spectroscopy than it is for infrared.
2. Glass cells and aqueous solutions may be used to obtain Raman spectra.
3. It is possible to purchase dual-purpose instruments: infrared/Raman spectrometers. However, dual-purpose instruments do not have available the same high specifications as those using a single technique.
4. The infrared and Raman spectra of a given sample usually differ considerably and hence each technique can provide additional, complementary information regarding the sample.

5. Often bands which are weak or inactive in the infrared, such as those due to the stretching vibrations of C=C, C≡C, C≡N, C–S, S–S, N=N and O–O functional groups, exhibit strong bands in Raman spectra. Also, in Raman spectra, skeletal vibrations often give characteristic bands of medium-to-strong intensity which in infrared spectra are usually weak. Although not always true, as a general rule, bands that are strong in infrared spectra are often weak in Raman spectra. The opposite is also true. (Bands due to the stretching vibrations of symmetrical groups/molecules may be observed by using Raman, i.e. infrared inactive bands may be observed by Raman. The reverse is also true – Raman inactive bands may be observed by using infrared spectroscopy.) For many molecules, Raman activity tends to be a function of the covalent character of bonds and so the Raman spectrum can reveal information about the backbone of the structure of a molecule. On the other hand, strong infrared bands are observed for polar groups.

6. Bands of importance to a particular study may occur in regions where they are overlapped by the bands due to other groups, hence, by making use of the other technique (infrared or Raman) it is often possible to observe the bands of importance in interference-free regions.

7. Raman spectrometers are usually capable of covering lower wavenumbers than infrared spectrometers, for example, Raman spectra may extend down to $100\,cm^{-1}$ or lower whereas most infrared spectra often stop at 400 or $200\,cm^{-1}$.

Separation, or even partial separation, of the individual components of a sample which is a mixture will result in simpler spectra being obtained. This separation may be accomplished by solvent extraction or by chromatographic techniques. Hence, combined techniques such as gas chromatography–mass spectroscopy, GC–MS, liquid chromatography–mass spectroscopy, etc. can be invaluable in the characterisation of samples.

Very early on, workers developing the techniques of infrared spectroscopy noticed that certain aggregates of atoms (functional groups) could be associated with definite characteristic absorptions, i.e. the absorption of infrared radiation for particular functional groups occurs over definite, and easily recognisable, frequency intervals. Hence, analysts may use these characteristic group frequencies to determine which functional groups are present in a sample. The infrared and Raman data given in the correlation tables and charts have been derived empirically over many years by the careful and painstaking work of very many scientists.

The infrared or Raman spectrum of any given substance is interpreted by the use of these known group frequencies and thus it is possible to characterise the substance as one containing a given type of group or groups. Although group frequencies occur within 'narrow' limits, interference or perturbation may cause a shift of the characteristic bands due to (a) the electronegativity of neighbouring groups or atoms, or (b) the spatial geometry of the molecule.

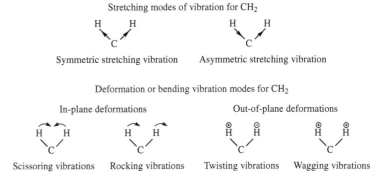

Figure 1.1

Functional groups sometimes have more than one characteristic absorption band associated with them. Two or more functional groups often absorb in the same region and can usually only be distinguished from each other by means of other characteristic infrared bands which occur in non-overlapping regions.

Absorption bands may, in the main, be regarded as having two origins, these being the fundamental vibrations of (a) functional groups, e.g. C=O, C=C, C≡N, –CH₂–, –CH₃, and (b) skeletal groups, i.e. the molecular backbone or skeleton of the molecule, e.g. C–C–C–C. Absorption bands may also be regarded as arising from *stretching* vibrations, i.e. vibrations involving bond-length changes, or *deformation* vibrations, i.e. vibrations involving bond-angle changes of the group. Each of these may, in some cases, be regarded as arising from symmetric or asymmetric vibrations. To illustrate this, the vibrational modes of the methylene group, CH₂ are given in Fig. 1.1. Any atom joined to two other atoms will undergo comparable vibrations, for example, any AX₂ group such as NH₂, NO₂.

The vibration bands due to the stretching of a given functional group occur at higher frequencies than those due to deformation. This is because more energy is required to stretch the group than to deform it due to the bonding force directly opposing the change.

Two other types of absorption band may also be observed: overtone and combination bands. Overtone bands are observed at approximately twice the frequency of strong fundamental absorption bands (overtones of higher order having too low an intensity to be observed). Combination bands result from the combination (addition or subtraction) of two fundamental frequencies.

As mentioned earlier, it is not the intention of this book to deal with the theoretical aspects of vibrational spectroscopy. However, as will be appreciated, some basic knowledge is of benefit. The theoretical aspects which should be borne in mind when using the group frequency approach for characterisation will be mentioned below in an easy, non-rigid and simple manner (there are many good books available dealing with the theory). A linear molecule (one where all the atoms are in a straight line in space, eg. carbon dioxide) consisting of N atoms has $3N - 5$ fundamental vibrations. A non-linear molecule with N atoms has $3N - 6$ fundamental vibrations. These give the maximum number of fundamental vibrations expected but some of these vibrations may be degenerate, i.e. have the same frequency, or be infrared or Raman inactive. In this simple approach, the molecule is considered to be isolated, in other words interactions between molecules and lattice vibrations are ignored. The vibrational frequency of a bond is expected to increase with increase in bond strength and is expected to decrease with increase in mass (strictly speaking reduced mass) of the atoms involved. For example, the stretching frequency increases in the order C–C < C=C < C≡C (triple bonds are stronger than double bonds which in turn are stronger than single bonds) and with regard to mass, the vibrational frequency decreases in the order H–F > H–Cl > H–Br > H–I. It should always be kept in mind that, strictly speaking, molecules vibrate as a whole and to consider separately the vibrations of parts of the molecule (groups of atoms) is a simplification of the true situation.

Many factors may influence the precise frequency of a molecular vibration. Usually it is impossible to isolate the contribution of one effect from another. For example, the frequency of the C=O stretching vibration in CH_3COCH_3 is lower than it is in CH_3COCl. There are several factors which may influence the C=O vibrational frequency: the mass difference between CH_3 and Cl; the associated inductive or mesomeric influence of Cl on the C=O group; the steric effect due to the size of the Cl atom, which affects the bond angle; and a possible coupling interaction between the C=O and C–Cl vibrations. The frequency of a vibration may also be influenced by phase (condensed phase, solution, gas) and may also be affected by the presence of hydrogen bonding.

When the atoms of two bonds are reasonably close to one another in a molecule, vibrational coupling may take place between their fundamental vibrations. For example, an isolated C–H bond has one stretching frequency but the stretching vibrations of the C–H bonds in the methylene group, CH_2, combine to produce two coupled vibrations of different frequencies, asymmetric and symmetric vibrations. Coupling may occur in polyatomic molecules when two vibrations have approximately the same frequency. The result of this coupling is to increase the frequency difference between the two vibrations, (i.e. the frequencies diverge).

Coupling may also occur between a fundamental vibration and the overtone of another vibration (or a combination vibration), this type of coupling being known as Fermi resonance. For example, the CH stretching mode of most aldehydes gives rise to a characteristic doublet in the region 2900–2650 cm^{-1} (3.45–3.77 μm) which is due to Fermi resonance between the fundamental C–H stretching vibration and the first overtone of the in-plane C–H deformation vibration. When the intensities of the two resulting bands are unequal, the stronger band has a greater contribution from the fundamental component than from the overtone (combination) component.

The intensity of an infrared absorption band is dependent on the magnitude of the dipole change during the vibration, the larger the change, the stronger the absorption band. In Raman spectroscopy, it is the change in polarisability which determines the intensity. Hence, if both infrared and Raman spectrometers are available, it is sometimes an advantage to switch from one technique to the other. An example of this is where the infrared spectrum of a sample gives weak bands for certain groups, or their vibrations may be infrared inactive, but, in either case, result in strong bands in the Raman spectra (For example, the C≡C stretching vibration of acetylene is infrared inactive as there is no dipole change whereas a strong band is observed in Raman.) Alternatively, it may be that strong, broad bands in the infrared obscure other bands which could be observed by Raman. Unfortunately, vibrational intensities have, in general, been overlooked or neglected in the analysis of vibrational spectra, infrared or Raman, even when they could provide valuable information.

The intensity of the band due to a particular functional group also depends on how many times (i.e. in how many places) that group occurs in the sample (molecule) being studied, the phase of the sample, the solvent (if any) being employed and on neighbouring atoms/groups. The intensity may also be affected by intramolecular/intermolecular bonding.

The intensities of bands in a spectrum may also be affected due to radiation being optically polarised. In spectral characterisation nowadays, the use of polarised radiation in both infrared and Raman is extensive. When a polarised beam of radiation is incident on a molecule, the induced oscillations are in the same plane as the electric vector of the incident electromagnetic wave so the resultant emitted radiation tends to be polarised in the same plane.

In Raman spectroscopy, the direction of observation of the radiation scattered by the sample is perpendicular to the direction of the incident beam. Polarised Raman spectra may be obtained by using a plane polarised source of electromagnetic radiation (e.g. a polarised laser beam) and placing a polariser between the sample and the detector. The polariser may be orientated so that the electric vector of the incident electromagnetic radiation is either parallel or perpendicular to that of the electric vector of the radiation falling on the detector. The most commonly used approach is to fix the polarisation of the incident beam and observe the polarisation of the Raman radiation in two different planes. The Raman band intensity ratio, given by the perpendicular polarisation intensity, $I_\perp$, divided by the parallel polarisation intensity, $I_\parallel$, is known as the depolarisation ratio, ρ.

$$\rho = \frac{I_\perp}{I_\parallel}$$

The symmetry property of a normal vibration can be determined by measuring the depolarisation ratio. If the exciting line is a plane polarised source (i.e. a polarised laser beam), then the depolarisation ratio may vary from near zero for highly symmetrical vibrations to a theoretical maximum of 0.75 for totally non-symmetrical vibrations. For example, carbon tetrachloride has Raman bands near $459\,cm^{-1}$ ($\sim21.79\,\mu m$), $314\,cm^{-1}$ ($\sim31.85\,\mu m$) and $218\,cm^{-1}$ ($\sim45.87\,\mu m$). The approximate depolarisation ratios of these bands are 0.01, 0.75 and 0.75 respectively, showing that the band near $459\,cm^{-1}$ ($\sim21.79\,\mu m$) is polarised (p) and the other two bands are depolarised (dp). Often depolarisation ratios are measured automatically by instruments at the same time as the Raman spectrum is recorded. This proves very useful for the detection of a weak Raman band overlapped by a strong band.

The vibrational frequencies, relative intensities and shapes of the absorption bands may all be used in the qualitative characterisation of a sample. The presence of a band at a particular frequency should not on its own be used as an indication of the presence of a particular functional group. Confirmation should always be sought from other bands or other analytical techniques if at all possible.

For example, if a sharp absorption is observed in the region 3100–$3000\,cm^{-1}$ (3.23–$3.33\,\mu m$), the sample may contain an aromatic or an olefinic component and the absorption observed may be due to the carbon–hydrogen ($=C-H$) stretching vibration. If bands are not observed in regions where other aromatic absorptions are expected, then aromatic components are absent from the sample. The suspected alkene is tackled in the same manner. By examining the absorptions observed, it is possible to determine the type of aromatic or alkene component in the sample. It may, of course, be that both groups are present, or indeed absent, the band observed being due to another functional group that absorbs in the same region, e.g. an alkane group with a strong adjacent electronegative atom or group.

It should be noted that the observation of a band at a position predicted by what is believed to be valid prior knowledge of the sample should not on its own be taken as conclusive evidence for the presence of a particular functional group.

Certain functional groups may not always give rise to absorption bands, even though they are present in the sample, since the particular energy transitions involved may be infrared inactive (due to symmetry). For example, symmetrical alkene groups do not have a $C=C$ stretching vibration band. Therefore, the absence of certain absorption bands from a spectrum leads one to conclude that (a) the functional group is not present in the sample, (b) the functional group is present but in too low a concentration to give a signal of detectable intensity, or (c) the functional group is present in the sample but is infrared inactive. In a similar way, the presence of an absorption band in the spectrum of a sample may be interpreted as indicating that (a) a given functional group is present (confirmed by other information), or (b) although more than one type of the given functional group is present in the sample their absorption bands all coincide, or (c) although more than one type of the given functional group is present, all but one have an infrared inactive transition.

The shape of an absorption band can give useful information, such as indicating the presence of hydrogen bonding.

The relative intensity of one band compared with another may, in some cases, give an indication of the relative amounts of the two functional groups concerned. The intensity of a band may also indicate the presence of certain atoms or groups adjacent to the functional group responsible for the absorption band.

These days, with modern instrumentation being so good, is not so essential to check the wavelength calibration of the spectrometer before running an infrared spectrum. This checking of the calibration may be done by examining a suitable reference substance (such as polystyrene film, ammonia gas, carbon dioxide gas, water vapour or indene) which has sharp bands, the positions of which are accurately known in the region of interest.

Purity is, of course, very important. In general, the more components a sample has, the more complicated the spectrum and hence the more difficult the analysis. Care should always be taken not to contaminate the sample or the cells used. The limits of detectability of substances vary greatly and, in general, depend on the nature of the functional groups they contain. Obviously, the parameters used for scanning the wavenumber range, e.g. resolution, number of scans, etc., are also important.

It should be noted that, when using a poorly-prepared sample, scattering of the incident radiation may result in what appears to be a gradual increase in absorption. In other words, a sloping base-line is observed.

Spurious Bands in Infrared and Raman Spectra

A spurious band is one which does not truly belong to the sample but results from either the sampling technique used or the general method of sample handling, or is due to an instrumental effect, or some other phenomenon. There are numerous reasons why spurious bands appear in spectra and it is extremely important to be aware of the possible sources of such bands and to be vigilant in the preparation of samples for study.

It should be obvious that incorrect conclusions may be drawn if the sample is contaminated so, if a solvent has been used in the extraction or separation of the sample, this solvent must be thoroughly removed. The presence of a contaminating solvent may be detected by examining regions of the spectrum in which the solvent absorbs strongly and hopefully the sample does not absorb. These bands are then used to verify the progress of subsequent solvent removal.

Certain samples may react chemically in the cell compartment even while the spectrum is being run and this may account for changes in spectra run at different times. Care should be taken that the sample does not react with the cell plates (or with the dispersive medium, or solvent, if used). For example, silicon tetrafluoride reacts with sodium chloride windows to form sodium silico-fluoride which has a band near $730\,cm^{-1}$ ($13.70\,\mu m$). A common error is to examine wet samples on salt plates (e.g. NaCl or KBr) which are, of course, soluble in water. Chemical and physical changes may also occur as a result of the sample preparation technique, e.g. due to melting of the sample in preparing a film or grinding of the sample for the preparation of discs or mulls.

One of the most common sources of false bands is the use of infrared cells which are contaminated, for example, by the previous sample studied – often it is extremely difficult for very thin sample cells to be cleaned thoroughly. Also, cell windows can become contaminated by careless handling. Some mulling agents, such as perfluorinated paraffins, are difficult to remove from cell windows if care is not taken.

It should always be borne in mind that some samples may decompose or react in a cell and, although the original substance(s) may be removed from the cell, the decomposition product remains to produce spurious bands in the spectra of subsequent samples. For example, silicon tetrachloride may leave deposits of silica on cell windows, resulting in a band near $1090–1075\,cm^{-1}$ ($9.17–9.30\,\mu m$), formaldehyde may form paraformaldehyde which may remain in the cell, producing a band at about $935\,cm^{-1}$ ($10.70\,\mu m$). Chlorosilanes hydrolyse in air to form siloxanes and hydrogen chloride. The siloxane may be deposited on the infrared cell windows and give a strong, broad band in the region $1120–1000\,cm^{-1}$ ($8.93–10.00\,\mu m$) due to the Si–O–Si group.

In addition to solute bands, traces of water in solvents such as carbon tetrachloride and chloroform may give rise to bands near $3700\,cm^{-1}$ ($2.70\,\mu m$), $3600\,cm^{-1}$ ($2.78\,\mu m$) and $1650\,cm^{-1}$ ($6.06\,\mu m$), this latter band being broad and weak. Amines may exhibit bands due to their protonated form if care is not taken in their preparation. In some instances, dissolved water and carbon dioxide in samples may form carbonates and hence result in CO_3^{2-} bands. Although not as common these days, stopcock greases (mainly silicones) can contaminate samples during chemical or sample preparation. Silicones have a sharp band at about $1265\,cm^{-1}$ ($7.91\,\mu m$) and a broad band in the region $1100–1000\,cm^{-1}$ ($9.09–10.00\,\mu m$). Some common salt crystals used for sample preparation may contain a trace of the meta-borate ion and hence have a sharp absorption line at about $1995\,cm^{-1}$ ($5.01\,\mu m$).

In some instances, the sample may not be as pure as expected, or it may have been contaminated during purification, separation or preparation, or it may have reacted with air, thus partly oxidising, etc. Also phthalates may leach out of plastic tubing during the use of chromatographic techniques and result in spurious bands. Silicon crystals often have a strong Si–O–Si band near $1100\,cm^{-1}$ ($9.09\,\mu m$) due to a trace of oxygen in the crystal.

It is also important not to lose information for a particular type of sample as a result of the sampling technique chosen. For example, hot pressing a polymer would alter the crystallinity or molecular orientation which could be of interest and would affect certain infrared bands.

The introduction to Inorganic Compounds and Coordination Complexes in Chapter 22 should also be read since this explains why certain differences may be observed in infrared and Raman spectra.

Due to the careless handling of cells, pressed discs, plates, films, internal reflection crystals, etc., spurious bands may be observed in spectra due to a person's fingerprints. These bands may be due to moisture, skin oils or even laboratory chemicals. Unfortunately, such carelessness is a common source of error. If an instrument experiences a sudden jolt, a sharp peak may be observed in the spectrum. Similarly, excessive vibration of the spectrometer may result in bands appearing in the spectrum.

It should be borne in mind that the Raman spectra of a sample may differ slightly when observed on different instruments. The reason for this is that scattering efficiency is dependent on the frequency of the radiation being scattered. In other words, the intensities of bands observed in Raman are partly dependent on the frequency of the excitation source so that the intensities of bands may differ 'significantly' if there are large differences in excitation frequencies (for example, when the instruments use visible and infrared radiations for excitation). Some instruments do not adequately compensate for changes in detector sensitivity over their spectral range and this too will have a bearing on the observations made. If the laser is unstable, its intensity fluctuates, an increase in noise may be observed and thus low intensity bands may be lost.

Although rare these days, if an interferometer is not correctly illuminated, errors in the positions of bands may be observed.

Spurious Bands at Any Position

Computer techniques   The computer manipulation of spectra is now a very common practice. Typical examples of such manipulations are to remove residual solvent bands, the addition of spectra, the flattening of base lines, the removal of bands associated with impurities, the accumulation of weak signals, etc. and the addition of spectral runs. Unfortunately, in the wrong hands (inexperienced or experienced), spectra can be so manipulated that they end up bearing little resemblance to the original recording and contain little, if any, useful information.

Although not so common these days, when recording a spectrum to magnetic disc, errors in software programmes have lead to spurious bands appearing in spectra or even bands disappearing from a recorded spectrum.

Regions of strong absorption by solvents   Insufficient radiation may reach the detector for proper intensity measurements to be taken when attempting to observe the spectrum of a solute in regions of strong solvent absorption with a solvent-filled cell in the reference beam. When using a difference technique, observations in regions of strong solvent absorptions are unpredictable and unreliable so it is important to mark clearly any such unusable regions of a spectrum in order that 'bands' in these regions cannot be misinterpreted later. It should be pointed out that nowadays, on modern spectrometers, spectral subtraction is computed electronically using the data collected when recording the spectrum of a sample.

Solvents should not damage the cell windows and should not react chemically with the sample. The spectral absorptions of a solute will be significantly distorted in a region where the solvent allows less than about 35% transmittance. Chart 1.1 indicates regions in which some common solvents should not be used. The cell path length is 0.1 mm unless indicated otherwise (*indicates a path length of 1 mm). Chart 1.3 indicates regions in the near infrared in which some common solvents should not be used. Of course, aqueous solutions may be used for Raman spectroscopy without problems being encountered, as water is a poor scatterer of radiation, see Chart 1.2. It should be borne in mind that in Raman a solvent may not have as strong an absorption as in infrared in a spectral region of interest. Of course, the opposite is also true.

Interference pattern   The spectra of thin unsupported films may exhibit interference fringes. For example, the spectra of thin polymeric films often have a regular interference pattern superimposed on the spectrum. Although possible, it is generally difficult to mistake such a wave pattern for absorption bands. When examined by reflection techniques, coatings on metals may also exhibit an interference pattern. The interference pattern can be a nuisance but can be relatively easily eliminated. The wave pattern observed may be used to determine film thickness (see page 266).

Christiansen effect   A spurious band on the high frequency side of a true absorption band may sometimes be observed when examining the mulls of crystalline materials if the particle size is of the same order of magnitude as the infrared wavelength being used.

Attenuated total reflectance, ATR, spectra   Bands may be observed when using attenuated total reflectance, ATR, due to surface impurities. Anomalous dispersions may be observed due to poorly-adjusted attenuated total reflectance samplers.

Chemical reaction   When a sample undergoes a chemical reaction, some bands may decrease in intensity and new bands, due to the product(s), may appear. Hence, some of the bands observed in the spectrum may vary in intensity with time. Although all the bands may belong to the sample, and in that sense are not truly spurious, they can nonetheless still be baffling.

Crystal orientation   In general, the infrared radiation incident on a sample is partially polarised so that the relative intensities of absorption bands may alter as a crystalline sample is rotated. In an orientated crystalline sample, a functional group may be fixed within its lattice in such a position that it will not interact with the incident radiation. These crystalline orientation effects can be dramatic, especially for thin crystalline films or single crystals.

Polymorphism   Differences are usually observed in the (infrared or Raman) spectra of different crystalline forms of the same substance. Therefore, it should be borne in mind that a different crystalline phase may be obtained after recrystallisation from a solvent. Also, in the preparation of a mull or disc, a change in the crystalline phase may occur.

Gaseous absorptions   These days, pollutant gases in the atmosphere, as well as carbon dioxide and water vapour, do not generally result in problems for modern spectrometers. When using older instruments, or single beam spectrometers, absorptions due to these gases may be superimposed on the observed spectrum.

Molten materials   The sudden crystallisation of a molten solid may result in a rapid drop in the transmittance which could be mistaken as an absorption band. Similarly, a phase change after crystallisation may result in absorbance changes.

Optical wedge   For older instruments, it is possible that an irregularity in their optical wedge may result in a small band or shoulder on the side of an absorption band.

Numerous laser emission frequencies   Some lasers used in Raman spectrometers produce a number of other emissions in addition to their base frequency which are of lesser intensity (i.e. the emission is not monochromatic). Of course, a sample can also reflect or scatter these additional radiations. As a result, spurious bands may be observed in Raman spectra at any position – the positions of bands and their intensities being dependent on the laser and the sample. The problem can be avoided by the use of a pre-monochromator or suitable filter.

Mains electricity supply   Bands due to electronic interference may be observed in Fourier transform spectra. Bands at frequencies related to that of the AC mains electricity supply may be observed. For example, a relatively strong line may be observed in Raman spectra at $100\,cm^{-1}$. Although such lines may be quite strong, they are easily recognised, for example, by observing that their position does not change when the scanning speed is altered. In order to avoid electronic interference, it is important that the detector and amplifier are screened.

Chart 1.1 Regions of strong solvent absorptions in the infrared

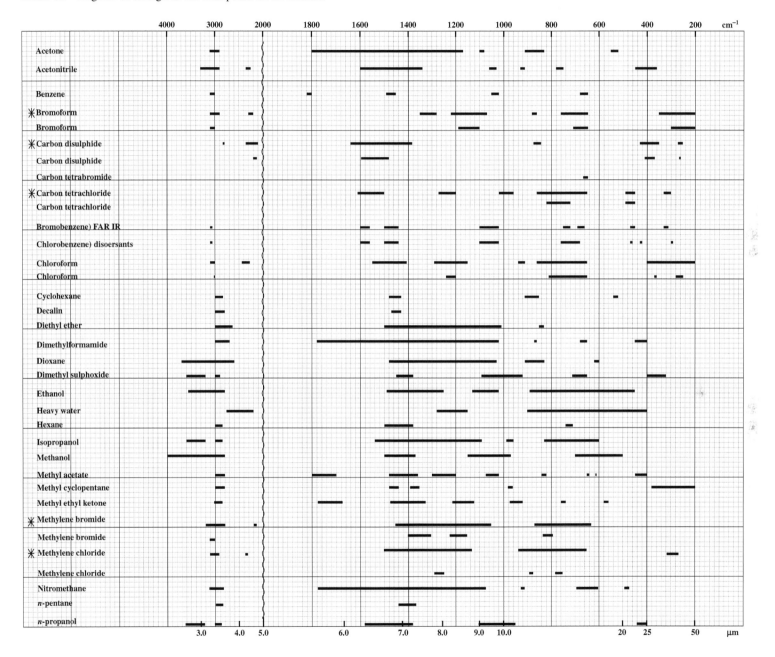

Chart 1.1 (*continued*)

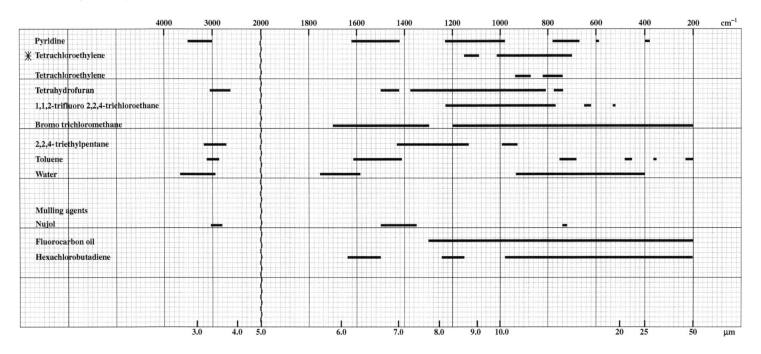

These days the stability of the Raman excitation radiation (i.e. the laser radiation source) is exceedingly good. As the intensity of the radiation is fairly constant, it allows the possibility of using Raman for quantitative analysis.

Fold-back   The maximum frequency that may be measured by an FT Raman spectrometer is governed by the frequency of the excitation radiation. However, radiation of a higher frequency than that of the maximum may still pass through the interferometer. As a result of this, the detector may observe electromagnetic interference due to this higher frequency which it cannot distinguish from that due to radiation that is below the maximum frequency by an equivalent amount. This fold-back below the maximum, by an amount equal to the difference in the frequencies, may therefore result in spurious bands appearing in Raman spectra. Most instruments these days have optical and electronic filters which try to overcome this effect but these devices do not always completely remove the problem.

Fluorescence   Many organic samples, and some inorganic, have fluorescent properties. The fluorescence of a sample, examined by Raman spectroscopy, may appear as a number of broad emissions over a large range. Although,

strictly speaking, such bands are not spurious since they do belong to the sample, they may nonetheless cause confusion. Obviously, if desired, such bands can be removed by computer, or other techniques.

Stray light   Stray light either entering a spectrometer or being generated from within, perhaps by poor optics, may result in spurious bands appearing in spectra. A common source of stray light is due to the sample compartment being left open.

Fluorescent lights   Due to the emissions of fluorescent room lights, sharp bands may be observed in the Raman spectrum.

Cosmic rays   In the observation of Raman spectra, cosmic ray interference may occur with charged coupled device (CCD) detectors. These detectors are sensitive to high energy photons and particles. The interference shows up as very sharp, intense spikes in the Raman spectra and so can easily be distinguished from true bands. There are programs available to remove these spikes.

Chart 1.2   Regions of strong solvent absorptions of the most useful solvents for Raman spectroscopy

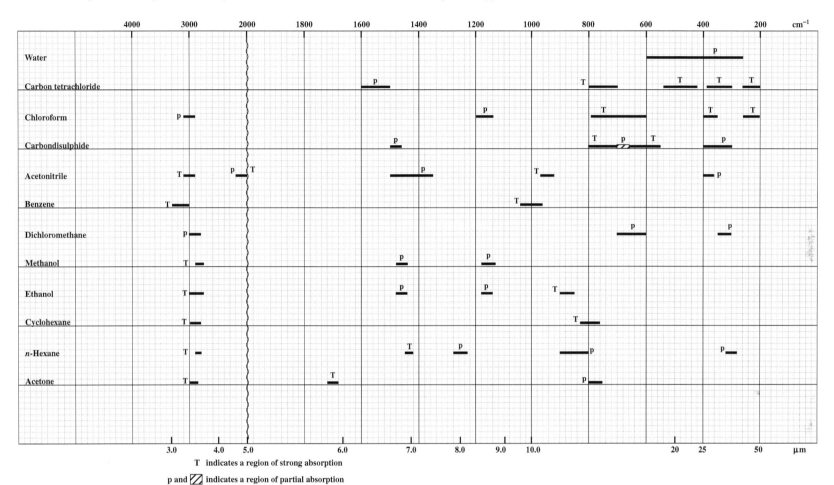

T indicates a region of strong absorption

p and ▨ indicates a region of partial absorption

Spurious Bands at Specific Positions

Table 1.1 gives the positions of some spurious bands and the reasons for their appearance.

Positive and Negative Spectral Interpretation

Both infrared and Raman spectra may be used as fingerprints of a sample. A bank of the infrared and Raman spectra of the constituents of the type of samples encountered in a given laboratory should be made or purchased. Such reference spectra are of great assistance in the interpretation of the spectrum of an unknown sample. It may often be the case that all that is required is a simple confirmation of a sample. This may easily be achieved by comparing the spectrum of the sample and that of the known reference material. If the absorption bands are the same (i.e. in wavelength, relative intensities and shapes), or nearly so, then it is reasonable to assume that the sample and reference are either identical or very similar in molecular structure.

Chart 1.3 Regions of strong solvent absorptions in the near infrared

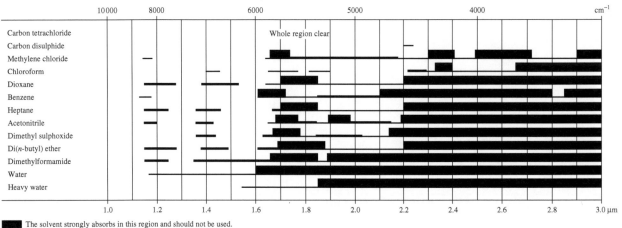

The solvent strongly absorbs in this region and should not be used.

Solutions having path lengths greater than 1 cm should not be used in this or the above region.

Solutions having path lengths greater than 2 cm should not be used in this or the above two regions.

In the interpretation of infrared and Raman spectra, there is no substitute for experience and, if possible, guidance from an expert in the field should be sought by the inexperienced.

The spectrum should be interpreted by (a) seeing which absorption bands are absent – negative spectral interpretation – and (b) examining those bands present – positive spectral interpretation.

Negative Spectral Interpretation

By examining a spectrum for the absence of bands in given regions, it is possible to eliminate particular functional groups and, hence, compounds containing these groups. In general, this type of interpretation is made by a search in a particular region where a given functional group always absorbs strongly. If no bands are observed in this region then this functional group may be excluded. For this purpose, Table 1.2 and the more detailed Chart 1.4 should be used. With a little experience, negative interpretation may be carried out at a glance.

Positive Spectral Interpretation

The technique of negative interpretation should, of course, be used in conjunction with the positive approach. It is important to be aware that correlation tables give the positions and intensities of bands characteristic of a large number of classes of compounds and groups. However, it may well be that bands appear in the spectrum of a particular sample which are not given in the tables. Assuming that these bands do belong to the sample and are not due to (a) solvent(s), (b) dispersive media, (c) air, (d) instrumental fault or (e) operator error, then correlations involving these bands may not as yet have been made, or the bands are not characteristic of the class of compound or group considered. It may well be, for example, that the band or bands have arisen due to solid-state effects, e.g. due to different crystalline modifications of the compound. In general, it is not necessary to identify every single (weak) band that appears in a spectrum in order to characterise a sample and be in a position to propose a molecular structure.

Regions for Preliminary Investigation

There are no rigid rules for the interpretation of infrared or Raman spectra. However, a few general hints may be given.

Preliminary Regions to Examine

It is usually advisable to tackle the bands at the higher-frequency end of the spectrum, the most intense bands being looked at first and associated bands,

Table 1.1 Spurious bands

Approximate band position cm^{-1}	Reason
3700–3600	Small traces of moisture in an organic solvent, such as carbon tetrachloride or hydrocarbon solvents, give rise to bands due to O–H vibrations in this region. A broad, weak band may also be observed near 1650 cm^{-1} (6.06 μm). Such bands are particularly noticeable using thick sample cells of long pathlength.
~3650	Occluded water in some fused silica windows gives rise to a sharp band.
3450–3300	Solid samples containing water have a band in this region and also a band near 1650 cm^{-1} (6.06 μm).
~3000	Contamination due to the use of plastic laboratory ware. Bands near 1450 cm^{-1} (6.90 μm) and 1380 cm^{-1} (7.25 μm) may also be observed. A band near 725 cm^{-1} (13.79 μm) may also be observed for polyethylene and polypropylene contamination. A band near 670 cm^{-1} (14.93 μm) is observed due to polystyrene contamination. Hydrocarbon oils, and also silicone oils (which also have a strong band near 1050 cm^{-1}), have a band in this region.
~2350	A band due to atmospheric carbon dioxide may be observed in older or poorly-adjusted instruments, for example, if the sample and reference beams of a double-beam spectrometer are not properly balanced. Also, a band near 665 cm^{-1} (15.04 μm) is observed.
~2325	Samples stored at low temperatures may exhibit a band due to dissolved carbon dioxide.
~2330	Absorptions due to gaseous nitrogen, N_2, may be observed in Raman spectra of samples.
2000–1280	Water vapour in air has many sharp, relatively strong bands, in this range. Water vapour often exhibits a sharp band near 1760 cm^{-1} (5.68 μm) which may account for shoulders seen on bands due to C=O stretching vibrations. Water vapour bands may be observed when using poorly balanced, double-beam instruments.
~1810	Spectroscopic grade chloroform has the trace of inhibitor, which is normally present, removed and therefore may oxidise to give phosgene on exposure to air and sunlight, so a band, due to the C=O group of phosgene, may be observed.
~1755	Phthalates, which are present as plasticisers in some polymeric materials, may leach out to contaminate samples and give a band at 1725 cm^{-1} (5.80 μm). Oxidation may convert the phthalate to phthalic anhydride which has a band at 1755 cm^{-1} (5.70 μm). Hence, dialkyl phthalate plasticiser present in plastic tubing attached to a chromatographic column may indirectly result in this band.
~1725	The phthalate plasticiser in flexible polyvinyl chloride tubing may dissolve in organic solvents and appear as a contaminant in samples.
~1650	Water present in many materials may result in this broad band. It may be difficult to remove all the water from some samples.
1615–1520	An alkali halide may react with a carboxylic acid or metal carboxylate to produce a salt and hence give rise to a spurious band due to the carboxylate anion. This may occur in the preparation of KBr discs or as an interaction with cell windows.
~1555	Absorptions due to gaseous oxygen, O_2, may be observed in Raman spectra of samples.
1450–1340	Nitrate formed by double decomposition (see also 825 cm^{-1}). This band is sometimes observed in the study of inorganic nitrates when using KBr discs/windows. It is due to the double decomposition reaction of potassium bromide with the nitrate to give potassium nitrate.
~1430	Although not a major problem these days, inorganic carbonate impurity in salts such as KBr may result in this band. This band occurs in the same region as that due to CH deformation vibrations.
~1380	Although not a major problem these days, potassium nitrate impurity in salts, such as KBr, may result in a band ~1380 cm^{-1} (7.25 μm). This band occurs in the same region as that due to CH_3 symmetric deformation vibrations.
~1265	Silicone stopcock grease is dissolved by aromatic and chlorinated solvents. Hence, the presence of a sharp band near 1265 cm^{-1} (7.91 μm) and a broad band in the 1110–1000 cm^{-1} (9.01–10.00 μm) range could be due to silicone stopcock grease.
~1100	Silica in contaminated cells (see also 475 cm^{-1}).
1100–1050	When preparing a sample for examination by a dispersive technique, it is possible to contaminate the sample with small amounts of powdered glass if the sample is ground between glass surfaces.
~1050	Usually due to the use of silicone oils and greases. Silicones have a strong broad band in this region and also bands at ~3000 and 1265 cm^{-1}.
~1000	This band is sometimes observed in the study of inorganic sulphates when using KBr discs/windows. It is due to the double decomposition reaction of potassium bromide with the sulphate to give potassium sulphate. There may also be a band in the region 670–580 cm^{-1} (14.93–17.24 μm).
~825	As above, the spectra of KBr discs containing inorganic nitrates may have a band due to potassium nitrate which is produced by double decomposition.

Table 1.1 *(continued)*

Approximate band position cm^{-1}	Reason
~790	Carbon tetrachloride vapour, being much heavier than air, may having escaped from a sample cell, remain in the instrument for some considerable time before being dispersed. Alternatively, having used a carbon tetrachloride solution, it may be that the cell has not been thoroughly cleaned before examination of the next sample. A weaker band near 765 cm^{-1} (13.07 μm) is also observed.
730–720	These days, polyethylene and polypropylene are widely used for laboratory ware and therefore may easily contaminate a sample. This band is usually split. A band due to the C–H stretching vibration would also be expected.
~670	Polystyrene containers used for mixing samples with KBr in mechanical vibrators may be abraded. Other bands due to styrene may also be observed (eg. ~3000 cm^{-1}, 1600 cm^{-1}).
670–580	Due to potassium sulphate through double decomposition (see 1000 cm^{-1}).
~665	Older, badly-balanced, double-beam instruments may exhibit bands due to atmospheric carbon dioxide, also a band at 2350 cm^{-1} (4.26 μm).
540–440	Broad band due to Si–O absorption.
~475	Due to silica (also ~1100 cm^{-1}).
~470	Due to carbon tetrachloride (other bands ~790 cm^{-1}).
~200	Due to silicone greases (see other bands ~1265 cm^{-1}).

Table 1.2 Negative spectral interpretation table

Absorption band absent in region cm^{-1}	μm	Type of vibration responsible for bands in this region	Type of group or compound absent
4000–3200	2.50–3.13	O–H and N–H stretching	Primary and secondary amines, organic acids and phenols
3310–3300	3.02–3.03	C–H stretching (unsaturated)	Alkynes
3100–3000	3.23–3.33	C–H stretching (unsaturated)	Aromatic and olefinic compounds
3000–2800	3.33–3.57	C–H stretching (aliphatic)	Methyl, methylene, methyne groups
2500–2000	4.00–5.00	X≡Y, X=Y=Z stretching[†]	Alkynes[‡], allenes[‡], cyanate, isocyanate, nitrile, isocyanides, azides, diazonium salts, ketenes, thiocyanates, isothiocyanates
1870–1550	5.35–6.45	C=O stretching	Esters, ketones, amides, carboxylic acids and their salts, acid anhydrides
1690–1620	5.92–6.17	C=C stretching	Olefinic compounds[‡]
1680–1610	5.92–6.21	N=O stretching	Organic nitrite compounds
1655–1610	6.04–6.21	–O–NO$_2$ asymmetric stretching	Organic nitrate compounds (the symmetric –O–NO$_2$ stretching vibration occurs at 1300–1255 cm^{-1} (7.69–7.97 μm)
1600–1510	6.25–6.62	–NO$_2$ asymmetric stretching	Organic nitro-compounds (the symmetric –NO$_2$ stretching vibration occurs at 1385–1325 cm^{-1} (7.22–7.55 μm)
1600–1450	6.25–6.90	C=C stretching	Aromatic ring system (normally four bands)
1490–1150	6.71–8.70	H–C–H bending	Methyl, methylene
1420–990	7.04–10.10	S=O stretching	Sulphoxides, sulphates, sulphites, sulphinic acids or esters, sulphones, sulphonic acids, sulphonates, sulphonamides, sulphonyl halides
1310–1020	7.63–9.80	C–O–C stretching	Ethers (aromatic, olefinic or aliphatic)
1225–1045	8.16–9.67	C=S stretching	Thioesters, thioureas, thioamides pyrothiones
1000–780	10.00–12.82	C=C–H deformation	Aliphatic unsaturation
900–670	11.11–14.93	C–H deformation	Substituted aromatics
850–500	11.76–20.00	C–X stretching[ξ]	Organohalogens
730–720	13.70–13.90	(CH$_2$)$_{n>3}$	Four or more consecutive methylene groups

[†] X, Y, and Z may represent any of the atoms C, N, O and S.
[‡] Band may be absent in the infrared due to symmetry of functional group but is a strong band in Raman.
[ξ] X may be Cl, Br or I.

Chart 1.4 Negative correlation chart. The absence of a band in the position(s) indicates the absence of group (or chemical class) specified. (Note the change of scale at 2000 cm^{-1} (5.0 μm).)

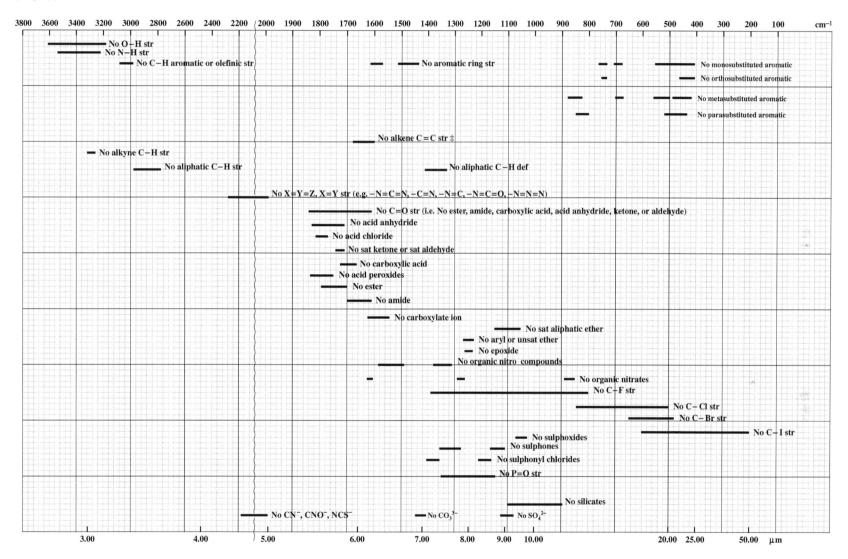

occurring in other regions, thus also being identified. In the light of the information gained, the region between 900 and 650 cm^{-1} (11.1 and 15.4 μm) can then be looked at. The origin of bands found in the so-called 'fingerprint' region 1350–900 cm^{-1} (7.4–11.1 μm) is usually difficult to decide on as the bands may arise in various ways, and similarly, below 650 cm^{-1} (above 15.4 μm), skeletal vibrations occur which are also often difficult to interpret. Hence these two regions are best avoided initially. Table 1.2 and Chart 1.4 may be used in reverse, i.e. to indicate the possible presence of a group which must then be confirmed.

Confirmation

It must be stressed again that the presence of a particular band should not, on its own, be used as an indication of the presence of a particular group. Confirmation should always be sought from the presence of other associated bands or from other independent techniques. For the interpretation of infrared spectra the correlation Charts 1.5 and 1.6 should be used first and then the tables and text of relevant chapters employed for the detailed confirmation and identification. Having positively identified the first band looked at, the next band is approached in a similar fashion. The interpretation of Raman spectra may be carried out in a similar fashion by making use initially of Chart 1.7.

Chemical Modification

Quite often it is helpful for identification purposes to modify the sample chemically and compare the spectra of the original and modified samples. Isotope exchanges may be helpful in the assignment of bands. Deuterium exchange is very useful and the most common. Labile hydrogen atoms are replaced by deuterium atoms. On comparing the spectra of the original and the deuterated sample, bands shifted in frequency by a factor of approximately $1/\sqrt{2}$ compared with the original may be associated with vibrations due to the substituted labile hydrogen.

Chemical reactions may also be helpful for assignment purposes, e.g.

(a) conversion of an acid to its salt or ester;
(b) conversion of an amine or amino acid to its hydrochloride;
(c) hydrogenation of unsaturated bonds;
(d) saponification of esters, this being particularly useful in the identification of the monomers of a polyester resin.

Collections of Reference Spectra

The most comprehensive collection of infrared spectra is that offered by Sadtler Research Laboratories[1] (a Division of Bio-Rad Laboratories). It consists of many thousands of spectra covering a wide variety of compounds and new additions are made periodically. The spectra are run under standard conditions. Spectra within the collection may be retrieved by the use of (a) an alphabetical index, name or synonym, (b) a molecular formula/structure index, (c) peak positions, or (d) a chemical class index. A pre-filter such as structure or physical properties may be applied to the search. Sadtler provide collections covering a broad range of pure and commercially available substances.

The total library available contains spectra of the following: (a) pure compounds and standards (b) dyes, pigments, coatings and paints, (c) fats, waxes, and derivatives, (d) fibre and textile chemicals, (e) starting materials and intermediates, (f) lubricants, (g) monomers, polymers (Vols I and II), plasticisers and additives, (h) natural resins, (i) perfumes, flavours and food additives, (j) petroleum chemicals, (k) pharmaceuticals, steroids and drugs (abused and prescription) (l) flame retardants, (m) polyols, (n) pyrolysates, (o) rubber chemicals, (p) solvents, (q) surface active agents (Vols I and II), (r) water treatment chemicals, (s) minerals and clays, (t) pollutants and toxic chemicals, (u) inorganic and organometallic compounds, (v) adhesives and sealants, (w) coating chemicals, (x) esters, (y) substances in the condensed phase and vapour phase, (z) agricultural chemicals and pesticides, etc. Sadtler have also published an atlas of near infrared spectra,[3] Raman spectra, ultraviolet–visible spectra, NMR spectra, and DTA data for materials. Many of the spectra in some of the collections are also referred to by trade names.

Sadtler offer nearly 200 000 digital infrared reference spectra in over fifty different collections and also publish handbooks and guides which cover the areas mentioned above. The Sadtler computer-based search system[4] and the other systems available from manufacturers such as Nicolet, Perkin Elmer, Bio Rad, etc., are all relatively easy to use. Sadtler also offer a computer-based system which contains both IR and NMR data, etc. Library search software packages, such as the Sadtler IR SearchMaster Software, the Spectrafile IR Search Software or the Spectra Calc Search Software, are frequently offered by FT-IR manufacturers in addition to specific search software formatted to operate with their particular data-stations/instruments/computer systems. Some of these search facilities may also cover a number of libraries not only of different suppliers but also of other techniques such as UV, NMR, MS etc. Obviously, such search software packages are dependent not only on the instrument but also on the user's interests. It should be borne in mind that the information retrieved from some search software may not cover certain aspects which may normally be available from the particular Sadtler library being searched, such as physical properties, molecular structure, *Chemical Abstracts* Service (CAS) Registry Number, common impurities, etc.

Aldrich[5-8] also produce a comprehensive, computer-based library of infrared spectra (and NMR spectra). The main classes of chemical covered by the Aldrich library are hydrocarbons, alcohols, phenols, aldehydes, ketones, acids, amides, amines, nitriles, aromatics, phosphorus and sulphur compounds, organometallics, inorganics, silanes, boranes, polymers, etc. The spectra are categorised by chemical functionality and arranged in order of increasing structural complexity. They are also indexed alphabetically, by molecular formula and by CAS number. The library also includes common organic substances, flavours, fragrances and substances of interest to forensic scientists.[7] An automobile (US) paint chip library is also available from Nicolet. Sigma[9] provide a computer-based library of FT-IR spectra which

Chart 1.5 Infrared – positions and intensities of bands. This chart may be used to identify the possible type of vibration responsible for a band at a given position. The range and position of the maximum absorption of a functional group is given in order of decreasing wavenumber. The information given in both the text and tables of relevant chapters may be used to confirm or eliminate a particular group. The relative intensities of bands are given

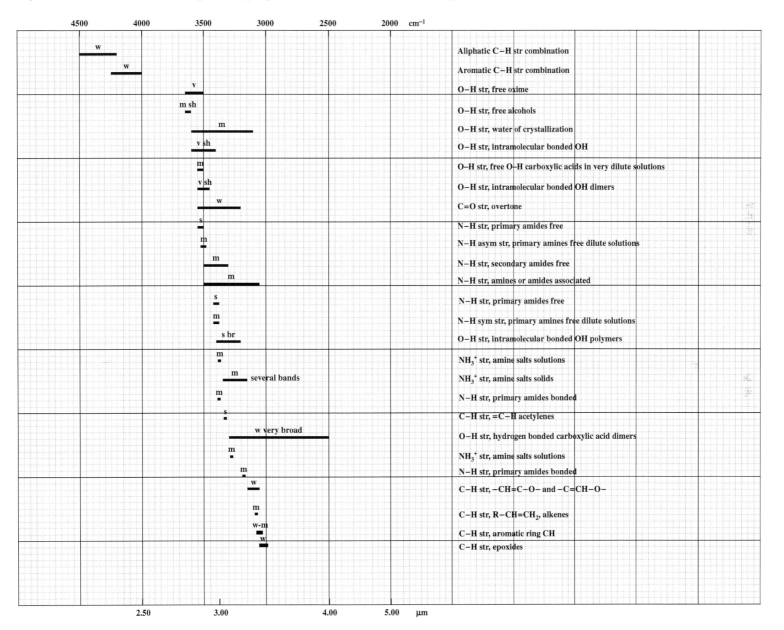

Chart 1.5 *(continued)*

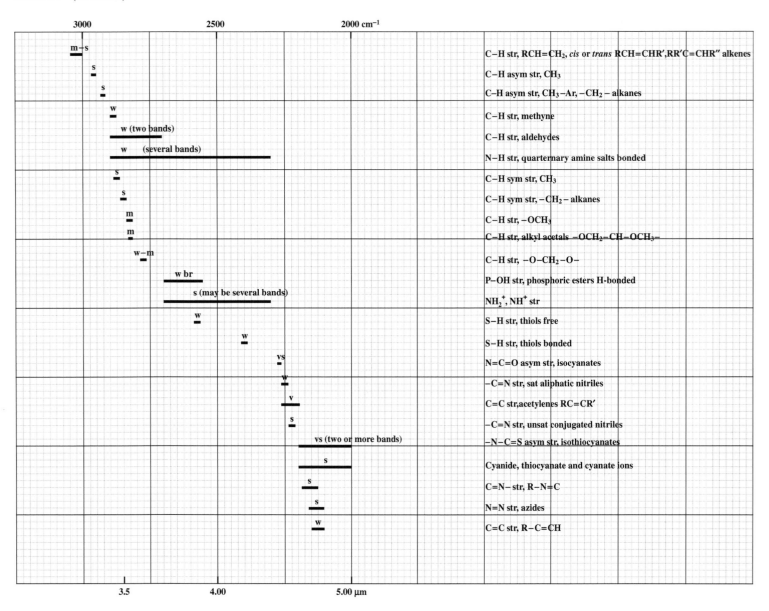

m–s			C–H str, RCH=CH$_2$, *cis* or *trans* RCH=CHR′, RR′C=CHR″ alkenes
s			C–H asym str, CH$_3$
s			C–H asym str, CH$_3$–Ar, –CH$_2$– alkanes
w			C–H str, methyne
w (two bands)			C–H str, aldehydes
w (several bands)			N–H str, quaternary amine salts bonded
s			C–H sym str, CH$_3$
s			C–H sym str, –CH$_2$– alkanes
m			C–H str, –OCH$_3$
m			C–H str, alkyl acetals –OCH$_2$–CH–OCH$_3$–
w–m			C–H str, –O–CH$_2$–O–
w br			P–OH str, phosphoric esters H-bonded
s (may be several bands)			NH$_2^+$, NH$^+$ str
w			S–H str, thiols free
w			S–H str, thiols bonded
vs			N=C=O asym str, isocyanates
w			–C≡N str, sat aliphatic nitriles
v			C≡C str, acetylenes RC≡CR′
s			–C≡N str, unsat conjugated nitriles
vs (two or more bands)			–N=C=S asym str, isothiocyanates
s			Cyanide, thiocyanate and cyanate ions
s			C=N– str, R–N=C
s			N=N str, azides
w			C=C str, R–C=CH

3000   2500   2000 cm^{-1}

3.5   4.00   5.00 μm

Chart 1.5 (*continued*)

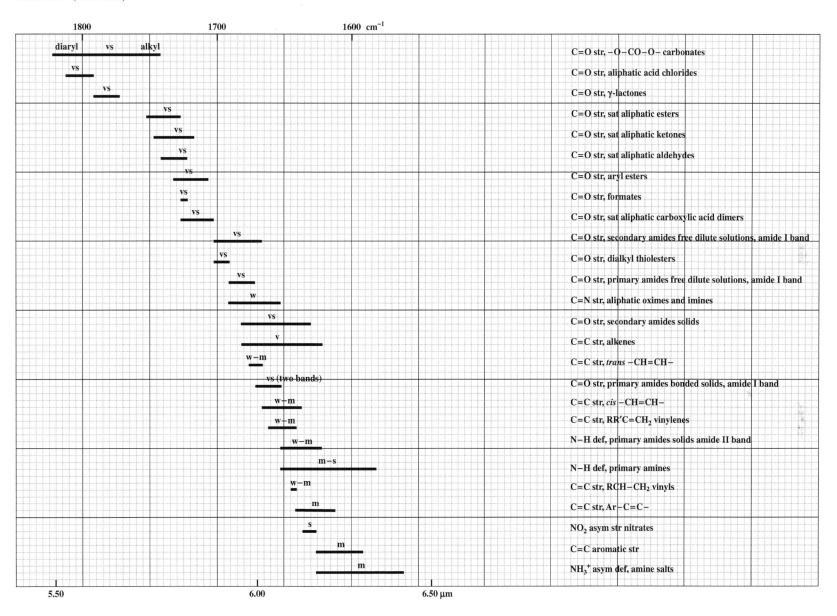

Chart 1.5 (*continued*)

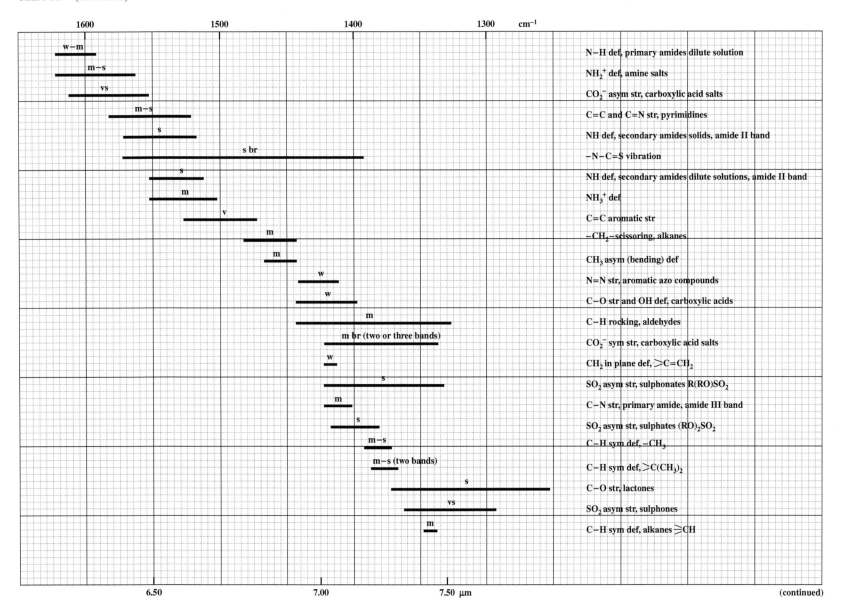

(continued)

Chart 1.5   (*continued*)

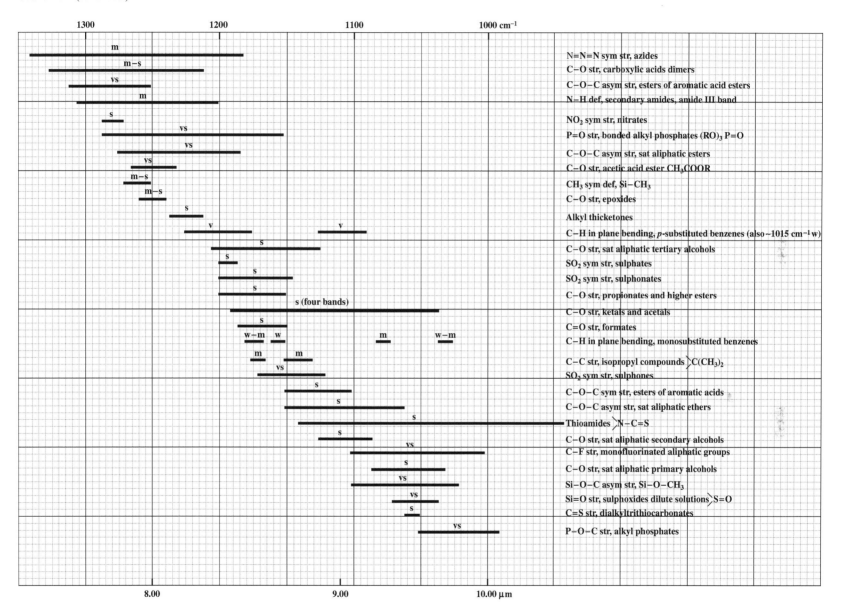

Chart 1.5   (*continued*)

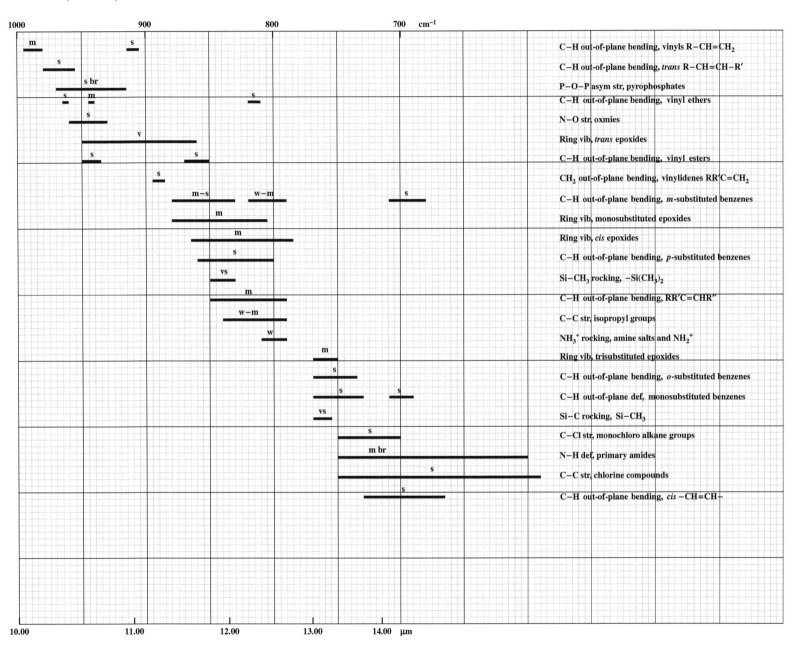

Chart 1.5   (*continued*)

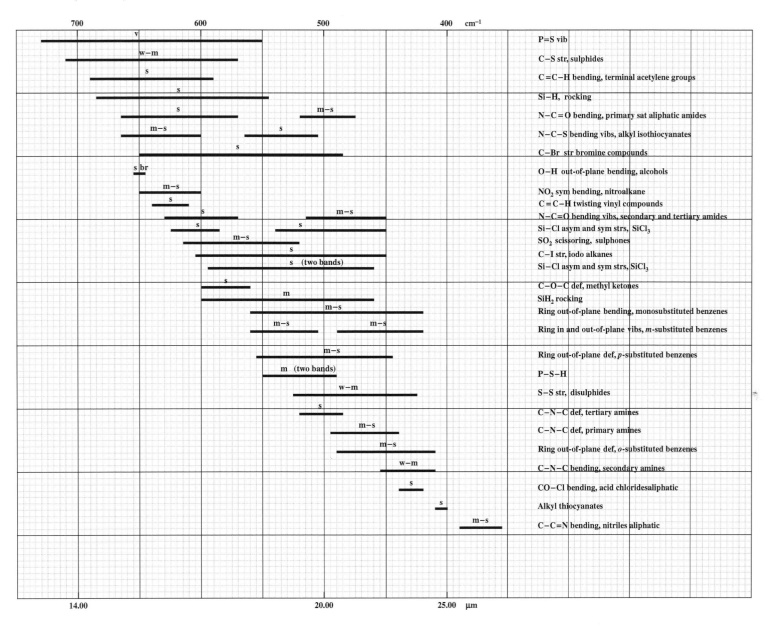

Chart 1.6 Infrared – characteristic bands of groups and compounds. The ranges of the main characteristic bands of groups or classes of chemical compound are indicated by either thick or fine lines. The thick lines indicate important band ranges which either are completely specific for that group or can be used in those ranges to distinguish the group from similar groups. The thin lines indicate other important band regions which should be borne in mind. The intensities of bands occurring in the region represented by thin lines are as given previously in the chart for similar groups unless specifically shown. (Note the change of scale at 2000 cm⁻¹ (5.0 μm).)

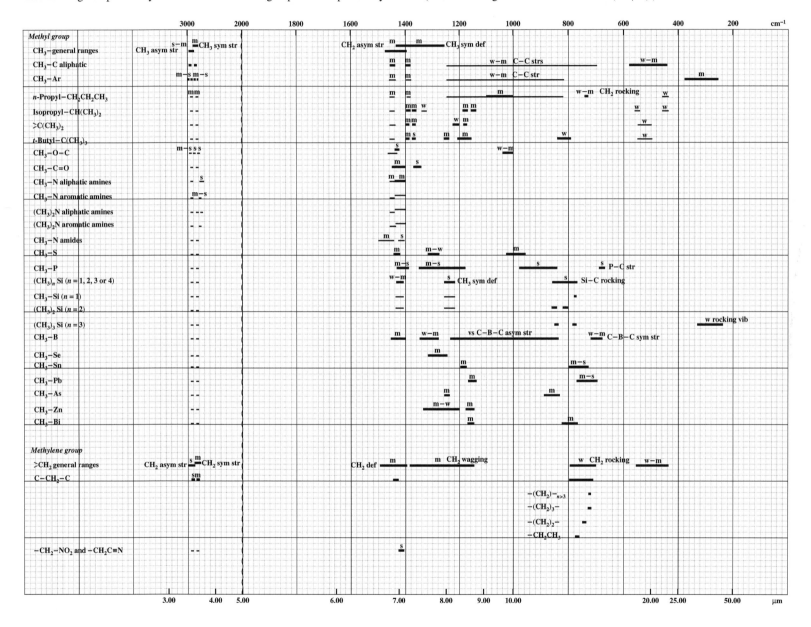

Chart 1.6   *(continued)*

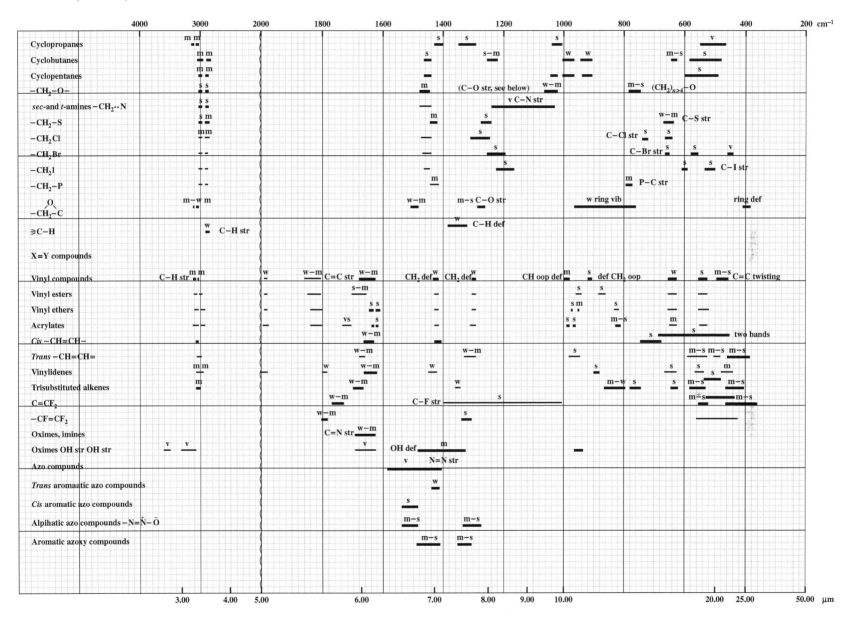

Chart 1.6 (*continued*)

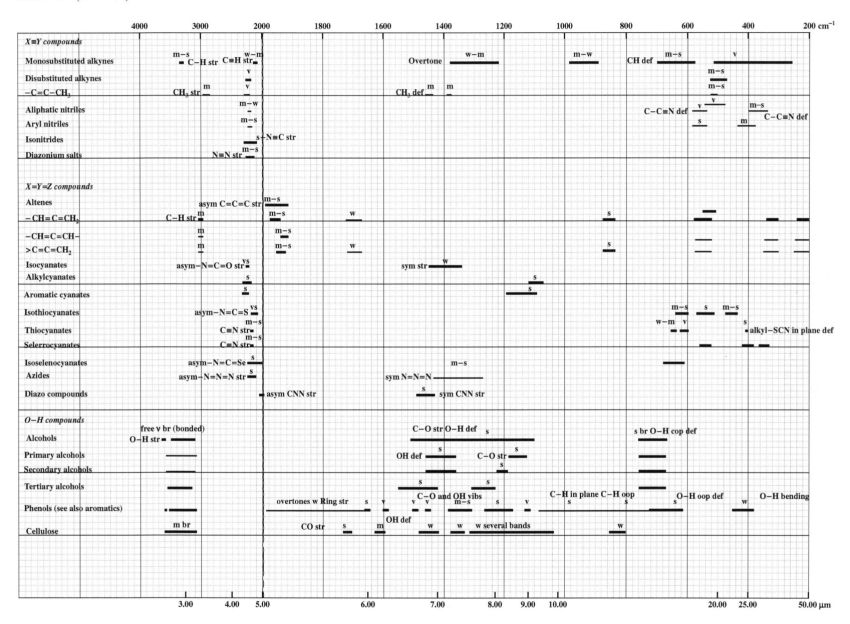

Chart 1.6 *(continued)*

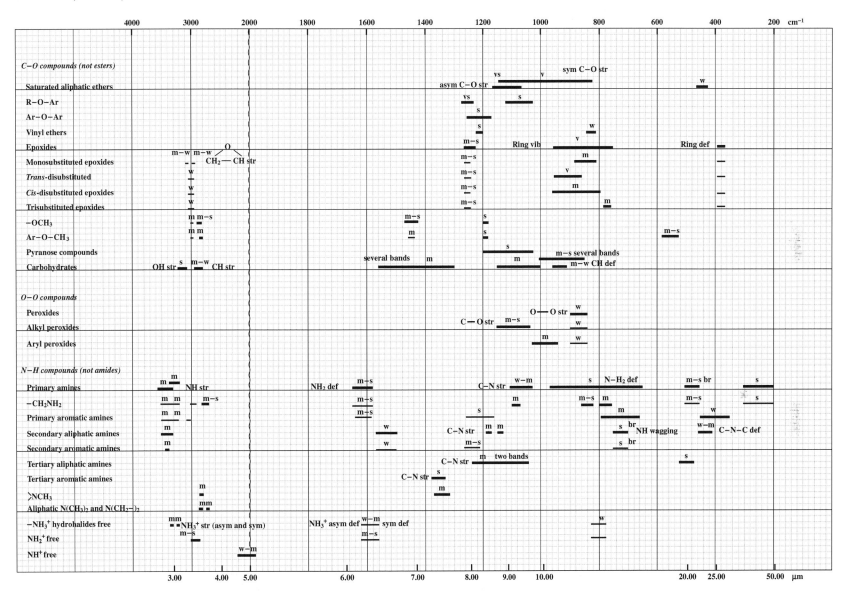

Chart 1.6 *(continued)*

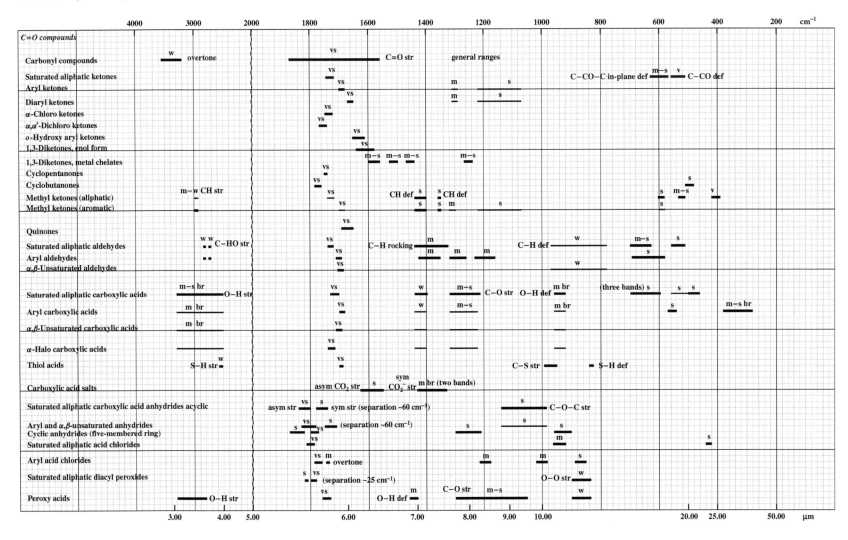

Chart 1.6 (*continued*)

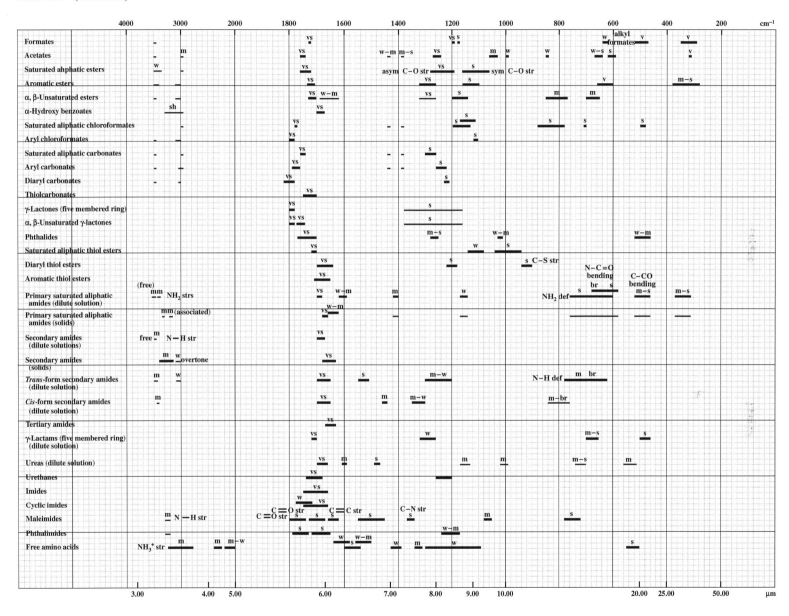

Chart 1.6   (*continued*)

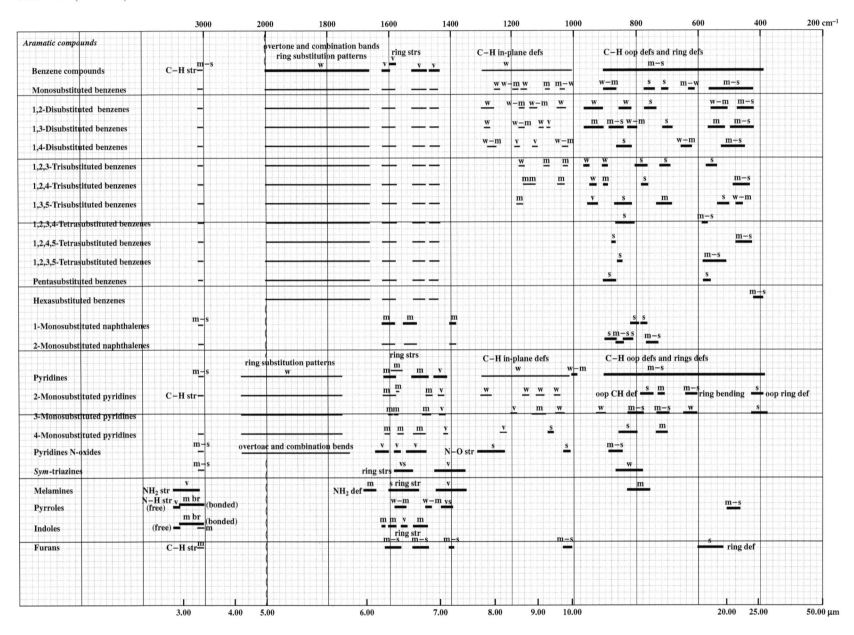

Chart 1.6 *(continued)*

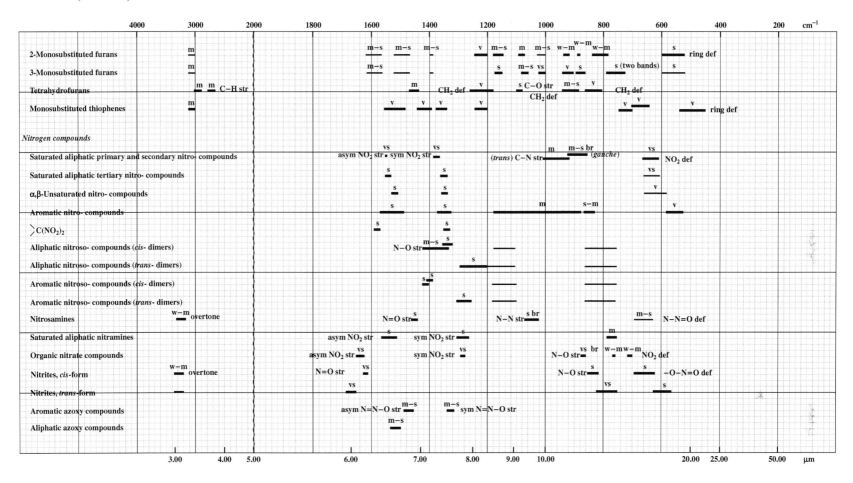

Chart 1.6 (*continued*)

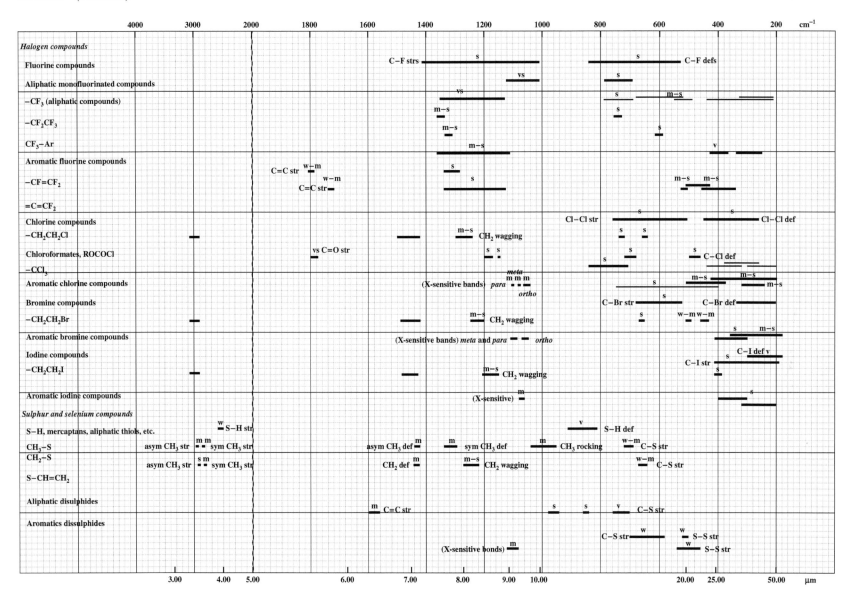

Chart 1.6   (*continued*)

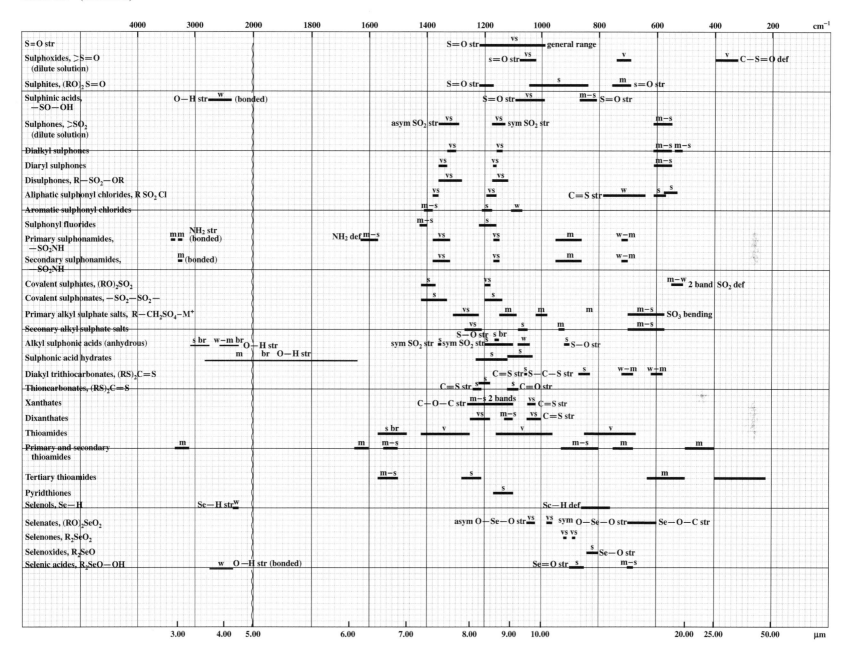

Chart 1.6 *(continued)*

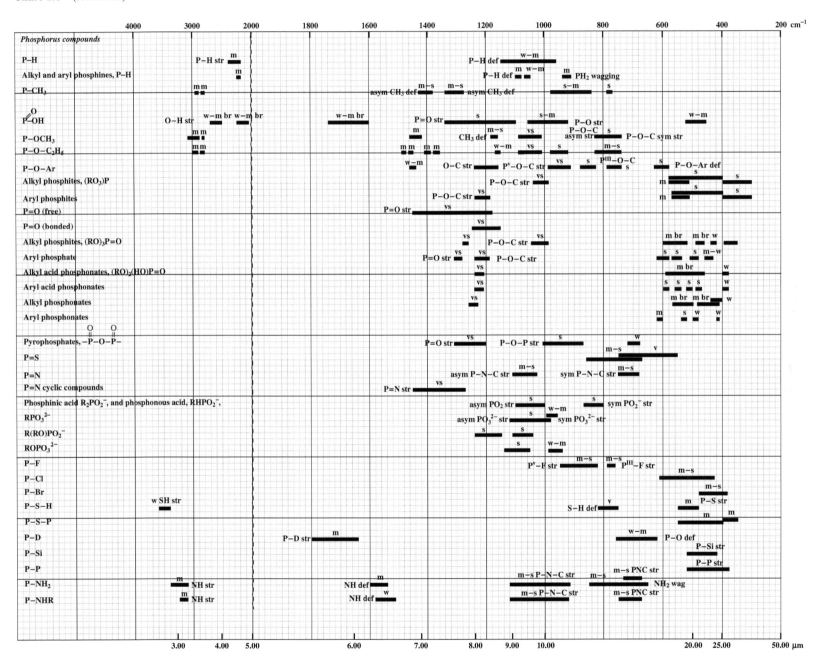

Chart 1.6 *(continued)*

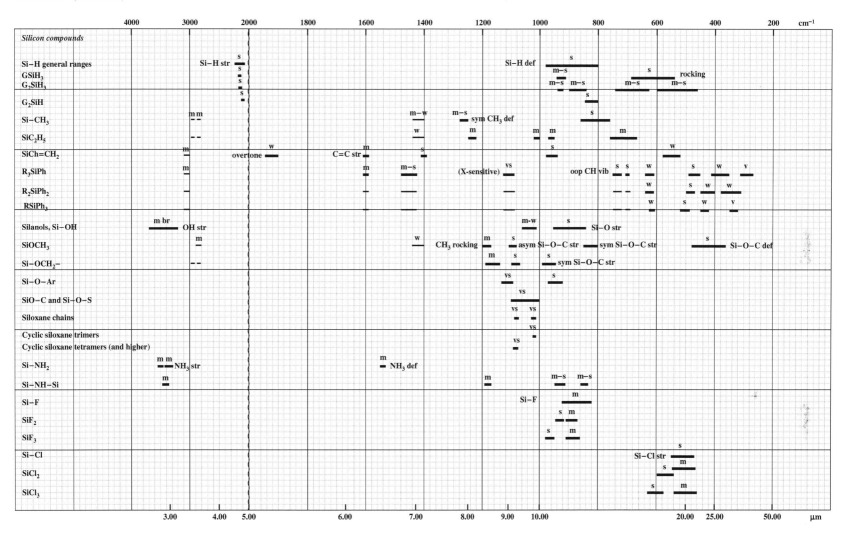

Chart 1.6 (*continued*)

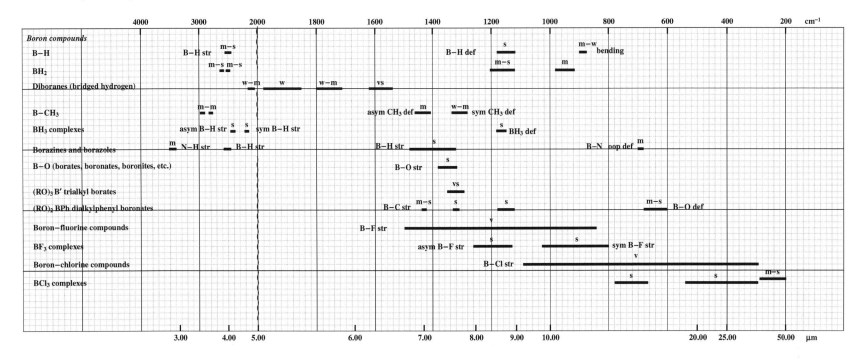

is devoted in the main to biochemical substances, for example, amino acids, enzymes, proteins, nucleotides, carbohydrates, steroids, etc.

Nicolet and Aldrich[8] have produced a computer-based search program for use with the Aldrich Library of FT-IR spectra.[5] The band positions and intensities from the spectrum of an unknown substance are entered into the program which then gives on the Aldrich FT-IR library reference numbers for spectra that match the spectral features of the unknown. Visual comparison may then confirm the identity of the unknown substance. Bio-Rad[2] provide a database of many thousands of spectra which cover many classes of substance such as polymers, surfactants, standards etc. The database may be searched at 4 cm^{-1} spacing and provides discrimination between several similar compounds. Sigma and Nicolet produce software designed to be used with the Sigma library of FT-IR spectra[9] which is aimed at identifying an unknown by entering band positions. The location in the library of matching spectra is given by the program. Digital forensic libraries are also offered by some suppliers.

There are numerous computer-based spectral libraries available.[1–15] Spectral libraries based on FT-IR and FT-Raman spectra have the advantage over those complied from digitised dispersive spectra in that the positions of band maxima are more precise and the signal-to-noise ratio is higher. In the main, the search packages available for these packages allow for library searches of unknown spectra. In addition, with most digital search packages it is possible to build one's own user library. Most FT-IR and FT-Raman instrument manufacturers either have their own collections of spectra or have the spectral libraries of others, such as Sadtler, Aldrich, etc, directly available to their customers. Of course, even though computer-based digital FT-IR libraries have become larger, more accurate in their representations of spectra and cheaper, there will still be a place for printed spectral libraries (hardcopies of spectra) for some time to come, although there is no doubt that computer-based digital libraries will eventually be the main medium used. The majority of the digital libraries available are also offered in printed form by suppliers.[1–15]

Spectral databases for FT-IR, NMR, and MS have been reviewed by Warr.[40] There are several reviews of computer methods used in the identification of unknowns[41–44] and a review of the use of computers in quantitative analysis.[45] In order to assist interpretation, there are computer programs which will, when a peak of interest has been highlighted, automatically locate and display

Chart 1.7 Raman – positions and intensities of bands. The range of the position of the maximum absorption of a functional group and its intensity are given in order of decreasing wavenumber. (Note the scale change at $2000\,cm^{-1}$.)

4500	4000	3500	3000	2500	2000 cm⁻¹

Free OH — w, sh OH str

Aliphatic alcohols (hydrogen bonded) — w OH str

OH intramolecular hydrogen bonded — w, sh OH str

Primary amines (dilute solution) — w, dp asym NH₂ str

Secondary amines — w NH str

Primary amines (dilute solution) — w–m, dp NH₂ sym. str

Primary amines (condensed phase) — w–m, br NH₂ str

Primary amides (hydrogen bonded) — w–m, asym NH₂ str

Alkyl acetylenes — w C≡C str

Secondary amides (hydrogen bonded) — w NH str

Carboxylic acids (associated) — w, br OH str

Vinyls, −CH=CH₂ — m asym ≡ CH₂ str (general range)

Aromatic compounds — m−s CH str

Cyclopropyl compounds — m−w asym CH₂ str

Cyclopropyl compounds — m−s sym CH₂ str

Epoxides — m CH asym str

Vinyls, −CH=CH₂ — m sym CH₂ str (general range)

trans (sat) −CH=CH (sat) — m CH str

P−OCH₃ — m−s asym CH₃ str

CH₃CO− — m−s asym CH₃ str

SCH₃ — m−s asym CH₃ str

cis-(sat)CH=CH-(sat) — m CH str

−OCH₃ — m−s CH₃ str

Methylesters — m−w CH₃ asym str

Epoxides — m−s sym CH

Alkanes — m−s CH₃ asym str

Methylesters — m−s CH₃ sym str

CH₃Si⊂ — m−s CH₃ asym str

CH₃OSi⊂ — m−s asym CH₃ str

Alkanes — m−s CH₂ asym str

Alkanes — m−s CH₃ sym str

Alkanes — m−s CH₂ sym str

4500	4000	3500	3000	2500	2000 cm⁻¹

Chart 1.7 (*continued*)

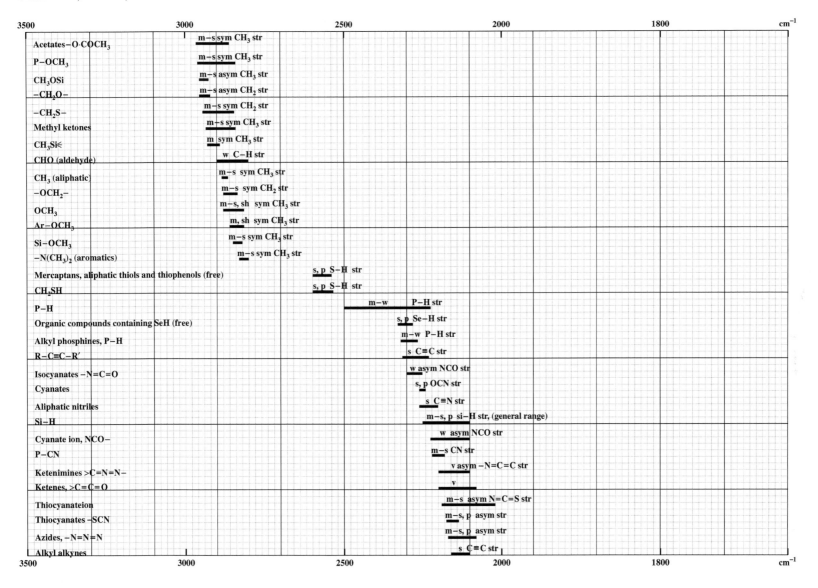

Chart 1.7 (*continued*)

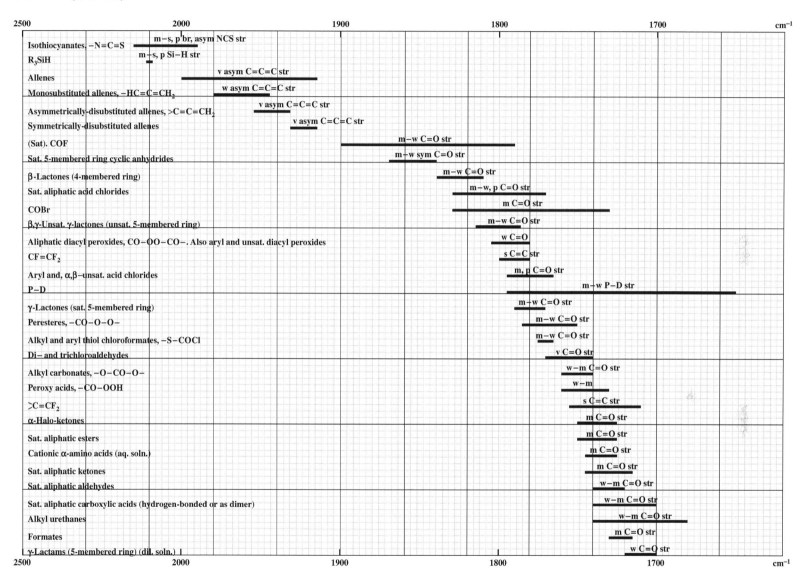

| | 2500 | 2000 | 1900 | 1800 | 1700 | cm⁻¹ |

Chart 1.7 (*continued*)

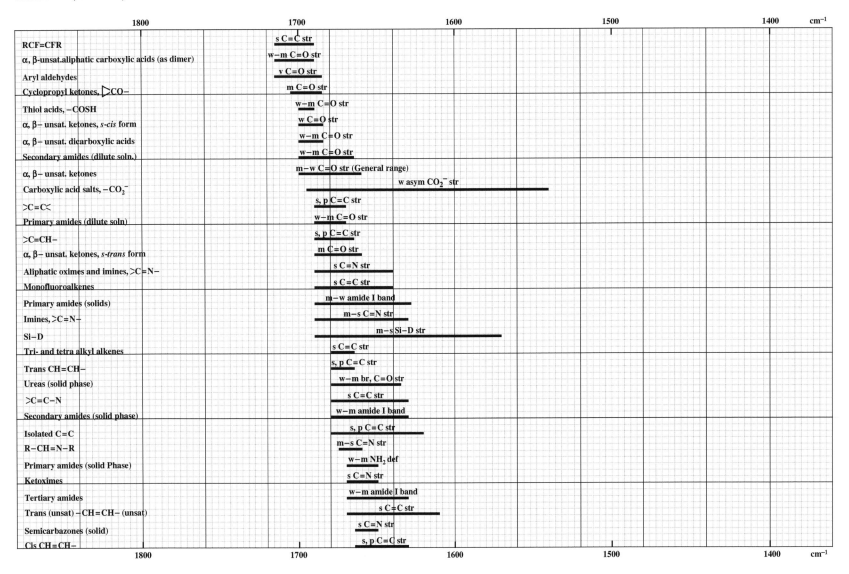

	1800	1700	1600	1500	1400	cm⁻¹

RCF=CFR — s C=C str

α, β-unsat.aliphatic carboxylic acids (as dimer) — w–m C=O str

Aryl aldehydes — v C=O str

Cyclopropyl ketones, ▷CO– — m C=O str

Thiol acids, –COSH — w–m C=O str

α, β– unsat. ketones, *s-cis* form — w C=O str

α, β– unsat. dicarboxylic acids — w–m C=O str

Secondary amides (dilute soln.) — w–m C=O str

α, β– unsat. ketones — m–w C=O str (General range)

Carboxylic acid salts, –CO₂⁻ — w asym CO₂⁻ str

>C=C< — s, p C=C str

Primary amides (dilute soln) — w–m C=O str

>C=CH– — s, p C=C str

α, β– unsat. ketones, *s-trans* form — m C=O str

Aliphatic oximes and imines, >C=N– — s C=N str

Monofluoroalkenes — s C=C str

Primary amides (solids) — m–w amide I band

Imines, >C=N– — m–s C=N str

Si–D — m–s Si–D str

Tri- and tetra alkyl alkenes — s C=C str

Trans CH=CH– — s, p C=C str

Ureas (solid phase) — w–m br, C=O str

>C=C–N — s C=C str

Secondary amides (solid phase) — w–m amide I band

Isolated C=C — s, p C=C str

R–CH=N–R — m–s C=N str

Primary amides (solid Phase) — w–m NH₂ def

Ketoximes — s C=N str

Tertiary amides — w–m amide I band

Trans (unsat) –CH=CH– (unsat) — s C=C str

Semicarbazones (solid) — s C=N str

Cis CH=CH– — s, p C=C str

Chart 1.7 *(continued)*

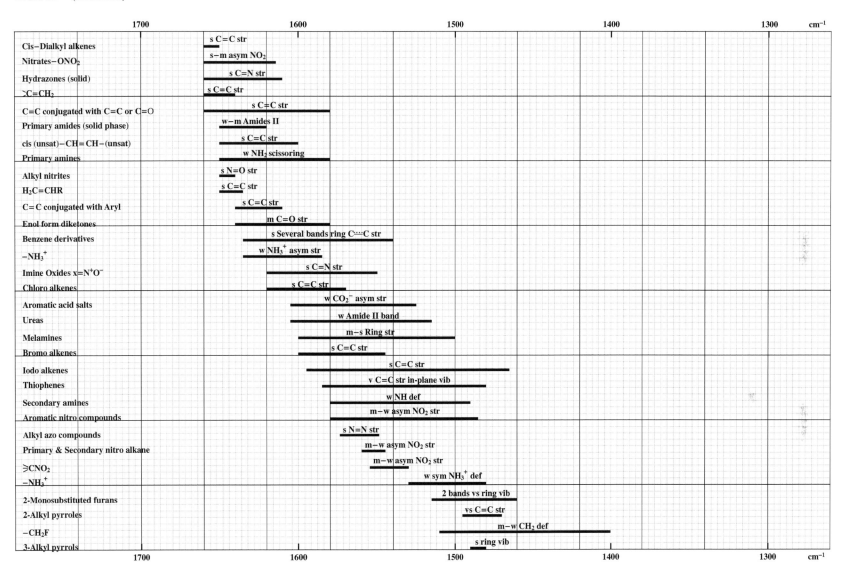

Chart 1.7 *(continued)*

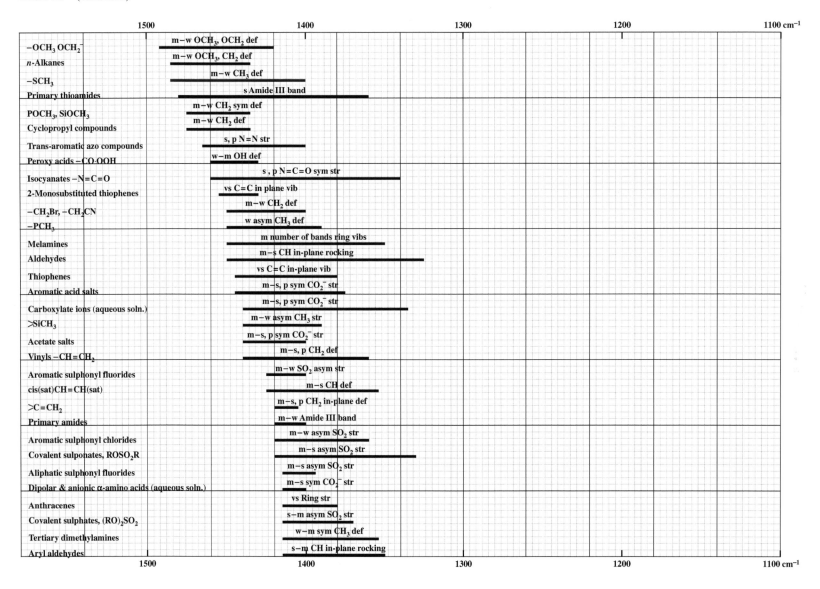

Chart 1.7 *(continued)*

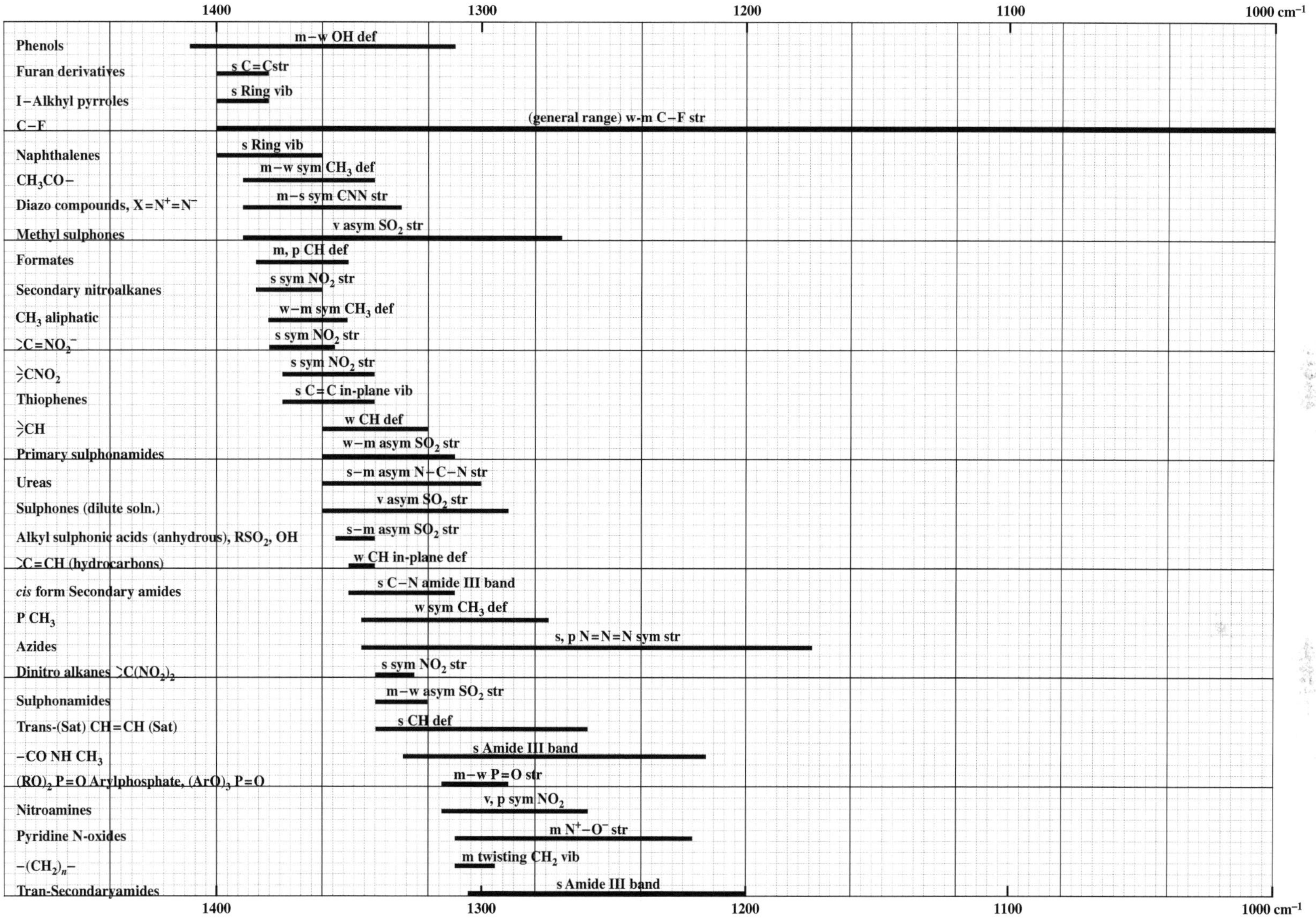

	1400	1300	1200	1100	1000 cm⁻¹

Phenols — m–w OH def

Furan derivatives — s C=C str

1–Alkyl pyrroles — s Ring vib

C–F — (general range) w-m C–F str

Naphthalenes — s Ring vib

CH_3CO- — m–w sym CH_3 def

Diazo compounds, $X=N^+=N^-$ — m–s sym CNN str

Methyl sulphones — v asym SO_2 str

Formates — m, p CH def

Secondary nitroalkanes — s sym NO_2 str

CH_3 aliphatic — w–m sym CH_3 def

$>C=NO_2^-$ — s sym NO_2 str

$>CNO_2$ — s sym NO_2 str

Thiophenes — s C=C in-plane vib

$>CH$ — w CH def

Primary sulphonamides — w–m asym SO_2 str

Ureas — s–m asym N–C–N str

Sulphones (dilute soln.) — v asym SO_2 str

Alkyl sulphonic acids (anhydrous), RSO_2, OH — s–m asym SO_2 str

$>C=CH$ (hydrocarbons) — w CH in-plane def

cis form Secondary amides — s C–N amide III band

P CH_3 — w sym CH_3 def

Azides — s, p N=N=N sym str

Dinitro alkanes $>C(NO_2)_2$ — s sym NO_2 str

Sulphonamides — m–w asym SO_2 str

Trans-(Sat) CH=CH (Sat) — s CH def

–CO NH CH_3 — s Amide III band

$(RO)_2$ P=O Arylphosphate, $(ArO)_3$ P=O — m–w P=O str

Nitroamines — v, p sym NO_2

Pyridine N-oxides — m N^+–O^- str

$-(CH_2)_n-$ — m twisting CH_2 vib

Tran-Secondary amides — s Amide III band

	1400	1300	1200	1100	1000 cm⁻¹

Chart 1.7 (*continued*)

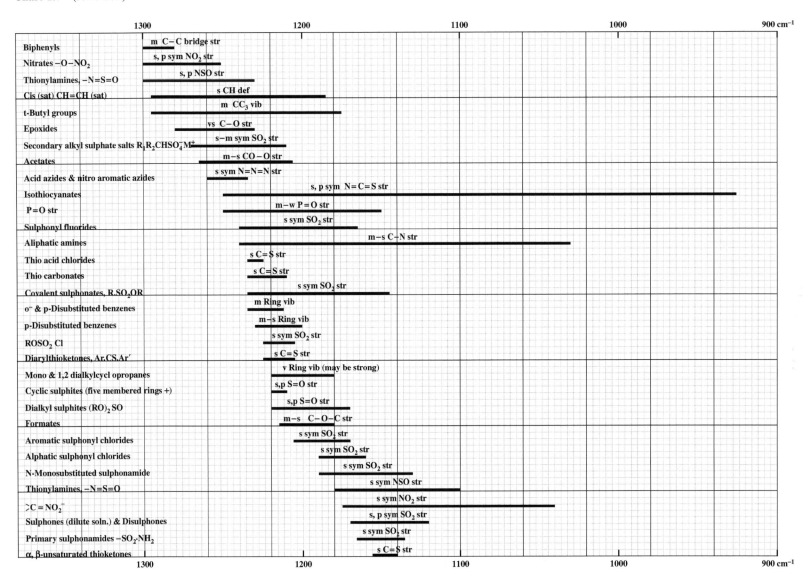

	1300	1200	1100	1000	900 cm⁻¹

Biphenyls — m C–C bridge str
Nitrates –O–NO₂ — s, p sym NO₂ str
Thionylamines, –N=S=O — s, p NSO str
Cis (sat) CH=CH (sat) — s CH def
t-Butyl groups — m CC₃ vib
Epoxides — vs C–O str
Secondary alkyl sulphate salts R₁R₂CHSO₄⁻M⁺ — s–m sym SO₂ str
Acetates — m–s CO–O str
Acid azides & nitro aromatic azides — s sym N=N=N str
Isothiocyanates — s, p sym N=C=S str
P=O str — m–w P=O str
Sulphonyl fluorides — s sym SO₂ str
Aliphatic amines — m–s C–N str
Thio acid chlorides — s C=S str
Thio carbonates — s C=S str
Covalent sulphonates, R.SO₂.OR — s sym SO₂ str
o– & p-Disubstituted benzenes — m Ring vib
p-Disubstituted benzenes — m–s Ring vib
ROSO₂ Cl — s sym SO₂ str
Diarylthioketones, Ar.CS.Ar′ — s C=S str
Mono & 1,2 dialkylcycl opropanes — v Ring vib (may be strong)
Cyclic sulphites (five membered rings +) — s,p S=O str
Dialkyl sulphites (RO)₂ SO — s,p S=O str
Formates — m–s C–O–C str
Aromatic sulphonyl chlorides — s sym SO₂ str
Alphatic sulphonyl chlorides — s sym SO₂ str
N-Monosubstituted sulphonamide — s sym SO₂ str
Thionylamines, –N=S=O — s sym NSO str
>C=NO₂⁻ — s sym NO₂ str
Sulphones (dilute soln.) & Disulphones — s, p sym SO₂ str
Primary sulphonamides –SO₂·NH₂ — s sym SO₂ str
α, β-unsaturated thioketones — s C=S str

	1300	1200	1100	1000	900 cm⁻¹

Chart 1.7 *(continued)*

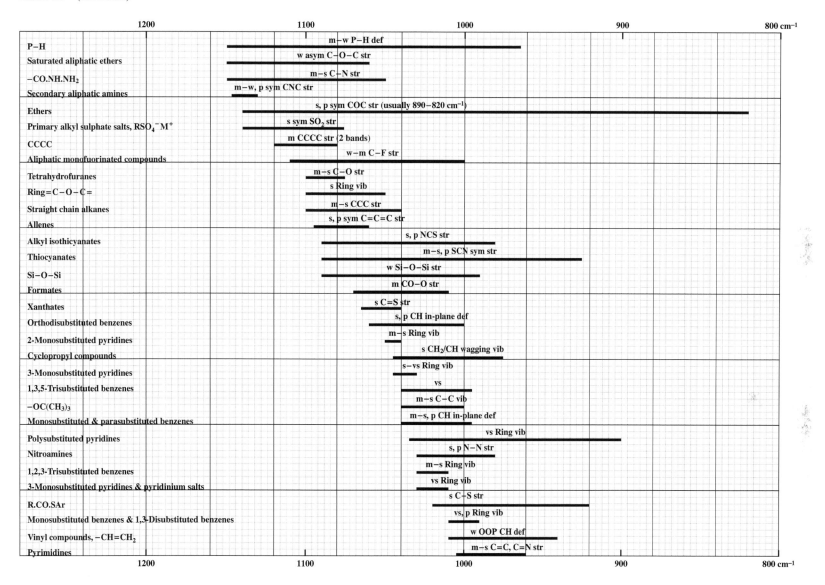

Chart 1.7   (*continued*)

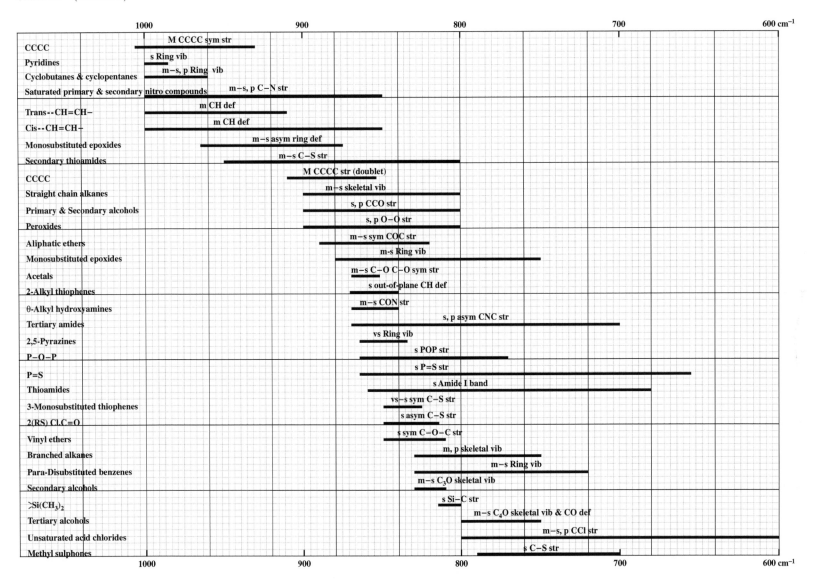

Chart 1.7 (*continued*)

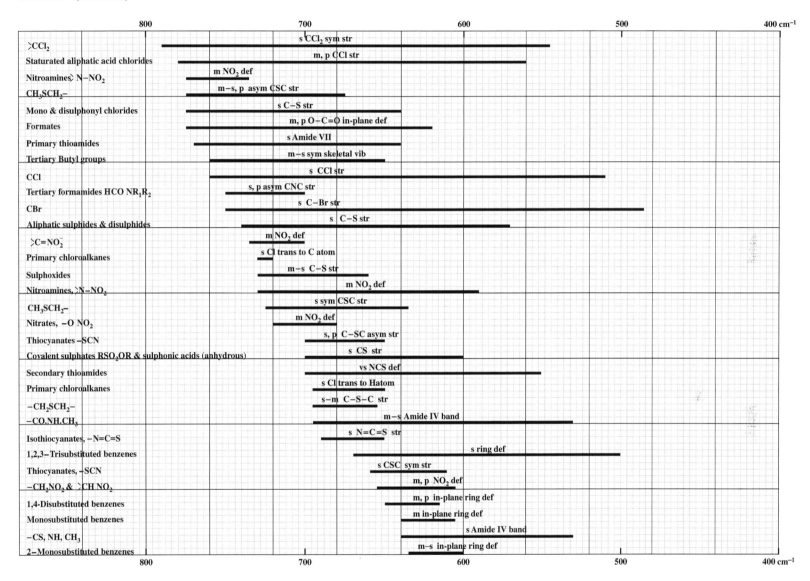

Chart 1.7 (*continued*)

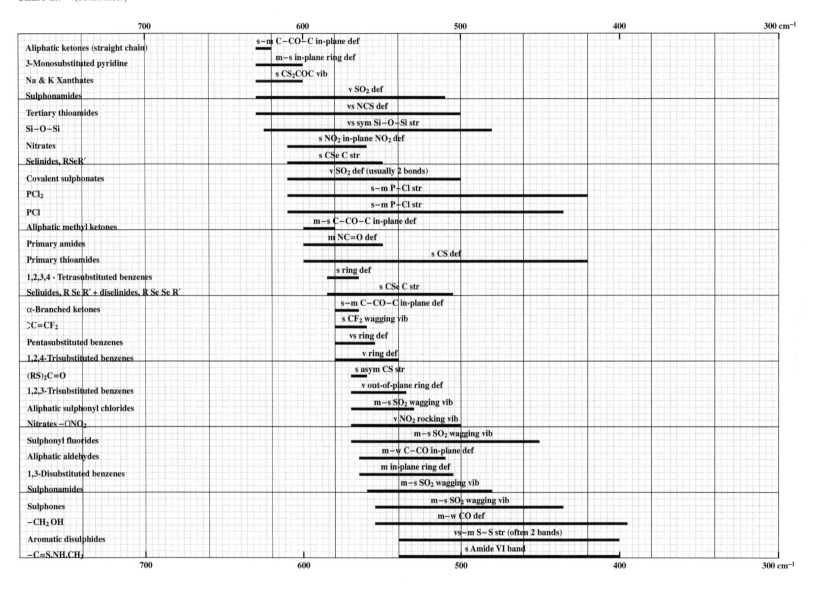

Chart 1.7 (*continued*)

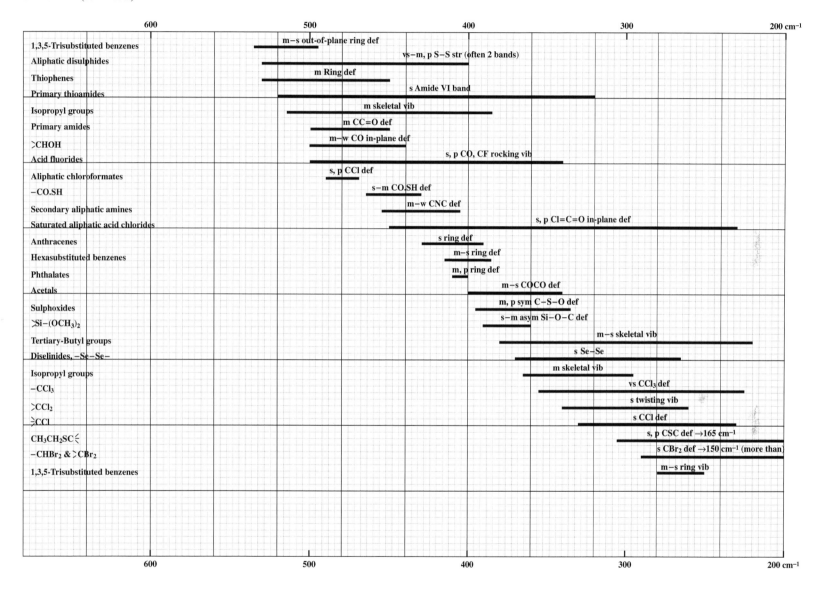

functional groups with that characteristic frequency, for example, Sadtler's IR Mentor and Nicolet's interpretation guide. Such computer programs can help a novice become familiar with infrared interpretation and will no doubt become more sophisticated with time. Normally, various algorithms based on absolute differences, squared differences, first-derivative squared differences, etc, are available for comparison purposes for the elucidation of unknown spectra.

There are numerous published libraries of infrared spectra.[1−33,35−39,53,54] The Coblentz Society[16] publish infrared spectra of numerous compounds, gases and vapours, halogenated hydrocarbons, plasticisers, and industrial chemicals. A large collection of spectra may be found in the Documentation of Molecular Spectroscopy (DMS)[18] system which also covers Raman and microwave spectroscopy. The American Petroleum Institute (API)[19] have published a large collection of spectra, mainly of hydrocarbons and compounds relevant to the petroleum industry. The Infrared Data Committee of Japan (IRDC)[20] published a collection similar to that of DMS. Mecke and Langenbucher[21] have published a small collection of infrared spectra of selected chemical compounds.

Some of the collections of spectra have in recent years been combined so that a more comprehensive collection can be obtained from a single source. For example, Sadtler have extended their polymer spectra by making the collection by Hummel available. However, the old spectra in non-digital form can still be obtained through reference libraries.

Other sources of spectra, generally of a more specialised nature, are available,[22−33,35−39,53,54] as is useful information regarding band positions and assignments.[34−39]

There are excellent books[46−52] available from which an introduction to various aspects of infrared spectroscopy may be obtained. A few of these are given in the Appendix. Of course, there is some degree of overlap of subject matter but the titles of the books generally indicate their contents. References included in the Appendix are, in general, of the review type. At the end of each chapter are given references of a more specialised nature pertinent to that chapter. It is intended that this book, rather than provide a complete bibliography or source of references, should act as a thorough guide to the newcomer to the field.

The use of computer programs to predict spectra from a knowledge of the molecular structure of the sample is still in its infancy. However, although a fair amount of work still needs to be done, there is no doubt that this type of approach will be of great importance to the analysts of the future. Certainly, the experience of a spectroscopist in the characterisation of infrared and Raman spectra will be essential for many years to come, just as is the ability of computer programs to search through libraries of spectra to find the best match to a sample's spectrum.

Final Comment

In the text and tables that follow in subsequent chapters, unless otherwise stated, the comments refer to infrared rather than Raman. Comments specifically aimed at Raman state that this is the case. The reason for this is that infrared is by far the more commonly used technique. Although, in general, the tables given in the chapters have been presented for specific classes of vibration, in some cases it was felt helpful and appropriate to include other types of vibration.

References

1. Sadtler Research Laboratories, 3316 Spring Garden Street, Philadelphia, PA 19104, USA, and PO Box 378, Hemel Hempstead, Herts HP2 7TF, UK.
2. 'FT-IR Digital Library', Bio-Rad, 237 Putnam Avenue, Cambridge, MA 02139, USA, and Hemel Hempstead, Herts HP2 7TF, UK.
3. *The Atlas of Near Infrared Spectra*, Sadtler Research Laboratories, Philadelphia, PA.
4. Sadtler SearchMaster Software Package, Sadtler Research Laboratories.
5. C. J. Pouchert, *Aldrich Library of Infrared Spectra*. 3rd edn, Aldrich Chemical Co. Inc., Milwaukee, WI, USA, 1981, and The Old Brickyard, Gillingham, Dorset SP8 4JL, UK.
6. C. J. Pouchert, *Aldrich Library of FT-IR Spectra*, 1st edn, Vols 1, 2, 1985 and (Vapour Phase) Vol. 3, 1989, Aldrich Chemical Co. Inc., Milwaukee, WI, USA, and The Old Brickyard, Gillingham, Dorset SP8 4JL, UK.
7. Aldrich Vapor-Phase FT-IR Library, as above.
8. Aldrich–Nicolet Digital FT-IR Data Base For Personal Computers, 1985, and Aldrich Chemical Co. Inc., Milwaukee, WI, USA, and Nicolet Analytical Instruments, 5225-1 Verona Road, PO Box 4508, Madison, WI 53711, USA.
9. R .J. Keller, *The Sigma Library of FT-IR Spectra*, Vols 1 and 2, 1986, Sigma Chemical Co., St. Louis, MO, USA.
10. The Sprouse Collections of infrared spectra of solvents, of polymers by ATR, of polymers by transmission and of polymer additives.
11. T. Mills, Georgia State Forensic Library (may be obtained through Nicolet or Stadler).
12. Toronto Forensic Library, Canadian Patents and Development Limited, Ontario Regional Laboratory, Health Protection Branch, of Health and Welfare (may be obtained through Nicolet or Stadler).
13. US Geological Survey Mineral Library.
14. Sadtler/Scholl Polymer Processing Chemicals Library, Sadtler.
15. D. O. Hummel and F. Scholl, *Infrared Analysis of Polymers Resins and Additives, An Atlas*, Vol. 1, part 2, Wiley, New York, 1971; Carl Hansen, Vols 2 and 3, 1984 (also available from Sadtler in digital form).
16. C. Carver (ed.), *Desk Book of Infrared Spectra*, 2nd edn, 1982, Gases and Vapours, 1980, Halogenated Hydrocarbons, 3rd edn, 1984, Plasticisers and Other Additives, 2nd edn, 1985, Regulated and Major Industrial Chemicals, 1983, The Coblentz Society, Kirkwood, MO.

17. *The Sprouse Collection of IR Spectra*, Elsevier, Amsterdam, 1990.
18. *DMS System*, Verlag Chemie GmbH, Weiheim/Bergstrasse, Germany.
19. American Petroleum Institute Research Project 44, Chemistry Department, Agricultural and Mechanical College Texas, College Station, TX 77843, USA.
20. Sanyo Shuppan Boeki Co. Inc., Hoyn Building, 8,2-Chome, Takara-cho, Chuo-ku, Tokyo, Japan.
21. R. Mecke and F. Langenbucher, *Infrared Spectra of Selected Chemical Compounds*, Heyden, London, 1966.
22. D. Welti, *Infrared Vapour Spectra*, Heyden, London, 1970.
23. D. Hansen (ed.), *The Sprouse Collection of Infrared spectra. Book 4, Common Solvents Condensed-Phase, Vapour Phase and Mass Spectra*, Elsevier, Amsterdam, 1990.
24. W. Karcher, R. J. Fordham, J. J. Dubois, P. G. J. M. Glade, and J. A. M. Ligthart (eds), *Spectral Atlas of Polycyclic Aromatic Compounds*, Reidel, Dordrecht, 1983.
25. *British Pharmacopoeia 1980*, 'Infrared Reference Spectra', HM Stationary Office 1980; Supplement 1, 1981; Supplement 2, 1982; Supplement 3, 1984.
26. V. C. Farmer (ed.), *The Infrared Spectra of Minerals*, Mineralogical Society, 1974.
27. K. G. R. Pachler, F. Matlok, and H. U. Gremlich, *Merk FT-IR Atlas*, VCH, New York, 1988.
28. Manufacturing Chemist Association Research Project, Chemistry Department, Agricultural and Mechanical College, Texas, USA.
29. A. F. Ardyukova, O. P. Shkurko, and V. F. Sedova, *Atlas of Spectra of Aromatic and Heterocyclic Compounds*, Nauka Sib. Otd, Novosibirsk, 1974.
30. *Infrared and Ultraviolet Spectra of Some Compounds of Pharmaceutical Interest*, revised edn, Association of Official Analytical Chemists, Washington, DC, 1972.
31. R. L. Davidovich, T. A. Kaidolova, T. F. Levchishina, and V. I. Sergineko, *Atlas of Infrared Absorption Spectra and X-ray Measurement Data for Complex Group IV and V Metal Fluorides*, Nauka, Moscow, 1972.
32. K. Dobriner, E. R. Katzenellenbogen, and R. N. Jones, *Infrared Absorption Spectra of Steroids, An Atlas*, Vol. I, Interscience, New York, 1953.
33. G. Roberts, B. S. Gallagher, and R. N. Jones, *Infrared Absorption Spectra of Steroids, An Atlas*, Vol. II, Interscience, New York, 1958.
34. K. Yamaguchi, *Spectral Data of Natural Products*, Elsevier, Amsterdam, 1970. (Contains spectra and data on infrared, ultraviolet, NMR, mass spectroscopy, etc.)
35. H. A. Szymanski, *Interpreted Infrared Spectra*, Plenum, New York, 1971.
36. H. A. Szymanski, *Infrared Band Hand Book*, Vols I–III, Plenum, New York, 1964, 1966, 1967, and *Correlation of Infrared and Raman Spectra of Organic Compounds*, Hertillon, 1969.
37. R. A. Nyquist, *The Interpretation of Vapour-Phase Spectra*, Sadtler, 1984.
38. R. A. Nyquist and R. O. Kagel, *Infrared Spectra of Inorganic Compounds* $(3800-45\,cm^{-1})$, Academic Press, New York, 1971.
39. L. Lang, S Holly, and P. Sohar (eds), *Absorption Spectra in the Infrared Region*, Butterworth, London, 1974.
40. W. A. Warr, *Chemom. Intell. Lab. Syst.*, 1991, **10**, 279.
41. M. Meyer, I. Weber, R. Sieler, and H. Hobart, *Jena Rev.*, 1990, **35**, 16.
42. W. O. George and H. A. Willis, (eds), *Computer Methods in UV–Visible and IR Spectroscopy*, Royal Soc. Chem., Cambridge, 1990.
43. H. J. Luinge, *Vib.Spectrosc.*, 1990, **1**, 3.
44. D. F. Averill *et al.*, *J. Chem.Inf. Comput. Sci.*, 1990, **30**, 133.
45. R. A. Cromcobe, M. L. Olson, and S. L. Hill, *Computerised Quantitative Infrared Analysis*, ASTM STP 934, G. L. McGlure (ed.), American Soc. For Testing Materials, Philadelphia, PA, 1987, 95–130.
46. F. F. Bentley, D. L. Smithson, and A. L. Rozek, *Infrared Spectra and Characteristic Frequencies* $\sim 700-300\,cm^{-1}$, Interscience, New York, 1968.
47. N. B. Colthurp, L. H. Daly, and S. E. Wiberley, *Introduction to Infrared and Raman Spectroscopy*, Academic Press, Boston, MA, 1990.
48. D. Lin-Vien, N. B. Colthurp, W. G. Fateley and J. G. Grasselli, *The Handbook of IR and Raman Characteristic Frequencies of Organic Molecules*, Academic Press, New York, 1991.
49. R. G. Messerschmidt and M. A. Harthcock (eds), *Infrared Microscopy*, Marcel Dekker, New York, 1988.
50. J. R. Durig (ed.), *Spectra and Structure*, Elsevier, Amsterdam, 1982.
51. J. R. Ferraro and K. Krishnan (eds), *Practical FTIR, Industrial and Laboratory Chemical Analysis*, Academic Press, New York, 1990.
52. J. L. Koenig, *Spectra of Polymers*, Amer. Chem. Soc., Dept. 31, 1155 Sixteenth St., N.W. Washington DC 20036.
53. V. A. Koptyug (ed.), *Atlas of Spectra of Organic Compounds in the Infrared, UV and Visible Regions*, Nos 1–32, Novosibirsk, 1987.
54. B. Schrader, *Raman/Infrared Atlas of Organic Compounds*, VCH, Weinheim, Germany, 1989.

2 Alkane Group Residues: C–H Group

Alkane Functional Groups

Residual alkane groups are found in a very large number of compounds and hence are an extremely important class.[1,22] Four types of vibration are normally observed, namely the stretching and the deformation of the C–H and the C–C bonds. The C–H vibration frequencies of the methyl and methylene groups fall in narrow ranges for saturated hydrocarbons. However, atoms directly attached to –CH_3 or –CH_2– may result in relatively large shifts in the absorption frequencies. In general, the effect of electronegative groups or atoms is to increase the C–H absorption frequency.

CH stretching vibrations 3000–2800 cm^{-1} (3.33–3.57 μm) result in bands of medium-to-strong intensity in both infrared and Raman spectra, as do the CH_3 and CH_2 deformation vibrations 1470–1400 cm^{-1} (6.80–7.14 μm). The CH_3 symmetric deformation vibration at ∼1380 cm^{-1} (∼7.25 μm), in general, gives medium-to-strong intensity bands in infrared spectra and weak-to-medium bands in Raman spectra except in the presence of an adjacent unsaturated group when the intensity is greatly increased. The C–C stretching bands of alicyclics and aliphatic residues, 1300–600 cm^{-1} (7.69–16.67 μm), are of weak-to-medium intensity in infrared spectra and medium-to-strong intensity in Raman spectra. The C–C deformation vibrations, 400–250 cm^{-1} (25.00–40.00 μm), are generally weak in infrared spectra and of strong-to-medium intensity in Raman spectra.

Alkane C–H Stretching Vibrations

For aliphatic hydrocarbons, with the exception of small ring compounds, the C–H stretching vibrations occur in the region 2975–2840 cm^{-1} (3.36–3.52 μm).[2-4] In strained ring systems[22-25], the frequency of the methylene C–H stretching vibration is increased, e.g. cyclopropanes absorb near 3050 cm^{-1} (3.28 μm). The CH_3 asymmetric stretching vibration occurs at 2975–2950 cm^{-1} (3.36–3.39 μm) and may easily be distinguished from the nearby CH_2 absorption at about 2930 cm^{-1} (3.41 μm). The symmetric CH_3 stretching absorption band occurs at 2885–2865 cm^{-1} (3.47–3.49 μm), and that of the methylene group at 2870–2840 cm^{-1} (3.49–3.52 μm).

The position of the CH_3 symmetric stretching vibration band may be altered due to an adjacent group, whereas the asymmetric stretching band is relatively insensitive, e.g. for the group –O–CH_3[5,6] the CH_3 symmetric stretching band occurs near 2850 cm^{-1} (∼3.51 μm) whereas the asymmetric stretching band generally occurs in the normal position (similarly for $\diagdown$N—CH_3).[6,7] A useful band for the identification of the OCH_3 and NCH_3 groups is that due to the CH_3 symmetric stretching vibration, which is sharp, of medium-to-strong intensity and is usually found in the region 2895–2815 cm^{-1} (3.45–3.55 μm). Correlations involving C–H stretching vibrations have been studied.[2] Information has also been derived from the intensities of these bands.[8,9] In the presence of a double bond adjacent to a methyl or a methylene group, the symmetric stretching vibration band splits into two. Methyl groups attached to unsaturated carbons, including aromatic groups, absorb in the range 3010–2905 cm^{-1} (3.32–3.45 μm) due to the asymmetric stretching vibration, the symmetric stretching band occurring in the region 2945–2845 cm^{-1} (3.41–3.53 μm). Electron-withdrawing groups directly attached to the CH_3 group result in the stretching vibrations occurring at slightly higher frequencies than those for saturated hydrocarbons. In polar molecules, a series of bands is observed between 2980 and 2700 cm^{-1} (3.36–3.70 μm) due to interactions between the fundamental vibrations of the methyl group and the overtones of their deformation vibrations. The n-propyl group has four medium-to-strong (overlapping) CH asymmetric stretching bands in the region 2990–2900 cm^{-1} (3.34–3.45 μm) and three overlapping bands of medium-to-strong intensity may be observed between 2940 and 2840 cm^{-1} (3.40–3.52 μm) due to the symmetric CH stretching vibrations. Most t-butyl compounds have three moderate-to-strong absorption bands in the region 2990–2930 cm^{-1} (3.34–3.41 μm) due to the asymmetric stretching vibrations. The symmetric stretching vibrations occur in the region 2950–2850 cm^{-1} (3.39–3.51 μm) with aromatic t-butyl compounds absorbing in the region 2915–2860 cm^{-1} (3.44–3.50 μm).

Alkane C–H *Deformation Vibrations*

The methyl groups of hydrocarbons give rise to two vibration bands, the asymmetric deformation band occurring at $1465–1440 \, cm^{-1}$ ($6.83–6.94 \, \mu m$) and the symmetric band at $1390–1370 \, cm^{-1}$ ($7.19–7.30 \, \mu m$). The former band is often overlapped by the $-CH_2-$ scissor vibration band occurring at $1480–1440 \, cm^{-1}$ ($6.76–6.94 \, \mu m$). The intensity of the methyl symmetric vibration band relative to the higher-frequency band (due to scissor $-CH_2-$ and/or asymmetric $-CH_3$ vibrations) may be used to indicate the relative number of methyl groups in the sample.

The presence of adjacent electronegative atoms or groups can alter the position of the methyl symmetric band significantly, its range being $1470–1260 \, cm^{-1}$ ($6.80–7.94 \, \mu m$), whereas the asymmetric band is far less sensitive, its range being $1485–1400 \, cm^{-1}$ ($6.73–7.14 \, \mu m$).[1,20] The position of the symmetric deformation band is dependent on the electronegativity of the element or group to which the methyl group is bonded and on its position in the Periodic Table. The more electronegative the element, the higher the frequency.[27–31]

The CH_3 symmetric deformation band is of medium intensity in infrared spectra and weak in Raman, unless directly adjacent to an unsaturated group, a carbonyl group or an aromatic.

Table 2.1 Alkane C–H stretching vibrations for alkane functional groups as part of a residual saturated hydrocarbon portion of the molecule (attached to a carbon atom)

Functional Groups	Region		Intensity		Comments
	cm^{-1}	μm	IR	Raman	
$-CH_3$ (aliphatic)	2975–2950	3.36–3.39	m–s	m	asym ⎱ frequency raised by electronegative
	2885–2865	3.47–3.49	m	m–s	sym ⎰ substituents
$-CH_2-$ (acyclic)	2940–2915	3.40–3.45	m–s	m–s	asym ⎱ frequency raised by electronegative
	2870–2840	3.49–3.52	m	m–s	sym ⎰ substituents
＼CH(acylic) ／	2890–2880	3.46–3.47	w	m	
Ar–CH₃	3000–2965	3.33–3.37	m–s	m–s	asym str, see refs 14, 15
	2955–2935	3.38–3.41	m–s	m–s	asym str, lower part of range for *ortho*-substituted compounds
	2930–2920	3.41–4.23	m–s	m–s	sym str
	2870–2860	3.48–3.50	m–w		def overtones
	2830–2740	3.53–3.65	w–m		def overtones. Fermi resonance enhanced.
(Unsat.)–CH₃	3010–2950	3.32–3.39	m–s	m–s	asym CH_3 str, usually $\sim 3000 \, cm^{-1}$
	2995–2905	3.34–3.44	m–s	m–s	asym CH_3 str (not acetylenes.)
	2945–2880	3.40–3.47	m–s	m–s	sym CH_3 str
CH_3Z, where $Z = -CR_3$, $-C(sat \; group)_3$, $-C(halogen)_3$, ＞CHOH, ＞CHCN	3035–2985	3.29–3.35	m–w	m–s	asym CH_3 str
	2975–2935	3.36–3.41	m–w	m–s	sym CH_3 str
Cyclopropanes, $-CH_2-$	3105–3070	3.22–3.26	m	m–w	asym str, see ref. 18
	3060–3020	3.27–3.31	v	m	
	3040–2995	3.29–3.34	m–w	m–s	sym str
	3020–3000	3.31–3.33	m	m	
Cyclobutanes, $-CH_2-$	3000–2975	3.33–3.36	m	m–w	asym str
	2925–2875	3.42–3.48	m	m	sym str
Cyclopentanes, $-CH_2-$	2960–2950	3.38–3.39	m	m	asym str
	2870–2850	3.48–3.51	m	m	sym str
Cyclohexanes, $-CH_2-$					(As for acyclic $-CH_2-$ groups, see ref. 19)

For the *n*-propyl group, the symmetric CH_3 deformation occurs near 1375 cm^{-1} ($\sim$7.30 μm) and the methylene rocking vibration occurs near 740 cm^{-1} ($\sim$13.51 μm). The *t*-butyl asymmetric deformations occur at 1495–1435 cm^{-1} (6.69–6.97 μm) and are of medium-to-strong intensity. The symmetric deformation bands are observed between 1420–1350 cm^{-1} (7.04–7.40 μm) although for most molecules the range is 1400–1370 cm^{-1} (7.14–7.30 μm). Hence most *t*-butyl groups have a strong band near 1365 cm^{-1} (7.32 μm) and a slightly weaker band near 1390 cm^{-1} (7.19 μm). The band normally found near 1380 cm^{-1} (7.25 μm) is split into two by resonance which occurs when two or three methyl groups are attached to a single carbon atom. The presence of a tertiary butyl group may be confirmed by its skeletal vibration bands which occur near 1255 cm^{-1} (7.97 μm) and 1210 cm^{-1}

(8.27 μm), whereas the corresponding bands for the isopropyl group are usually found near 1170 cm^{-1} (8.55 μm) and 1145 cm^{-1} (8.73 μm).

Methyl rocking vibrations[10] are generally weak and not very useful for assignment purposes even though they are mass sensitive. For *n*-alkanes, a band due to the CH_2 wagging vibration occurs near 1305 cm^{-1} (7.66 μm), the intensity of this band being less than the band at $\sim$1460 cm^{-1} (6.85 μm) while being dependent on the number of CH_2 groups present.

The CH_2 wagging, rocking and twisting vibrations which occur in the region 1430–715 cm^{-1} are usually of weak intensity in infrared spectra and of medium intensity in Raman. The bands due to $\diagdown$CH deformation are weak in both infrared and Raman spectra.

Table 2.2 Alkane C–H deformation vibrations for alkane functional groups as part of a residual saturated hydrocarbon portion of molecule (attached to a carbon atom)

Functional Groups	Region		Intensity		Comments
	cm^{-1}	μm	IR	Raman	
–CH$_3$ (aliphatic)	1465–1440	6.83–6.94	m	m	asym ⎫ Frequency raised by
	1390–1370	7.19–7.30	m–s	w–m	sym (characteristic of C–CH$_3$) ⎬ electronegative substituents
$\diagdown$C(CH$_3$)$_2$	1385–1335	7.22–7.49	m–s	w–m	Two bands of almost equal intensity
–C(CH$_3$)$_3$	1475–1435	6.78–6.97	m	w–m	asym CH$_3$ def vib
	1420–1375	7.04–7.27	m	w	CH$_3$ sym bending vib
	1395–1350	7.17–7.41	m–s	w	CH$_3$ sym bending vib. Often $\sim$1365 cm^{-1}.
CH$_3$Z, where Z = –CR$_3$, –C(sat group)$_3$, –C(halogen)$_3$, $\diagdown$CHOH, $\diagdown$CHCN	1465–1430	6.83–6.99	m–w	w	asym CH$_3$ def vib
$\diagdown$CH$_2$	1410–1350	7.09–7.41	m–s	w	CH$_3$ sym bending vib
	1480–1440	6.76–6.94	m	m	Scissor vib
$\diagdown$CH	1360–1320	7.35–7.58	w	w	
–(CH$_2$)$_n$–	1485–1445	6.73–6.92	m	w–m	def vib
	1305–1295	7.66–7.72	–	m	Not usually observed in IR. Intensity increases with *n*.
(Unsat.)–CH$_3$	1480–1430	6.76–6.99	v	m	asym CH$_3$ def vib, usually medium intensity
	1470–1400	6.80–7.14	v	m	asym CH$_3$ def vib, usually medium intensity (not acetylenes)
	1405–1355	7.12–7.38	m–s	s	sym CH$_3$ def vib
Cyclopropanes	1420–1400	7.04–7.14	s	w–m	
	1365–1295	7.33–7.22	s	w	
Cyclobutanes	1450–1440	6.90–6.94	s	m	
	1360–1250	7.35–8.00	w–m	m	
	1245–1220	8.21–8.20	s–m	w	

As mentioned previously, in the spectra of hydrocarbons, the methylene deformation band is found in the region 1480–1440 cm^{-1} (6.76–6.94 μm), but in the presence of adjacent unsaturated groups this band is found near 1440 cm^{-1} (6.94 μm). With an adjacent chlorine, bromine, iodine, sulphur, or phosphorus atom, or a nitrile, nitro-, or carbonyl group, this band occurs at 1450–1405 cm^{-1} (6.90–7.12 μm).

Alkane C–C Vibrations: Skeletal Vibrations

The skeletal vibrations of alkane residues are often weak in infrared and usually of weak-to-medium intensity in Raman spectra. Of the skeletal vibrations, the C–C stretching absorptions occur in the region 1260–700 cm^{-1} (7.94–14.29 μm) and are normally weak and of little use in assignments. Dimethyl quaternary carbon compounds have a characteristic absorption near 1180 cm^{-1} (8.48 μm). The C–C deformation bands occur below 600 cm^{-1} (16.67 μm)[11,17] and these also are weak. Straight-chain alkanes have two bands, one at 540–485 cm^{-1} (18.52–20.62 μm) and the other near 455 cm^{-1} (21.98 μm). The former band is usually slightly more intense than the second band and tends to the higher frequency end of the range as the length of the chain increases. An exception is n-pentane which has only one band, near 470 cm^{-1} (21.28 μm). Branched alkanes not containing methyl or ethyl groups have at least one band in the region 570–445 cm^{-1} (17.54–22.47 μm). Alkanes with three or more branches absorb near 515 cm^{-1} (19.42 μm). Straight-chain

Table 2.3 Alkane C–C skeletal vibrations for alkane functional groups as part of a residual saturated hydrocarbon portion of the molecule (attached to a carbon atom)

| Functional Groups | Region | | Intensity | | Comments |
	cm^{-1}	μm	IR	Raman	
$\diagdown$C(CH$_3$)$_2$	1175–1165	8.51–8.58	m	w	C–C str. If no hydrogen on central carbon then one band at ~1190 cm^{-1}
	1150–1130	8.90–8.85	v	w	Rocking vib
	1060–1040	9.43–9.62	w		
	955–900	10.47–11.11		m, p	
	840–790	11.90–12.66	w	m, p	
	495–490	20.20–20.41	w	m, p	
	320–250	31.25–40.00	–	m	
–C(CH$_3$)$_3$	1255–1245	7.98–8.03	m	m	
	1225–1165	8.17–8.58	m	m	
	~1000	~10.00	w–m	m–s	
	930–925	10.75–10.81	m	m	
	360–270	27.78–37.04	–	m	
$\diagdown$C–CH$_3$	~970	~10.31	m	w	CH$_3$ rocking vib
(Unsat.)–CH$_3$	1130–1000	8.85–10.00	w–m	w	Rocking vib
	1060–900	9.43–11.11	w–m	w	Rocking vib
	245–120	40.82–83.33			Torsional vib
CH$_3$Z, where Z = –CR$_3$, –C(sat group)$_3$, –C(halogen)$_3$, $\diagdown$CHOH, $\diagdown$CHCN	1080–960	9.26–10.42	w	w	CH$_3$ rocking vib
$\diagdown$C–CH$_2$–CH$_3$	~925	~10.81	m	w	CH$_3$ rocking vib
–CH(C$_2$H$_5$)$_2$	510–505	19.61–19.80	w		

(continued overleaf)

Table 2.3   (*continued*)

Functional Groups	Region		Intensity		Comments
	cm^{-1}	μm	IR	Raman	
Straight-chain alkanes	1175–1120	8.51–8.93	m	m	Doublet
	1100–1040	9.09–9.62	–	m–s	CCC str. May be strong in Raman
	900–800	11.11–12.50	–	m–s	May be strong in Raman
	540–485	18.52–20.62	w	m	} not *n*-pentane
	~455	~21.98	w	m	
	~300	~33.33	–	w, br	
Branched alkanes	1175–1165	8.51–8.58	m	w	
	1170–1140	8.55–8.77	m	w	
	1060–1040	9.43–9.62	–	m	
	950–900	10.53–11.11	–	m	
	830–800	12.05–12.50	–	m, p	
	570–445	17.54–22.47	w–m	m	At least one band
	470–440	21.28–22.73	w–m	m	
	320–250	31.25–40.00	–	m–s	
Monobranched alkanes	570–540	17.86–18.52	w–m	m	
	470–440	21.28–22.73	w–m	m	
Dibranched alkanes not possessing CH$_3$ or C$_2$H$_5$	555–535	18.02–18.69	w	m	
3,3-Dibranched alkanes	~530	~18.87	w	m	
2,2-Dibranched alkanes	~490	~20.41	w	m	
Alkanes with three or more branches	~515	~19.42	w	m	
–(CH$_2$)$_n$–	1305–1295	7.66–7.72	–	s	Twisting vib
–(CH$_2$)$_n$–(*n*>3)	725–720	13.79–13.89	w–m	–	} Rocking vib; splits into two components in the crystalline phase } Usually very weak in Raman
–(CH$_2$)$_3$–	735–725	13.61–13.79	w–m	w	
–(CH$_2$)$_2$–	745–735	13.42–13.61	w–m	w	
–CH$_2$–	785–770	12.74–12.99	w–m	–	
CCCC	1120–1090	8.93–9.17	–	m	CCCC sym str
	1110–1080	9.01–9.26	–	m	CCCC asym str
	1005–930	9.95–10.75	–	m	CCCC sym str
	910–855	10.99–11.70	–	m	Doublet
C C C C (C)	1255–1200	7.97–8.33	m–w	m	Two bands
	750–650	13.33–15.38	–	s, p	
Methyl benzenes	1070–1010	9.34–9.90	m	w	Rocking vib
	390–260	25.64–38.46	m		In-plane bending vib of aromatic C–CH$_3$ bond
Ethyl benzenes	565–540	17.70–38.46	m–s		In-plane bending vib of =C–C–C group
Isopropyl benzenes	545–520	18.35–19.23	m–w		In-plane bending vib of =C–C–C group
Propyl and butyl benzenes	585–565	17.09–17.70	m		Two bands
Cyclopropanes	1200–1180	8.50–8.47	s–m	v	May be strong in Raman
	1050–1000	9.52–10.00	w–m	v	Often ~1020 cm^{-1}
	960–900	10.42–11.11	s	v	Ring vib
	870–850	11.49–11.76	v	s–m	Ring vib. Often absent but may be strong
	540–500	18.52–20.00	v		

Table 2.3 (*continued*)

Functional Groups	Region		Intensity		Comments
	cm^{-1}	μm	IR	Raman	
Saturated aliphatic cyclopropanes	470–460	21.28–21.74	s		
Cyclobutanes	1000–960	10.00–10.42	w	s–m, p	Ring vib. (CH$_2$ scissoring vib, ~1445 cm^{-1})
	930–890	10.75–11.24	m–w	m–w	See ref. 16
	780–700	12.82–14.29	s		
	640–625	15.63–16.00	m–w	m	Ring def vib
	580–490	17.24–20.41	s		
	180–140	55.56–71.43	w	w	Ring puckering vib
Alkyl cyclobutanes	580–530	17.24–18.87	s		
Cyclopentanes	1000–960	10.00–10.42	w	s	
	930–890	10.75–11.24	w	s–m	
	595–490	16.81–20.41	s		
Saturated aliphatic cyclopentanes	585–530	17.09–18.87	s		
Cyclohexanes	1055–1000	9.48–10.00	w	v	See ref. 19
	1015–950	9.86–10.53	w	s	
	~900	~11.11	s	m	
	570–435	17.54–22.99	v	m	

Table 2.4 C–H stretching vibrations for alkane residues attached to atoms other than saturated carbon atoms (excluding olefines)

Functional Groups	Region		Intensity		Comments
	cm^{-1}	μm	IR	Raman	
Methyl groups					
–O–CH$_3$	3030–2950	3.30–3.39	w–m	m–s	asym CH$_3$ str. Aromatic compounds 3005–2965 cm^{-1}
	2985–2920	3.35–3.42	w–m	m–s	asym CH$_3$ str. May extend up to 3015 cm^{-1}. For ethers usually 2850 cm^{-1}.
	2880–2815	3.47–3.55	m, sh	m–s	sym CH$_3$ str, sharp. May extend to 2960 cm^{-1} (overtone, see refs 5,6)
Ar–O–CH$_3$	3005–2965	3.33–3.37	w–m	m–s	asym CH$_3$ str. Usually 2985 cm^{-1}.
	2975–2935	3.36–3.41	w–m	m–s	asym CH$_3$ str. Usually 2950 cm^{-1}.
	2860–2815	3.50–3.55	m	m	sh, usually well separated, sym CH$_3$ str. Usually 2850 cm^{-1}.
–SCH$_3$	3040–2980	3.29–3.36	w–m	m–s	asym CH$_3$ str
	3030–2935	3.30–3.41	m	m–s	sym CH$_3$ str
	3000–2840	3.33–3.52	m	m–s	sym CH$_3$ str
CH$_3$SCH$_2$–	3000–2980	3.33–3.36	w–m	m–s	asym CH$_3$ str
	2980–2960	3.36–3.38	m	m–s	asym CH$_3$ str
	2975–2945	3.36–3.44	m	m–s	asym CH$_2$ str
	2930–2910	3.41–3.44	s	m–s	sym CH$_3$ str
	2915–2855	3.43–3.50	m	m–s	sym CH$_2$ str

(*continued overleaf*)

Table 2.4 (*continued*)

Functional Groups	Region cm^{-1}	Region µm	Intensity IR	Intensity Raman	Comments
RSCH$_3$	2995–2955	3.34–3.38	m	m–s	asym CH$_3$ str
	2900–2865	3.45–3.49	m	m–s	sym CH$_3$ str
N–CH$_3$ (amines and imines)	2820–2760	3.55–3.62	s	m–s	sym CH$_3$ str, general range, see refs 6, 7
N–CH$_3$ (aliphatic amines)	2805–2780	3.56–3.60	s	m–s	sym CH$_3$ str, $\diagdown$NCH$_2$– band may also occur in this region
N–CH$_3$ (aromatic amines)	2820–2810	3.55–3.56	s	m–s	sym CH$_3$ str
–N(CH$_3$)$_2$ (aliphatics)	2825–2810	3.54–3.56	s	m–s	sym CH$_3$ str, see ref. 7
	2775–2765	3.60–3.62	s	m–s	
–N(CH$_3$)$_2$ (aromatics)	2830–2800	3.53–3.57	s	m–s	sym CH$_3$ str
Amides, CH$_3$NH·CO–	3000–2940	3.33–3.40	m–s	m–s	asym CH$_3$ str
	2990–2900	3.34–3.45	m–s	m–s	asym CH$_3$ str
	2920–2825	3.42–3.54	m–s	m–s	sh, sym CH$_3$ str
CH$_3$–CO–	3045–2965	3.28–3.37	w–m	m–s	asym CH$_3$ str
	3010–2960	3.32–3.38	w–m	m–s	asym CH$_3$ str
	2970–2840	3.37–3.52	m	m–s	sym CH$_3$ str
CH$_3$–CO– (unsat group or Ar)	3030–2970	3.30–3.37	m–w	m–s	asym CH$_3$ str. Overlapped by ring CH str bands.
	3000–2930	3.33–3.41	m–w	m–s	asym CH$_3$ str
	2950–2850	3.39–3.51	m	m–s	sym CH$_3$ str
Acetates, –O·CO·CH$_3$	3050–2980	3.28–3.36	m–w	m–s	asym CH$_3$ str
	3030–2950	3.30–3.39	m–w	m–s	asym CH$_3$ str
	2960–2860	3.38–3.50	m	m–s	sym CH$_3$ str
Thioactetates, –S·CO·CH$_3$	3010–2990	3.32–3.34	m–w	m	asym CH$_3$ str
	3000–2980	3.33–3.36	m–w	m	asym CH$_3$ str
	2930–2910	3.41–3.44	m	m	sym CH$_3$ str
Acetamides, CH$_3$—CO—N$\diagup\diagdown$	3010–2970	3.32–3.37	w	m	asym CH$_3$ str
	3000–2980	3.33–3.36	w–m	m	asym CH$_3$ str
	2945–2855	3.40–3.50	m	m	sym CH$_3$ str
Methyl esters, CH$_3$O·CO·–	3050–2980	3.28–3.36	m–w	m–w	asym CH$_3$ str
	3030–2950	3.30–3.39	m–w	m–w	asym CH$_3$ str
	3000–2940	3.33–3.40	m–w	m–w	sym CH$_3$ str
CH$_3$O·CS–	3040–2990	3.29–3.34	m–w	m–w	asym CH$_3$ str
	3010–2985	3.32–3.35	m–w	m–w	asym CH$_3$ str
	2960–2920	3.38–3.42	m–w	m–w	sym CH$_3$ str
CH$_3$O·SO–	3040–2990	3.29–3.34	m–w	m–w	asym CH$_3$ str
	3025–2975	3.31–3.36	m–w	m–w	asym CH$_3$ str
	2965–2915	3.37–3.43	m–w	m–w	sym CH$_3$ str
Z–CH$_3$, Z=CN, NH$_2$, –NHCO	3060–2950	3.27–3.39	m–s	m–w	asym CH$_3$ str
	3045–2900	3.28–3.45	m–s	m–w	asym CH$_3$ str
	2945–2785	3.40–3.59	m–s,	m	sym CH$_3$ str

Table 2.4   (*continued*)

Functional Groups	Region		Intensity		Comments
	cm^{-1}	μm	IR	Raman	
Z–SO$_2$–CH$_3$, Z=R, Ar, ArNH–	3050–2940	3.28–3.40	w–m	m–w	asym CH$_3$ str
	3045–2975	3.28–3.36	w–m	m–w	asym CH$_3$ str
	2950–2900	3.39–3.45	w–m	m	sym CH$_3$ str
CH$_3$Si	3000–2930	3.33–3.41	w–m	m–w	asym CH$_3$ str
	2975–2925	3.36–3.42	w–m	m–w	asym CH$_3$ str
	2930–2890	3.41–3.46	w–m	m	sym CH$_3$ str
Methylene and other groups					
–CHO (aldehyde)	2900–2800	3.45–3.57	w	w	C–H str
	2775–2695	3.63–3.71	w		Overtone
X–CH$_2$– (X = halogen)	~3050	~3.28	w	m	C–H str
–CH$_2$–O–	2940–2915	3.40–3.43	m–s	m–s	asym CH$_2$ str
	2870–2840	3.48–3.52	m–s	m–s	sym CH$_2$ str
–O–CH$_2$–O–	2820–2710	3.55–3.69	m	m–w	C–H str
–CH$_2$NH$_2$ and –CH$_2$CN	2945–2915	3.40–3.43	m	m–s	asym CH$_2$ str
	2890–2850	3.46–3.51	s	m–s	sym CH$_2$ str
–CH$_2$–S–	2985–2920	3.35–3.43	m	m	asym CH$_2$ str
	2945–2845	3.40–3.51	m	m	sym CH$_2$ str
Cyclopropyl compounds	3115–3065	3.21–3.26	m	m–w	asym CH$_2$ str ⎫
	3100–3050	3.23–3.28	m	m–w	asym CH$_2$ str ⎪ Only two bands observed
	3080–3000	3.25–3.33	m–s	m–s	sym CH$_2$ str ⎬ due to overlap
	3060–2970	3.27–3.37	m	m	CH str ⎪
	3040–2995	3.29–3.34	m–s	m–s	sym CH$_2$ str ⎭
Methylene dioxy compounds 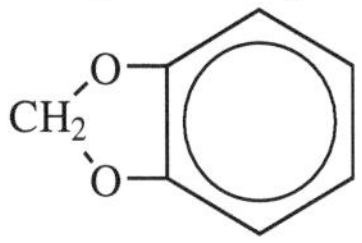	~2780	~3.60	m	m–w	sym CH$_2$ str also a band at ~925 cm^{-1}
(epoxides) 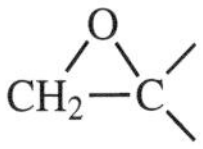	3075–3030	3.25–3.03	w	s–m	asym C–H str, see ref. 18
(epoxides) 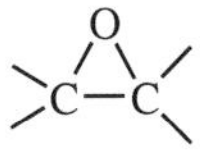	3000–2990	3.33–3.34	w	s	C–H str
Aziridinyl compounds 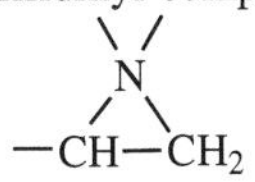	3100–3060	3.23–3.27	m–s	m	asym CH$_2$
	3000–2945	3.33–3.40	m–s	m–s	sym CH$_2$
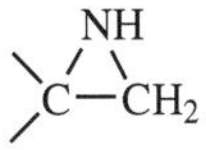	~3050	~3.28	m–s	m	asym CH$_2$ str

(*continued overleaf*)

Table 2.4 (*continued*)

Functional Groups	Region		Intensity		Comments
	cm^{-1}	μm	IR	Raman	
Ethyl groups	3000–2960	3.33–3.37	m–s	m	asym CH$_3$ str. Most commonly found in range 2990–2960 cm^{-1} (ref. 26)
	2990–2940	3.34–3.40	m–s	m–s	asym CH$_3$ str. Most commonly found in range 2980–2940 cm^{-1}
	2970–2900	3.37–3.35	m	m–s	asym CH$_2$ str. Most commonly found in range 2970–2920 cm^{-1}
	2970–2840	3.37–3.52	m–s	m–s	sym CH$_3$ str. Most commonly found in range 2940–2960 cm^{-1}
	2890–2840	3.46–3.52	m	m–s	sym CH$_2$ str. Most commonly found in range 2890–2850 cm^{-1}
Et·CO·–	2940–2860	3.40–3.50	m	m–s	sym CH$_3$ str.
	2940–2820	3.40–3.55	m	m–s	sym CH$_2$ str.
EtO– (ethers)	2995–2975	3.34–3.36	m–s	m	asym CH$_3$ str.
	2990–2940	3.34–3.40	m–s	m	asym CH$_3$ str.
	2990–2840	3.34–3.52	m–s	m–s	sym CH$_2$ str.
	2950–2920	3.39–3.42	m–s	m	asym CH$_2$ str.
	2940–2880	3.40–3.47	m–s	m–s	sym CH$_3$ str.
EtO·CO·– (esters)	2995–2975	3.34–3.36	w–m	m–s	asym CH$_3$ str.
	2985–2960	3.35–3.38	w–m	m–s	asym CH$_3$ str.
	2960–2930	3.38–3.41	w–m	m–s	asym CH$_2$ str.
	2930–2890	3.41–3.46	w–m	m–s	sym CH$_3$ str.
	2910–2860	3.44–3.50	w–m	m–s	sym CH$_2$ str.
Isopropyl compounds	3005–2985	3.33–3.50	m–s	m	asym str, usually below 3000 cm^{-1}
	2940–2860	3.40–3.50	m–s	m–s	sym str
–CHF$_2$	3005–2975	3.33–3.36	m–s	m	asym str
–CHCl$_2$	3015–2985	3.32–3.35	m–s	m	asym str
P–O–CH$_3$	3050–2990	3.28–3.34	w	m–s	asym CH$_3$ str
	3020–2950	3.31–3.39	m	m–s	asym CH$_3$ str
	2960–2840	3.38–3.52	w–m	m–s	sym CH$_3$ str
Si–O–CH$_3$	2990–2960	3.34–3.38	m–s	m–s	asym CH$_3$ str
	2955–2925	3.38–3.42	m–s	m–s	asym CH$_3$ str
	2850–2820	3.51–3.55	m	m–s	sym CH$_3$ str
t-Butyl cation, (CH$_3$)$_3$C$^+$	~2830	~3.53	s	m	CH$_3$ str
	~2500	~4.00	w	m	CH$_3$ str
Isopropyl cation, (CH$_3$)$_2$CH$^+$	~2730	~3.66	s	m	CH$_3$ str

paraffins have two characteristic bands at 1150–1130 cm^{-1} (8.70–8.85 μm) and 1090–1055 cm^{-1} (9.17–9.48 μm), both due to C–C stretching and CH$_3$ rocking vibrations.

Cyclopropane derivatives[12,13,21] have a band of variable intensity at 540–500 cm^{-1} (18.52–20.00 μm). An exception is that of vinylcyclopropane which has a strong absorption at 455 cm^{-1} (21.98 μm), and other unsaturated cyclopropanes also have a medium-intensity absorption in this region. Saturated aliphatic cyclopropyl compounds have a medium-to-weak band at about 1045 cm^{-1} (9.57 μm), a medium-intensity band at about 1020 cm^{-1} (9.80 μm), and a strong band at 470–460 cm^{-1} (21.28–21.74 μm). Cyclopentanes absorb strongly at 595–490 cm^{-1} (16.81–20.41 μm), alkyl monosubstituted cyclopentanes absorbing in the higher-frequency half of this range, 585–530 cm^{-1} (17.09–18.87 μm). Cyclohexane derivatives have bands of variable intensity in the region 570–435 cm^{-1} (17.54–22.99 μm).

Table 2.5   C–H deformation and other vibrations for alkane residues attached to atoms other than saturated carbon atoms

Functional Groups	Region		Intensity		Comments
	cm^{-1}	μm	IR	Raman	
–O–CH$_3$	1485–1445	6.73–6.92	m–w	m–w	asym CH$_3$ def vib
	1475–1435	6.78–6.97	m	m–w	asym CH$_3$ def vib
	1460–1420	6.85–7.04	m	m–w	sym CH$_3$ def vib
	1235–1155	8.10–8.66	w–m	w	Rocking CH$_3$ vib, usually ~1200 cm^{-1}.
	1190–1100	8.40–9.09	w–m		Rocking CH$_3$/CO vib (overlapped by C–O–C vib. strong at 1200–1040 cm^{-1})
	1025–855	9.76–1170	v		Rocking CO/CH$_3$ vib, unsat. compounds 995–895 cm^{-1}, aromatic compounds 1055–995 cm^{-1}
	580–340	17.24–29.41	w–m		CO def vib, usually w–vw, unsat. compounds 530–330 cm^{-1}, aromatic compounds 370–270 cm^{-1}
	265–185	37.74–54.05			CH$_3$ torsional vib
	210–110	47.62–90.90			CH$_3$ torsional vib
–OC(CH$_3$)$_3$	1200–1155	8.33–8.66	s	m–s	C–O str
	1040–1000	9.62–10.00	w–m	m–s	C–C vib almost always observed
	920–820	10.87–12.20	w–m		skeletal vib
	770–720	13.00–13.89	w–m		t-Bu sym skeletal vib
CH$_3$–CO–	1465–1415	6.83–7.07	m–w	m–w	asym CH$_3$ def vib (not amides)
	1440–1410	6.94–7.04	m–w	m–w	asym CH$_3$ def vib
	1390–1340	7.19–7.46	m–s	m–w	sym CH$_3$ def vib
	1155–1015	8.66–9.85	w–m	w	Rocking vib. May be of variable intensity
	1070–900	9.35–11.11	w	w	Rocking vib. May be of variable intensity
	270–130	37.04–76.92			Torsional vib.
CH$_3$–CO– (unsat group or Ar)	1470–1410	6.80–7.09	m–w	m–w	asym CH$_3$ def vib (unsat group 1440–1410 cm^{-1})
	1450–1390	6.90–7.19	m–w	m–w	asym CH$_3$ def vib
	1365–1345	7.33–7.43	m–s	m–w	sym CH$_3$ def vib
	1100–1020	9.09–9.80	w–m	w	Rocking vib. May be of variable intensity (Ar 1095–1045 cm^{-1})
	1040–975	9.62–10.62	w–m	w	Rocking vib. May be of variable intensity
	225–185	44.44–54.05			Torsional vib.
Methyl esters, CH$_3$O·CO·–	1485–1435	6.73–6.97	m–s	m–w	asym CH$_3$ def vib
	1465–1435	6.83–6.97	m–s	m–w	asym CH$_3$ def vib
	1460–1420	6.85–7.04	m–w	m–w	sym CH$_3$ def vib
	1220–1150	8.20–8.70	v	w	Rocking vib. Often weak-to-medium intensity
	1190–1120	8.40–8.93	v	w	Rocking vib. Often weak-to-medium intensity
	290–160	34.48–62.50			Torsional vib.
CH$_3$O·CS·–	1475–1435	6.78–6.97	m–s	m–w	asym CH$_3$ def vib
	1465–1435	6.83–6.97	m–s	m–w	asym CH$_3$ def vib
	1430–1420	6.99–7.04	m–w	m–w	sym CH$_3$ def vib
	1200–1150	8.33–8.70	v	w	Rocking vib. Often weak-to-medium intensity
	1165–1120	8.58–8.93	v	w	Rocking vib. Often weak-to-medium intensity
	290–210	34.48–47.62			Torsional vib.
CH$_3$O·SO·–	1485–1445	6.73–6.92	m–s	m–w	asym CH$_3$ def vib
	1460–1430	6.85–6.99	m–s	m–w	asym CH$_3$ def vib
	1460–1420	6.85–7.04	m–w	m–w	sym CH$_3$ def vib
	1220–1170	8.20–8.55	v	w	Rocking vib. Often weak-to-medium intensity

(*continued overleaf*)

Table 2.5   (*continued*)

Functional Groups	Region cm⁻¹	Region μm	Intensity IR	Intensity Raman	Comments
	1190–1140	8.40–8.77	v	w	Rocking vib. Often weak-to-medium intensity
	290–160	34.48–62.50			Torsional vib.
Amides $CH_3NH \cdot CO \cdot -$ and thioamides $CH_3NH \cdot CS \cdot -$	1480–1420	6.76–7.04	m–s	m–w	asym CH_3 def vib
	1475–1410	6.78–7.09	m–s	m–w	asym CH_3 def vib
	1425–1375	7.02–7.27	m–s	m–w	sym CH_3 def vib
	1190–1100	8.40–9.09	w	w	Rocking vib
	1165–1035	8.58–9.66	w	w	Rocking vib
	260–200	38.46–50.00			Torsional vib.
Acetamides $CH_3-CO-N\langle$	1480–1420	6.76–7.04	m–w	m–w	asym CH_3 def vib
	1460–1420	6.85–7.04	m–w	m–w	asym CH_3 def vib
	1375–1355	7.27–7.38	m–s	m–w	sym CH_3 def vib
	1130–1030	8.85–9.71	w–m	w	Rocking vib.
	1090–940	9.17–10.64	w–m	w	Rocking vib. May be of variable intensity
$Z-CH_3$, $Z=-CN$, $-NH_2$, $-NHCO$, $-NCO$, $-NCS$, $-NO_2$, $-NHSO_2$, $-NHCS$, $-N_3$,	1485–1425	6.73–8.03	m	m–w	asym CH_3 def vib
	1475–1415	6.78–7.07	m	m–w	asym CH_3 def vib
	1445–1375	6.92–7.27	w	m–w	sym CH_3 def vib
	1200–1100	8.33–9.09	w	w	Rocking vib.
	1165–1025	8.58–9.76	w	w	Rocking vib.
	260–145	38.46–68.97			Torsional vib.
$Z-SO_2-CH_3$, $Z=R$, Ar, $ArNH-$, NH_2, Halogen	1470–1400	6.80–7.14	m	m–w	asym CH_3 def vib
	1460–1400	6.85–7.14	m	m–w	asym CH_3 def vib
	1380–1290	7.25–7.75	m–s	m–w	sym CH_3 def vib
	1035–955	9.66–10.47	m–w	w	Rocking vib. (MeSH $\sim 1065\,cm^{-1}$)
	985–895	10.15–11.17	w	w	Rocking vib.
$-SO-CH_3$	1440–1410	6.94–7.09	m	m–w	asym CH_3 def vib
	1430–1400	6.99–7.14	m	m–w	asym CH_3 def vib
	1320–1290	7.58–7.75	m–s	m–w	sym CH_3 def vib
	1025–945	9.76–10.58	m–w	w	Rocking vib
	960–895	10.42–11.17	w	w	Rocking vib
$-SCH_3$	1485–1420	6.73–7.04	m	m–w	asym CH_3 def vib
	1470–1415	6.80–7.07	m	m–w	asym CH_3 def vib
	1460–1400	6.85–7.14	m–s	m–w	sym CH_3 def vib
	1340–1290	7.46–7.75	m–w	m–w	def vib
	1220–1150	8.20–8.70	w	w	Rocking CH_3 vib
	1190–1120	8.40–8.93	w	w	Rocking CH_3/CS vib
	1100–1120	9.09–8.93	w	w	Rocking CS/CH_3 vib
	1030–950	9.71–10.53	m–w	w	Rocking vib
	390–250	25.64–40.00			CS def vib
	290–160	34.48–62.50			CH_3 torsional vib
CH_3SCH_2-	1455–1425	6.87–7.02	w–m	m–w	asym CH_3 def vib

Table 2.5 (*continued*)

Functional Groups	Region cm^{-1}	Region μm	Intensity IR	Intensity Raman	Comments
	1440–1410	6.94–7.09	w–m	m–w	asym CH$_3$ def vib
	1435–1375	6.97–7.27	m	m	CH$_2$ def vib
	1330–1310	7.52–7.63	w	m	sym CH$_3$ def vib
	1305–1195	7.66–8.37	m	m	CH$_2$ wagging vib
	1280–1120	7.81–8.93	m	m	CH$_2$ twisting vib
	1035–965	9.66–10.36	w	w	CH$_3$ rocking vib
	970–910	10.31–10.99	w	w	CH$_3$ rocking vib
	890–740	11.24–13.51	w	w	CH$_2$ rocking vib
	775–675	12.90–14.81	w–m	s–m	asym CSC vib
	725–635	13.79–15.75	w–m	s	sym CSC vib
	420–320	23.81–31.25	w	m	Skeletal vib
	290–210	34.48–47.62	w		Skeletal vib
	220–160	45.45–62.50			Torsional vib
	180–110	55.56–90.90			Torsional vib
	105–45	95.24–222.22			Torsional vib
Ethyl groups	1480–1420	6.76–7.04	m–w	m–w	CH$_2$ def vib Most common range 1470–1440 cm^{-1}
	1475–1455	6.78–6.87	m–w	m–w	asym CH$_3$ def vib. Most common range 1475–1455 cm^{-1}
	1465–1435	6.83–6.97	m	m–w	asym CH$_3$ def vib. Most common range 1465–1445 cm^{-1}
	1390–1360	7.19–7.35	m–s	m–w	sym CH$_3$ def vib. Most common range 1385–1370 cm^{-1}
	1365–1295	7.33–7.72	m–w	m–w	CH$_2$ wagging vib. Most common range 1360–1320 cm^{-1}
	1290–1200	7.75–8.33	w	m–w	CH$_2$ twisting vib. Most common range 1285–1215 cm^{-1}
	1190–1060	8.40–9.43	w–m	w	CH$_2$ rocking vib. Most common range 1150–1070 cm^{-1}
	1090–1005	9.17–9.95	w	m–s	C–C str. Most common range 1090–1025 cm^{-1}
	1000–880	10.00–11.36	w	w	CH$_3$ rocking vib. Most common range 980–890 cm^{-1}
	835–715	11.98–13.99	w–m	w	CH$_2$ rocking vib. Most common range 790–730 cm^{-1}
	490–290	20.41–34.48	w–m	m	Skeletal vib. Most common range 470–440 cm^{-1}
	335–125	29.85–80.00			CH$_3$ torsional vib. Most common range 270–180 cm^{-1}
	150–90	66.67–111.11			Et torsional vib. Most common range 150–90 cm^{-1}
EtO– (ethers)	1495–1455	6.69–6.87	w	m–w	CH$_2$ def vib. (unsat. and aromatic ethers 1490–1470 cm^{-1})
	1480–1450	6.76–6.90	w	m–w	asym CH$_3$ def vib
	1465–1425	6.83–7.02	w	m–w	asym CH$_3$ def vib
	1400–1370	7.14–7.30	m–s	m–w	sym CH$_3$ def vib
	1380–1310	7.25–7.63	m–w	m–w	CH$_2$ wagging vib
	1310–1260	7.63–7.94	w	m–w	CH$_2$ twisting vib
	1195–1135	8.37–8.81	w–m	w	CH$_3$ rocking vib (aromatic ethers 1175–1145 cm^{-1})
	1160–1080	8.62–9.26	w–m	w	CH$_3$ rocking vib (unsat. and aromatic ethers 1130–1110 cm^{-1})
	1100–1030	9.09–9.71	m	m–w	CO/CC str (unsat. ethers 1100–1060 cm^{-1})
	940–810	10.64–12.35	m	m–w	CC/CO str (unsat. ethers 900–840 cm^{-1}, aromatic ethers 935–835 cm^{-1})
	825–785	12.12–12.74	w–m	w	CH$_2$ rocking vib (unsat. ethers 835–765 cm^{-1}, Ar ethers 840–740 cm^{-1})
	530–410	18.87–24.39	w–m		COC def vib (unsat. and Ar ethers 470–370 cm^{-1})
	470–320	21.28–31.25	w–m		OCC def vib (unsat. ethers 440–340 cm^{-1}, Ar ethers 340–240 cm^{-1})
	260–200	38.46–50.00			CH$_3$ torsional vib

(*continued overleaf*)

Table 2.5 (*continued*)

Functional Groups	Region		Intensity		Comments
	cm^{-1}	μm	IR	Raman	
	200–100	50.00–100.00			Et torsional vib
EtCO–	1445–1405	6.92–712	m	m–w	CH$_2$ def vib
	1380–1300	7.25–7.69	m–w	m–w	CH$_2$ wagging vib
EtS–	1445–1415	6.92–7.07	m	m–w	CH$_2$ def vib
	1310–1250	7.63–8.00	m–s	m–w	CH$_2$ wagging vib
Isopropyl groups	1485–1430	6.73–6.99	m–s	m–w	asym def vib (see ref. 25)
	1400–1360	7.14–7.35	w–m	m–w	sym def vib
	1190–1150	8.40–8.70	m	w	
	1160–1070	8.62–9.35	v	w	
	1120–1040	8.93–9.62	v	w	
	1000–940	10.00–10.64	w	w	
	905–765	11.05–13.07	w	m	CC$_2$ str
	515–385	19.42–25.97	w	m	Skeletal vib. Usually 480–400 cm^{-1}
	410–310	24.39–32.26	w	m	Skeletal vib
	365–275	27.40–36.36	w	m	Skeletal vib
t-Butyl groups	1495–1450	6.69–6.70	m–s	m	asym CH$_3$ def vib
	1475–1455	6.78–6.87	m–s	m	asym CH$_3$ def vib
	1470–1435	6.80–6.97	m–s	m	asym CH$_3$ def vib
	1395–1355	7.17–7.38	m	m	sym CH$_3$ def vib
	1370–1360	7.30–7.35	m–s	m	sym CH$_3$ def vib
	1295–1175	7.72–8.51	w	m	Skeletal CC$_3$ vib
	1215–1105	8.23–9.05	w	w	Rocking vib, usually 1185–1125 cm^{-1}
	1085–980	9.22–10.20	w–m	w	Rocking vib
	1050–890	9.52–11.24	w	w	Rocking vib(three bands)
	890–710	11.24–14.08	w–m	m	Skeletal vib
	520–350	19.23–28.57	w–m	m	Skeletal vib
	415–255	24.10–39.22	w–m	m	Skeletal vib
	380–220	26.32–45.45	w–m	m–s	Skeletal vib
–O–CH$_2$– (esters)	1475–1460	6.78–6.85	m–s	m–w	CH$_2$ sym def vib
	∼1030	∼9.71	w–m	w	Not always observed
Esters (acyclic)	1470–1435	6.80–6.97	m–s	m–w	CH$_2$ sym def vib
Esters (cyclic, small rings)	1500–1470	6.67–6.80	m	m–w	sym def vib, several bands
Acetates –O–CO–CH$_3$	1465–1415	6.83–7.08	m–w	m–w	asym def vib
	1460–1400	6.85–7.14	m–w	m–w	asym def vib
	1390–1340	7.19–7.46	m–s	m	sym def vib
	1080–1020	9.26–9.80	w–m	w	Rocking vib. Often variable intensity
	1025–930	9.76–10.75	w	w	Rocking vib. Often variable intensity
	220–110	45.45–90.90			Torsional vib
Thioacetates –OCSCH$_3$	1450–1420	6.90–7.04	w–m	m–w	asym def vib
	1430–1410	6.99–7.09	w–m	m–w	asym def vib
	1365–1345	7.33–7.43	m	m–w	sym def vib
	1140–1100	8.77–9.09	w–m	w	Rocking vib. Often variable intensity
	1065–935	9.39–10.70	w–m	w	Rocking vib. Often variable intensity
EtO·CO·– (esters)	1490–1460	6.71–6.85	w	m–w	CH$_2$ def vib
	1475–1445	6.78–6.92	w	m–w	asym CH$_3$ def vib
	1465–1435	6.83–6.97	w	m–w	asym CH$_3$ def vib

Table 2.5 (*continued*)

Functional Groups	Region		Intensity		Comments
	cm^{-1}	μm	IR	Raman	
	1400–1370	7.14–7.30	m–s	m–w	sym CH$_3$ def vib
	1385–1335	7.22–7.49	w–m	m	CH$_2$ wagging vib
	1330–1240	7.52–8.06	w	m–w	CH$_2$ twisting vib
	1195–1135	8.37–8.81	w–m	w	CH$_3$ rocking vib
	1150–1080	8.70–9.26	w–m	w	CH$_3$ rocking vib
	1100–1020	9.09–9.80	w–m	w–m	CO/CC str
	940–840	10.64–11.90	w–m	w–m	CC/CO str
	825–775	12.12–12.90	w–m	w	CH$_2$ rocking vib
	370–250	27.03–40.00	w–m		COC def vib
	395–305	25.31–32.79			OCC def vib
	280–210	35.71–47.62			CH$_3$ torsional vib
	200–120	50.00–83.33			Et torsional vib
–CO–CH$_3$ (ketones)	1450–1400	6.90–7.14	s	m–w	asym def vib
	1360–1355	7.35–7.38	s	m–w	sym def vib
–CO–CH$_2$– (small-ring ketones)	1475–1425	6.78–7.02	s	m–w	asym def vib, several bands
–CO–CH$_2$– (acyclic ketones)	1435–1405	6.97–7.12	s	m–w	asym def vib
–CH$_2$–COOH	~1200	~8.33	m	m	CH$_2$ def vib
Acetyl acetonates	1415–1380	7.07–7.25	s	m–w	asym def vib
	1360–1355	7.35–7.38	s	m–w	sym def vib
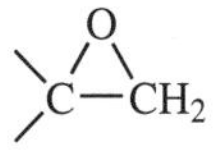 C—CH$_2$ with O bridge (epoxides)	~1500	~6.67	w–m	m–w	asym bending vib
–CHO (aldehydes)	1440–1325	6.94–7.55	m–s	m–w	CH def vib
CHOH (secondary alcohols) (free)	1410–1350	7.09–7.41	w		CH def vib
	1300–1200	7.69–8.33	w	m–w	CH def vib
Secondary alcohols (bonded)	1440–1400	6.94–7.14	w	m–w	CH def vib
	1350–1285	7.41–7.78	w	m–w	CH def vib
–(CH$_2$)$_n$–O–, (n > 4)	745–735	13.42–13.61	m–s	m–w	CH$_2$ def vib
N—CH$_3$	1440–1390	6.94–7.19	m	m–w	sym def vib, usually moves to higher wavenumbers for hydrohalides
N—CH$_3$ (amine hydrochlorides)	1475–1395	6.78–7.17	m	m–w	sym def vib
N—CH$_3$ (amino acid hydrochlorides)	1490–1480	6.71–6.76	m	m–w	sym def vib
N—CH$_3$ (amides)	1420–1405	7.04–7.12	s	m–w	sym def vib (asym def 1500–1450 cm^{-1})
N—CH$_2$— (amides, lactams)	~1440	~6.94	m	m–w	
N—CH (amines) and groups	1350–1315	7.41–7.61	w	m	CH def vib

(*continued overleaf*)

Table 2.5   (*continued*)

Functional Groups	Region cm^{-1}	Region μm	Intensity IR	Intensity Raman	Comments
with $-O-CH$ such as acetals orthoformates and peroxides					
N$-CH_2-$(ethylenediamine complexes)	1480–1450	6.76–6.90	s	m–w	sym def vib, two bands
	1400–1350	7.14–7.41	m–s		
$-CH_2-NO_2$	1425–1415	7.02–7.07	s	m–w	sym def vib
$-CH_2-CN$	1450–1405	6.90–7.12	m–s	m–w	sym def vib. CH$_2$ wagging vib at 1365–1230 cm^{-1} weak-to-medium band
$-CH_2-C=C\overset{/}{\underset{\backslash}{}}$ and $-CH_2-C\equiv C-$	1445–1430	6.92–6.99	m	m–s	Conjugation to CH$_2$ decreases wavenumber
X$-CH_2-$, (X=halogen, X ≠ F)	1460–1385	6.85–7.22	m	m	(Strong band at 1315–1215 cm^{-1} due to CH$_2$ wagging vib for Cl, ~1230 cm^{-1} for Br and ~1170 cm^{-1} for I)
$-CH_2-S-$	1435–1410	6.97–7.09	m	m	CH$_2$ def vib
	1305–1215	7.66–8.23	s	m	CH$_2$ wagging vib
Cyclopropyl compounds	1475–1435	6.78–6.97	m–w	m–w	CH$_2$ def vib
	1440–1410	6.94–7.09	m	m–w	CH$_2$ def vib
	1420–1240	7.04–8.06	s	m–w	CH def vib
	1220–1180	8.20–8.47	m	s	Ring breathing vib
	1195–1155	8.37–8.66	m–s	v	CH$_2$ torsional vib, may be strong in Raman.
	1170–1090	8.55–9.17	w		CH$_2$ torsional vib
	1105–1035	9.05–9.66	w–m	m–w	CH$_2$/CH wagging vib
	1070–1010	9.35–9.90	w–m	m–w	CH$_2$/CH wagging vib
	1045–975	9.57–10.26	m–w	s	CH$_2$/CH wagging vib, usually at ~1020 cm^{-1}
	985–825	10.15–12.12	m–s	v	asym ring def vib
	905–815	11.05–12.27	w	s, p	sym ring def vib
	870–790	11.49–12.66	v	w–m	CH$_2$ rocking vib
	815–755	12.27–13.25	w	w	CH$_2$ rocking vib
Aziridinyl compounds, $\overset{\backslash\,/}{\underset{-CH-CH_2}{N}}$	1485–1455	6.73–6.87	m–w	m–w	CH$_2$ def vib
	1465–1425	6.83–7.02	m	m–w	CH$_2$ def vib
	1285–1185	7.78–8.44	w	s	ring def vib
	1260–1160	7.94–8.62	m	m	CH$_2$ torsional vib
	1195–1105	8.37–9.05	w	m–w	CH$_2$ torsional vib
	1145–1095	8.73–9.13	w	m–w	CH$_2$ wagging vib
	1105–1025	9.05–9.76	w	m–w	CH$_2$ wagging vib
	925–885	10.81–11.30	w	m–s	asym ring def vib
	890–820	11.24–12.20	w	s	sym ring def vib
	840–790	11.90–12.66	w	w	CH$_2$ rocking vib
	800–730	12.50–13.70	w	w	CH$_2$ rocking vib
$-CHF_2$	1445–1345	6.92–7.43	m–s	m–w	CH def vib
	1345–1205	7.43–8.30	m–s	m–w	CH def vib

Table 2.5 (*continued*)

Functional Groups	Region		Intensity		Comments
	cm^{-1}	μm	IR	Raman	
⟩CHCl$_2$	1310–1200	7.63–8.50	m–s	m	2 bands, CH def vib
t-Butyl cation (CH$_3$)$_3$C$^+$	~1455	~6.87	w	m–w	CH$_3$ def vib
	~1300	~7.69	m	s	asym C–C–C str
	~1290	~7.75	m	s	CH$_3$ def vib
	~1070	~9.34	m–w	w	in-plane CH$_3$ rocking vib
	~960	~10.42	m–w	w	in-plane CH$_3$ rocking vib
Isopropyl cation (CH$_3$)$_2$CH$^+$	~1490	~6.71	s–m	m–w	CH in-plane def vib
	~1260	~7.94	v	m–s	asym C–C–C str
	~1175	~8.51	w	w	
	~940	~10.64	vw	w	
F–CH$_3$	~1475	~6.78	m	m–w	sym def vib
Cl–CH$_3$	~1355	~7.38	m	m–w	sym def vib
	~1015	~9.85	m	w	Rocking vib
–CH$_2$Cl	1450–1410	6.90–7.09	m–s	m	asym def vib
	1315–1215	7.60–8.23	m–s	m–w	CH$_2$ wagging vib
	1280–1145	7.81–8.73	m	m–w	CH$_2$ twisting vib
	990–780	10.10–12.82	w–m	w	CH$_2$ rocking vib
Br–CH$_3$	~1305	~7.61	m	m–w	sym def vib
–CH$_2$Br	1300–1200	7.69–8.33	m–s	m–w	CH$_2$ wagging vib. Unsat.CH$_2$Br 1240–1200 cm^{-1}
	1245–1105	8.03–9.05	m	m–w	CH$_2$ twisting vib
	945–715	10.58–13.99	w	w	CH$_2$ rocking vib
I–CH$_3$	~1250	~7.98	m	m–w	sym def vib
–CH$_2$I	1275–1050	7.84–9.52	m–s	m–w	CH$_2$ wagging vib. Rotational isomerism results in up to 80 cm^{-1} band separation
P–CH$_3$	1320–1280	7.58–7.81	m–w	m–w	sym def vib
	960–830	10.42–12.05	m	w	Rocking vib
P–CH$_2$	1440–1405	6.94–7.12	m	m–w	CH$_2$ def vib
S–CH$_2$–	1460–1410	6.85–7.09	s	m–w	sym def vib(strong band at 1305–1215 cm^{-1} due to CH$_2$ wagging vib)
Se–CH$_3$	~1280	~7.81	m	m–w	sym def vib
B–CH$_3$	1460–1405	6.85–7.12	m	m–w	asym def vib
	1440–1410	6.94–7.09	w	m–w	asym def vib
	1285–1250	7.78–8.00	m–s	m–w	sym def vib
	890–790	11.24–12.66	m		
	870–765	11.49–13.07	m–s	w	Rocking vib
⟩SiCH$_3$	1440–1410	6.94–7.09	w	m–w	asym CH$_3$ def vib
	1440–1390	6.94–7.19	w	m–w	asym CH$_3$ def vib
	1290–1240	7.75–8.06	m, sh	m–w	sym CH$_3$ def vib
	890–790	11.24–12.66	s–m	w	Rocking CH$_3$ vib
	870–740	11.49–13.51	m–s	w	Rocking CH$_3$ vib
Si–OCH$_3$	1475–1450	6.78–6.90	w	m–w	asym CH$_3$ def vib
	1470–1450	6.80–6.90	w	m–w	asym CH$_3$ def vib

(*continued overleaf*)

Table 2.5   (*continued*)

Functional Groups	Region		Intensity		Comments
	cm^{-1}	μm	IR	Raman	
	1465–1435	6.83–6.97	w	m–w	sym CH$_3$ def vib
	1200–1170	8.33–8.55	w–m	w	Rocking CH$_3$ vib
	1185–1135	8.44–8.81	w–m		Rocking CH$_3$/CO vib
	1095–1045	9.13–9.57			Rocking CO/CH$_3$ vib
	345–295	28.99–33.90			CO def vib
	230–150	76.92–66.67			CH$_3$ torsional vib
Sn–CH$_3$	1200–1180	8.33–8.48	m	m–w	sym def vib
	~770	~12.99	m–s	w	Rocking vib
Pb–CH$_3$	1170–1155	8.55–8.66	m	m–w	sym def vib ⎫
	770–700	12.99–14.29	m–s	w	Rocking vib ⎪
As–CH$_3$	1265–1240	7.91–8.07	m	m–w	sym def vib
	~860	~11.63	m	w	Rocking vib
Ge–CH$_3$	1240–1230	8.07–8.13	m	m–w	sym def vib
	~820	~12.20	m	w	Rocking vib
SbCH$_3$	1215–1195	8.23–8.37	m	m–w	sym def vib
	~800	~12.50	m	w	Rocking vib ⎬ CH$_3$– metal groups
Bi–CH$_3$	1165–1145	8.58–8.73	m	m–w	sym def vib ⎪ have strong band
	~790	~12.50	m	w	Rocking vib ⎪ at 900–700 cm^{-1} due
Zn–CH$_3$	1340–1200	7.46–8.33	m–w	m–w	asym def vib ⎪ to CH$_2$ rocking
	1190–1150	8.40–8.70	m	m–w	sym def vib
Be–CH$_3$	~1220	~8.26	m–w	m–w	asym def vib
	~1080	~9.26	m	m–w	sym def vib
Al–CH$_3$	1100–1020	9.09–9.20	m	m–w	sym def vib
Ga–CH$_3$	~1220	~8.20	m–w	w	asym def vib ⎪
	~1100	~9.09	m		sym def vib ⎭
In–CH$_3$	1140–1100	8.77–9.09	m	m–w	sym def vib
Hg–CH$_3$	~1180	~8.47	w–m	w	
	790–700	12.66–14.29	m–s	w	
P–O–CH$_3$	1475–1445	6.78–6.92	m	w	asym CH$_3$ def vib
	1470–1435	6.80–6.97	m	m–w	asym CH$_3$ def vib
	1470–1420	6.80–7.04	m	m–w	sym CH$_3$ def vib
	1190–1140	8.40–8.77	m–s	w	Rocking CH$_3$ vib
	1090–1010	9.17–9.90	m–w	m–w	Rocking CH$_3$/CO vib
	500–450	20.00–22.22			CO def vib
	270–170	37.03–58.82			CH$_3$ torsional vib
	200–170	50.00–58.82			CH$_3$O torsional vib
P–OCH$_2$CH$_3$	1480–1470	6.76–6.80	m–w	m–w	OCH$_2$ def vib
	1450–1435	6.90–6.97	m	m–w	CH$_3$ def vib
	~1395	~7.17	w–m	m–w	OCH$_2$ wagging vib
	~1370	~7.30	m	m–w	CH$_3$ sym def vib
	~1160	~8.62	m–w	w	CH$_3$ rocking vib
	~1100	~9.09	m–w	w	CH$_3$ rocking vib
–CH$_2$–SO$_2$–	~1250	~8.00	m	m–w	sym def vib
–CH$_2$– metal (metal=Cd, Hg, Zn, Sn)	1430–1415	6.99–7.07	m	m–w	CH$_2$ def vib

Methyl-substituted benzenes have an absorption band of medium intensity in the range $390-260\,\mathrm{cm}^{-1}$ $(25.64-38.46\,\mu\mathrm{m})$ which is due to the in-plane bending of the aromatic $C-CH_3$ bond. Ethyl-substituted benzenes have a medium-to-strong absorption at $565-540\,\mathrm{cm}^{-1}$ $(17.70-18.52\,\mu\mathrm{m})$ and isopropyl benzenes have a medium-intensity absorption band at $545-520\,\mathrm{cm}^{-1}$ $(18.35-19.23\,\mu\mathrm{m})$. Both these variations are due to the in-plane bending of the $=C-C-C$ group.

For propyl and butyl benzenes, two bands of medium intensity close together, usually not completely resolved, are observed at $585-565\,\mathrm{cm}^{-1}$ $(17.09-17.70\,\mu\mathrm{m})$. Mono branched alkanes have bands of medium intensity at $570-445\,\mathrm{cm}^{-1}$ $(17.54-22.47\,\mu\mathrm{m})$ and $470-440\,\mathrm{cm}^{-1}$ $(21.28-22.73\,\mu\mathrm{m})$.

References

1. N. Sheppard and D. M. Simpson, *Quart. Rev.*, 1953, **7**, 19.
2. H. J. Bernstein, *Spectrochim. Acta*, 1962, **18**, 161.
3. D. C. McKean *et al.*, *Spectrochim. Acta*, 1973, **29A**, 1037.
4. M. T. Forel *et al.*, *J. Opt. Soc. Am.*, 1960, **50**, 1228.
5. H. B. Henbest *et al.*, *J. Chem. Soc.*, 1957, 1462.
6. F. Dalton *et al.*, *J. Chem. Soc.*, 1960, 2927.
7. R. D. Hill and G. D. Meakins, *J. Chem. Soc.*, 1958, 761.
8. A. S. Wexler, *Appl. Spectrosc. Rev.*, 1968, **1**, 29.
9. S. Higuchi *et al.*, *Spectrochim. Acta*, 1972, **28A**, 1335.
10. J. Van Schooten *et al.*, *Polymer*, 1961, **2**, 357.
11. F. F. Bentley and E. F. Wolfarth, *Spectrochim. Acta*, 1959, **15**, 165.
12. K. H. Ree and F. A. Miller, *Spectrochim. Acta*, 1971, **27A**, 1.
13. N. C. Craig *et al.*, *Spectrochim. Acta*, 1972, **28A**, 1175.
14. G. M. Badger and A. G. Moritz, *Spectrochim. Acta*, 1959, **15**, 672.
15. A. B. Dempster *et al.*, *Spectrochim. Acta*, 1972, **28A**, 373.
16. H. E. Ulery and J. R. McCienon, *Tetrahedron*, 1963, **19**, 749.
17. A. S. Gilbert *et al.*, *Spectrochim. Acta*, 1976, **32A**, 931.
18. C. J. Wurrey and A. B. Nease, *Vib. Spectra Struct.*, 1978, **7**, 1.
19. T. C. Rounds and H. L. Strauss, *Vib. Spectra Struct.*, 1978, **7**, 237.
20. N. B. Colthup, *Appl. Spectrosc.*, 1980, **34**, 1.
21. G. Schrumpf, *Spectrochim. Acta*, 1983, **39A**, 505.
22. C. J. Pouchert, *The Aldrich Library of FT-IR Spectra*, Aldrich Chemical Co., Milwaukee, WI, 1985.
23. T. Woldbaek *et al.*, *Spectrochim Acta.* 1985, **41A**, 43
24. P. M. Green *et al.*, *J. Raman Spectrosc*, 1986, **17**, 355
25. G. Schrumpf, *J. Raman Spectrosc*, 1986, **17**, 183 & 433
26. S. Konaka *et al.*, *J. Mol. Struct.*, 1991, **244**, 1.
27. P. Kbeboe *et al.*, *Spectrochim Acta*, 1985, **41A**, 53.
28. P. Kbeboe *et al.*, *Spectrochim Acta*, 1985, **41A**, 1315.
29. A. Piart-Goypiron *et al.*, *Spectrochim Acta*, 1993, **49A**, 103
30. P. Derreumaux *et al.*, *J. Mol. Struct*, 1993, **295**, 203.
31. J. R. Hill *et al.*, *J. Phys. Chem*, 1991, **85**, 3037.

3 Alkenes, Oximes, Imines, Amidines, Azo Compounds: C=C, C=N, N=N Groups

Alkene Functional Group, $\diagup$C=C$\diagdown$

The most useful bands are those resulting from the C=C stretching and the C–H out-of-plane deformation vibrations, the latter bands being the strongest observed in the infrared spectra of alkenes[1] (see Charts 1.5, 1.6 and 3.1). The symmetry of the molecule and its interactions, if any, affect the change in the dipole moment and hence the intensity of the bands in the infrared. For example, for compounds which have a symmetrical configuration, the C=C stretching vibration is infrared inactive, whereas in Raman spectra this band is strong and easily recognised. In the infrared, the intensity of the C=C stretching band decreases markedly as the symmetry of the alkene molecule increases. Symmetrical vinylene compounds, Z–C=C–Z, in the *trans* configuration have C_{2h} symmetry and as a consequence the CH, C=C, and the C–Z stretching vibrations, the CH and CZ in-plane deformations and the CH and CZ wagging vibrations are all infrared inactive but their vibrations are all observable in Raman spectra.

Alkene $\diagup$C=C$\diagdown$ *Stretching Vibrations*

Non-conjugated alkenes have a weak C=C stretching absorption band in the range $1680–1620\,\text{cm}^{-1}$ ($5.95–6.17\,\mu\text{m}$). This band is absent for symmetrical molecules. Therefore, it is not surprising to find that olefins which have terminal double bonds have the most intense absorptions. Vinyl, vinylidene, and *cis*-disubstituted olefins tend to absorb at the lower end of the range given, below $1665\,\text{cm}^{-1}$ (above $6.01\,\mu\text{m}$), whereas *trans*-disubstituted, tri-, and tetrasubstituted olefins absorb at the higher wavenumbers.

In conjugated systems, the C=C stretching vibration frequency is lower than that of an isolated C=C group.[2-4] Often there is the same number of bands as there is of double bonds, e.g. with two double bonds, two bands of different intensities are observed due to the C=C–C=C symmetric and asymmetric stretching. For conjugated dienes without a centre of symmetry, two absorption bands are normally observed, one at about $1650\,\text{cm}^{-1}$ ($6.06\,\mu\text{m}$) and another more intense band near $1600\,\text{cm}^{-1}$ ($6.25\,\mu\text{m}$). The presence of this latter band may be used to confirm the presence of conjugation. For dienes with a centre of symmetry, only one C=C stretching band is observed in their infrared and Raman spectra. In the infrared, the asymmetric C=C stretching vibration band, which is of weak intensity, occurs near $1600\,\text{cm}^{-1}$($6.25\,\mu\text{m}$). In Raman spectra, it is the symmetric stretching band which is observed, this being strong and occurring at $1640\,\text{cm}^{-1}$ ($6.10\,\mu\text{m}$), the asymmetric band being Raman inactive. Different rotational isomers are possible for dienes; hence the intensities of the asymmetric and symmetric stretching bands are dependent on the conformational structure. Obviously steric effects have a bearing on the population and structure of the isomers and hence on the intensity of the bands observed. Alkenes conjugated to aromatic rings exhibit a strong absorption near $1625\,\text{cm}^{-1}$ ($6.15\,\mu\text{m}$).[3,37,38] In this case, the aromatic C=C ring absorption is at about $1590\,\text{cm}^{-1}$ ($6.28\,\mu\text{m}$). In poly-conjugated systems, a series of weak bands is observed at $2000–1660\,\text{cm}^{-1}$ ($5.00–6.02\,\mu\text{m}$), similar to that of aromatic compounds.

The effect of electronegative substituents such as chlorine etc., attached directly to alkene groups, is generally to lower the C=C stretching vibration frequency. Fluorine, on the other hand, increases this frequency. In alkene strained-ring compounds, the frequency of the C=C stretching vibration is decreased[5-10] – the smaller the ring, the lower the frequency. Information on the integrated intensity of the band due to the C=C stretching vibration is also available.[15,22]

Alkene C–H *Stretching Vibrations*

In general, bands due to both alkene and aromatic C–H stretching occur above $3000 \, cm^{-1}$ (below $3.33 \, \mu m$). Although alkane C–H stretching vibrations generally occur below $3000 \, cm^{-1}$, it must be noted that small-ring paraffins and alkanes substituted with electronegative atoms or groups also absorb above $3000 \, cm^{-1}$. The $=CH_2$ stretching vibration of vinyl and vinylidine groups occurs at $3095–3075 \, cm^{-1}$ ($3.24–3.25 \, \mu m$) and the $=CH$ stretching vibration at $3050–3000 \, cm^{-1}$ ($3.28–3.33 \, \mu m$), whilst their symmetric stretching vibration occurs near $2975 \, cm^{-1}$ ($3.36 \, \mu m$), although this is unfortunately often overlapped by alkane absorptions. The $=C–H$ stretching vibrations generally result in strong bands in the Raman spectra and bands of medium intensity in the infrared.

Alkene C–H *Deformation Vibrations*

The deformation vibrations of C–H may be either perpendicular to or in the same plane as that containing the carbon–carbon double bond and the other bonds:

The arrows indicate the vibrational motions of a single C—H

The absorption bands due to the out-of-plane vibrations occur mainly at $1000–800 \, cm^{-1}$ ($10.00–12.50 \, \mu m$) and have strong-to-medium intensities. These bands are important in the characterisation of alkenes,[11–13] e.g. for hydrocarbons:

(a) Vinyl groups, $-CH=CH_2$, absorb strongly[14,39] in the regions $995–980 \, cm^{-1}$ ($10.05–10.20 \, \mu m$) and $915–905 \, cm^{-1}$ ($10.93–11.05 \, \mu m$), the overtones of these bands being found near $1980 \, cm^{-1}$ ($5.05 \, \mu m$) and $1830 \, cm^{-1}$ ($5.46 \, \mu m$) respectively. For the nitrile compound, the first band occurs at $960 \, cm^{-1}$ ($10.42 \, \mu m$) and for the corresponding isothiocyanate and thiocyanate this band occurs near $940 \, cm^{-1}$ ($10.64 \, \mu m$)

(b) Vinylidene groups, $>C=CH_2$, absorb strongly at $895–885 \, cm^{-1}$ ($11.17–11.30 \, \mu m$).

(c) *Trans*-disubstituted alkenes, $-CH=CH-$, absorb strongly at $980–955 \, cm^{-1}$ ($10.20–10.47 \, \mu m$).

(d) *Cis*-disubstituted alkenes, $-CH=CH-$, absorb strongly at $730–650 \, cm^{-1}$ ($13.70–15.38 \, \mu m$).

(e) Trisubstituted alkenes, $>C=CH-$, absorb at $850–790 \, cm^{-1}$ ($11.76–12.66 \, \mu m$).

The $=CH_2$ out-of-plane deformation vibration is not mass sensitive for non-hydrocarbon olefins but it is sensitive to electronic changes. Groups that withdraw electrons mesomerically from the $=CH_2$ group, e.g.

$$-CO-\overset{\overset{\text{O}}{\|}}{C}-CH=CH_2 \text{ and } CNCH=CH_2,$$

tend to raise the frequency and those which donate electrons mesomerically lower the frequency relative to that of the hydrocarbon olefin.

For vinylidene compounds[14,15] with halogens directly bonded to the $>CH=CH_2$ group, the out-of-plane deformation vibration frequency is decreased. This shift in frequency becomes greater with increase in the electronegativity of the halogen atom and appears to have an approximately additive effect. Oxygen atoms directly bonded to the vinylidene group also tend to decrease the $=CH_2$ out-of-plane vibration frequency.

For *cis*-vinylenes, the in-plane CH deformation may be found in the range $1425–1265 \, cm^{-1}$ ($7.02–7.91 \, \mu m$) (but is usually in the region $1400–1290 \, cm^{-1}$) and $1295–1185 \, cm^{-1}$ ($7.72–8.44 \, \mu m$). For *trans*-vinylenes, these bands occur at $1340–1260 \, cm^{-1}$ ($7.46–7.94 \, \mu m$) (but usually in the region $1330–1215 \, cm^{-1}$) and $1305–1265 \, cm^{-1}$ ($7.66–7.91 \, \mu m$). For *symmetrical trans*-1,2-disubstituted vinylenes, the out-of-plane CH deformation vibration is infrared inactive but Raman active ($1000–910 \, cm^{-1}$ ($10.00–10.99 \, \mu m$)). In the Raman spectra of the *cis*-isomers, this is a weak band and occurs at $1000–850 \, cm^{-1}$ ($10.00–11.76 \, \mu m$).

Alkene *Skeletal Vibrations*[15–19]

For unbranched 1-alkenes, strong bands are observed near $635 \, cm^{-1}$ ($15.75 \, \mu m$) and $550 \, cm^{-1}$ ($18.18 \, \mu m$) and these have been assigned to ethylenic twisting vibrations.

All *cis*-alkenes have two, well-separated, strong bands at $630–570 \, cm^{-1}$ ($15.87–17.54 \, \mu m$) and $500–460 \, cm^{-1}$ ($20.00–21.74 \, \mu m$) and in general have weak bands or no bands in the region $455–370 \, cm^{-1}$ ($21.98–27.03 \, \mu m$), whereas all *trans*-alkenes have medium-to-strong absorption bands, usually only one, in this latter region. For example, unbranched *cis*-2-alkenes absorb in the regions $590–570 \, cm^{-1}$ ($16.95–17.54 \, \mu m$) and $490–465 \, cm^{-1}$ ($20.41–21.51 \, \mu m$) whereas unbranched *trans*-2-alkenes have absorptions at

Chart 3.1   Infrared – band positions of alkenes

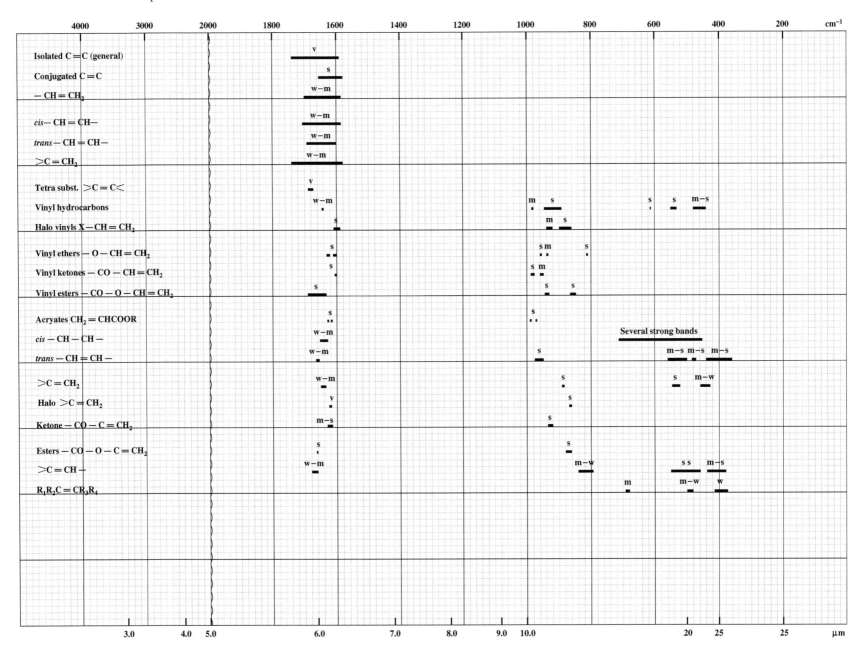

Table 3.1 Alkene C=C stretching vibrations

Functional Groups	Region		Intensity		Comments
	cm^{-1}	μm	IR	Raman	
Isolated C=C	1680–1620	5.95–6.17	w–m	s, p	May be absent for sym compounds
C=C conjugated with aryl	1640–1610	6.10–6.21	m	s	*Ortho* substitution increases frequency
C=C conjugated with C=C or C=O	1660–1580	6.02–6.33	s–m	s	C=C–C=C usually ~1600 cm^{-1}. See ref. 10
Conjugated, CH$_2$=CH–C≡C–	1620–1610	6.17–6.21	s	s	Conjugated with C=C see ref. 21
Dienes and trienes	1670–1610	5.99–6.21	m–w	s	sym C=C str (often ~1640 cm^{-1}) usual range, but may occur up to 1700 cm^{-1}. Trienes sometimes one band only and may have shoulder on 1650 cm^{-1}
	1610–1550	6.21–6.45	m	m–w	asym C=C str
Polyenes	1660–1580	6.02–6.33	m–w	s	br, often more than one band. In Raman, overtone bands may easily be observed
Vinyls					
Vinyl group, –CH=CH$_2$	1645–1640	6.08–6.10	w–m	s, p	Hydrocarbons
Halo- or cyano-vinyls	1620–1580	6.17–6.27	s	s	Fluoro- ~1650 cm^{-1}. (For 3,3-difluoroalkenes refs: 43, 44)
Vinyl ether, –O–CH=CH$_2$	1660–1630	6.02–6.54	s	s	Usually a doublet in region 1640–1610 cm^{-1}, see ref. 13
	1620–1610	6.17–6.21	s	s	
–S–CH=CH$_2$	1590–1580	6.29–6.33	s	s	Also strong bands in Raman at ~1390 and ~1280 cm^{-1}
Vinyl ketone, –CO–CH=CH$_2$	1625–1615	6.15–6.19	s–m	s	(For dichlorovinyl ketones, see ref. 36)
Vinyl ester, CH$_2$=CHOCOR	1700–1645	5.88–6.08	s–m	s	
Acrylates, CH$_2$=CHCOOR	1640–1635	6.10–6.12	s–m	s	
	1625–1620	6.16–6.17	s–m	s	
⟩Si–CH=CH$_2$	1630–1580	6.13–6.33	v	s	
Vinylenes					
cis–CH=CH–	1665–1630	6.01–6.13	m	s, p	Hydrocarbons. Absorbs more strongly than *trans* isomers for symmetrical compounds. Non-hydrocarbons 1680–1630 cm^{-1}
cis (unsat)–CH=CH– (unsat)	1650–1600	6.06–6.25	m	s	
trans –CH=CH–	1680–1665	5.95–6.02	w–m	s, p	Hydrocarbons. In general, *trans* isomers absorb at higher wavenumbers than the equivalent *cis* isomer. Non-hydrocarbons 1680–1650 cm^{-1}
trans (unsat)–CH=CH– (unsat)	1670–1610	5.99–6.21	m	s	

(*continued overleaf*)

Table 3.1 (*continued*)

Functional Groups	Region		Intensity		Comments
	cm^{-1}	μm	IR	Raman	
Vinylidenes					
(Sat.)$_2$C=CH$_2$	1675–1625	5.97–6.15	w–m	s, p	Hydrocarbons 1660–1640 cm^{-1}
Halo- and cyano-substituted	1630–1620	6.13–6.17	v	s	Difluoro-substituted ~1730 cm^{-1}
$\diagdown$C=CH$_2$ $\diagup$					
–CO–C=CH$_2$, ketones	~1630	~6.14	m–s	s	
–CO–O–C=CH$_2$, esters	1675–1670	5.97–5.99	s	s	
α,β-unsaturated amines, CH$_2$=CN$\diagup\diagdown$	1700–1660	5.88–6.02	m	s	
Trisubstituted alkenes					
$\diagdown$C=CH— $\diagup$	1690–1665	5.92–6.01	m–s	s, p	Adjacent C=O decreases frequency and increases intensity
CH$_2$=CF–	1650–1645	6.06–6.08	m	s	
CF$_2$=CF–	1800–1780	5.56–5.62	m	s	See ref. 19
$\diagdown$C=CF$_2$ $\diagup$	1755–1735	5.70–5.76	m	s	
$\diagdown$C=C–N$\diagup\diagup\diagdown$	1680–1630	5.95–6.13	m–s	s	More intense than normal C=C str band
Tetrasubstituted alkenes					
$\diagdown$C=C$\diagup$ $\diagup\diagdown$	1690–1670	5.92–5.99	w	s, p	May be absent for symmetrical compounds
Internal double bonds					
Cyclopropene	~1655	~6.04	w–m	s, p	Polyfluorinated compound ~1945 cm^{-1}. Monosubstituted compound ~1790 cm^{-1}, disubstituted compound ~1900–1860 cm^{-1}
Cyclopropenones	1865–1840	5.36–5.43	s	s	Mainly C=O and C=C str
	1660–1600	6.02–6.25	s	s	Mainly C=C and C=O str. Ring str ~880 cm^{-1}
Cyclobutene	~1565	~6.39	w–m	s	See ref. 9. Polyfluorinated compound ~1800 cm^{-1}. Monosubstituted compound ~1640 cm^{-1}, disubstituted compound ~1675 cm^{-1}
Cyclopentene	~1610	~6.21	w–m	s	See ref. 18. Polyfluorinated compound ~1770 cm^{-1}. Monosubstituted compound 1670–1640 cm^{-1}, disubstituted compound 1690–1670 cm^{-1}. Raman sym ring str a band ~900 cm^{-1}.

Table 3.1 (*continued*)

Functional Groups	Region cm^{-1}	Region μm	Intensity IR	Intensity Raman	Comments
Cyclohexene	~1645	~6.08	w–m	s	Polyfluorinated compound ~1745 cm^{-1}. Raman strong band ~820 cm^{-1} due to ring sym str
Cycloheptene	~1650	~6.06	w–m	s	
1,2-Dialkylcyclopropenes	1900–1860	5.26–5.38	w–m	s	
1,2-Dialkylcyclobutenes	~1675	~5.97	w–m	s	
1,2-Dialkylcyclopentenes	1690–1670	5.92–5.65	w–m	s	
1-Alkylcyclopentenes	1675–1665	5.97–6.01	w	s, p	
1,2-Dialkylcyclohexenes	1685–1675	5.93–5.63	w–m	s	
3,4 Dihydroxy-3 cyclobutene 1,2-dione	~1515	~6.60	w–m		

Exocyclic double bonds:

$$\text{>C=C(CH}_2)_n$$

$n = 2$	1780–1730	5.62–5.78	m	s	
$n = 3$	~1680	~5.95	m	s	Shift to lower frequency as ring size increases
$n = 4$	~1655	~6.04	m	s	
$n = 5$	~1650	~6.06	m	s	
Alkyl-substituted fulvenes 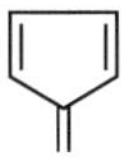	~1645	~6.08	m	s	Aromatic groups on the exo double bond lower frequency to ~1600 cm^{-1}
Benzofulvenes	~1630	~6.13	m	s	
A–CH–CH$_2$	1650–1565	6.23–6.39	v	s	A = heavy element, or group involving heavy element, directly attached to C=C, see ref. 26
A = see comments C=C π-interaction with metal	1580–1500	6.23–6.67		s	E.g. Pt(C$_2$H$_4$) see Chapter 22 and refs: 23–25

Table 3.2 Alkene C–H vibrations

Functional Groups	Region cm^{-1}	Region μm	Intensity IR	Intensity Raman	Comments
Vinyls Vinyls, –CH=CH$_2$ (general ranges)	3150–3000	3.17–3.33	m	m	asym CH$_2$ str
	3070–2930	3.26–3.41	m	m	sym CH$_2$ str
	3110–2980	3.22–3.36	m	m, p	CH str
	1440–1360	6.94–7.35	m	m–s, p	CH$_2$ def vib
	1330–1240	7.52–8.06	m	m	CH def vib
	1180–1010	8.47–9.90	m–w	m	CH in-plane def vib
	1010–940	9.90–10.64	s	w	Out-of-plane CH vib
	980–810	10.20–12.35	s	w	Out-of-plane CH$_2$ vib

(*continued overleaf*)

Table 3.2   (*continued*)

Functional Groups	Region cm^{-1}	Region μm	Intensity IR	Intensity Raman	Comments
	720–410	13.89–24.39	w	w	CH$_2$ twisting vib
	600–250	16.67–40.00		w	C=C skeletal vib
	200–40	50.00–250.00			Torsional vib
Vinyl hydrocarbon compounds, –CH=CH$_2$	3095–3070	3.23–3.26	m	m	CH str of CH$_2$
	3030–2995	3.30–3.34	m	s, p	CH str of CH
	1985–1970	5.04–5.08	w	–	Overtone
	1850–1800	5.41–5.56	w	–	Overtone
	1420–1410	7.04–7.09	m	s–m, p	CH$_2$ in-plane def vib, scissoring
	1300–1290	7.69–7.75	w	m	CH in-plane def vib
	995–980	10.05–10.20	m–s	w	CH out-of-plane def vib. Overtone ~1980 cm^{-1}
	915–905	10.93–11.05	s	w	CH$_2$ out-of-plane def vib, insensitive to conjugation, see ref. 21 Overtone ~1830 cm^{-1}
	690–610	14.49–16.39	w	w	CH wagging vib
	635–620	15.75–16.13	w	w	C–H out-of-plane def vib
Vinyl halogen compounds	945–935	10.58–10.83	m–s	w	CH out-of-plane def (nitrile-substituted compound 960 cm^{-1})
	905–865	11.05–11.56	s	w	CH$_2$ out-of-plane def vib (nitrile-substituted compounds 960 cm^{-1})
Vinyl ethers –O–CH=CH$_2$	970–960	10.31–10.42	s	w	CH out-of-plane def vib, see ref. 13
	945–940	10.58–10.64	m	w	CH out-of-plane def vib. Raman ~845 cm^{-1} COC str
	825–810	12.12–12.35	s	w	CH$_2$ out-of-plane def vib
Vinyl ketones, –COCH=CH$_2$	995–980	10.05–10.20	s	w	CH out-of-plane def vib
	965–955	10.36–10.47	m	w	CH$_2$ out-of-plane def vib
Vinyl esters, CH$_2$=CHOCOR	950–935	10.53–10.70	s	w	CH out-of-plane def vib
	870–850	11.49–11.76	s	w	CH$_2$ out-of-plane def vib
Acrylates, CH$_2$=CHCOOR	990–980	10.10–10.20	s	w	out-of-plane def vib
	970–960	10.31–10.42	s	w	out-of-plane def vib
Vinyl amides –(CO)NR–CH=CH$_2$	980–965	10.20–10.36	s	w	
	850–830	11.77–12.05	s	w	
⟩Si—CH=CH$_2$	1010–990	9.90–10.10	s–m	w	Out-of-plane CH def vib
	980–940	10.20–10.64	s–m	w	Out-of-plane CH$_2$ def vib
(Sat)–CH=CH$_2$	1000–980	10.00–10.20	s	w	
	965–905	10.36–11.05	s	w	
(Unsat)–CH=CH$_2$	1000–960	10.00–10.42	s	w	
	950–870	10.53–11.49	s–m	w	
Vinylidenes Hydrocarbons, ⟩C=CH$_2$	3095–3075	2.53–2.67	m–w	m	CH asym str
	2985–2970	3.35–3.37	m–w	s, p	CH sym str. General range 3040–3010 cm^{-1}
	1800–1750	5.56–5.71	w	–	overtone
	1420–1405	7.04–7.12	w	m–s, p	CH$_2$ in-plane def vib, scissoring vib
	1320–1290	7.58–7.75	w	m	CH$_2$ in-plane def vib
	895–885	11.17–11.30	s	w	CH$_2$ out-of-plane def vib. Overtone ~1780 cm^{-1}
	715–680	13.99–14.70	w	w	CH$_2$ def vib

Table 3.2 (*continued*)

Functional Groups	Region		Intensity		Comments
	cm^{-1}	μm	IR	Raman	
	560–420	17.86–23.81			Skeletal vib, out-of-plane –C=C def vib
	470–370	21.28–27.03		w	Skeletal vib
Mono- and dihalogen- substituted	890–865	11.24–11.56	s	w	CH$_2$ out-of-plane def vib (difluoro- at ~805 cm^{-1})
$\diagdown$C=CH$_2$					
Cyano-substituted $\diagdown$C=CH$_2$	960–895	10.42–11.17	s	w	CH$_2$ out-of-plane def vib (dicyano ~985 cm^{-1})
–CO–C=CH$_2$ (ketones and esters)	~930	~11.07	s	w	CH$_2$ out-of-plane def vib
–CO–O–C=CH$_2$ (esters)	880–865	11.36–11.56	s	w	CH$_2$ out-of-plane def vib
(Unsat)$_2$–C=CH$_2$	940–890	10.64–11.24	s	w	CH$_2$ wagging vib
	750–630	13.33–15.87	w	w	CH$_2$ twisting vib
	560–460	17.86–21.74			Skeletal vib
	470–340	21.28–29.41		w	Skeletal vib
Vinylenes					
cis-CH=CH-(hydrocarbons)	3040–3010	3.29–3.32	m	m	CH str.
	1425–1355	7.02–7.38	w	–	CH in-plane def vib
	1295–1200	7.72–8.33	w	s–m	CH sym rocking vib
	980–880	10.20–11.36	w	m	Out-of-plane CH def vib
	730–650	13.70–15.38	s	w	CH out-of-plane def vib, conjugation increases frequency range to 820 cm^{-1}. General range 730–650 cm^{-1}.
	630–620	15.87–16.13	s		Usually strong
	675–435	14.81–22.99	m–s	w	Skeletal vib
	490–250	20.41–40.00			Torsional vib
	310–175	32.26–57.14			
cis-(Sat)-CH=CH-(Sat')	3090–3010	3.31–3.32	m	m	CH str. For unsat. groups 3080–3030 cm^{-1}
	3040–2980	3.29–3.36	m	m	CH str. For unsat. groups 3030–2980 cm^{-1}
	1425–1355	7.02–7.38	w	m–s	CH def vib. (Unsat. conj. groups 1410–1290 cm^{-1})
	1295–1185	7.72–8.44	w	s	CH def vib. (Unsat. conj. groups 1290–1200 cm^{-1})
	1000–850	10.00–11.76	w–m	m–w	CH wagging vib. (Unsat. conj. groups 1000–910 cm^{-1})
	790–650	12.66–15.38	s–m	w	CH wagging vib. (Unsat. conj. groups 790–710 cm^{-1})
	590–440	16.95–22.73	m–s		–C=CH def vib. (Unsat. conj. groups 675–435 cm^{-1})
	490–320	20.41–31.25	m–s		Torsional vib. (Unsat. conj. groups 410–320 cm^{-1})
	310–220	32.26–45.45			(For unsat. conj. groups 295–175 cm^{-1})
Halogen-substituted *cis*-CH=CH-	780–770	12.82–12.99	s	m	
trans-CH=CH-(hydrocarbons)	3040–3010	3.29–3.32	m	s	CH str
	1340–1260	7.46–7.94	v	w	CH in-plane def vib, sometimes absent
	1305–1215	7.66–8.23	v		CH def vib
	1000–910	10.00–10.99	–	m	CH def vib
	980–955	10.20–10.47	s	w	CH out-of-plane def vib (usually ~965 cm^{-1}), conjugation increases frequency slightly and polar groups decrease it significantly (e.g. for *trans–trans* system, may be ~1000 cm^{-1})
	630–430	15.87–23.26		w	Skeletal C=C vib
	455–250	21.98–40.00	m–s		Torsional vib
	340–200	29.41–50.00			

(*continued overleaf*)

Table 3.2   (*continued*)

Functional Groups	Region cm^{-1}	Region μm	Intensity IR	Intensity Raman	Comments
trans–(Sat)–CH=CH–(Sat)	3065–3015	3.26–3.32	m	m	CH str. For unsat. groups 3095–3015 cm^{-1}
	3050–3000	3.28–3.33	m	m	CH str. For unsat. groups 3000–2990 cm^{-1}
	1340–1300	7.46–7.69	v	s	CH def vib. (Unsat. conj. groups 1330–1260 cm^{-1})
	1305–1260	7.66–7.94	v	s	CH def vib. (Unsat. conj. groups 1260–1215 cm^{-1})
	1000–910	10.00–10.99	v	m	CH wagging vib. (Unsat. conj. groups 1000–940 cm^{-1})
	850–750	11.76–13.33	m–w	w	CH wagging vib. (Unsat. conj. groups 900–760 cm^{-1})
	620–440	16.13–22.73		w	–C=CH def vib. (Unsat. conj. groups 550–430 cm^{-1})
	410–250	24.39–40.00			Torsional vib. (Unsat. conj. groups 450–250 cm^{-1})
	310–230	32.26–43.48			(For unsat. conj. groups 340–200 cm^{-1})
Halogen-substituted *trans*–CH=CH–	~930	~10.75	s	w	CH out-of-plane def vib
trans–CH=CH–conjugated with C=C or C=O	~990	~10.10	s	w	CH out-of-plane def vib
trans–CH=CH–O–(ethers)	940–920	10.64–10.87	s	w	
Trisubstituted alkenes					
＞C=CH—(hydrocarbons)	3040–3010	3.29–3.32	m	m	CH str
	1680–1600	5.95–6.25	w	–	Overtone
	1350–1340	7.41–7.46	m–w	w	CH in-plane def vib.
	850–790	11.76–12.66	m–w	w	CH out-of-plane def vib, electronegative groups at lower end of frequency range
	525–485	19.05–20.62		w	C=C–C skeletal vib
Cyclic alkenes	3090–2995	3.24–3.34	m	m	=C–H str, ring-strain dependent: highest frequencies for smallest rings. Normally more than one band.
	780–665	12.82–15.04	m	w	CH out-of-plane def vib
Dienes	990–965	10.10–10.36	s	m	*trans* isomer CH def vib
	~720	~13.88	s	m	*cis* isomer CH def vib
Trienes	~990	~10.10	s	m	*trans–cis–trans* CH def vib
	~960	~10.42	m	m	*cis–trans–trans* CH def vib
	~720	~13.89	m	m	CH def vib
Polyenes	990–970	10.10–10.31	s	m	Doublet CH def vib
CH$_2$=CH–M (M=metal)	1425–1385	7.02–7.19	w	m–s	CH$_2$ def vib, see ref. 26
	1265–1245	7.91–8.03	w–m	w	CH rocking vib
	1010–985	9.90–10.15	m	w	CH out-of-plane vib
	960–940	10.42–10.64	s	w	CH$_2$ out-of-plane vib
Cyclopentadienyl derivatives	3110–3020	3.22–3.31	m	m	CH str
	1445–1440	6.92–6.94	m	s–m	C=C str
	1115–1090	8.97–9.17	m–s	s–m	C=C str
	1010–990	9.91–10.10	s	m	In-plane CH def vib
	830–700	12.05–14.29	s	w	Out-of-plane CH def vib
Fulvenes	1665–1605	6.01–6.23	m–s	s	C=C str. Strong intensity due to exo C=C dipole
	1370–1340	7.30–7.46	m–s	m–s	Ring vib. Characteristic of unsaturated five-membered ring
	~765	~13.07	s	w	CH out-of-plane def vib
Benzofulvenes	~790	~12.66	s	w	CH out-of-plane def vib

Table 3.3 Alkene skeletal vibrations

Functional Groups	Region		Intensity		Comments
	cm⁻¹	μm	IR	Raman	
R–CH=CH₂	690–610	14.49–16.39	v	w	Ethylenic twisting vib, see ref. 20 (exception is propene ~578 cm⁻¹)
	600–380	16.67–26.32	m–s	w	Ethylenic twisting vib
	485–445	20.62–22.47	m–s		Torsional vib. 200–70 cm⁻¹
$\searrow$Si—CH=CH₂	540–410	18.52–24.39	v	w	Twisting CH₂ vib
	410–250	24.39–40.00	m–s	w	
	150–70	66.67–142.86			Torsional vib
cis-Alkenes	670–455	14.93–21.98	s		Two bands
trans-Alkenes	455–370	21.98–27.03	m–s		Usually one band
Unbranched cis- R–CH=CH–CH₃	590–570	16.95–17.54	s	w	
	490–460	20.41–21.74	s		
Unbranched trans- R–CH=CH–CH₃	420–385	23.81–25.97	s		
	325–285	30.77–35.09	s		
cis-R₁CH=CHR₂	630–570	15.87–17.54	s		
	500–460	20.00–21.74	s		
trans-R₁CH=CHR₂	580–515	17.24–19.42	m–s		
	500–480	20.00–20.83	m–s	w	
	455–370	21.98–27.03	m–s		
R₁R₂C=CH₂	560–530	17.86–18.87	s		
	470–435	21.28–22.99	m–w	w	
R₁R₂C=CHR₃	570–515	17.54–19.42	s	w	Rocking motion, may have medium intensity
	525–470	19.05–21.28	s		Probably out-of-plane bending vib
	450–395	22.22–25.32	m–s		
R₁R₂C=CR₃R₄	690–675	14.49–14.81	m	s–m	C–C str
	510–485	19.61–20.62	m–w	m	Skeletal vib
	425–385	23.53–25.97	w	w	Skeletal vib
Aryl olefins	~550	~18.18	m		

420–385 cm^{-1} (23.81–25.97 µm) and 325–285 cm^{-1} (30.77–35.09 µm). For C=C conjugated to an aromatic group, an absorption band near 550 cm^{-1} (18.18 µm) is observed.

Cyclobutene derivatives have a ring breathing vibration at 1000–950 cm^{-1} (10.00–10.53 µm) of variable intensity, whereas in the case of oxocarbon compounds this band occurs in the region 750–550 cm^{-1} (13.33–18.18 µm).

Oximes, $\diagup$C=N–OH, Imines, $\diagup$C=N– , Amidines, $\diagup$N–C=N– , etc.

The N–H stretching vibration of the group C=N–H occurs in the region 3400–3300 cm^{-1} (2.94–3.03 µm). The frequency of the vibration is decreased in the presence of hydrogen bonding. In Raman spectra, the band due to the C=N stretching vibration is of strong intensity whereas in infrared it is generally of weak intensity. For oximes and imines,[27,28,30,31,40–42] the C=N stretching band occurs in the region 1690–1620 cm^{-1} (5.92–6.17 µm), the infrared band being weak in the case of aliphatic oximes and occurring at the higher-frequency end of the range given. For α, β-unsaturated and aromatic oximes[28,32] this band is of medium intensity and occurs in the lower-frequency half of the range. The closeness of this band to that due to the C=C stretching vibration often presents difficulties. Conjugated cyclic systems containing C=N have a band of variable intensity, due to the stretching vibration, in the region 1660–1480 cm^{-1} (6.02–6.76 µm), e.g. pyrrolines absorb at 1660–1560 cm^{-1} (6.02–6.41 µm). As the ring size of cyclic imines decreases, the frequency of the C=N stretching vibration decreases. Protonation of the imine group to form salts results in a 30 cm^{-1} increase in the C=N stretching frequency.

The O–H stretching vibration[30,31] for oximes in a dilute solution using non-polar solvents occurs in the region 3650–2570 cm^{-1} (2.78–2.79 µm), a strong absorption being observed. If hydrogen bonding occurs, this band appears at 3300–3130 cm^{-1} (3.03–3.20 µm). In general, oximes have a strong band near 930 cm^{-1} (10.75 µm) due to the stretching vibration of the N–O bond, the general range for this band being 1030–870 cm^{-1} (9.71–11.49 µm).

Amidines[29] absorb strongly at 1685–1580 cm^{-1} (5.93–6.33 µm), due to the C=N stretching vibration, the band being found as low as 1515 cm^{-1} (6.60 µm) for amidines in solution.

Table 3.4 Oximes, imines, amidines, etc.: C=N stretching vibrations

Functional Groups	Region		Intensity		Comments
	cm^{-1}	µm	IR	Raman	
Aliphatic oximes and imines, $\diagup$C=N–	1690–1640	5.92–6.10	w	s	
α,β-Unsaturated and aromatic oximes and imines	1650–1620	6.06–6.17	m	s	
Conjugated cyclic systems (oximes and imines)	1660–1480	6.02–6.76	v	s	
R$_1$ $\diagup$C=N–H R$_2$	1650–1640	6.06–6.10	s	s	sh
Ar $\diagup$C=N–H R$_2$	1635–1620	6.12–6.17	m	s	
R$_2$C=N–R	1665–1645	6.01–6.08	m–w	s	
RCH=N–R$_2$	1690–1630	6.92–6.13	v	s	(Schiff bases)
Ar–CH=N–Ar	1645–1605	6.08–6.23	v	s	Often two bands, see ref. 32
R(RO)C=N–H	~1655	~6.04	v	s	
Ar(RO)C=N–H	1645–1630	6.08–6.13	v	s	
Ar(RO)C=N–	1700–1630	5.88–6.13	v	s	

Table 3.4 (*continued*)

Functional Groups	Region		Intensity		Comments
	cm^{-1}	μm	IR	Raman	
RHC=N–OH	1670–1645	5.99–6.08	m–w	s	
	1640–1630	6.10–6.13	m–w	s	
R$_2$C=N–OH	1670–1650	5.99–6.06	m–w	s	
Quinone oximes,	1560–1520	6.37–6.58	s	s	
O=⟨benzene⟩=N—OH					
Guanidines, N—C=N—, —N—	1690–1550	5.92–6.45	s	s	
	1050–990	9.52–10.10		s	sym CN$_3$ str
Guanidine hydrochlorides mono-substituted	~1660	~6.02	s	s	strong band in Raman observed at 1050–990 cm^{-1} due to C–CN$_3$ str
	~1630	~6.14	s	s	
Guanidine hydrochlorides di-substituted	~1680	~6.00	s	s	
	~1595	~6.27	s	s	
Guanidine hydrochlorides tri-substituted	~1635	~6.12	s	s	Only one band
Azines, C=N—N=C	1670–1635	5.99–6.12	s	w	asym C=N–N=C str. Compounds with centre of symmetry have active bands only for IR asym str or Raman sym str. Compounds with no centre of symmetry have asym and sym bands active in both IR and Raman
	1625–1600	6.15–6.25	w	s	sym C=N–N=C str. For aryl azines, C=N str 1635–1605 cm^{-1} and 1565–1535 cm^{-1}
Benzamidines, φ—C=N—, —N—	1630–1590	6.14–6.29	m	s	
Hydrazones, C=N—N	1645–1610	6.08–6.21	m–w	s	
Semicarbazones, G$_2$C=N—N—CO— and G$_2$C=N—N—CS—	1655–1640	6.04–6.10	m–w	s	Conjugation lowers C=N str to 1630–1610 cm^{-1}
Hydrazoketones, —CO—C—N—N—	1600–1530	6.25–6.54	vs	s	
Amidines and guanidines, N—C=N—	1685–1580	5.93–6.33	v	s	
Imino ethers, –O–C=N–	1690–1645	5.92–6.08	v	s	Usually strong doublet due to rotational isomerism
–S–C=N–	1640–1605	6.10–6.23	v	s	
Imine oxides, C=N$^+$–O$^-$	1620–1550	6.17–6.45	s	s	N–O str 1280–1065 cm^{-1}

Table 3.5 Oximes, imines, amidines, etc.: other bands

Functional Groups	Region cm^{-1}	Region μm	Intensity IR	Intensity Raman	Comments
Oximes	3650–3500	2.74–2.86	v	w	Free O–H str, dilute solution
	3300–3130	3.03–3.20	v	w	Associated O–H str
	1475–1315	6.78–7.60	m	m–w	O–H def vib
	960–930	10.42–10.75	s	m	N–O str
Quinone oximes, O=⟨⟩=N—OH	3540–2700	2.82–3.70	s	w	br, associated O–H str
	1670–1620	5.99–6.17	s	w–m	C=O str
	1560–1520	6.37–6.58	s	s	C=N str
Imines	3400–3300	2.94–3.03	v	m	Free N–H str
	3400–3100	2.94–3.23	m	m	associated N–H str
–N–D	2600–2400	3.85–4.15	w–m	m	Free N–D str

Table 3.6 Azo compounds

Functional Groups	Region cm^{-1}	Region μm	Intensity IR	Intensity Raman	Comments
Alkyl azo compounds	1575–1555	6.35–6.43	v	s	N=N str
α,β-Unsaturated azo compounds	~1500	~6.67	v	s–m	
trans-Aromatic azo compounds	1465–1380	6.94–7.25	w	s, p	N=N str
cis-Aromatic azo compounds	~1510	~6.62	s	w–m	N=N str
Aliphatic azoxy compounds, –N=N$^+$–O$^-$	1530–1495	6.54–6.69	m–s	m	} Electron-withdrawing group on N–O nitrogen increases frequency
	1345–1285	7.43–7.78	m–s	m	
Aromatic azoxy compounds, –N=N$^+$–O$^-$	1490–1410	6.71–7.09	m–s	v	asym N=N–O str. In Raman, trans form s, for cis form w
	1340–1315	7.46–7.60	m–s		sym N=N–O str
Azothio compounds, –N=N$^+$–S$^-$	1465–1445	6.83–6.92	w	s	N=N str
	1070–1055	9.35–9.48	w		N–S str
Diazirines, \C/ /N=N\	~1620	~6.17	w	s	N=N str
Diazoketones, –CO–CN$_2$–	2100–2055	4.76–4.87	s	m	
	1650–1600	6.06–6.25	s	m–s	
	1390–1330	7.19–7.52	s		

N-Unsubstituted amidine hydrochlorides have a strong band at 1710–1675 cm^{-1} (5.85–5.97 μm) and a weak band at 1530–1500 cm^{-1} (6.54–6.67 μm). *N,N*-Disubstituted amidine hydrochlorides have a medium-intensity band at 1590–1530 cm^{-1} (6.29–6.54 μm) due to the deformation vibration of the =NH$_2$ group. Substituted amidines absorb strongly at 1700–1600 cm^{-1} (5.88–6.25 μm).

Azo Compounds, –N=N–

Azo compounds[33–35] are difficult to identify by infrared spectroscopy because no significant bands are observed for them, the azo group being non-polar in nature. In addition, the weak absorption of the azo group occurs in the same region as the absorptions of aromatic compounds, the *cis* form having much stronger bands normally than the *trans* form. However, in Raman spectra, the N=N stretching band is generally of strong intensity.

Aromatic azo compounds in the *trans* form absorb at $1465–1380\,cm^{-1}$ ($6.83–7.25\,\mu m$) and in the *cis* form, near $1510\,cm^{-1}$ ($6.62\,\mu m$). Aromatic compounds which are in the *trans* form absorb at the lower frequency end of the range given if they are substituted with strong electron donors. In Raman spectra, a strong band is observed near $590\,cm^{-1}$ ($16.95\,\mu m$) due to C–N stretching and C–N=N deformation vibrations.

References

1. N. Sheppard and D. M. Simpson, *Quart. Rev.*, 1952, **6**, 1.
2. J. L. H. Allan *et al.*, *J. Chem. Soc.*, 1955, 1874.
3. J. H. Wotiz *et al.*, *J. Am. Chem. Soc.*, 1950, **72**, 5055.
4. A. A. Petrov and G. I. Semenov, *J. Gen. Chem. Moscow*, 1959, **29**, 3689.
5. K. B. Wiberg and B. J. Nist, *J. Am. Chem. Soc.*, 1961, **83**, 1226.
6. S. Pinchas *et al.*, *Spectrochim. Acta*, 1965, **25**, 783.
7. J. Shabati *et al.*, *J. Inst. Petroleum*, 1962, **48**, 13.
8. J. B. Miller, *J. Org. Chem.*, 1960, **25**, 1279.
9. E. M. Suzuki and J. W. Nibler, *Spectrochim. Acta*, 1974, **30A**, 15.
10. K. Noack, *Spectrochim. Acta*, 1962, **18**, 697 and 1625.
11. W. J. Potts and R. A. Nyquist, *Spectrochim. Acta*, 1959, **15**, 679.
12. E. M. Popov and G. I. Kajan, *Opt. Spectrosc.*, 1962, **12**, 102.
13. E. M. Popov *et al.*, *Opt. Spectrosc.*, 1962, **12**, 17.
14. J. Overend and J. R. Scherer, *J. Chem. Phys.*, 1960, **32**, 1720.
15. G. P. Ford *et al.*, *J. Chem. Soc. Perkin Trans.*, 1974, 1569.
16. F. F. Bentley and E. F. Wolfarth, *Spectrochim. Acta*, 1959, **15**, 165.
17. P. M. Sverdlov, *Akad. Nauk SSSR Doklady*, 1957, **112**, 706.
18. J. L. Lauer *et al.*, *J. Chem. Phys.*, 1959, **30**, 1489.
19. D. E. Mann *et al.*, *J. Chem. Phys.*, 1957, **27**, 51.
20. J. R. Scherer and W. J. Potts, *J. Chem. Phys.*, 1959, **30**, 1527.
21. A. A. Petrov and G. I. Semenov, *J. Gen. Chem. Moscow*, 1957, **27**, 2974; 1958, **28**, 73.
22. A. S. Wexler, *Appl. Spectrosc. Rev.*, 1968, **1**, 29.
23. D. M. Adams and J. Chatt, *Chem. Ind.*, 1960, 149.
24. D. B. Powell and N. Sheppard, *J. Chem. Soc.*, 1960, 2519.
25. D. W. Wertz *et al.*, *Spectrochim. Acta*, 1973, **29A**, 1439.
26. D. B. Powell *et al.*, *Spectrochim. Acta*, 1974, **30A**, 15.
27. J. Fabian *et al.*, *Bull. Soc. Chim. France*, 1956, 287.
28. D. Hadzi, *J. Chem. Soc.*, 1956, 2725.
29. J. Fabian *et al.*, *Bull. Soc. Chim. France*, 1956, 287.
30. D. Hadzi and L. Premru, *Spectrochim. Acta*, 1967, **23A**, 35.
31. M. St. C. Flett, *Spectrochim. Acta*, 1957, **10**, 21.
32. J. D. Margerum and J. A. Sousa, *Appl. Spectrosc.*, 1965, **19**, 91.
33. L. E. Clougherty *et al.*, *J. Org. Chem.*, 1957, **22**, 462.
34. R. Von Kubler, *Z. Electrochem.*, 1960, **64**, 650.
35. K. J. Morgan, *J. Chem. Soc.*, 1961, 2151.
36. G. A. Gavrilova *et al.*, *Izv. Acad. Nauk SSSR Ser. Khim.*, 1978, **1**, 84.
37. R. A. Nyquist, *Appl. Spectrosc.*, 1986, **40**, 196.
38. H. Yashida, Y. Furukawa and M. Tasumi, *J. Mol. Struct.*, 1989, **194**, 279.
39. I. S. Ignatyev *et al.*, *J. Mol. Struct.*, 1981, **72**, 25.
40. K. Hashiguchi *et al.*, *J. Mol. Spectrosc.*, 1984, **105**, 81.
41. Y. A. Matatsu, Y. Hamada and M. Tsuboi, *J. Mol. Spectrosc.*, 1987, **123**, 276.
42. V. M. Kolb *et al.*, *J. Org. Chem.*, 1987, **52**, 3003.
43. G. A. Giurgis *et al.*, *J. Phys. Chem. A*, 2000, **104(19)**, 4383.
44. J. R. Durig *et al.*, *J. Phys. Chem. A*, 2000, **104(4)**, 741.

4 Triple Bond Compounds: $-C\equiv C-$, $-C\equiv N$, $-N\equiv C$, $-N\equiv N$ Groups

Alkyne Functional Group, $-C\equiv C-$

Two bands due to stretching vibrations may be observed, one due to the $-C\equiv C-$ group and the other to the $\equiv C-H$ group.[1,2] Information is also available on band intensities.[3,4,25,26,35] For symmetrical disubstituted alkynes, the $-C\equiv C-$ stretching vibration is infrared inactive but it is strong and easily identified in Raman spectra.

Alkyne $C\equiv C$ Stretching Vibrations

In infrared spectra, this band is weak,[3,4,13] for monosubstituted alkynes[2,5] occurring in the region $2150-2100\,\text{cm}^{-1}$ ($4.65-4.76\,\mu\text{m}$) and for disubstituted alkynes[1,3] in the region $2260-2190\,\text{cm}^{-1}$ ($4.43-4.57\,\mu\text{m}$). For disubstituted alkynes, two bands are often observed, due to Fermi resonance, in the region $2310-2190\,\text{cm}^{-1}$ ($4.33-4.57\,\mu\text{m}$).

For central $-C\equiv C-$ the band is usually weak and occurs at $2260-2190\,\text{cm}^{-1}$ ($4.43-4.57\,\mu\text{m}$). The $C\equiv C$ band is completely absent for simple acetylenes where there is a high degree of symmetry. Hence, as with alkenes, alkynes with a terminal triple bond have the most intense band due to $C\equiv C$ stretching vibrations and as the triple bond is moved to an internal position its intensity becomes less. Conjugation[6–11] increases both the intensity[3,4,26,35] and the frequency of the $C\equiv C$ stretching vibration. Information on cyclic acetylenes is also available.[14–17] In the Raman spectra of disubstituted alkynes, there are often two bands, near $2310\,\text{cm}^{-1}$ ($4.33\,\mu\text{m}$) and $2230\,\text{cm}^{-1}$ ($4.48\,\mu\text{m}$). The additional band has been attributed to an overtone/combination band enhanced by Fermi resonance.

Alkyne $C-H$ Vibrations

For monosubstituted alkynes,[2,18] strong bands are observed at $3340-3300\,\text{cm}^{-1}$ ($2.99-3.03\,\mu\text{m}$) due to the $C-H$ stretching vibration (this is a weak band in Raman spectra), and at $730-575\,\text{cm}^{-1}$ ($13.70-17.39\,\mu\text{m}$) due to the $C-H$ deformation vibration (alkyl monosubstituted alkynes $640-625\,\text{cm}^{-1}$ ($15.63-16.00\,\mu\text{m}$)). Care must be taken since the $C-H$ stretching absorption occurs in the same region as those for $N-H$, which fortunately are usually much broader. The position of the band due to the $\equiv C-H$ stretching vibration is generally not sensitive to molecular structure changes, exceptions being acetylenes with halogen atoms directly bonded to the triple bond. Phase changes alter the position of the $\equiv C-H$ stretching vibration band significantly, in solid-phase spectra the band being up to $50\,\text{cm}^{-1}$ lower ($0.05\,\mu\text{m}$ higher) than in dilute solution in inert solvents. An increase in wavenumber of similar magnitude is observed for vapour-phase spectra as compared with liquid-phase spectra.

For monosubstituted alkynes, the CH in-plane and out-of-plane deformation vibrations result in characteristic bands of medium-to-strong intensity at $730-620\,\text{cm}^{-1}$ ($13.70-16.13\,\mu\text{m}$) and $700-575\,\text{cm}^{-1}$ ($14.29-17.39\,\mu\text{m}$) respectively. In Raman spectra, these bands tend to be of weak intensity. The separation of these bands is less for saturated groups attached to the carbon than for unsaturated or carbonyl groups.

A band of variable intensity and uncertain origin is sometimes observed in the region $1740-1630\,\text{cm}^{-1}$ ($5.76-6.14\,\mu\text{m}$). The hydrogen bonding of acetylenes[19] and their formation of complexes with nitrogen-containing compounds[20] have been studied.

For disubstituted alkynes of the type $-C\equiv C-(CH_2)_2-$ a characteristic band due to the CH_2 wagging vibration is usually observed in the range $1340-1325\,\text{cm}^{-1}$ ($7.46-7.55\,\mu\text{m}$).

Alkyne Skeletal Vibrations

Monosubstituted acetylenes have skeletal vibrations occurring at $370-220\,\text{cm}^{-1}$ ($27.03-45.45\,\mu\text{m}$) and $290-140\,\text{cm}^{-1}$ ($34.48-71.43\,\mu\text{m}$).

Table 4.1 Alkyne $C\equiv C$ stretching vibrations

Functional Groups	Region		Intensity		Comments
	cm^{-1}	μm	IR	Raman	
Monosubstituted alkynes, $-C\equiv CH$	2150–2100	4.65–4.76	w–m	s, p	See ref. 2. Vapour phase higher: 2165–2135 cm^{-1}
Disubstituted alkynes	2260–2190	4.43–4.57	v	s, p	Intensity decreases as symmetry of molecule increases.
$R-C\equiv CR'$	2240–2190	4.46–4.57	m–w	s	Also medium intensity band at 2325–2285 cm^{-1}
Conjugated alkynes (see comments)	2270–2200	4.41–4.55	m	s, p	Conjugated with $C=C$, $C\equiv C$
	2125–2035	4.71–4.91	w	s	
Conjugated alkynes (see comments)	~2250	~4.43	s	s	Conjugated with COOH or COOR
$CH_2X-C\equiv CH$, X = halogen	2135–2125	4.68–4.71	m	s	
$-C\equiv C-Cl$	2270–2190	4.41–4.56	m	s, p	Strong band due to C–Cl str 760–430 cm^{-1}
$-C\equiv C-Br$	2250–2150	4.44–4.65	m	s, p	Strong band due to C–Br str 690–350 cm^{-1}
$-C\equiv C-I$	2220–2120	4.50–4.72	m	s, p	C–I str 660–310 cm^{-1}
$M-C\equiv C-H$, M=P, As, Sb, Ge, Sn, SiH$_3$	2055–2015	4.87–4.96	w–m	s, p	
$M-C\equiv C-CH_3$, (M as above)	2200–2170	4.55–4.61	s	s, p	

Table 4.2 Alkynes: other bands

Functional Groups	Region		Intensity		Comments
	cm^{-1}	μm	IR	Raman	
Monosubstituted alkynes, $-C\equiv CH$	3340–3280	2.99–3.05	m–s	w	sh, CH str
	1375–1225	7.27–8.17	w–m		CH wagging vib overtone
	1020–905	9.80–11.05	w	m–w	$C-C\equiv C$ str
	970–890	10.31–11.24	m–w	m–w	
	730–575	13.70–17.39	m–s	w	CH def vib, two bands if molecule has axial symmetry 730–620 cm^{-1} and 700–575 cm^{-1} (Fluoro ~580 cm^{-1})
	370–220	27.03–45.45	w	m–w	$-C\equiv CH$ skeletal vib
	290–140	34.48–71.43	w		$-C\equiv CH$ skeletal vib
Alkyl monosubstituted acetylenes	640–625	15.63–16.00	s	w	$C\equiv C-H$ bending vib
	355–335	28.17–29.85	v	m–w	$C-C\equiv CH$ def vib
$-C\equiv CH$	510–260	19.61–38.46	v		Non-alkyl substituent
$R-C\equiv C-CH_3$	520–495	19.23–20.20	m–s		
$R-C\equiv C-C_2H_5$	495–480	20.20–20.83	s–m	s	br
$R-C\equiv C-(CH_2)_2CH_3$	475–465	21.05–21.51	m		
$R_2N-CH_2C\equiv C-H$	~2100	~4.76	w–m	s	
	935–895	10.70–11.17	m	m–w	
	665–645	15.04–15.50	m–s	w	
	345–320	28.99–31.25	v	m–w	
$-C\equiv C(CH_2)_2-$	1340–1325	7.46–7.55	m	m–w	CH$_2$ wagging vib
$(\alpha,\beta\text{-Unsat})-C\equiv CH$	3340–3280	2.99–3.05	m	w	CH str. ($C\equiv C$ str 2125–2095 cm^{-1})
	700–620	14.29–16.13	m–s	w	CH def vib (aromatic compounds 660–630 cm^{-1})
	630–610	15.87–16.39	m–s	w	
	340–240	29.41–41.67		m–w	CH out-of-plane def vib. (aromatic compounds 370–320 cm^{-1})
	240–150	41.67–66.67			
$CH_2X-C\equiv C-H$, X = halogen	675–650	14.81–15.38	m	w	$C\equiv C-H$ def vib
	640–635	15.63–15.75	m	w	$C\equiv C-H$ def vib

(*continued overleaf*)

Table 4.2   (*continued*)

Functional Groups	Region		Intensity		Comments
	cm^{-1}	μm	IR	Raman	
	~310	~32.26			
	190–155	52.63–64.52			
H–C≡C–(substituted benzenes)	660–630	15.15–15.87	m–s	m–w	CH def vib
	630–610	15.87–16.39	m–s	m–w	CH def vib
	370–320	27.03–31.25	v		
C≡C–X (X=Cl, Br or I)	185–160	54.05–62.50	v		C≡C–X bending vib
–C≡C–Cl	470–370	21.78–27.03			C≡C skeletal vib
	435–125	22.99–80.00			≡C–Cl def vib
	360–260	27.78–38.46	w	m–w	C≡C skeletal vib
	190–90	52.63–111.11			≡C–Cl def vib
–C≡C–Br	470–320	21.78–31.50			C≡C skeletal vib
	375–125	26.67–80.00			≡C–Cl def vib
	360–260	27.78–38.46	w	m–w	C≡C skeletal vib
	170–70	58.82–142.86			≡C–Cl def vib
Z–C≡C–Cl, Z=CN, CHO, CH$_3$	580–540	17.24–18.52	s	s	C–Cl str
Z–C≡C–Br, Z=CN, CHO, CH$_3$	475–395	21.05–35.32	s	s	C–Br str
Z–C≡C–I, Z=CN, CHO, CH$_3$	405–360	24.69–27.78	s	s	C–I str
M–C≡C–H, M=P, As, Sb, Ge, Sn, SiH$_3$	3305–3280	3.03–3.05	m	m	
	710–675	14.08–14.81	m–s	w	
	665–575	15.04–17.39	m–s	w	

All alkyl monosubstituted acetylenes have an absorption of variable intensity in the region 355–335 cm^{-1} (28.17–29.85 μm) due to the skeletal deformations of the C–C≡CH group. Monosubstituted acetylenes in which the substituent is not an alkyl group absorb in the region 510–260 cm^{-1} (19.61–38.46 μm) as a result of deformation vibrations. Methyl- and ethyl-substituted acetylenes absorb strongly at 520–495 cm^{-1} (19.23–20.20 μm) and 495–480 cm^{-1} (20.20–20.83 μm) respectively. Benzenes substituted with –C≡C–[12] absorb at about 550 cm^{-1} (18.18 μm).

All acetylenic compounds absorb at 970–890 cm^{-1} (10.31–11.24 μm) due to the ≡C–C stretching vibration.

Nitriles, –C≡N

Nitrile-containing compounds normally have a sharp absorption in the region 2260–2200 cm^{-1} (4.43–4.55 μm). Care must be taken since acetylenic derivatives also absorb in this general region (due to the C≡C stretching vibration), as do compounds with cumulative double bonds. In infrared spectra, the C≡N stretching band may be of variable intensity (it may be very weak to very strong). In Raman spectra, the band is of medium-to-strong intensity.

For saturated aliphatic nitriles,[21,22] the band due to the stretching vibration of the –C≡N group occurs near 2250 cm^{-1} (4.44 μm) and for aryl and conjugated nitriles near 2230 cm^{-1} (4.48 μm).[23–25,39–43] The intensity of this band varies considerably. For example, oxygen atoms on neighbouring carbon atoms,

$$-O-\overset{|}{\underset{|}{C}}-C≡N,$$ tend to reduce the intensity of the band, for instance, cyanohy-

drins, $\diagdown C(OH)CN$, have no observable C≡N absorption whereas conjugation to the C≡N group appears to increase the intensity of the band. The intensity is reduced by electron-withdrawing atoms or groups, e.g. oxygen or chlorine atoms. Normally, medium-to-strong bands are observed for relatively small molecules not containing oxygen atoms. Aromatic nitriles with electron-donating substituents on the ring tend to have a more intense C≡N stretching band than those with electron-accepting groups. Solvents may also affect the intensity of this band.[31] The position of the band is about the same for dimers as for monomers.

In general, all aliphatic nitriles have a medium-to-strong band at 390–340 cm^{-1} (25.64–29.41 μm) in their infrared and Raman spectra due to the C–C≡N deformation.[27] Saturated primary aliphatic nitriles have medium-to-strong bands at 580–555 cm^{-1} (17.24–18.02 µm) and 560–525 cm^{-1} (17.86–19.05 µm) due to the C–C–CN in-plane deformation vibration. These two bands may be assigned to rotational isomers, the first band to the isomer where the C$\equiv$N group is *trans* to a carbon atom and the second band to that where the C$\equiv$N group is *trans* to a hydrogen atom. Aliphatic nitriles exhibit a very strong band in their Raman spectra at 200–160 cm^{-1} (50.00–62.50 µm).

Aromatic nitriles have two bands, one strong at 580–540 cm^{-1} (17.24–18.52 µm) and one of medium intensity at 430–380 cm^{-1} (23.26–26.32 µm). The former band is due to the combination of the out-of-plane aromatic ring-deformation vibration and the in-plane deformation vibration of the $-C\equiv N$ group. The latter band is due to the in-plane bending of the aromatic ring C–CN bond.

Inorganic cyanides[28] in the solid phase absorb over a wide range, 2250–2000 cm^{-1} (4.44–5.00 µm), as do coordination complexes: 2150–1980 cm^{-1} (4.65–5.05 µm).[28–30] For nitrile complexes with iodine monochloride, the band due to the C$\equiv$N stretching vibration is slightly higher by about 10 cm^{-1} (lower by 0.02 µm) than for the corresponding normal nitrile compound, the bands being broader and slightly stronger than usually observed. On the other hand, the coordination of nitriles to metal ions (R–C$\equiv$N $\rightarrow$ M) results in the band due to the C$\equiv$N stretching vibration being of greater intensity and occurring at a higher wavenumber, 2360–2225 cm^{-1} (4.23–4.47 µm), than for the uncoordinated nitrite compound. The cyanide ion absorbs at 2200–2070 cm^{-1} (4.55–4.83 µm).

Isonitriles, $-N\equiv C$

Alkyl and aryl isonitriles have strong absorptions in the regions 2175–2130 cm^{-1} (4.60–4.69 µm) and 2150–2110 cm^{-1} (4, 72–4.74 µm) respectively.[33,34] The intensity of the band is very sensitive to changes in the substituent. Isonitriles have a characteristic band, not found for nitriles, near 1595 cm^{-1} (6.25 µm).

Nitrile N-oxides, $-C\equiv N \rightarrow O$

Aryl nitrile N-oxides absorb strongly at 2305–2285 cm^{-1} (4.34–4.38 µm), due to the C$\equiv$N stretching vibration, and at 1395–1365 cm^{-1} (7.17–7.33 µm) due to the N–O stretching vibration.[35]

Table 4.3 Nitrile, isonitrile, nitrile N-oxide, and cyanamide C$\equiv$N stretching vibrations

Functional Groups	Region cm^{-1}	Region µm	Intensity IR	Intensity Raman	Comments
Saturated aliphatic nitriles	2260–2230	4.42–4.48	m	s, p	n-Alkyl 2250 cm^{-1}
α,β-Unsaturated nitriles	2250–2200	4.44–4.50	m–s	m–s, p	Polynuclear aromatics 2225–2210 cm^{-1}
Aryl nitriles	2240–2220	4.46–4.50	m–s	m–s, p	
α-Halogen-substituted nitriles	2280–2240	4.39–4.46	w–m	s, p	
β-Halogen-substituted nitriles	2260–2250	4.42–4.44	m–s	s, p	
ROCH$_2$CN	2260–2245	4.42–4.45	w	s, p	
RCO·CN	2225–2210	4.49–4.52	s	s, p	
NHR·CO·CHR·CN	2270–2255	4.41–4.43	s	s, p	
ROCOCH$_2$CN	~2260	~4.42	w	s, p	
$\diagdown$N—CH=C—CN $\diagup$	2210–2185	4.52–4.58	m–s	s, p	
(Sat·ring)·CN	2245–2230	4.45–4.48	s	s, p	
Aliphatic isonitriles	2175–2130	4.60–4.69	s	s, p	Conjugation lowers range to 2125–2105 cm^{-1}
R·CO·CH$_2$NC	2170–2160	4.61–4.63	s	s, p	
Aryl isonitriles	2150–2100	4.65–4.76	s	s, p	
Aryl nitrile N-oxides	2305–2285	4.34–4.38	s	s, p	
Thiocyanates, S–C$\equiv$N	2175–2135	4.60–4.68	m–s	s, p	
Cyanamides	2225–2200	4.49–4.55	s	s, p	
Cyanoguanidines	2210–2175	4.52–4.60	s	s, p	Often multiple peaks
$-$CF$_2$–C$\equiv$N	2280–2270	4.39–4.41	m–s	s, p	

Table 4.4 Nitrile, isonitrile, nitrile *N*-oxide, and cyanamide C≡N deformation vibrations

Functional Groups	Region cm⁻¹	Region μm	Intensity IR	Intensity Raman	Comments
Aliphatic nitriles	390–340	25.64–29.41	m–s	s, p	C–C≡N def vib
	200–160	50.00–62.50		s	
Primary aliphatic nitriles	580–555	17.24–18.02	m–s	m	C–C≡N in-plane def vib, C≡N *trans* to carbon atom
	560–525	17.86–19.05	m–s	m	C–C≡N in-plane def vib, C≡N *trans* to hydrogen atom
Secondary aliphatic nitriles	580–550	17.24–18.18	v	m	C–C≡N in-plane def vib
	545–530	18.35–18.87	v	m	C–C≡N in-plane def vib, C≡N *trans* to two hydrogen atoms
	565–535	17.70–18.69	v	m	C–C≡N in-plane def vib, C≡N *trans* to one hydrogen atom
Tertiary aliphatic nitriles	~575	~17.39	m–s	m	C–C–CN in-plane def vib, C≡N *trans* to three hydrogen atoms
	~595	~16.81	s	m	C–C–CN in-plane def vib, C≡N *trans* to a carbon atom and two hydrogen atoms
α,β-Unsaturated nitriles	285–220	35.09–45.45	m	v	C–CN def vib
	245–150	40.82–66.67		m, p	
Aromatic nitriles	580–540	17.24–18.52	s	m	combination of C≡N in-plane bending vib and out-of-plane bending vib of aromatic ring
	430–380	23.26–26.32	m	m–w	in-plane bending vib of aromatic C–CN bond
Aryl nitrile *N*-oxides	1395–1365	7.17–7.33	s	m	N–O str

Table 4.5 Diazonium compounds

Functional Groups	Region cm⁻¹	Region μm	Intensity IR	Intensity Raman	Comments
Diazonium salts	2300–2130	4.35–4.69	m–s	m–s	N≡N see refs: 31 and 32 in Chapter 5

Cyanamides, ⟍N–C≡N

Cyanamides absorb more strongly at lower frequencies than might be expected for the C≡N stretching mode. This is due to the presence of the resonance
⟍N–C≡N ⟷ ⟍N⁺=C=N⁻ which reduces the force constant.

The **C≡N** stretching band is found to be strong in both infrared and Raman spectra, the range being 2225–2210 cm⁻¹ (4.49–4.53 μm). The same resonance effect is found for cyanoguanidines

Diazonium Salts, Aryl–N≡N⁺X⁻

Diazonium salts[36–38] have a strong absorption in the region 2300–2130 cm⁻¹ (4.35–4.69 μm) which is due to the stretching vibration of the N≡N group. This band is dependent on the nature of the ring substituents but is less dependent on the nature of the anion, a shift of about 40 cm⁻¹ at most being observed for different anions. Aryl diazonium salts may be represented by the resonance structures

Electron-donating groups at *ortho* or *para* positions tend to increase the contribution of the second structure and hence tend to decrease the frequency of the N≡N stretching vibration, whereas electron-withdrawing groups have the opposite effect.

References

1. N. Sheppard and D. M. Simpson, *Quart. Rev*, 1952, **6**, 1.
2. E. A. Gastilovich and D. N. Shigorin, *Usp. Khim.*, 1973, **42**, 1358.

3. A. S. Wexler, *Appl. Spectrosc. Rev.*, 1968, **1**, 29.

4. T. L. Brown, *J. Chem. Phys.*, 1962, **38**, 1049.

5. R. A. Nyquist and W. J. Potts, *Spectrochim. Acta*, 1960, **16**, 419.

6. A. D. Allen and C. D. Cook, *Can. J. Chem.*, 1963, **41**, 1084.

7. A. A. Petrov *et al.*, *J. Gen. Chem. Moscow*, 1957, **27**, 2081.

8. A. A. Petrov and G. I. Semenov, *J. Gen. Chem. Moscow*, 1957, **27**, 2974.

9. A. A. Petrov and G. I. Semenov, *J. Gen. Chem. Moscow*, 1958, **28**, 73.

10. T. V. Yakovlera *et al.*, *Opt. Spectrosc.*, 1962, **12**, 106.

11. J. L. H. Allan *et al.*, *J. Chem. Soc.*, 1955, 1874.

12. J. C. Evans and R. A. Nyquist, *Spectrochim. Acta*, 1960, **16**, 918.

13. P. N. Daykin *et al.*, *J. Chem. Phys.*, 1962, **37**, 1087.

14. F. Sondheimer *et al.*, *J. Am. Chem. Soc*, 1962, **84**, 270.

15. F. Sondheimer and R. Wolovsky, *J. Am. Chem. Soc.*, 1962, **84**, 260.

16. N. A. Domnin and R. C. Kolinsky, *J. Gen. Chem. Moscow*, 1961, **33**, 1682.

17. G. Eglington and A. R. Galbraith, *J. Chem. Soc.*, 1959, 889.

18. J. C. D. Brand *et al.*, *J. Chem. Soc.*, 1960, 2526.

19. R. West and C. S. Kraihanel, *J. Am. Chem. Soc.*, 1961, **83**, 765.

20. E. A. Gastilovich *et al.*, *Opti. Spectrosc.*, 1961, **10**, 595.

21. A. Hidalgo, *Anales Real Soc. Espanola Fis. Chem. Madrid*, 1962, **58A**, 71.

22. J. P. Jesson and H. W. Thompson, *Spectrochim. Acta*, 1958, **13**, 217.

23. R. Heilmann and J. Bonner, *Compt. Rend.*, 1959, **248**, 2595.

24. D. A. Long and W. O. George, *Spectrochim. Acta*, 1964, **20**, 1799.

25. T. L. Brown, *Chem. Rev.*, 1958, **58**, 581.

26. H. W. Thompson and G. Steel, *Trans. Faraday Soc.*, 1956, **52**, 1451.

27. J. Hidalgo, *Compt. Rend.*, 1959, **249**, 395.

28. K. Nakamoto, *Infrared Spectra of Inorgnic and Coordination Compounds*, Wiley, New York, 1986.

29. J. P. Fackler, *J. Chem. Soc.*, 1962, 1957.

30. H. A. Brune and W. Zeil, *Z. Naturforsch*, 1961, **16A**, 1251.

31. G. L. Cadlow *et al.*, *Proc. R. Soc. London*, 1960, **A254**, 17.

32. L. van Haverbeke and M. A. Herman, *Spectrochim. Acta*, 1975, **31A**, 959.

33. I. Ugi and R. Meyer, *Chem. Ber.*, 1960, **93**, 239.

34. I. Ugi and C. Steinbruckner, *Chem. Ber.*, 1961, **94**, 2797 and 2802.

35. S. Califano *et al.*, *J. Chem. Phys.*, 1957, **26**, 1777.

36. L. A. Kazitsyna *et al.*, *Dokl. Acad. Nauk. SSSR*, 1963, **151**, 573.

37. L. A. Kazitsyna *et al.*, *J. Phys. Chem. Moscow*, 1960, **34**, 404.

38. R. H. Nuttall *et al.*, *Spectrochim. Acta*, 1961, **17**, 947.

39. G. Schrumpf, *Spectrochim. Acta*, 1983, **39A**, 505.

40. D. A. Compton *et al.*, *Spectrochim. Acta*, 1983, **39A**, 541.

41. T. Satto, M. Yamakawa and M. Takasuka, *J. Mol. Spectrosc.*, 1981, **90**, 359.

42. J-X. Han *et al.*, *J. Mol. Spectrosc.*, 1999, **198(2)**, 421.

43. F. Winther *et al.*, *J. Mol. Struct.*, 2000, 517–518, 265–270.

5 Cumulated Double-bond Compounds: X=Y=Z Group

Often resonance hybrids are possible for compounds of this type: X=Y=Z, X^+-Y^-≡Z, etc. The asymmetric stretching vibration of the cumulated double-bond group X=Y=Z gives rise to a band in the range $2275-1900\,cm^{-1}$ ($4.40-5.26\,\mu m$) which is in approximately the same region as the band due to the triple bond X≡Y, $2300-2000\,cm^{-1}$ ($4.35-5.00\,\mu m$).

The symmetric stretching vibration is generally weak and not very useful. It occurs in the region $1400-1100\,cm^{-1}$ ($7.14-9.09\,\mu m$). It can be seen that some compounds dealt with in this chapter could, in fact, be considered as triple-bond compounds (depending on the triple-bond character) and therefore could equally well have been dealt with in the previous chapter, e.g. thiocyanates.

Allenes, $\diagup C=C=C\diagdown$ [1,19-21]

Monosubstituted allenes have a medium-to-strong absorption in the region $1980-1945\,cm^{-1}$ ($5.05-5.14\,\mu m$) which is due to the asymmetric stretching vibration of the C=C=C group. For polar substituents, this band is in the higher-frequency portion of this range, and also, with strong polar groups such as carbonyls or nitriles, the band is observed to consist of two peaks. Asymmetrically- and symmetrically-disubstituted allenes absorb at $1955-1930\,cm^{-1}$ ($5.12-5.18\,\mu m$) and $1930-1915\,cm^{-1}$ ($5.18-5.22\,\mu m$) respectively.[20] Tri- and tetrasubstituted allenes absorb in the region $2000-1920\,cm^{-1}$ ($5.00-5.21\,\mu m$).

Mono- and asymmetrically-substituted allenes absorb strongly at $875-840\,cm^{-1}$ ($11.43-11.90\,\mu m$) due to the out-of-plane deformation vibrations of the $=CH_2$ group. The overtone of this band occurs near $1700\,cm^{-1}$ ($5.88\,\mu m$). Symmetrically-disubstituted allenes absorb near $870\,cm^{-1}$ ($11.49\,\mu m$) due to the CH deformation vibrations. Trisubstituted allenes absorb strongly at $880-840\,cm^{-1}$ ($11.36-11.90\,\mu m$).

The C=C=C symmetric stretching band is of medium or weak intensity, or absent. It occurs in the region $1095-1060\,cm^{-1}$ ($9.13-9.43\,\mu m$) and is not a useful band in making assignments.

Bands due to the C–H stretching vibrations of $C=C=CH_2$ occur near $3050\,cm^{-1}$ ($3.28\,\mu m$) and $2990\,cm^{-1}$ ($3.34\,\mu m$), the first band being at a slightly lower frequency than the corresponding band for vinyl and vinylidene groups.

The C=C=C bending vibration near $355\,cm^{-1}$ ($28.17\,\mu m$) is of strong intensity in Raman spectra.

Isocyanates, –N=C=O, and Cyanates

Due to the asymmetric stretching vibration of the –N=C=O group, a band, sometimes with shoulders, occurs at $2300-2250\,cm^{-1}$ ($4.35-4.44\,\mu m$) which is a useful band for characterisation,[2,7,18,36] except for methyl isocyanate which absorbs near $2230\,cm^{-1}$ ($4.48\,\mu m$). This band is slightly broader than the corresponding band observed for thiocyanates and is not affected by conjugation. However, for α,β-unsaturated compounds the C=C stretching vibration is affected and moves from its normal position to $1670-1630\,cm^{-1}$ ($5.99-6.14\,\mu m$). In Raman spectra, the asymmetric NCO stretching band is weak or not observed. The symmetric –N=C=O stretching vibration band occurs at $1460-1340\,cm^{-1}$ ($6.85 - 7.46\,\mu m$). It is weak and not usually of use for assignment purposes since it is often overlapped by aliphatic absorption bands which occur in the same region.

Aliphatic and aryl isocyanate trimers,[3] i.e. isocyanurates, have a strong band due to the carbonyl stretching vibration in the region $1715-1680\,cm^{-1}$ ($5.83-5.95\,\mu m$). Aromatic isocyanate dimers[3] have a strong similar band at $1785-1775\,cm^{-1}$ ($5.60-5.63\,\mu m$). A band of variable intensity in the region

Table 5.1 Allenes

Functional Groups	Region cm^{-1}	Region µm	Intensity IR	Intensity Raman	Comments
Allenes	2000–1915	5.00–5.22	m–s	v	asym C=C=C str
	1095–1060	9.31–9.43	w	s, p	sym C=C=C str
	~355	~28.17	w–m	s	C=C=C bending vib. For haloallenes 625–590 and 550–485 cm^{-1}.
Monosubstituted allenes, –HC=C=CH$_2$	1980–1945	5.05–5.14	m–s	w	asym C=C=C str
	~1700	~5.88	w		=CH$_2$ wagging vib overtone
	875–840	11.43–11.90	s	w	=CH$_2$ out-of-plane def vib
Symmetrically-disubstituted allenes, –CH=C=CH–	1930–1915	5.18–5.22	m–s	v	asym C=C=C str
Asymmetrically-disubstituted allenes, >C=C=CH$_2$	1955–1930	5.12–5.18	m–s	v	asym C=C=C str
	~1700	~5.88	w		=CH$_2$ wagging vib overtone
	875–840	11.43–11.90	s	w	=CH$_2$ out-of-plane def vib
Tri- and tetrasubstituted allenes	2000–1920	5.00–5.21	m–s	v	asym C=C=C str
Methyl, ethyl, propyl, and butyl allenes	~555	~18.02	m	s	C=C=C bending vib
	550–520	18.18–19.23	w–m	s	C=CH bending vib
	355–305	28.17–32.79	w–m	s	C=C=C bending vib
	~200	~50.00	w–m		C=C=C bending vib
Cyclopropyl allenes, ▷=C=C<	~2020	~4.95	m	v	asym C=C=C str

650–580 cm^{-1} (15.39–17.24 µm) may be observed due to the NCO bending vibration. This band is often broad and of medium intensity.

Cyanates have a strong band in the region 2260–2240 cm^{-1} (4.42–4.46 µm). The C–O–CN stretching vibration results in a strong band in the region 1125–1080 cm^{-1} (8.89–9.26 µm) for alkyl compounds and 1190–1110 cm^{-1} (8.40–9.01 µm) for aromatics.

Note that the symmetric NCO band for isocyanates occurs above 1200 cm^{-1} (below 8.33 µm) and hence can easily be distinguished from cyanates.

Isothiocyanates, –N=C=S[6–12]

Due to the asymmetric stretching vibration of the –N=C=S group, a very strong band in the region 2150–1990 cm^{-1} (4.65–5.03 µm) is observed. For aliphatic compounds, this band is usually a broad doublet, although it may sometimes have a shoulder which appears at 2225–2150 cm^{-1} (4.49–4.65 µm). Alkyl isothiocyanates[11] absorb in the region 2140–2080 cm^{-1} (4.67–4.81 µm) whereas aryl derivatives[12,35] tend to absorb in the region 2100–1990 cm^{-1} (4.76–5.03 µm). For alkyl compounds, the symmetric

stretching vibration gives rise to a band of variable intensity in the region 1250–1080 cm^{-1} (8.00–9.26 µm) whereas for aryl isothiocyanates, a strong-to-medium intensity band is observed at 940–925 cm^{-1} (10.64–10.81 µm), this being a weak band in Raman spectra. Most alkyl isothiocyanates have absorptions at 640–600 cm^{-1} (15.63–16.67 µm) and at 565–510 cm^{-1} (17.70–19.61 µm) which are of strong-to-medium intensity. These bands have been assigned to the in-plane and out-of-plane deformation vibrations of the –NCS group. A medium-to-strong band is also usually observed at 470–440 cm^{-1} (21.28–22.73 µm). In Raman spectra, alkyl isothiocyanates have strong, polarised bands at 1090–980 cm^{-1} (9.17–10.20 µm).

Thiocyanates, –S–C≡N

(Rather than include this section in the previous chapter, it was felt that it would best be treated here together with isothiocyanate compounds.)

A sharp band of medium-to-strong intensity is observed in the region 2175–2135 cm^{-1} (4.60–4.68 µm) due to the C≡N stretching vibration.[4–7,9,10,13] The absorption due to aryl derivatives is found in the upper end of this frequency range while that for alkyl derivatives[4] is in the lower half of the range.

All aliphatic thiocyanates have a strong band at $405-400 \, \text{cm}^{-1}$ ($24.69-25.00 \, \mu\text{m}$) which is due to the in-plane deformation vibration of the $-$SCN group. Primary aliphatic thiocyanates have a weak-to-medium intensity band at $650-640 \, \text{cm}^{-1}$ ($15.38-15.63 \, \mu\text{m}$) due to the stretching vibration of the S$-$CN bond and a band of medium-to-strong intensity near $620 \, \text{cm}^{-1}$ ($16.13 \, \mu\text{m}$) due to the C-S stretching vibration (where the carbon is the α-carbon). Secondary aliphatic thiocyanates have a band of variable intensity at $610-600 \, \text{cm}^{-1}$ ($16.39-16.67 \, \mu\text{m}$) due to the S$-$CN bond stretching vibration and, in addition, as many as three bands may be observed due to different molecular configurations: one near $655 \, \text{cm}^{-1}$ ($15.27 \, \mu\text{m}$), another at $640-630 \, \text{cm}^{-1}$ ($15.63-15.87 \, \mu\text{m}$), and one near $575 \, \text{cm}^{-1}$ ($17.39 \, \mu\text{m}$). As with alkyl halides, cyanides, etc., different rotational isomers are possible. In Raman spectra, the C$-$S stretching vibration bands are of medium-to-strong intensity.

Simple inorganic thiocyanates[14-16] absorb strongly near $2050 \, \text{cm}^{-1}$ ($4.90 \, \mu\text{m}$), this band usually being the predominant one in the region $5000-650 \, \text{cm}^{-1}$ ($2-15 \, \mu\text{m}$). A weak symmetrical stretching band is observed at $1090-925 \, \text{cm}^{-1}$ ($9.17-10.81 \, \mu\text{m}$) but in Raman spectra this band is of medium intensity and is polarised.

Selenocyanates and Isoselenocyanates[17]

Aromatic selenocyanates have a medium-to-strong sharp band near $2160 \, \text{cm}^{-1}$ ($4.63 \, \mu\text{m}$) whilst the corresponding isoselenocyanates have a strong, broad band, usually with two peaks, in the region $2200-2000 \, \text{cm}^{-1}$ ($4.55-5.00 \, \mu\text{m}$). The symmetric $-$N$=$C$=$Se stretching vibration band of isoselenocyanates occurs in the region $675-605 \, \text{cm}^{-1}$ ($14.85-16.53 \, \mu\text{m}$). Selenocyanates have a band of medium intensity at $545-520 \, \text{cm}^{-1}$ ($19.23-18.35 \, \mu\text{m}$) due to the stretching vibration of the Se$-$CN bond, another band at about $420-400 \, \text{cm}^{-1}$ ($23.81-40.00 \, \mu\text{m}$) due to the in-plane vibration of the Se$-$C$\equiv$N group, and a band at about $360 \, \text{cm}^{-1}$ ($27.78 \, \mu\text{m}$) due to the out-of-plane vibration of the group.

Alkyl isoselenocyanates absorb in the regions $2185-2100 \, \text{cm}^{-1}$ ($4.58-4.76 \, \mu\text{m}$) and $560-500 \, \text{cm}^{-1}$ ($17.86-20.00 \, \mu\text{m}$). For isoselenocyanates where the nitrogen atom is not bound to a carbon atom (e.g. to an atom of Si, Sn, Ge, etc.), the asymmetric $-$N$=$C$=$Se stretching vibration band occurs at about $2140 \, \text{cm}^{-1}$ ($4.67 \, \mu\text{m}$), a single band being observed. For isoselenocyanato-phosphates, $(\text{RO})_2\text{P}=\text{ONCSe}$, and thiophosphates, $(\text{RO})_2\text{P}=\text{SNCSe}$,[22] the N$=C=$Se asymmetric stretching vibration band occurs in the region $1975-1960 \, \text{cm}^{-1}$ ($5.06-5.10 \, \mu\text{m}$).

Isoselenocyanato- complexes have a strong band at $430-370 \, \text{cm}^{-1}$ ($23.26-27.03 \, \mu\text{m}$).

Azides, $-$N$=$N$^+$$=N^-$

Resonance is possible for these compounds:

$$-\text{N}=\text{N}^+=\text{N}^- \longleftrightarrow -\text{N}^--\text{N}^+\equiv\text{N}.$$

Organic azides[23-27,35] have a strong band in the region $2170-2080 \, \text{cm}^{-1}$ ($4.60-4.81 \, \mu\text{m}$) due to the asymmetric stretching vibration of the N$=$N$=$N group and in Raman spectra this band is of medium-to-strong intensity. This band is relatively insensitive to conjugation and to changes in the electronegativity of the adjacent group. A weak band at $1345-1175 \, \text{cm}^{-1}$ ($7.43-8.51 \, \mu\text{m}$) is also observed due to the symmetric stretching of the NNN group, this band being of strong intensity in Raman spectra. This band is not observed for ionic azides,[28] which have their strong absorption in the range $2170-2030 \, \text{cm}^{-1}$ ($4.61-4.93 \, \mu\text{m}$). Information is also available for inorganic azides.[29,30]

Diazo Compounds, $\text{\textbackslash}$C$=$N$^+$$=N^-$

Diazo compounds may be represented by resonance hybrids:

$$\text{C}=\text{N}^+=\text{N}^- \longleftrightarrow \text{C}^--\text{N}^+\equiv\text{N}^-$$

Diazo compounds with the group $-$CH$=$N$^+$$=N^-$ have a strong absorption in the region $2050-2035 \, \text{cm}^{-1}$ ($4.88-4.91 \, \mu\text{m}$) and disubstituted compounds, $\text{\textbackslash}$C$=$N$^+$$=N^-$, absorb at $2035-2000 \, \text{cm}^{-1}$ ($4.91-5.00 \, \mu\text{m}$).

Diazoketones and diazoesters, $\text{\textbackslash}$CO$-$C$=$N$^+$$=N^-$, have their carbonyl stretching frequencies slightly decreased from that expected for an ordinary ketone or ester. Similarly, the stretching vibration frequency of the C$=$N$^+$$=N^-$ group for these compounds is increased (probably due to coupling), indicating that there is an increase in the proportion of triple-bond character. Monosubstituted diazoketones, $-$CO$-$CH$=$N$^+$$=N^-$ absorb at $2100-2080 \, \text{cm}^{-1}$ ($4.76-4.81 \, \mu\text{m}$) and disubstituted diazoketones, $-$CO$-$C$=$N$^+$$=N^-$, absorb at $2075-2050 \, \text{cm}^{-1}$ ($4.82-4.88 \, \mu\text{m}$), the frequency of the carbonyl stretching absorption being lowered to $1650-1645 \, \text{cm}^{-1}$ ($6.06-6.08 \, \mu\text{m}$) for aliphatic compounds and to $1630-1605 \, \text{cm}^{-1}$ ($6.14-6.23 \, \mu\text{m}$) for aromatic compounds.

Table 5.2 X=Y=Z groups (except allenes)

Functional Groups	Region cm^{-1}	μm	Intensity IR	Raman	Comments
Isocyanates –N=C=O	2300–2250	4.35–4.44	vs	w	asym NCO str, (see refs 35 and 36) br. Aryl isocyanates 2285–2265 cm^{-1}
	1460–1340	6.85–7.46	w–m	s, p	sym NCO str
	650–580	15.39–17.24	m, br	w	def vib
Cyanates	2260–2240	4.42–4.46	s	s, p	–OCN str
	1190–1080	8.40–9.26	s		COCN str. (C–O str 1125–1080 cm^{-1})
Cyanate ion, NCO$^-$	2225–2100	4.49–4.76	s	w–m	asym NCO str
	1335–1290	7.49–7.75	s	m–s, p	sym NCO str
	1295–1180	7.72–8.47	w		Combination band
	650–600	15.38–16.67	s	w	NCO bending vib
Isothiocyanates –N=C=S	2150–1990	4.65–5.03	vs	m, p	br. asym NCS str, usually a doublet, in range 2125–2085 cm^{-1} vs, br
	1250–925	8.00–10.81	v	s, p	sym NCS (see text)
	690–650	14.49–15.39	–	s	
	645–575	15.50–17.39	m–w	w	
Alkyl isothiocyanates	1250–1080	8.00–9.26	v	w	sym NCS str
	1090–980	9.17–10.20	–	s, p	
	640–600	15.63–16.67	m–s	w	br, –NCS in-plane def vib
	565–510	17.70–19.61	m–s	w	–NCS out-of-plane def vib
	470–440	21.28–22.73	m–s		
Aryl isothiocyanates	940–925	10.64–10.81	m–s	w	sym NCS str
Thiocyanates, –SC≡N	2175–2135	4.60–4.68	m–s	m–s, p	asym str. Aryl at upper end of frequency range 2175–2160 cm^{-1}, alkyl at lower end 2160–2135 cm^{-1}
	1090–925	9.17–10.81	w	m–s, p	sym str
	700–670	14.29–14.93	w	s, p	C–S–C asym str
	660–610	15.15–16.39	m–s	s	C–S–C sym str
	515–450	19.42–22.22	w–m		SCN bending vib
	420–400	23.81–25.00	w–m		In-plane def vib
Alkyl thiocyanates	405–400	24.69–25.00	s		SCN group in-plane def vib
Primary aliphatic thiocyanates	650–640	15.38–15.63	w–m	s	S–CN str
	~620	~16.13	m–s	s	CS–S str (absent for MeSCN)
	~405	~24.69	s		
Secondary aliphatic thiocyanates	~655	~15.27	w	s	C$_\alpha$–S str
	640–630	15.63–15.87	w	s	C$_\alpha$–S str
	610–600	16.39–16.00	v	s	C–SN str
	~575	~17.39	m	s	C$_\alpha$–S str
	~405	~24.69	s		
Thiocyanate ion	2190–2020	4.57–4.95	s	m–s	asym N=C=S str
	~950	~10.53	w		Bending vib overtone
	~750	~13.33	w	m	sym str
	~470	~21.28	s	w	Bending vib
Coordinated thiocyanate ions, NCS–metal	730–690	13.70–14.49	m–s	m–s	See refs 15, 16 and 34, and Chapter 22
Coordinated isothiocyanate ions, SCN–metal	860–780	11.63–12.82	m–s	m–s	See Chapter 22

(continued overleaf)

Table 5.2 (*continued*)

Functional Groups	Region cm^{-1}	Region μm	Intensity IR	Intensity Raman	Comments
Alkyl selenocyanates, –SeCN	~2150	~4.65	s	s, p	
	545–520	18.35–19.23	m–s	s	Se–CN str
	420–400	23.81–25.00	w		
	365–360	27.39–27.78	w		
Aromatic selenocyanates, –SeCN	~2160	~4.63	s	s, p	sh
	420–400	23.81–25.00	w		
	~350	~28.57	w		SeCN bending vib
Alkyl isoselenocyanates	2185–2100	4.58–4.76	s	s	
	560–500	17.86–20.00	m–s	s	
Aromatic isoselenocyanates	2200–2000	4.55–5.00	s	s	br, doublet
Azides –N=N=N	2170–2080	4.61–4.81	vs–s	m–s, p	asym str (sometimes a doublet)(–CO–N$_3$··, ~2150 cm^{-1})
	1345–1175	7.43–8.51	m–w	s, p	sym str
Metal azides and azide ion	680–410	14.71–24.39	w		N=N=N bending vib
Acid azides and nitro-aromatic azides	2240–2170	4.46–4.61	s	m–s	asym N=N=N str
	2155–2140	4.64–4.67	s	s	asym N=N=N str
	1710–1690	5.85–5.92	s	w–m	C=O str for acid azides
	1260–1235	7.94–8.10	m	s	sym N=N=N str
Diazo compounds $\diagdown$C=N$^+$=N$^-$	2050–2000	4.88–5.00	vs	v	br, asym str CNN
	1390–1330	7.19–7.52	s	m–s	sym str CNN
Diazoketones and diazoesters, –CO–C=N$^+$=N$^-$	2075–2050	4.82–4.88	s	m–s	(Ketones: C=O str, 1650–1600 cm^{-1} and strong band, 1390–1330 cm^{-1} – may be doublet; alkylketones: C=O str, ~1645 cm^{-1})
Diazonium salt, Ar–N≡N$^+$X$^-$	2300–2230	4.35–4.69	m–s	m–s	N≡N str, see refs: 31 and 32
Ketenes, $\diagdown$C=C=O	2200–2080	4.45–4.81	m–s	v	often found near 2150 cm^{-1}
	~1130	~8.85	v	m–s, p	sym C=C=O str. Range 1420–1120 cm^{-1} usually s–m (Aromatics: IR intensity w, Raman s, p).
R$_3$SiCH=C=O	2115–2085	4.73–4.80	s	w	asym C=C=O str
	1295–1265	7.72–7.90	w	s	sym C=C=O str
Ketenimines, $\diagdown$C=C=N–	2170–2000	4.61–5.00	s	v	asym C=C=N str
	~1235	~8.10	m	s	sym C=C=N str
	1190–1080	8.40–9.26	s		COCN str
Aliphatic Carbodi-imines R–N=C=N–R	2155–2130	4.64–4.70	vs	w	asym N=C=N str, see ref. 33
	~1460	~6.85	w	s, p	sym N=C=N str
Aryl carbodi-imines Ar–N=C=N–Ar	2145–2135	4.66–4.68	vs	m–s, p	C=N str doublet due to Fermi resonance band at ~2110 cm^{-1} usually being the stronger
	2115–2105	4.73–4.75	vs	m–s, p	
Thionylamines –N=S=O	1300–1230	7.69–8.13	v	s, p	NSO asym str
	1180–1100	8.48–9.09	v	s	NSO sym str

Carbodi-imides, $-N=C=N-$

Aliphatic compounds have a very intense band in the region 2155–2130 (4.64–4.70 μm) and a weaker band at 1580–1340 cm^{-1} (6.33–7.46 μm). Aromatic compounds have a very intense band near 1240 cm^{-1} (8.07 μm) and near 1210 cm^{-1} (8.26 μm) and a weaker band being observed at 1680–1380 cm^{-1} (5.95–7.25 μm). The symmetrical stretching band is weak and occurs near 1460 cm^{-1} (6.85 μm) but in Raman spectra this is a strong band.

References

1. L. M. Sverdlov and M. G. Borisov, *Opt. Spectrosc.*, 1960, **9**, 227.
2. E. A. Nicol *et al.*, *Spectrochim. Acta*, 1974, **30**, 1717.
3. B. Taub and C. E. McGinn, *Dyestuffs*, 1958, **42**, 263.
4. R. P. Hirschmann *et al.*, *Spectrochim. Acta*, 1964, **20**, 809.
5. K. Kottke *et al.*, *Pharmazie*, 1973, **28**, 736.
6. N. S. Ham and J. B. Willis, *Spectrochim. Acta*, 1960, **16**, 279.
7. G. D. Caldow and H. W. Thompson, *Spectrochim. Acta*, 1958, **13**, 212.
8. E. Svatek *et al.*, *Acta Chem. Scand.*, 1959, **13**, 442.
9. A. Foffani *et al.*, *R. C. Acad. Lincei*, 1960, **29**, 355.
10. E. Lieber *et al.*, *Spectrochim. Acta*, 1959, **13**, 296.
11. R. N. Knisely *et al.*, *Spectrochim. Acta*, 1967, **23A**, 109.
12. P. Kristián *et al.*, *Coll. Czech. Chem. Comm.*, 1964, **29**, 2507.
13. R. A. Cummins, *Austral. J. Chem.*, 1964, **17**, 838.
14. C. Pecile *et al.*, *R. C. Acad. Lincie*, 1960, **28**, 189.
15. A. Turco and C. Pecile, *Nature*, 1961, **191**, 66.
16. A. Tramer, *J. Chem. Phys.*, 1962, **59**, 232.
17. W. J. Franklin *et al.*, *Spectrochim. Acta*, 1974, **30A**, 1293.
18. R. P. Hirschmann *et al.*, *Spectrochim. Acta*, 1965, **21**, 2125.
19. J. H. Wotiz and D. E. Mancuso, *J. Org. Chem.*, 1957, **22**, 207.
20. A. A. Petrov *et al.*, *Opt. Spectrosc.*, 1959, **7**, 170.
21. M. G. Borisov and L. M. Sverdlov, *Opt. Spectrosc.*, 1963, **15**, 14.
22. T. Gabrio and G. Barnikow, *Z. Chem.*, 1969, **9**, 183.
23. E. Mantica and G. Zerbi, *Gazz. Chim. Ital.*, 1960, **90**, 53.
24. E. Lieber and A. E. Thomas, *Appl. Spectrosc.*, 1961, **15**, 144.
25. E. Lieber and E. Oftedahl, *J. Org. Chem.*, 1959, **24**, 1014.
26. E. Lieber *et al.*, *Anal. Chem.*, 1957, **29**, 916.
27. W. R. Carpenter, *Appl. Spectrosc.*, 1963, **17**, 70.
28. J. I. Bryant and G. C. Turrell, *J. Chem. Phys.*, 1962, **37**, 1069.
29. H. A. Papazian, *J. Chem. Phys.*, 1961, **34**, 1614.
30. W. Dobramsyl *et al.*, *Spectrochim. Acta*, 1975, **31A**, 905.
31. L. A. Kazitsyna *et al.*, *J. Phys. Chem. Moscow*, 1960. **34**, 850.
32. L. A. Kazitsyna *et al.*, *Dokl. Akad. Nauk. SSSR*, 1963, **151**, 573.
33. G. D. Meakins and R. J. Moss, *J. Chem. Soc.*, 1957, 993.
34. J. Lewis *et al.*, *J. Chem. Soc.*, 1961, 4590.
35. A. El Shahawy and R. Gaufres, *J. Chim. Phys. Phys.-Chim. Biol.*, 1978, **75**, 196.
36. R. A. Nyquist *et al.*, *Appl. Spectrosc.*, 1992, **46**, 841 & 972.

6 Hydroxyl Group Compounds: O–H Group

Alcohols, R–OH

Bands due to O–H stretching and bending vibrations and C–O stretching vibrations are observed.

Alcohol O–H Stretching Vibrations

In the infrared, the O–H stretching band[1-5,32] is of medium-to-strong intensity,[6] although it may be broad (see below). However, in Raman spectra, the band is generally weak.

Unassociated hydroxyl groups absorb strongly in the region $3670-3580 \, \text{cm}^{-1}$ ($2.73-2.80 \, \mu\text{m}$).[1,2,27-29] However, free hydroxyl groups only occur in the vapour phase or in very dilute solutions in non-polar solvents. The band due to the free hydroxyl group[28] is sharp and its relative intensity increases in the following order:

aromatic alcohols < tertiary alcohols < secondary alcohols

< primary alcohols

In very dilute solution in non-polar solvents, the normal O-H absorptions of alcohols are:

primary aliphatic alcohols	$3645-3630 \, \text{cm}^{-1}$ ($2.74-2.75 \, \mu\text{m}$)
secondary aliphatic alcohols	$3640-3620 \, \text{cm}^{-1}$ ($2.75-2.76 \, \mu\text{m}$)
tertiary aliphatic alcohols	$3625-3610 \, \text{cm}^{-1}$ ($2.76-2.77 \, \mu\text{m}$)
$\text{R—OH}\cdots\text{O=C}$	$3600-3450 \, \text{cm}^{-1}$ ($2.78-2.90 \, \mu\text{m}$)

The relative intensity of the band due to the hydroxyl stretching vibration decreases with increase in concentration, with additional broader bands appearing at lower frequencies $3580-3200 \, \text{cm}^{-1}$ ($2.73-3.13 \, \mu\text{m}$). These bands are the result of the presence of intermolecular bonding, the amount of which increases with concentration. The precise position of the O–H band is dependent on the strength of the hydrogen bond.[4] In some samples, intramolecular hydrogen bonding[4-8] may occur, the resulting hydroxyl group band which appears at $3590-3400 \, \text{cm}^{-1}$ ($2.79-2.94 \, \mu\text{m}$) being sharp and unaffected by concentration changes.

For solids, liquids, and concentrated solutions, a broad band is normally observed at about $3300 \, \text{cm}^{-1}$ ($3.00 \, \mu\text{m}$). Polyhydric alcohols in dilute solution in non-polar solvents normally have a sharp band at about $3600 \, \text{cm}^{-1}$ ($2.78 \, \mu\text{m}$) and a broader band at $3550-3450 \, \text{cm}^{-1}$ ($2.82-2.90 \, \mu\text{m}$). Hydroxyl groups which are hydrogen-bonded to aromatic ring π-electron systems absorb at $3580-3480 \, \text{cm}^{-1}$ ($2.79-2.87 \, \mu\text{m}$).[30]

Overtone bands of carbonyl stretching vibrations also occur in the region $3600-3200 \, \text{cm}^{-1}$ ($2.78-3.13 \, \mu\text{m}$) but are, of course, of weak intensity. Bands due to N–H stretching vibrations may also cause confusion. However, these bands are normally sharper than those due to intermolecularly hydrogen bonded O–H groups.[30,31]

Alcohol C–O Stretching Vibrations

The absorption region for the alcohol C–O group due to its stretching vibration is $1200-1000 \, \text{cm}^{-1}$ ($8.33-10.00 \, \mu\text{m}$). Hydrogen bonding has the effect of decreasing the frequency of this band slightly: saturated primary alcohols absorb strongly in the region $1090-1000 \, \text{cm}^{-1}$ ($9.17-10.00 \, \mu\text{m}$); secondary alcohols absorb at $1125-1085 \, \text{cm}^{-1}$ ($8.90-9.22 \, \mu\text{m}$); tertiary alcohols absorb strongly at $1205-1125 \, \text{cm}^{-1}$ ($8.30-8.90 \, \mu\text{m}$). The COC stretching band at $1090-1000 \, \text{cm}^{-1}$ ($9.17-10.00 \, \mu\text{m}$) is characteristic of primary alcohols. These ranges, which are given for pure liquids, should be extended slightly for solution spectra. In general, the presence of unsaturation and chain branching both lower the C–O stretching vibration frequency. Care must be taken since esters, carboxylic acids, acid anhydrides, and ethers all absorb strongly in the general range $1300-1000 \, \text{cm}^{-1}$ ($7.69-10.00 \, \mu\text{m}$) due to the C–O stretching vibration.

Table 6.1 Hydroxyl group O–H stretching vibrations

Functional Groups	Region		Intensity		Comments
	cm^{-1}	μm	IR	Raman	
Free O–H	3670–3580	2.73–2.80	v	w	sh, OH str
Hydrogen-bonded O–H (intermolecular), —Ö—H H H₂O' H₂O' H₂O [dimer] [polymer]	3550–3230	2.82–3.10	m–s	w	Usually broad but may be sharp, frequency is concentration-dependent
Hydrogen-bonded O–H (intramolecular), H H O O	3590–3400	2.79–2.94	v	w	Usually sharp, frequency is concentration-independent
Chelated O–H,	3200–2500	3.13–4.00	v	w	Usually broad, frequency concentration-independent
–OD	2780–2400	3.60–4.17	v	w	O–D str
OH of enol form of β-diketones	2700–2500	3.71–4.00	v	w	br, chelated OH
Intramolecular-bonded *ortho*-phenols	3200–2500	3.13–4.00	m	w	Free phenols ∼3610 cm^{-1}
Carboxylic acids, –COOH	3300–2500	3.03–4.00	w–m	w	br, O–H str, hydrogen-bonded, sometimes number of weak bands in region 2700–2500 cm^{-1}. Band is concentration-dependent
OH of water of crystallization	3600–3100	2.78–3.23	w	w	In solid-state spectra
	1630–1600	6.13–6.25	m	w	def
OH of water in dilute solution	∼3760	∼2.66	w–m	w	In non-polar solvents
Free oximes, C=N—OH	3600–3570	2.78–2.79	w–m	w	sh
Oximes, hydrogen-bonded	3300–3150	3.03–3.17	m	w	br
Free hydroperoxides, –O–O–H	3560–3530	2.82–2.83	m	w	
Peracids, –CO–O–OH	∼3280	∼3.05	m	w	
Tropolones	∼3100	∼3.23	w–m	w	
Phosphorus acids, P OH	2700–2560	3.70–3.91	m	w	br

Primary and secondary alcohols have a band of medium intensity in their infrared spectra at 900–800 cm^{-1} (11.11–12.50 μm) due to C–C–O stretching vibration. In Raman spectra, this is a strong band. For tertiary alcohols, this band occurs at 800–750 cm^{-1} (12.50–13.33 μm) and is of strong intensity in Raman spectra.

Alcohol O–H Deformation Vibrations

The in-plane O–H deformation vibration gives rise to a medium-to-strong band in the region 1440–1260 cm^{-1} (6.94–7.93 μm). In concentrated solutions, this band is very broad, extending over approximately 1500–1300 cm^{-1} (6.67–7.69 μm). On dilution, the band becomes weaker and is eventually replaced by a sharp, narrow band at about 1260 cm^{-1} (7.93 μm). In the presence of hydrogen bonding, the O–H deformation vibration is lowered in frequency. (Bands due to CH, deformation vibrations may also be present in this region.) In Raman spectra, the COH bending vibration band is generally of weak-to-medium intensity. This can be used to advantage, since other bands which would otherwise be difficult to observe may be seen by the use of Raman spectroscopy.

Table 6.2 Hydroxyl group O–H deformation vibrations

Functional Groups	Region		Intensity		Comments
	cm^{-1}	μm	IR	Raman	
Primary and secondary alcohols	1440–1260	7.41–7.94	m–s	m–w	In-plane O–H def vib , br
Secondary alcohols, ＼CHOH ／	1430–1370	6.99–7.30	m–s	m–w	In-plane O–H def vib coupled with CH wagging vib , br. In dilute soln., moves to 1310–1250 cm^{-1}
Tertiary alcohols	1410–1310	7.09–7.63	m	m–w	In-plane O–H def vib , br. Hydrogen-bonded: near 1410 cm^{-1}, dilute soln:, 1320 cm^{-1}.
Alcohols	710–570	14.08–17.54	m, br	w	O–H out-of-plane def vib
Phenols	1410–1310	7.09–7.63	s	m–w	O–H def vib and C–O str combination
Carboxylic acids	960–875	10.41–11.42	m	m–w	O–H out-of-plane def vib , br diffuse
Deuterated carboxylic acids	~675	~14.81	s	m–w	O-D in-plane def vib

Table 6.3 Alcohol C–O stretching vibrations, deformation and other bands

Functional Groups	Region		Intensity		Comments
	cm^{-1}	μm	IR	Raman	
Primary alcohols, –CH$_2$–OH	2990–2900	3.34–3.45	w–m	m–s	asym CH$_2$ str
	2935–2840	3.41–3.52	w–m	m–s	sym CH$_2$ str
	1480–1410	6.76–7.09	w–m	m	CH$_2$ def vib
	1390–1280	7.19–7.81	w–m	m–w	CH$_2$ wagging – alcohol OH def vib may obscure
	1300–1280	7.69–7.81	w–m	m	CH$_2$ twisting vib, may be obscured by OH def vib
	1090–1000	9.17–10.00	s	s–m	CCO str, characteristic band
	900–800	11.11–12.50	m	s	CCO str
	960–800	10.42–12.50	w–m	m	CH$_2$ twisting vib
	710–570	14.08–17.54	w–m	w	br, OH out-of-plane def vib
	555–395	18.01–25.32	w–m	m–w	C–O def vib
RCH$_2$CH$_2$OH	~1050	~9.52	s	m–s	Ethanol ~1065 cm^{-1}. CCO str
R$_1$R$_2$CHCH$_2$OH	~1035	~9.66	s	m–s	CCO str
R$_1$R$_2$R$_3$CCH$_2$OH	~1020	~9.80	s	m–s	CCO def vib
(Unsat group) –CH$_2$CH$_2$OH	~1015	~9.85	s	m–s	vinyl or aryl substituted
Secondary alcohols, ＼CH—OH ／	1400–1330	7.14–7.52	w	m–w	CH wagging vib
	1350–1290	7.41–7.75	w	m	CH def vib
	1150–1075	8.70–9.30	s	m–s	C–O str, often shows multiple bands due to coupling
	900–800	11.11–12.50	m	s, p	CCO str
	660–600	15.15–16.67	m, br	w	OH out-of-plane def vib
	500–440	20.00–22.73	w	m–w	CO in-plane def vib
	390–330	25.64–30.30	m	w	CO out-of-plane def vib
RH$_2$C＼ ／CHOH H$_3$C	~1085	~9.22	s	m–s	(IsoPropyl alcohol ~1100 cm^{-1}) Each additional alkyl group increases wavenumber by ~15 cm^{-1}
(Unsat group) –CH$_2$CH(OH)CH$_3$	~1070	~9.35	s	m–s	

Table 6.3 (*continued*)

Functional Groups	Region cm⁻¹	Region µm	Intensity IR	Intensity Raman	Comments
[(Unsat group) CH₂]₂–CHOH	~1010	~9.90	s	m–s	
R(Aryl)CHOH	1350–1260	7.41–7.94	m–s	m–w	OH def vib
	1075–1000	9.30–10.00	s	w	CO str
(Aryl–CH₂)₂CHOH	~1050	~9.52	s	m–s	
Aromatic and α,β-unsaturated secondary alcohols	1085–1030	9.22–9.71	s	m–s	
>C=CH—CHOH	1080–1020	9.26–9.80	s	m–s	CCO str
Tertiary alcohols >C–OH	1210–1100	8.26–9.09	s	m–s	CC–O str
	800–750	12.50–13.33	m	s, p	C–O def vib
	~360	~27.78		m–w	CCO bending vib
Saturated tertiary alcohols —C–OH	1210–1100	8.26–9.09	s	m–s	
RCH₂(CH₃)₂COH	~1135	~8.81	s	m–s	(*t*-Butyl alcohol ~1150 cm⁻¹)
CH₃(R₁CH₂)(R₂CH₂)COH	~1120	~8.93	s	m–s	Each additional alkyl group increases the wavenumber by ~15 cm⁻¹
(Unsat group) –CH₂(CH₃)₂COH	~1120	~8.93	s	m–s	
[(Unsat group) –CH₂]₂CH₃COH	~1060	~9.43	s	m–s	
[(Unsat group) –CH₂]₃COH	~1010	~9.90	s	m–s	
α-Unsaturated and cyclic tertiary alcohols	1125–1085	8.90–9.22	s	m–s	see refs 9 and 10
	1060–1020	9.43–9.80	s	m–s	
Alicyclic secondary alcohols (three- or four-membered rings)	1060–1020	9.43–9.80	s	m–s	
Alicyclic secondary alcohols (five- or six-membered rings)	1085–1030	9.22–9.71	s	m–s	
Phenols	1260–1180	7.94–8.48	s	m–w	O–H def vib and C–O str combination

The O–H out-of-plane vibration gives a broad band in the region 710–570 cm⁻¹ (14.08–17.54 µm). The position of this band is dependent on the strength of the hydrogen bond – the stronger the hydrogen bond, the higher the wavenumber. Bonded primary and secondary alcohols have two bands: one near 1420 cm⁻¹ (7.04 µm) and the other near 1330 cm⁻¹ (7.52 µm). As mentioned, in dilute solution, these bands shift to lower frequencies ~1385 cm⁻¹ (7.22 µm) and ~1250 cm⁻¹ (8.00 µm). Bonded tertiary alcohols absorb near 1410 cm⁻¹ (7.04 µm) and in dilute solution near 1320 cm⁻¹ (7.58 µm). In Raman spectra, the CCO and CCC skeletal vibration bands are in general of medium-to-strong intensity.

In the far infrared spectroscopic region,[11–13] aliphatic alcohols in a cyclohexane solvent exhibit a characteristic strong band at 220–200 cm⁻¹ (45.45–50.00 µm) due to the torsional motion of the O–H group.[13] This band is insensitive to steric effects but becomes broad with increase in concentration,

Table 6.4 Phenols: O–H stretching vibrations

Functional Groups	Region cm^{-1}	Region μm	Intensity IR	Intensity Raman	Comments
Unassociated	3620–3590	2.76–2.79	m	w	(In dilute solution) sh
Associated	3250–3000	3.08–3.33	v	w	(In solution) br, concentration and solvent dependent
Ortho-substituted, OH ⬡–X where X=C=O	3200–2500	3.13–4.00	m	w	Intramolecular hydrogen bonded
X=F	3635–3630	~2.75	m	w	Dilute solution
X=Cl	3600–3550	2.78–2.82	m	w	Dilute solution
X=Br	3550–3540	~2.82	m	w	Dilute solution
X=I	3540–3525	2.82–2.84	m	w	Dilute solution
X=NO$_2$	3275–3235	3.05–3.09	m	w	
X=OR	3595–3470	2.78–2.88	m	w	
X=alkene	3600–3585	2.78–2.79	m	w	
X=NH$_2$	~3660	~2.73	m	w	
X=SMe	~3445	~2.90	m	w	
Ortho-t-butyl phenols (dilute solutions)	3650–3640	2.74–2.75	m	w	
	3615–3605	~2.77	m	w	

Table 6.5 Phenols: interaction of O–H deformation and C–O stretching vibrations

Functional Groups	Region cm^{-1}	Region μm	Intensity IR	Intensity Raman	Comments
Associated	1410–1310	7.09–7.63	m	m–w	COH bending vib
	1260–1180	7.94–8.48	s	w	CO str
Unassociated (dilute solution)	1360–1300	7.35–7.69	m	m–w	COH bending
	1225–1150	8.17–8.70	s	w	CO str
o-Alkyl phenols (solution)	~1320	~7.58	m	m–w	COH bending vib
	1255–1240	7.97–8.07	s	w	CO str
	1175–1160	8.51–8.62	s	w	
	~750	~13.33	m	w	
m-Alkyl phenols (solution)	1285–1270	7.78–7.87	s	m–w	
	1190–1180	8.40–8.48	s	w	
	1160–1150	8.62–8.70	s	w	
	820–770	12.20–12.99	m–s	–	
p-Alkyl phenols (solution)	1260–1245	7.94–8.03	s	m–w	
	1175–1165	8.51–8.58	s	w	
	835–815	11.98–12.97	m	w	

Table 6.6 Phenols: other bands

Functional Groups	Region		Intensity		Comments
	cm⁻¹	μm	IR	Raman	
Phenols	~1660	~6.02	s	m–s	usually a doublet, aromatic ring C=C str
	~1110	~9.01	v	w	aromatic C–H def vib
	720–600	13.89–16.67	s	w	br, O–H out-of-plane bending vib (hydrogen bonding), see ref. 26
	450–375	22.22–26.67	w	m–w	in-plane bending vib of aromatic C–OH bond

eventually disappearing. In benzene solution, a band is observed at about $300\,cm^{-1}$ ($33.33\,\mu m$) which is believed to be due to an alcohol–benzene complex which is formed.

Phenols

In the absence of intramolecular hydrogen bonding and in the case of a dilute solution in a non-polar solvent[14–16] (i.e. in the additional absence of intermolecular hydrogen bonding), phenols have an absorption band at 3620–$3590\,cm^{-1}$ (2.76–$2.79\,\mu m$) due to the O–H stretching vibration.[17,18]

If strong intramolecular hydrogen bonding does occur, for example, to a carbonyl group, then a relatively sharp band is found at about $1200\,cm^{-1}$ ($3.13\,\mu m$). If, on the other hand, hydrogen bonding is inhibited by the presence of large groups in the *ortho* positions,[19–21] the absorption occurs in the region 3650–$3600\,cm^{-1}$ (2.74–$2.78\,\mu m$). Phenols without bulky *ortho* groups, whether in concentrated solutions or as solids or in the pure liquid phase, have a broad absorption at 3400–$3230\,cm^{-1}$ (2.94–$3.10\,\mu m$).

Medium-to-strong bands are observed at 1255–$1240\,cm^{-1}$ (7.97–$8.07\,\mu m$), 1175–$1150\,cm^{-1}$ (8.51–$8.70\,\mu m$), and 835–$745\,cm^{-1}$ (11.98–$13.42\,\mu m$) for alkyl phenols.[22,23] In addition, *o*-phenols usually have a band near $1320\,cm^{-1}$ ($7.58\,\mu m$) and *m*-alkyl phenols one at $1185\,cm^{-1}$ ($8.44\,\mu m$).[24] The three main bands may be attributed to the C–O stretching and the O–H in-plane and out-of-plane deformation vibrations.

The C–O stretching vibration for *p*-monosubstituted phenols,[24] i.e. the strongest band in the region 1300–$1200\,cm^{-1}$ (7.69–$8.33\,\mu m$), increases in frequency with the electron-withdrawing ability of the substituent.

In the solid phase, or in cases where strong hydrogen bonding may occur, a broad absorption at 720–$600\,cm^{-1}$ (13.89–$16.67\,\mu m$) is observed due to the out-of-plane deformation of the O–H group. In dilute solution, i.e. in the unassociated state, this absorption occurs near $300\,cm^{-1}$ ($33.33\,\mu m$). In

the presence of hydrogen bonding, a characteristic weak absorption, due to the in-plane bending of the ring C–OH bond, is observed at 450–$375\,cm^{-1}$ (22.22–$26.67\,\mu m$).[25,26] In the absence of hydrogen bonding, this band may be shifted by about 20–$40\,cm^{-1}$ (1.00–$2.20\,\mu m$). For monosubstituted phenols,[25] the position of this weak band is influenced by the nature of the substituent. In the case of electron-accepting or almost neutral groups, such as alkyl groups,[25] the band is found above $400\,cm^{-1}$ (below $25.00\,\mu m$) whereas with electron-donating substituents the band occurs below $400\,cm^{-1}$ for solid samples.

References

1. I. Motoyama and C. H. Jarboe, *J. Phys. Chem.*, 1966, **70**, 3226.
2. J. H. van der Maas and E. T. G. Lutz, *Spectrochim Acta*, 1974, **30A**, 2005.
3. J. S. Cook. and I. H. Reece, *Austral. J. Chem.*, 1961, **14**, 211.
4. U. Liddel, *Ann. N Y. Acad. Sci.*, 1957, **69**, 70.
5. S. Siggia, *et al.*, in *Chemistry of the Hydroxyl Group*, Part 1, S. Patai, (ed.), Interscience, London, 1971, p. 311.
6. A. S. Wexler, *Appl. Spectrosc. Rev.*, 1968, **1**, 29.
7. A. O. Diallo, *Spectrochim. Acta*, 1972, **28A**, 1765.
8. L. T. Pitzner and R. H. Atalla, *Spectrochim. Acta*, 1975, **31A**, 911.
9. A. A. Petrov and G. I. Semenov, *J. Gen. Chem.* USSR AQOTU, 1957, **27**, 2974.
10. J. C. Richer and P. Belanger, *Can. J. Chem.*, 1966, **44**, 2057.
11. J. E. Chamberlain *et al.*, *Nature*, 1975, **255**, 319.
12. W. F. Passchier *et al.*, *Chem. Phys. Lett.*, 1970, **4**, 485.
13. S. M. Craven, *US Natl Tech, Inform. Service AD Rep.*, 1971, No. 733, p. 666.
14. M. St C. Fiett, *Spectrochim. Acta*, 1957, **10**, 21.
15. N. A. Putnam, *J. Chem. Soc.*, 1960, 5100.
16. K. U. Ingold, *Can. J. Chem.*, 1960, **38**, 1092.
17. T. Cairns and G. Eglinton. *J. Chem. Soc.*, 1965, 5906.
18. Z. Yoshida and E. Osawa, *J. Am. Chem. Soc.*, 1966, **88**, 4019.
19. L. J. Bellamy *et al.*, *J. Chem. Soc.*, 1961, 4762.
20. A. W. Baker *et al.*, *Spectrochim Acta*, 1964, **20**, 1467 and 1477.

21. K. U. Ingold and D. R. Taylor, *Can. J. Chem.*, 1961, **39**, 471 and 481.
22. D. D. Shrewsbury, *Spectrochim. Acta*, 1960, **16**, 1294.
23. J. H. S. Green *et al.*, *Spectrochim. Acta*, 1972, **28A**, 33.
24. J. H. S. Green *et al.*, *Spectrochim. Acta*, 1971, **27A**, 2199.
25. R. J. Jakobsen, *Wright Air Development Division Tech. Rep.*, 1960, No. 60–204.
26. V. Bekarek and K. Pragerova, *Coll. Czech. Chem. Commun.*, 1975, **40**, 1005.
27. N. S. Sundra, *Spectrochim. Acta*, 1985, **41A**, 1449.
28. E. T. G. Lutz and J. H. van der Mass, *Spectrochim Acta*, 1986, **42A**, 755.
29. R. A. Nyquist, *The Interpretation of Vapour-Phase Spectra*, Sadtler, 1985.
30. S. Chakravarty *et al.*, *Spectrochim. Acta*, 1993, **49A**, 543.
31. R. A. Shaw. *et al.*, *J. Am. Chem. Soc.*, 1990, **112**, 5401.
32. R. Laenen *et al.*, *J. Phys. Chem. A*, 1999, **103(50)**, 10708.

7   Ethers: G_1-O-G_2 Group

The mass and bond strength for the C–O group is similar to that of C–C and therefore, as expected, there is a close similarity in their band positions. However, the change in dipole moment of the C–O group is much larger and therefore the intensity band due to the C–O stretching vibration is considerably greater.

Ethers have characteristic, strong absorption bands in the range $1270–1060\,cm^{-1}$ (7.94–9.43 μm) which may be associated with the C–O–C asymmetric stretching vibration.[1] Carboxylic esters and lactones also absorb strongly in this region. For saturated aliphatic ethers, this band may be found at $1150–1060\,cm^{-1}$ (8.70–9.43 μm), usually within the range $1140–1110\,cm^{-1}$ (8.77–9.01 μm). In the case of branched-chain aliphatic ethers, two peaks may be observed. Benzyl ethers absorb at about $1090\,cm^{-1}$ (9.17 μm) and cyclic ethers absorb at $1270–1030\,cm^{-1}$ (7.87–9.71 μm). Aryl ethers absorb strongly in the region $1270–1230\,cm^{-1}$ (7.87–8.13 μm). Alkyl aryl ethers have two strong absorptions, the most intense of which is at $1310–1210\,cm^{-1}$ (7.63–8.26 μm), the other being at $1120–1020\,cm^{-1}$ (8.93–9.80 μm), these bands being due to the asymmetric and symmetric vibrations of the group C–O–C respectively.

The asymmetric C–O–C stretching vibration frequency depends on the group directly bonded to the oxygen atoms and decreases in the following order:

$$C_6H_5- > CH_2{=}CH- > R_3C- > RCH_3CH- > RCH_2- > C_6H_5CH_2-$$

For aliphatic ethers, a weak band is observed, usually in the region $930–900\,cm^{-1}$ (10.75–11.11 μm) but sometimes found as high as $1140\,cm^{-1}$ (8.77 μm) when it is usually strong. This band is due to the symmetric stretching vibration of the C–O–C group and may be absent for symmetric ethers due to symmetry factors (see below).

Vinyl ethers usually absorb very close to $1205\,cm^{-1}$ (8.30 μm) in the range $1225–1200\,cm^{-1}$ (8.16–8.33 μm) due to the asymmetric $=$C–O stretching vibration. The C$=$C stretching vibration results in a band which appears as a strong doublet, the stronger portion being at $1620–1610\,cm^{-1}$ (6.17–6.21 μm), the other peak being near $1640\,cm^{-1}$ (6.10 μm). The doubling is due to the presence of rotational isomerism which is the result of rotation being restricted about the $=$C–O bond. The stronger band is due to the more stable, planar, *trans-* form and the weaker band is due to the *gauche-* form, the *cis-* form being sterically inhibited.

As a rough approximation, the asymmetric stretching vibration of the C–O bond occurs at about $1130\,cm^{-1}$ (8.85 μm) when the carbon is fully saturated and at about $1200\,cm^{-1}$ (8.33 μm) when it is unsaturated. This may be either aromatic or olefinic unsaturation.

Often the symmetric and asymmetric C–O–C absorption bands are well separated, by about $200\,cm^{-1}$ (1.7 μm). On simple theoretical grounds these bands would be expected to occur closer together. The large difference is due to coupling. When a central atom is attached to two groups of similar mass by bonds of similar order, coupling of vibrations may occur, e.g. coupling occurs for CH_3OCH_3 whereas there is no coupling for CH_3OH. In general, it has been found that the separation of coupled frequencies is a maximum if the bond angle between the central atom and the two attached groups is 180° and a minimum if it is 90°. For symmetrical ethers, e.g. diethyl ether, due to the presence of coupling, the C–O–C asymmetric stretching vibration band, which is of strong intensity, occurs at about $1110\,cm^{-1}$ (9.01 μm) and the symmetric stretching vibration band is weak or absent.

In general, the asymmetric stretching frequency is lowered for molecules with electron-withdrawing groups since the electron density of the C–O bond is reduced. The opposite is true of electron-donating groups. Any group which increases the double-bond character of the C–O group tends to increase the stretching vibration frequency of this bond and this may be the result of either electronic induction or resonance.

As a result of resonance, aromatic ethers have a contribution from $=O^+-$, e.g.

which tends to increase the force constant of the C–O (aromatic carbon–oxygen bond) and hence increases the C–O stretching vibration frequency as compared with aliphatic compounds. Electron-donating groups

Table 7.1 Ether C–O stretching vibrations

Functional Groups	Region cm⁻¹	Region μm	Intensity IR	Intensity Raman	Comments
Saturated aliphatic ethers, C–O–C	1150–1060	8.70–9.43	vs	w	asym C–O–C str
	1140–820	8.77–12.20	v	s, p	sym C–O–C str, usually weak. Raman band usually 890–820 cm⁻¹ and also strong band at 500–400 cm⁻¹.
Alkyl-aryl ethers, =C–O–C	1310–1210	7.63–8.26	vs	w	asym =C–O–C str
	1120–1020	8.93–9.80	s	s, p	sym C–O–C str
Vinyl ethers, CH₂=CH–O–	1225–1200	8.16–8.33	s	w	asym C–O–C str, usually ~1205 cm⁻¹
	850–810	11.76–12.35	w	s	sym COC str
–CH₂–O–CH₂–	1140–1085	8.77–9.22	s	s	usually ~1120 cm⁻¹
CH₃·CO·CH₂–X, X=halogen	1180–1040	8.47–9.62	vs	s	asym COC str
	970–890	10.31–11.24	v	m–s	sym COC str
Ar–O–CH₂–O–Ar	1265–1225	7.91–8.17	s	w	=C–O str, may be as high as ~1205 cm⁻¹
	1050–1025	9.52–9.76	s	s	
Cyclic ethers	1270–1030	7.87–9.71	s	s	sym C–O–C str, frequency decreases with increase in ring size.
Trimethylene oxides (four-membered ring)	1035–1020	9.66–9.80	s–m	vs	sym C–O–C str. Raman vs band ~1030 cm⁻¹ due to ring vib, also m at ~1140 and 930 cm⁻¹.
	990–930	10.10–10.75	s	m	sym COC str
Cyclic ethers (five-membered ring)	1080–1060	9.26–9.43	s	m	asym COC str
	920–905	10.87–11.05	m	s	sym COC str
Cyclic ethers (six-membered ring)	1110–1090	9.01–9.17	s	m	asym COC str
	820–805	12.20–12.42	m	s	sym COC str
Acyclic diaryl ethers, =C–O–C=	1250–1170	8.00–8.55	s	w	sym COC str
Ring =C–O–C=	1200–1120	8.33–8.93	s	m	ring vib
	1100–1050	9.09–8.70	s	s	ring vib
Oxirane compounds: Epoxides,	1280–1230	7.81–8.13	m–s	vs	C–O str
	950–815	10.53–12.27	v	m–w	ring vib
	880–750	11.36–13.33	m–s	m–s	ring vib
Monosubstituted epoxides, –CH–CH₂	880–775	11.36–12.90	m–s	m–s	ring vib
Trans-epoxides	950–860	10.53–11.63	v	m–s	ring vib
Cis-epoxides	865–785	11.56–12.74	m–s	m–s	ring vib
Trisubstituted epoxides	770–750	12.99–13.33	m–s	–	ring vib
Ketals and acetals,	1190–1140	8.40–8.77	s	v	C–O–C–O–C vib (see ref. 5),
	1145–1125	8.73–8.89	s	v	C–O–C–O–C vib
	1100–1060	9.09–9.43	s	m, p	C–O–C–O–C vib, strongest band
	1060–1035	9.43–9.65	s	v	C–O–C–O–C vib, sometimes observed
Acetals	1115–1105	8.96–9.02	s	m, p	C–H def vib (perturbed by C–O group), as for ketals
	870–850	11.49–11.76	s	m–s	sym C–O–C–O str

Table 7.1 (*continued*)

Functional Groups	Region		Intensity		Comments
	cm⁻¹	μm	IR	Raman	
Phthalans	915–895	10.93–11.17	s	m	
Aromatic methylene dioxy compounds, 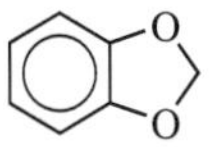	1265–1235	7.90–8.10	s	w–m	C–O str
Pyranose compounds	1200–1030	8.33–9.70	s	w–m	C–O str

Table 7.2 Ethers: other bands

Functional Groups	Region		Intensity		Comments
	cm⁻¹	μm	IR	Raman	
Aliphatic ethers, –OCH₃	2995–2955	3.34–3.38	m	m–s	asym –CH₃ str
	2900–2840	3.45–3.52	m	m–s	sym –CH₃ str
	2835–2815	3.53–3.55	m–w	m–s	
	1470–1435	6.80–6.97	m	m	asym and sym –CH₃ def vib
	1200–1185	8.33–8.45	m–w	w	Rocking vib.
–OCH₂–	2955–2920	3.38–3.43	m	m–s	asym CH₂ str
	2880–2835	3.47–3.53	m	m–s	sym CH₂ str (almost equal in intensity to asym str)
	1475–1445	6.78–6.92	m	m–w	CH₂ def vib
	1400–1360	7.14–7.35	m	m–w	Wagging vib.
–O–CH₂–O	~2780	~3.60	m	m	CH str (range 2820–2710 cm⁻¹)
Aliphatic ethers	~430	~23.26	w	s, p	C–O–C def vib
R–O–Ar	1310–1210	7.63–8.26	m	w	
	1050–1010	9.52–9.90	m	m	
Methyl aromatic ethers, =C–O–CH₃	2830–2815	3.53–3.55	m	m	CH₃ str
	580–505	17.24–19.80	m–s	m–w	C–O–C def vib
Ar–O–CH₂–O–Ar	1375–1350	7.27–7.41	v	m	C–H def vib
	940–920	10.64–10.87	v		
Vinyl ethers	3150–3000	3.18–3.33	w	m	C–H str, a number of bands
	1660–1635	6.02–6.12	m	s	C=C str, *gauche-* form
	1620–1610	6.17–6.21	s	s	C=C str, *trans-* form
	~1320	~7.58	v	m–w	=CH rocking vib
	970–960	10.31–10.42	m	w	=CH wagging vib, *trans-* form
	820–810	12.20–12.35	s	w	=CH₂ wagging vib
Epoxides	3075–3030	3.25–3.30	w	s	C–H str, one or two bands, see ref. 6
	~370	~27.03	w–m	m–s	Ring def vib
Monosubstituted epoxides 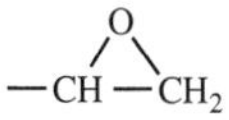	3075–3040	3.25–3.29	m	s	asym CH str
	3035–2975	3.29–3.36	m	m–s	CH str
	3010–2970	3.32–3.37	m	m–s	sym CH str
	1500–1430	6.67–6.99	m–w	m–w	CH₂ def vib
	1445–1375	6.92–7.27	m	m–s	CH def vib
	1265–1245	7.91–8.03	m	m–s	Ring str
	1210–1140	8.26–8.77	m	m–w	CH₂ twisting def vib

(*continued overleaf*)

Table 7.2 (*continued*)

Functional Groups	Region		Intensity		Comments
	cm^{-1}	μm	IR	Raman	
	1140–1120	8.77–88.93	m	m	CH$_2$ wagging vib
	1110–1040	9.01–9.62	m	m–w	CH wagging vib
	965–875	10.36–11.43	s	m–s	asym ring def vib
	880–810	11.36–12.34	s	s	sym ring def vib
	800–750	12.50–13.33	w	w	CH$_2$ rocking def vib
Aromatic ethers	1310–1230	7.63–8.13	s	w	X-sensitive band
Phenoxy, Ph–O–	765–750	13.10–13.33	s	w	C–H out-of-plane def vib, ring def vib
	695–690	14.39–14.49	s	w	C–H out-of-plane def vib, ring def vib
Benzyloxy, Ph–CH$_2$–O–	745–730	13.42–13.70	s	w	C–H out-of-plane def vib, ring def vib
	700–695	14.29–14.39	s	w	
Acetals	2830–2820	3.53–3.55	w	m–w	
	660–600	15.15–16.67	s	m	COCO def vib
	540–450	18.52–22.22	s	m–s	COCO def vib
	400–320	25.00–31.25	s	m–s	COCO def vib
Aromatic methylene dioxy compounds,	2950–2750	3.40–3.64	m	m–s, p	C–H str, two bands
	1485–1350	6.73–7.60	v	m	Several bands
	940–915	10.58–10.93	s		

at *ortho* or *para* positions on the ring tend to reduce this frequency relative to a similar *meta*-substituted compound. The reverse is true of electron-attracting groups.

The CH$_3$–O group for aliphatic ethers may be distinguished from the group CH$_3$–C since the former absorbs at 1470–1440 cm^{-1} (6.80–6.94 μm) due to both the CH$_3$ symmetric and asymmetric deformation vibrations, whereas the latter group absorbs at 1385–1370 cm^{-1} (7.22–7.30 μm) due to the symmetric deformation vibration of the CH$_3$ group. The OCH$_3$ group can usually be distinguished by its CH$_3$ symmetric stretching vibration band which occurs in the region 2830–2815 cm^{-1} (3.53–3.55 μm). Aromatic compounds with methoxy groups have an absorption in the region 580–505 cm^{-1} (17.24–19.80 μm), of medium-to-strong intensity, due to the in-plane deformation vibration of the C–O–C groups.

Cyclic ethers (five membered ring) often have several bands of medium intensity in the region 1080–800 cm^{-1} (9.26–12.50 μm).

Epoxides[2–4,7] absorb near 1250 cm^{-1} (8.00 μm) due to the C–O stretching vibration and near 370 cm^{-1} (27.03 μm) due to their ring deformations. The CH$_2$ and CH of epoxy rings have their stretching bands in the regions 3005–2990 cm^{-1} (3.33–3.34 μm) and 3050–3025 cm^{-1} (3.28–3.31 μm).

In the case of acetals and ketals, the C–O stretching vibration band is split into three: 1190–1140 cm^{-1} (8.40–8.77 μm), 1145–1125 cm^{-1} (8.73–8.89 μm), and 1100–1060 cm^{-1} (9.09–9.43 μm). The vibration modes may be considered as similar to the asymmetric C–O stretching vibration of ethers. A fourth band at 1060–1035 cm^{-1} (9.43–9.65 μm) which is due to the symmetric vibration may sometimes be observed. In addition, acetals have a characteristic, strong band in the region 1115–1105 cm^{-1} (8.96–9.02 μm) due to a C–H deformation vibration being perturbed by the neighbouring C–O groups. This band may be used to distinguish between acetals and ketals. Acetals have three characteristic deformation bands in their Raman spectra at 600–550 cm^{-1} (16.67–18.18 μm), 540–450 cm^{-1} (18.52–22.22 μm) and 400–320 cm^{-1} (25.00–31.25 μm).

References

1. A. R. Katritzky and N. A. Coats, *J. Chem. Soc.*, **1959**, 2062.
2. A. J. Durbetaki, *J. Org. Chem.*, 1961, **26**, 1017.
3. H. von Hoppff and H. Keller, *Helv. Chim. Acta*, 1959, **42**, 2457.
4. J. Bomstein, *Anal. Chem.*, 1958, **30**, 544.
5. B. Wladislaw *et al.*, *J. Chem. Soc. B*, **1966**, 586.
6. C. J. Wurrey and A. B. Nease, *Vib. Spectra Struct.*, 1978, **7**, 1.
7. R. A. Nyquist, *Appl. Spectrosc.*, 1986, **40**, 275

8 Peroxides and Hydroperoxides: –O–O–Group

Peroxides[1–3,5] and hydroperoxides[2–4] have two main structural units, the C–O and O–O groups. The band due to the C–O stretching vibration occurs in the region $1300-1000\,\mathrm{cm^{-1}}$ ($7.69-10.00\,\mu\mathrm{m}$). Electron-withdrawing substituents attached to the C–O group tend to reduce the frequency of this absorption band.

All peroxides have a band at $900-800\,\mathrm{cm^{-1}}$ ($11.11-12.50\,\mu\mathrm{m}$) due to their O–O stretching vibration. The Raman band is of strong intensity and easily identified, whereas it is usually weak and often difficult to observe in infrared. For symmetrical peroxides this O–O stretching vibration is infrared inactive, although as a result of environmental interaction it may still be observed. Tertiary peroxides and tertiary hydroperoxides have a strong band in the region $920-800\,\mathrm{cm^{-1}}$ ($10.87-12.50\,\mu\mathrm{m}$) which is believed to be due to the skeletal vibration of the group

$$
\begin{array}{c}
\quad\ \ \mathrm{C} \\
\quad\ \ | \\
\mathrm{C-C-O} \\
\quad\ \ | \\
\quad\ \ \mathrm{C}
\end{array}
$$

For organic peroxides, the range of the O–O stretching vibration, determined by Raman studies[6], was originally quoted as $950-700\,\mathrm{cm^{-1}}$ ($10.52-14.29\,\mu\mathrm{m}$). However, in recent studies[7], the range has been given as $875-845\,\mathrm{cm^{-1}}$ ($11.43-11.83\,\mu\mathrm{m}$), which is clearly much smaller.

Symmetrical aliphatic diacyl peroxides, –CO–O–O–CO–, have two strong infrared bands in the region $1820-1780\,\mathrm{cm^{-1}}$ ($5.49-5.62\,\mu\mathrm{m}$) due to the stretching vibrations of the C=O groups. Similarly, symmetrical aromatic diacyl peroxides have two strong bands in the region $1820-1760\,\mathrm{cm^{-1}}$ ($5.50-5.88\,\mu\mathrm{m}$), the position of these bands being dependent on the nature and position of the aromatic substituents.

Metal peroxide compounds absorb in the region $900-800\,\mathrm{cm^{-1}}$ ($11.11-12.50\,\mu\mathrm{m}$) due to the O–O stretching vibration.

Ozonides have a medium-intensity absorption at $1065-1040\,\mathrm{cm^{-1}}$ ($9.39-9.62\,\mu\mathrm{m}$) due to the stretching vibration of the C–O bond. This band is

Table 8.1 Peroxides and hydroperoxides

Functional Groups	Region cm⁻¹	Region μm	IR	Raman	Comments
Peroxides	900–800	11.11–12.50	w	s, p	O–O str
Alkyl peroxides	1150–1030	8.70–9.71	m–s	m–w	C–O str
Aryl peroxides	~1000	~10.00	m	m–w	C–O str
Peracids, peroxides of the type G·CO·OO·H	~3450	~2.90	m	m–w	O–H str
	1785–1755	5.60–5.70	vs	w	C=O str
	~1175	~8.51	m–s	w–m	C–O str
Aliphatic diacyl peroxides, –CO–OO–CO–	1820–1810	5.50–5.52	vs	w	C=O str
	1805–1780	5.54–5.62	vs	w	C=O str
	1300–1050	7.69–9.52	m–s	w–m	C–O str
Aryl and unsaturated diacyl peroxides	1805–1780	5.54–5.62	vs	w	C=O str
	1785–1735	5.60–5.76	vs	w	C=O str
	1300–1050	7.69–9.52	m–s	w–m	C–O str
Ozonides,	1065–1040	9.39–9.62	m	w–m	C–O str
	900–700	11.11–14.29	w	s	O–O str
–OCH₃	2995–2980	3.34–3.56	m	s–m	asym CH str
	2970–2920	3.37–3.42	m	s–m	sym CH str
	1470–1435	6.80–6.97	m–s	m	CH def vib
	1200–1185	8.33–8.44	m–w	w	Rocking vib
–OCH₂	2955–2920	3.38–3.42	m	s–m	asym CH str
	2880–2835	3.47–3.53	m	s	sym CH str
	1473–1445	6.78–6.92	m–s	m	CH def vib
	1400–1360	7.14–7.35	m–s	m	Wagging vib

often not of use for assignment purposes since alcohols and ethers also absorb in this region.

References

1. D. Swern (ed.), *Organic Peroxides*, Vol. 2, Interscience, New York, 1971, pp. 683–697.

2. H. A. Szymanski, *Prog. Infrared Spectrosc.*, 1967, **3**, 139.
3. W. P. Keaveney *et al.*, *J. Org. Chem.*, 1967, **32**, 1537.
4. M. A. Kovner *et al.*, *Opt. Spectrosc.*, 1960, **8**, 64.
5. M. E. Bell and J. Laane, *Spectrochim. Acta*, 1972, **28A**, 2239.
6. P. A. Budinger *et al.*, *Anal. Chem.*, 1981, **53**, 884.
7. V. Vacque *et al.*, *Spectrochim. Acta*, 1997, **53A(1)**, 55.

9 Amines, Imines, and Their Hydrohalides

Amine Functional Groups

Amine N–H *Stretching Vibrations*

As solids or liquids, in which hydrogen bonding may occur, primary aliphatic amines[1–5,18–22] absorb in the region 3450–3160 cm^{-1} (2.90–3.16 μm). This is a broad band of medium intensity which may show structure depending on the hydrogen-bond polymers formed. In dilute solution in non-polar solvents, two bands are observed for primary amines due to the N–H asymmetric and symmetric vibrations. In the aliphatic case,[1,2] they are in the range 3550–3250 cm^{-1} (2.82–3.08 μm) whereas in the aromatic case[6–9] they are of medium intensity,[15] one at 3520–3420 cm^{-1} (2.84–2.92 μm) and the other at 3420–3340 cm^{-1} (2.92–2.99 μm). In the condensed phase, for example, as liquids, α-saturated primary amines may exhibit a broad, symmetrical doublet of weak-to-medium intensity[18–20] at 3200–3160 cm^{-1} (3.13–3.16 μm). Various empirical relationships[3,10] between the bands have been proposed one of which is $\nu_{sym} = 345.5 + 0.876\nu_{asym}$ where the two N–H bonds of the primary amine are equivalent. For primary amines in the solid phase, the two bands are usually observed at approximately 100 cm^{-1} lower (0.09 μm higher) than for dilute non-polar solvent solutions.

Secondary amines[5,11,12,20] have only one N–H stretching band which is usually weak and occurs in the range 3500–3300 cm^{-1} (2.86–3.03 μm). In the solid and liquid phases, a band of medium intensity may be observed at 3450–3300 cm^{-1} (2.90–3.03 μm) for secondary aromatic amines.[12] As a result of hydrogen bonding, bands due to the N–H stretching vibrations may, in some solvents, be found as low as 3100 cm^{-1} (3.23 μm). In general, bands due to the N–H stretching vibration are sharper and weaker than, and do not occur in as wide a range as, those due to the O–H stretching vibration.

It is sometimes useful to convert tertiary amines into their hydrochlorides and then examine the resulting spectra for the presence of a band due to the N–H stretching vibration,[13,14] a technique which may also be found useful for distinguishing between imines and amines.

N–H stretching vibrations result in bands of weak-to-medium intensity in the Raman spectra of amines.[22] (For peptides[19] see Chapter 23.)

Amine N–H *Deformation Vibrations*

Primary amines[20–22] have a medium-to-strong absorption band in the 1650–1580 cm^{-1} (6.06–6.33 μm) region and secondary amines have a weaker band at 1580–1490 cm^{-1} (6.33–6.71 μm). Primary aromatic amines[20] normally absorb at 1615–1580 cm^{-1} (6.19–6.33 μm). Care must be taken since aromatic ring absorptions also occur in this general region.

Amines often exhibit a number of peaks when examined as pressed discs, due to a reaction with the dispersing agent and the formation of amine hydrohalides. Hydrogen bonding has the effect of moving the N–H deformation band to higher frequencies. This shift is dependent on the strength of the hydrogen bond. Primary amines have a broad absorption of weak-to-medium intensity at 895–650 cm^{-1} (11.17–15.40 μm) which alters in shape and position depending on the amount of hydrogen bonding present.

Secondary aliphatic amines have an absorption in the range 750–700 cm^{-1} (13.33–14.29 μm).

Amine C–N *Stretching Vibrations*

The C–N stretching absorption of primary aliphatic amines is weak and occurs in the region 1090–1020 cm^{-1} (9.17–9.77 μm). Secondary aliphatic amines have two bands of medium intensity at 1190–1170 cm^{-1} (8.40–8.55 μm) and 1145–1130 cm^{-1} (8.73–8.85 μm).

For aromatic and unsaturated amines =C–N, two bands are observed at 1360–1250 cm^{-1} (7.36–8.00 μm) and 1280–1180 cm^{-1} (7.81–8.48 μm) due to conjugation of the electron pair of the nitrogen with the ring imparting double-bond character to the C–N bond, primary and secondary aromatic amines absorbing strongly in the first region. The C–N band for tertiary aromatic amines is found at 1380–1265 cm^{-1} (7.25–7.91 μm).

Imines with aliphatic groups attached to the nitrogen atom have a band near $1670\,cm^{-1}$ ($5.99\,\mu m$), with aromatic groups attached, this band is near $1640\,cm^{-1}$ ($6.10\,\mu m$) and, with extended conjugated groups, it is near $1620\,cm^{-1}$ ($6.17\,\mu m$).

Amine $>$N–CH$_3$ and $>$N–CH$_2$– Absorptions

A band of medium-to-strong intensity due to the stretching vibration of the C–H bond of the N–C–H group occurs near $2820\,cm^{-1}$ ($3.55\,\mu m$). This band is lower in frequency and more intense than ordinary alkyl bands and is therefore easily identified. Aliphatic amines with –N(CH$_3$)$_2$ have two bands, one near $2820\,cm^{-1}$ ($3.55\,\mu m$) and the other near $2770\,cm^{-1}$ ($3.61\,\mu m$).

Other Amine Bands

Primary aromatic amines (e.g. anilines) have a weak-to-medium intensity band at $445–345\,cm^{-1}$ ($22.47–28.99\,\mu m$) which is probably due to the in-plane deformation vibration of the aromatic ring-amine bond. For monosubstituted aminobenzenes with electron-donating substituents, this band is observed below $400\,cm^{-1}$ (above $25.00\,\mu m$), whereas with electron-accepting or alkyl substituents in the ring, this band is above $400\,cm^{-1}$.

Primary alkyl amines have a strong absorption in the vicinity of $200\,cm^{-1}$ ($50.00\,\mu m$). It has been suggested that this band is due to torsional oscillations about the C–N bond and that the band which occurs at $495–445\,cm^{-1}$ ($20.20–22.47\,\mu m$) is an overtone of this band.

Amine Hydrohalides,[13,14] –NH$_3^+$, $>$NH$_2^+$, $-$NH$^+$ and Imine Hydrohalides, $>$C=NH$^+$–

Amine Hydrohalide N–H$^+$ Stretching Vibrations

In the solid phase, amine hydrohalides containing –NH$_3^+$ have an absorption of medium intensity at $3350–3100\,cm^{-1}$ ($2.99–3.23\,\mu m$) due to stretching

Table 9.1 Amine N–H stretching vibrations

Functional Groups	Region cm^{-1}	Region μm	Intensity IR	Intensity Raman	Comments
Primary amines, –NH$_2$ (dilute solution spectra)	3550–3330	2.82–3.00	w–m	w, dp	asym NH$_2$ str
	3450–3250	2.90–3.08	w–m	w–m, dp	sym NH$_2$ str
Primary amines (condensed phase spectra)	3450–3160	2.90–3.16	w–m	w–m	br.
Primary aromatic amines (dil. soln.)	3520–3420	2.84–2.92	m	m–w	asym NH$_2$ str
	3420–3340	2.92–2.99	m	m–w	sym NH$_2$ str
Secondary aliphatic amines, $>$NH	3500–3300	2.86–3.03	w	w	
Secondary aromatic amines	3450–3400	2.90–2.94	m	w	Greater intensity than aliphatic compounds
N–D (free)	2600–2400	3.85–4.15	w	w–m	
–NH·CH$_3$ (condensed phase)	3315–3215	3.02–3.11	w–m	w–m	br, NH str
–NH·CH$_3$ (dilute solutions)	3440–3350	2.91–2.99	w–m	w–m	Much narrower band than for condensed phase
(α-unsat or Ar)–NH·CH$_3$ (condensed phase)	3440–3360	2.91–2.98	m–s	w–m	NH str
(α-unsat or Ar)–NH·CH$_3$ (dilute solutions)	3480–3420	2.87–2.92	m–s	w–m	NH str
Diamines (condensed phase) (see ref. 16)	3360–3340	2.98–2.99	w–m	w–m	asym N–H str
	3280–3270	3.05–3.06	w–m	w–m	sym N–H str
Imines, C=NH (see ref. 17)	3400–3300	2.94–3.03	m	w	
o-Alkyl hydroxylamines RONH$_2$	3255–3235	3.07–3.10	m	w–m	NH$_2$ str

Table 9.2 Amine N–H deformation vibrations

Functional Groups	Region		Intensity		Comments
	cm^{-1}	μm	IR	Raman	
Saturated primary amines	1650–1580	6.06–6.33	m–s	w	br, scissor vib. Aromatic amines at lower end of frequency range 1640–1580 cm^{-1}. Amides 1640–1580 cm^{-1}, thioamides 1650–1590 cm^{-1}, sulphonamides 1575–1550 cm^{-1}.
	1295–1145	7.72–8.73	w	m–w	NH$_2$ rocking/twisting vib. Aromatic amines 1120–1020 cm^{-1}, amides 1170–1080 cm^{-1}, thioamides 1305–1085 cm^{-1}, sulphonamides 1190–1130 cm^{-1}
	895–650	11.17–15.40	m–s	w	Usually br, N–H out-of-plane bending vib , usually multiple bands (for non-hydrogen bonded amines, band may be sharp). Aromatic amines 720–520 cm^{-1}, amides 730–610 cm^{-1}, thioamides 710–580 cm^{-1}, sulphonamides 710–650 cm^{-1}
Primary aliphatic amines, R–CH$_2$NH$_2$ and R$_1$R$_2$R$_3$CNH$_2$	850–810	11.76–12.35	m–s	w	
	795–760	12.58–13.10	m	w–m	
Primary aliphatic amines, R$_1$\ R$_2$/ CHNH$_2$	850–750	11.76–13.33	s	w	br
	~795	~12.58	s	w–m	
Secondary amines	1580–1490	6.33–6.71	w–m	w	May be masked by an aromatic band at 1580 cm^{-1}
	750–700	13.33–14.29	s	w	br, N–H wagging vib
Secondary aliphatic amines: R$_1$–CH$_2$–NH–CH$_2$–R$_2$ and R$_1$\ R$_4$\ R$_2$–C–NH–C–R$_5$ R$_3$/ R$_6$/	750–710	13.33–14.08	s	w	
R$_1$R$_2$CH–NH–RHR$_3$R$_4$	735–700	13.61–14.29	s	w	
(α-Sat.)NH·CH$_3$	1580–1480	6.33–6.76	w–m	w	NH def vib. –CO·NH·CH$_3$ 1600–1500 cm^{-1}, –SO$_2$·NH·CH$_3$ 1575–1550 cm^{-1}, –CS·NH·CH$_3$ 1570–1500 cm^{-1}
Imines, \ /C=N–H	1590–1500	6.29–6.67	m	w	N–H bending vib
o-Alkyl hydroxylamines RONH$_2$	1595–1585	6.27–6.31	m	w	

vibrations. Depending on the amount of hydrogen bonding, a number of bands may appear in this region. Also in the solid phase, amines with >NH$_2^+$, –NH$^+$ and C=NH$^+$– have a broad absorption of medium intensity at 2700–2250 cm^{-1} (3.70–4.44 μm). In dilute solution using non-polar solvents, the stretching vibrations of –NH$_3^+$ result in two bands, one near 3380 cm^{-1} (2.96 μm) and the other near 3280 cm^{-1} (3.05 μm), the stretching vibrations of >NH$_2^+$ result in strong bands at 3000–2700 cm^{-1} (3.33–3.70 μm), those of –NH$^+$ result in a weak-to-medium intensity band at 2200–1800 cm^{-1} (4.55–5.56 μm) while for >C=NH$^+$– a strong absorption is the result at 2700–2330 cm^{-1} (3.70–4.29 μm). Quaternary salts have no characteristic absorption bands.

Amine Hydrohalide N–H$^+$ Deformation Vibrations

Amine –NH$_3^+$ groups have medium-to-strong absorptions near 1600 cm^{-1} (6.25 μm) and 1500 cm^{-1} (6.67 μm) due to asymmetric and symmetric deformation vibrations. Secondary amine hydrohalides have only one band which

Table 9.3 Amine C–N stretching vibrations

Functional Groups	Region cm⁻¹	Region µm	Intensity IR	Intensity Raman	Comments
Primary aliphatic amines	1240–1020	8.06–9.80	w–m	m–s, dp	General range
Primary aliphatic amines, –CH₂–NH₂	1100–1050	9.09–9.52	m	m–s, dp	
Primary aliphatic amines, ＞CH–NH₂	1140–1080	8.77–9.26	w–m	m, dp	
	1045–1035	9.57–9.66	w	m, dp	
Primary aliphatic amines, —CNH₂	1240–1170	8.07–8.55	w–m	m–w	
	1040–1020	9.62–9.80	w	m	
Secondary aliphatic amines	1190–1170	8.40–8.55	m	m	General range. Asym CNC str
	1145–1130	8.73–8.85	m	m–w, p	General range. Sym CNC str
Secondary aliphatic amines, –CH₂–NH–CH₂⁻	1145–1130	8.73–8.85	m–s	m	
Secondary aliphatic amines, ＞CH–NH–CH＜	1190–1170	8.40–8.55	m	m	
Tertiary aliphatic amines	1240–1030	8.06–9.71	m	m–s	General range, doublet. Also Raman asym C–N str band at 835–750 cm⁻¹
	1040–1020	9.62–9.80	w	m	CN str
Tertiary aliphatic amines, —CH₂ –CH₂–N —CH₂	1210–1150	8.25–8.70	m	m–w	
Tertiary dimethyl amines, (CH₃)₂N–CH₂–	~1270	~7.87	m	m–w	
	~1190	~8.40	m	m	
	~1040	~9.62	m–w	m	
Tertiary diethyl amines, (C₂H₅)₂N–C＜	~1205	~8.30	m	m–w	
	~1070	~9.35	m		
Primary and secondary aromatic amines	1360–1250	7.36–8.00	s	m–w	X-sensitive band
ArNHR	1360–1250	7.35–8.00	s–m	m–w	C_Ar–N str
	1280–1180	7.81–8.48	m	m–w	C_R–N str
Tertiary aromatic amines, N＜	1380–1265	7.25–7.91	s–m	m–w	
Imines, ＞C=N–	1690–1630	5.92–6.14	v	m–s	C=N str
R–CH=N–R	1675–1660	5.97–6.02	m–w	m–s	C=N str
Ar–CH=N–Ar	1650–1645	6.06–6.08	m–s	m–s	C=N str
R–CH=N–Ar	1660–1630	6.02–6.14	m–s	m–s	C=N str

Table 9.4 Amines: other vibrations

Functional Groups	Region		Intensity		Comments
	cm^{-1}	μm	IR	Raman	
−CH$_2$NH$_2$	2945–2915	3.40–3.43	m	m−s	CH$_2$ asym str
	2890–2850	3.46–3.51	m	m−s	CH$_2$ sym str
	1470–1430	6.80–6.99	m	m−w	CH$_2$ def vib
	1385–1335	7.22–7.49	m−w	m−w	CH$_2$ wagging vib
	1335–1245	7.49–8.03	m−w	m−w	CH$_2$ twisting vib
	1285–1145	7.78–8.73	m−w	m	CH$_2$/NH$_2$ twisting vib
	945–835	10.58–11.98	w	w	CH$_2$ rocking vib
	895–795	11.17–12.58	s−m	w−m	br, NH$_2$ wagging vib. When unassociated becomes sharp at 800–740 cm^{-1}, also affected by temperature.
	465–315	21.51–31.75	w	w−m	CN def vib
Primary aliphatic amines	495–445	20.20–22.47	m−s		broad
	350–210	28.57–47.62	s		broad
Primary aromatic amines	445–345	22.47–28.99	w		in-plane bending vib of aromatic −NH$_2$ bond
	300–160	33.33–62.50			torsional vib
Secondary aliphatic amines (Sat.)NH·CH$_3$	455–405	21.98–24.69	w−m	m−w	C−N−C def vib
	2990–2940	3.34–3.40	w−m	m	asym CH$_3$ str
	2975–2925	3.36–3.42	w−m	m	asym CH$_3$ str
	2925–2785	3.42–3.59	m−s	m	sym CH$_3$ str
	1580–1480	6.33-6.76	w	w	NH in-plane def vib
	1485–1455	6.73–6.87	w−m	m−w	asym CH$_3$ def vib
	1475–1445	6.78–6.92	w−m	m−w	asym CH$_3$ def vib
	1445–1375	6.92–7.27	w−m	m−w	sym CH$_3$ def vib
	1180–1100	8.47–9.09	w−m	m	CH$_3$ rocking vib and CN str. Ar·NH·CH$_3$ 1155–1125 cm^{-1}.
	1150–1020	8.70–9.80	w−m	w−m	CH$_3$ rocking vib and CN str. Ar·NH·CH$_3$ 1080–1030 cm^{-1} and −SO$_2$·NH·CH$_3$ 1085–1055 cm^{-1}.
	1070–920	9.35–10.87	w−m	w	CH$_3$ rocking vib and CN str. Ar·NH·CH$_3$ 1050–920 cm^{-1}.
	750–700	13.33–14.29	m	s, p	very br. Ar·NH·CH$_3$ and −SO$_2$·NH·CH$_3$ 670–600 cm^{-1}, −CO·NH·CH$_3$ 795–675 cm^{-1}, −CS·NH·CH$_3$ 720–610 cm^{-1}
	410–310	24.39–32.26		m−w	CNC skeletal vib. Ar·NH·CH$_3$ and −SO$_2$·NH·CH$_3$ 410–310 cm^{-1}, −CO·NH·CH$_3$ 370–260 cm^{-1}, −CS·NH·CH$_3$ 330–200 cm^{-1}
	260–200	38.46–50.00			CH$_3$ torsional vib
	130–70	76.92–142.86			NH·CH$_3$ torsional vib
Tertiary dimethyl amines, −N(CH$_3$)$_2$	3020–2960	3.31–3.38	m−s	w−m	asym CH$_3$ str
	3020–2960	3.31–3.38	m−s	w−m	asym CH$_3$ str
	2975–2925	3.36–3.42	m−s	w−m	asym CH$_3$ str
	2975–2925	3.36–3.42	m−s	w−m	asym CH$_3$ str
	2925–2785	3.42–3.59	m−s	m−s	sym CH$_3$ str
	2900–2770	3.45–3.61	m−s	m−s	sym CH$_3$ str
	1490–1440	6.71–6.94	m−s	w−m	asym CH$_3$ def vib
	1490–1440	6.71–6.94	m−s	w−m	asym CH$_3$ def vib
	1470–1420	6.80–7.04	m−s	w−m	asym CH$_3$ def vib
	1470–1420	6.80–7.04	m−s	w−m	asym CH$_3$ def vib
	1445–1375	6.92–7.27	m−s	w−m	sym CH$_3$ def vib
	1415–1355	7.07–7.38	m−s	w−m	sym CH$_3$ def vib

(*continued overleaf*)

Table 9.4   (*continued*)

Functional Groups	Region		Intensity		Comments
	cm^{-1}	µm	IR	Raman	
	1300–1200	7.69–8.33	w	m–w	CH$_3$ rocking vib and asym CCN str.
	1200–1130	8.33–8.84	w	m–w	asym CCN str and CH$_3$ rocking vib
	1180–1050	8.47–9.52	w	w–m	CH$_3$ rocking vib
	1100–1020	9.09–9.80	w	w–m	CH$_3$ rocking vib
	1070–940	9.35–10.64	w	w–m	CH$_3$ rocking vib
	980–820	10.20–12.20			sym CCN str
	525–395	19.05–25.32			CCN def vib
	410-310	24.39–32.26		m–w	CCN wagging vib
	375–225	26.67–44.44			CCN rocking vib
	295–195	33.90–51.28			CH$_3$ torsional vib
	240–130	41.67–76.92			CH$_3$ torsional vib
	170–70	58.82–142.86			CCN torsional vib
Tertiary aliphatic amines	510–480	19.61–20.83	s		
$\diagdown$C=C−N$\diagdown$	1680–1630	5.95–6.14	m–s	s	C=C str usually more intense than normal C=C str band
o-Alkyl hydroxylamines, RONH$_2$	855–840	11.70–11.90	m	s	C–O–N str. In aqueous solution, band moves to higher wavenumbers

Table 9.5   Amine and imine hydrohalide N–H$^+$ stretching vibrations

Functional Groups	Region		Intensity		Comments
	cm^{-1}	µm	IR	Raman	
−NH$_3$$^+$	3350–3100	2.99–3.23	m	m	br, solid phase spectra
$\diagdown$NH$_2$$^+$, −NH$^+$, $\diagdown$NC=NH$^+$−	2700–2250	3.70–4.44	m	m	br, sometimes a group of sharp bands, solid phase spectra
−NH$_3$$^+$	~3380	~2.96	m	m	asym str, dilute solution spectra
	~3280	~3.05	m	m	sym str, dilute solution spectra
−CH$_2$NH$_3$$^+$	3235–3030	3.09–3.30	m, br	m	asym NH$_3$ str
	3115–2985	3.21–3.35	m, br	m	asym NH$_3$ str
	3010–2910	3.32–3.44	m	m	sym NH$_3$ str
$\diagdown$NH$_2$$^+$	3000–2700	3.33–3.70	m–s	m	Dilute solution spectra, two bands
−NH$^+$	2200–1800	4.55–5.56	w–m	m	Dilute solution spectra
$\diagdown$C=NH$^+$−	2700–2330	3.70–4.29	m–s	m	Dilute solution spectra, overtone bands occur at 2500–2300 cm^{-1}
Ammonium salts, NH$_4$$^+$	3300–3030	3.03–3.30	s	m	br

Table 9.6 Amine and imine hydrohalide N–H$^+$ deformation and other vibrations

Functional Groups	Region cm^{-1}	Region μm	Intensity IR	Intensity Raman	Comments
–NH$_3^+$	~2500	~4.00	w		Overtone (sometimes absent)
	~2000	~5.00	w		Overtone (sometimes absent)
	1635–1585	6.15–6.31	m	w	asym NH$_3^+$ def vib
	1585–1560	6.31–6.41	m	w	asym NH$_3^+$ def vib
	1530–1480	6.54–6.76	w	w	sym NH$_3^+$ def vib
	1280–1150	7.81–8.70	w	w	NH$_3^+$ rocking vib
	1135–1005	8.81–9.95	w–m	w	NH$_3^+$ rocking vib /CN str vib
	1100–930	9.09–10.75	w–m	w	NH$_3^+$ rocking vib /CN str vib
⟩NH$_2^+$	~2000	~5.00	w		Overtone (sometimes absent)
	1620–1560	6.17–6.41	m–s	w	
	~800	~12.50	w	w	NH$_2^+$ rocking vib
–CH$_2$NH$_3^+$	2960–2900	3.38–3.45	m	m	asym CH$_2$ str
	2920–2800	3.42–3.57	m	m	sym CH$_2$ str
	1635–1585	6.12–6.31	m–s	m	asym NH$_3^+$ def vib
	1615–1560	6.19–6.41	m–s	m	asym NH$_3^+$ def vib
	1520–1480	6.58–6.76	w	m	sym NH$_3^+$ def vib
	1280–1150	7.81–8.70	w	w	NH$_3^+$ rocking vib
	1135–1005	8.81–9.95	w–m	w	NH$_3^+$ rocking vib /CN str vib
	1100–930	9.09–10.75	w–m	w	NH$_3^+$ rocking vib /CN str vib
	535–425	18.69–23.53	w–m		NH$_3^+$ twisting vib /CCN def vib
	370–250	27.02–40.00			CCN twisting vib /NH$_3^+$ def vib
Imine, ⟩C=N$^+$–H	2200–1800	4.55–5.56	m	w	One or more bands
	~1680	~5.95	m	s	C=N$^+$ str
Ammonium salts, NH$_4^+$	1430–1390	6.99–7.19	s	w	N–H def vib

is near to 1600 cm^{-1} (6.25 μm). Unfortunately, aromatic ring C=C stretching vibrations also give rise to bands in this general region so that care must be exercised in interpretations.

Amine and Imine Hydrohalides: Other Bands

Other relevant bands have, of course, been discussed in the previous section dealing with uncharged amines and this should be referred to.

Primary amine hydrohalides have a number of sharp bands in the region 2800–2400 cm^{-1} (3.57–4.15 μm), and a band around 2000 cm^{-1} (5.00 μm) which is believed to be a combination band involving NH$_3^+$ deformation vibrations. Secondary amine hydrohalides have two sharp bands, at about 2500 cm^{-1} (4.00 μm) and 2400 cm^{-1} (4.15 μm). Primary amine hydrohalides

also absorb in the region 1280–1005 cm^{-1} (7.81–9.95 μm) due mainly to the rocking vibration of the NH$_3^+$ group. Most amine hydrohalides have in addition a medium-to-strong band at about 1120 cm^{-1} (8.93 μm).

Imine hydrohalides have one or more bands of medium intensity in the region 2200–1800 cm^{-1} (4.55–5.56 μm) which may be used to distinguish them clearly from amine hydrohalides.

References

1. H. J. Bernstein, *Spectrochim, Acta*, 1962, **18**, 161.
2. W. J. Orville-Thomas *et al.*, *J. Chem. Soc.*, 1958, 1047,
3. L. J. Bellamy and R. L. Williams, *Spectrochim. Acta*, 1957, **9**, 341.
4. E. V. Titov and M. V. Poddubnaya, *Teor. Eksp. Khim.*, 1972, **8**, 276.

5. J. E. Stewart, *J. Chem. Phys.*, 1959, **30**, 1259.
6. S. F. Mason, *J. Chem. Soc.*, **1958**, 3619.
7. P. J. Kreuger and W. H. Thompson, *Proc. R. Soc. London*, 1957, **A243**, 143.
8. P. J. Kreuger, *Nature*, 1962, **194**, 1077.
9. A. Bryson, *J. Am. Chem. Soc.*, 1960, **82**, 4862.
10. A. I. Finkekhtejn, *Opt. Spectrosc.*, 1966, **12**, 454.
11. L. K. Dyall and J. E. Kemp, *Spectrochim. Acta*, 1966, **22**, 467.
12. A. G. Moritz, *Spectrochim. Acta*, 1960, **16**, 1176.
13. B. Chenon and C. Sandorfy, *Can. J. Chem.*, 1958, **36**, 1181.
14. L. Segal and F. V. Eggeston, *Appl. Spectrosc.*, 1961, **15**, 112.
15. A. S. Vexler, *Appl. Spectrosc. Rev.*, 1968, **1**, 29.
16. M. E. Baldwin, *Spectrochim. Acta*, 1962, **18**, 1455.
17. J. Fabian, *Bull. Soc. Chim. Fr.*, 1956, 1499.
18. C. Lawrence *et al.*, *Spectrochim. Acta*, 1981, **38A**, 791.
19. R. H. Collins *et al.*, *J. Mol. Struct.*, 1990, **216**, 53.
20. C. J. Pouchert, *The Aldrich Library of FTIR Spectra*, The Aldrich Co., 1985.
21. K. Ohno *et al.*, *J. Mol. Struct.*, 1992, **268**, 41.
22. J. R. Durig *et al.*, *J. Raman Spectrosc.*, 1989, **20**, 311

10 The Carbonyl Group: C=O

Introduction

The carbonyl group is contained in a large number of different classes of compounds, e.g. aldehydes, ketones, carboxylic acid, esters, amides, acid anhydrides, acid halides, lactones, urethanes, lactams, etc., for which a strong absorption band due to the C=O stretching vibration[1,2] is observed in the region $1850–1550 \, cm^{-1}$ ($5.41–6.45 \, \mu m$). Because of its intensity[3,10] in the infrared and the relatively interference-free region in which it occurs, this band is reasonably easy to recognize. In Raman spectra, the CO stretching band is much less intense than in infrared.

The frequency of this carbonyl stretching vibration is dependent on various factors:

1. The structural environment of the C=O group.
   (a) The more electronegative an atom or group directly attached to a carbonyl group,[7,9] the greater is the frequency.
   (b) Unsaturation[11–17] in the α,β- position tends to decrease the frequency, except for amides which are little influenced, by $15–40 \, cm^{-1}$ from that expected without the conjugation, further conjugation having little effect on the frequency.
   (c) Hydrogen bonding[18–26] to the C=O results in a decrease in the frequency of $40–60 \, cm^{-1}$, this being independent of whether the bonding is inter- or intramolecular.
   (d) In situations where ring strain occurs, the greater the strain, i.e. the smaller the ring, the greater is the frequency.[21,27–34]
2. The physical state of the sample. In the solid phase, the frequency of the vibration is slightly decreased compared with that in dilute non-polar solutions.[35,36] The presence of hydrogen bonding is an important contributing factor to this decrease in frequency.

In cases where more than one of these influences is present, the net shift in the position of the band due to the C=O stretching vibration appears to be the result of an approximately additive process, although this does not always hold in cases where hydrogen bonding to the C=O group is present.

If the double-bond character of the carbonyl group is increased (i.e. the force constant of the bond is increased) by its neighbouring groups, then the frequency of the stretching vibration is increased (i.e. the wavelength is decreased). If the presence of an adjacent group results in resonance hybrids, such as II (below), making a greater contribution, then this will tend to decrease the double-bond character of the carbonyl group and hence decrease the frequency of the carbonyl stretching vibration:

On the other hand, an electron-accepting group tends, through the inductive effect, to increase the double-bond character and hence increases the frequency of the vibration:

Hence, for a particular group, these two opposite effects determine the frequency of the vibration and it is therefore possible, in general, to give an approximate order for the C=O bond stretching vibration frequency for different groups:

$$RCOO^- < RCONH_2 < (RCOOH)_2 < RCOR'$$
$$\text{dimer}$$

$$< RCHO < RCOOR' < RCOOH < RCOOCOR'$$
$$\text{monomer}$$

Hydrogen bonding tends to decrease the double-bond character of the carbonyl group, therefore shifting the absorption band to lower frequency:

For example, the C=O stretching vibration band of aliphatic carboxylic acids as monomers appears near $1760 \, cm^{-1}$ ($5.68 \, \mu m$), but as dimers (which are predominant in liquid and solid samples) the band occurs near $1700\,cm^{-1}$ ($5.88\,\mu m$). For this hydrogen-bonded dimer, a characteristic broad band is observed at about $920\,cm^{-1}$ ($10.87\,\mu m$) due to its out-of-plane bending vibration.

Due to hydrogen-bonding in solvents such as chloroform, the band due to the carbonyl stretching vibration of ketones may be split by about $5-10\,cm^{-1}$ ($0.02-0.04\,\mu m$).

If a carbonyl group is part of a conjugated system, then the frequency of the carbonyl stretching vibration decreases, the reason being that the double-bond character of the $C{=}O$ group is less due to the π-electron system being delocalized.

For *meta-* and *para*-substituted aromatic carbonyl compounds, a linear relationship exists between the carbonyl absorption frequency and the Hammett reactivity constant.[2,5,18,37,38] A relationship between the carbonyl stretching vibration frequency of aromatic carbonyl compound and the pK[39,40] of the corresponding aromatic carboxylic acid has been demonstrated. Correlations with other parameters, such as electronegativities,[41] ionization potentials, Taft σ^* values, half-wave potentials, etc., have also been made.[42,43] For aromatic compounds with *ortho-* substituents, a combination of factors may be important, such as chelation, steric effects, and field effects (dipole interactions through space).

In detailed spectral studies of carbonyl compounds in which conjugation with an olefinic group occurs, geometrical isomerism must be taken into account.[13,16,17] α,β-Unsaturated carbonyl compounds have a contribution from the $\overset{\beta}{C^+}-\overset{\alpha}{C}{=}C-O^-$ form in addition to the form $\overset{\beta}{C}{=}\overset{\alpha}{C}-C{=}O$. Some partial double-bond character exists between the $C{=}O$ and the α,β-unsaturated $C{=}C$ bond. Hence, geometrical isomerism about this 'single' bond is possible, resulting in s-*trans* and s-*cis* forms, where the s indicates restricted rotation about a single bond:

s-*trans*   s-*cis*

If the R groups are different, then various s-*trans* and s-*cis* forms may exist, e.g.

Different s-*trans* forms

In the case of two olefinic groups conjugated to a carbonyl group, various configurations are possible such as

s-*trans*, s-*trans*   s-*trans*, s-*cis*

α-Dicarbonyl[44,45] compounds may exist in two configurations, cisoid and transoid:

cisoid   transoid

In the cisoid conformation, a degree of interaction between the dipoles of the two carbonyl groups would be expected which would result in an increase in the carbonyl character or possibly result in enolization. However, for acyclic α-dicarbonyl compounds, no such interaction is observed: the carbonyl stretching vibration frequency is virtually the same as for the equivalent monocarbonyl compound. This can be explained if the α-dicarbonyl substances exist in the more energetically-favoured transoid conformation. For the symmetric stretching vibration, the dipole interactions of the two carbonyls would be cancelled. The symmetric stretching mode is infrared inactive since there would be no net change in the dipole moment during vibration. In the case of cyclic α-dicarbonyls, the two $C{=}O$ groups are held, depending on the ring size, more-or-less rigidly in the cisoid conformation. This results in these cyclic compounds with smaller rings having a marked tendency to enolization.

The great difference between the spectra of a carboxylic acid and its salt may be useful when doubt exists as to whether or not a $C{=}O$ band should be attributed to a carboxyl group.

In general, for the $C{=}O$ stretching vibration band, acids absorb more strongly than ketones, aldehydes, or amides. The intensity of the $C{=}O$ absorptions of ketones and aldehydes is approximately the same, whereas that of amides may vary greatly.

A relatively small number of compounds containing only one carbonyl group has more than one band due to the carbonyl stretching vibration, examples being benzoyl chloride,[46] cyclopentanone,[47–49] cyclopent-2-enone, ethylene carbonate and certain α,β-unsaturated lactones (five- and six-membered rings)[50,51] and lactams. It would seem that Fermi resonance is responsible for this doubling of the carbonyl band.[52,72,170]

Fermi resonance occurs if the energy associated with a combination or an overtone band coincides approximately with that for a fundamental energy level of a different vibration. This may be thought of as a to-and-fro transfer of energy between the two levels. Fermi resonance results in two bands of similar intensity almost equidistant from the position at which the fundamental and combination bands would have occurred. These doublets are, of course, concentration-independent but may depend on temperature and solvent polarity.

With the exception of thioacids, the carbonyl stretching vibration frequency of thiol compounds[53,54] is found approximately 40 cm^{-1} lower (0.15 μm higher) than that of the corresponding oxygen compound. Similarly, dithiol carbonates have bands which are about 80 cm^{-1} (0.35 μm higher) than for the corresponding –O–CO–O–compound.

Ketones, $\diagdown$C=O

Ketone C=O Stretching Vibrations

Ketones and aldehydes have almost identical carbonyl absorption frequencies. Aldehydes usually absorb at about 10 cm^{-1} higher (0.03 μm lower) than the corresponding ketone.

Saturated aliphatic ketones[43,122,172,173,175] and cyclic ketones (six-membered rings and greater) in the pure liquid and solid phases absorb strongly in the range 1725–1705 cm^{-1} (5.80–5.86 μm). In dilute solution in non-polar solvents, the absorption occurs at 1745–1715 cm^{-1} (5.73–5.83 μm). Therefore, in general, in the solid phase, the frequency of the C=O stretching vibration is 10–20 cm^{-1} lower than that observed in dilute solutions using non-polar solvents. In non-polar solvents, aryl ketones[42,55] absorb at 1700–1680 cm^{-1} (5.88–5.95 μm), diaryl ketones at 1670–1600 cm^{-1} (5.99–6.25 μm), α,β-unsaturated ketones ($\diagdown_\beta$C=$\overset{\alpha}{C}$–C=O) at 1700–1660 cm^{-1} (5.88–6.02 μm), α-halo-ketones at 1750–1725 cm^{-1} (5.71–5.80 μm),[4,56,57] and α,α'-dihalo-ketones at 1765–1745 cm^{-1} (5.66–5.73 μm).[58–60] α-Chloro-ketones absorb at the higher frequencies if the chlorine atom is near the oxygen and at the lower values if away from it.[4] In the case of α,β-unsaturation, the C=C stretching vibration frequency is also reduced. The aromatic band near 1600 cm^{-1} (6.25 μm) usually appears as a doublet and the band near 1500 cm^{-1} (6.67 μm) can be very weak.

α-Diketones have a very strong symmetric C=O stretch at about 1720 cm^{-1} (5.81 μm). When the carbonyl groups are fixed in the *cis* position by a six-membered ring two bands are observed one at approximately 1760 cm^{-1} (5.68 μm) and the other at about 1730 cm^{-1} (5.78 μm). If the *cis* configuration is held by a five-membered ring then these strong bands occur at about 1775 cm^{-1} (5.63 μm) and 1760 cm^{-1} (5.68 μm).

Enolized β-diketones[44,61,63,67] have a very strong band in the region of 1610 cm^{-1} (6.21 μm) (the band due to the C=C stretching vibration being at 1520–1500 cm^{-1}). For α-diketones, a single band is observed at a slightly higher frequency than that expected for the single ketone.

Unsymmetrical *para*-substituted benzils have two bands at 1690–1660 cm^{-1} (5.92–6.02 μm). *Ortho*-hydroxy or *ortho*-amino-aryl ketones[64] exhibit a strong band in the region 1655–1610 cm^{-1} (6.04–6.21 μm) due to the carbonyl stretching vibration. The presence of intramolecular hydrogen bonding causes this frequency to be lower than might otherwise be expected.

As mentioned previously, the band due to the C=O stretching vibration is shifted from its expected position by a number of parameters, these influences being approximately additive in their effect. The approximate magnitude of these shifts is given in Table 10.1.

Methyl and Methylene Deformation Vibrations in Ketones

For the group –CH$_2$–CO–, the methylene scissoring vibration occurs in the range 1435–1405 cm^{-1} (6.97–7.12 μm).[65] This is lower than that for CH$_2$ in aliphatic hydrocarbons which occurs in the range 1480–1440 cm^{-1} (6.76–6.94 μm). For methyl groups adjacent to carbonyl groups, the symmetrical C–H bending vibration has a lower frequency, 1360–1355 cm^{-1} (7.35–7.38 μm), than that for aliphatic hydrocarbons, 1390–1370 cm^{-1} (7.19–7.30 μm). Ketones with the structure –CH$_2$–CO–CH$_2$– have a medium-intensity band at 1230–1100 cm^{-1} (8.13–9.09 μm) due to the asymmetric stretching vibration of the backbone. For methyl ketones, this band is near 1170 cm^{-1} (8.55 μm).

Ketone Skeletal and Other Vibrations

A band of medium-to-strong intensity due to the C–C stretching vibration may be found at 1325–1115 cm^{-1} (7.55–8.95 μm) for aliphatic ketones[65] and

Table 10.1 Influence on C=O stretching vibration for ketones and aldehydes

	Wavenumber shift (cm^{-1})	Wavelength shift (μm)
α,β-Unsaturation	−30	−0.11
α-Halogen	+20	−0.07
α,α'-Dihalogen	+40	−0.15
α,α-Dihalogen	+20	−0.07
Solid phase	−20	+0.07

at 1225–1075 cm^{-1} (8.16–9.30 μm) for aromatic ketones. However, this band is not normally used for assignment purposes.

Due to the in-plane deformation of the C–CO–C group, aliphatic ketones have a strong absorption at 630–620 cm^{-1} (15.87–16.13 μm)[66] which is shifted to lower frequencies, 580–565 cm^{-1} (17.24–17.70 μm), if α-branching occurs.

Aliphatic ketones have an absorption at 540–510 cm^{-1} (18.52–19.61 μm) which is due to C–C=O deformation. This band is shifted to 560–550 cm^{-1} (17.86–18.18 μm) if α-branching occurs. Small-ring cyclic ketones absorb strongly at 505–480 cm^{-1} (19.80–20.83 μm).[66]

With the exception of acetone and α-branched compounds, methyl ketones have prominent bands at about 1355 cm^{-1} (7.38 μm) and at about 1170 cm^{-1}

Chart 10.1 Infrared – band positions of carbonyl groups including carboxylic acids and their salts etc. All these bands are very strong

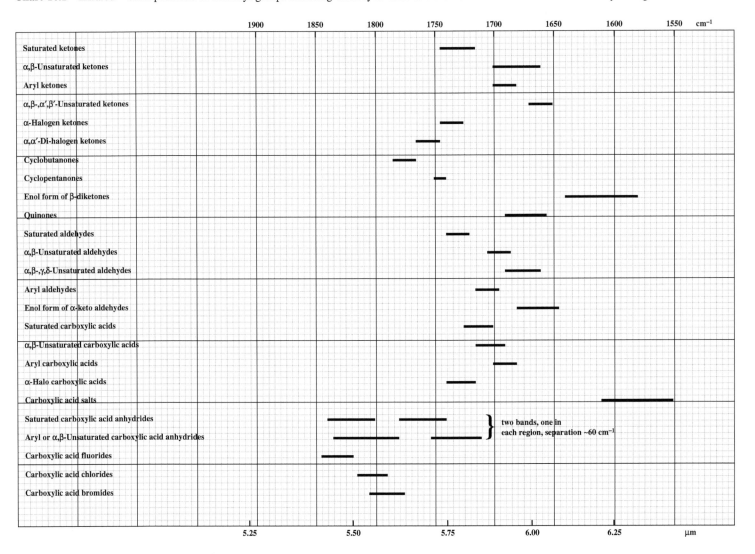

Chart 10.1   (*continued*)

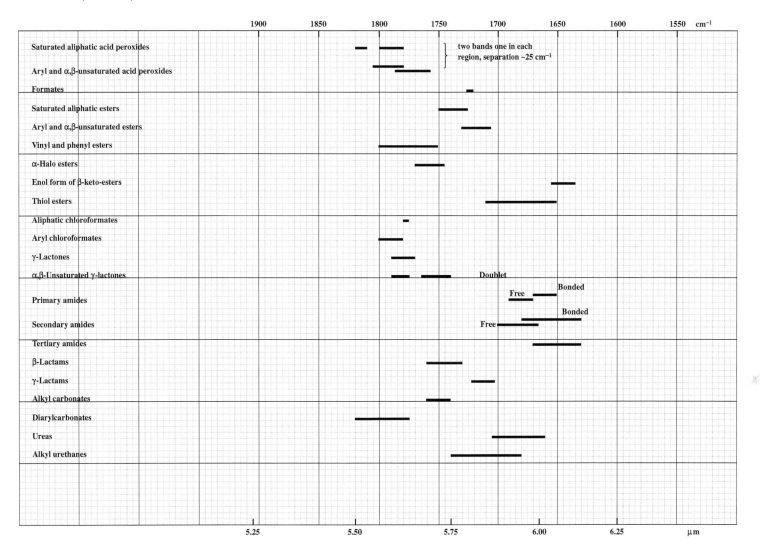

Table 10.2 Ketone C=O stretching vibrations

Functional Groups	Region		Intensity		Comments
	cm^{-1}	μm	IR	Raman	
Saturated aliphatic ketones	1745–1715	5.73–5.83	vs	m	sat. methyl ketones 1730–1700 cm^{-1}
Aryl ketones	1700–1680	5.88–S.95	vs	v	
Diaryl ketones	1670–1600	5.99–6.25	vs	v	see ref. 55
o-Hydroxy diaryl ketones	1655–1635	6.04–6.12	vs	m	
α,β-Unsaturated ketones, $\overset{\beta}{\diagdown}\text{C=}\overset{\alpha}{\text{C}}\text{—C=O}$	1700–1660	5.88–5.02	vs	m–w	General range
	1690–1660	5.92–6.02	vs	m	s-*trans*- form (C=C str, 1645–1615 cm^{-1})
	1700–1685	5.88–5.93	vs	w	s-*cis*- form (C=C str, 1625–1615 cm^{-1})
α,β-Unsaturated, β-amino ketones, *cis*- form	1640–1600	6.10–6.25	s	m	Intramolecular hydrogen bonding occurs. *Trans*- form has no hydrogen bonding and carbonyl band occurs in normal range
α,β-, α′,β′-Di-unsaturated ketones, C=C—CO—C=C	1680–1650	5.95–6.06	vs	m–w	
α-Halo-ketones	1750–1725	5.71–5.80	vs	m	
α,α′-Dihalo-ketones	1765–1745	5.66–5.73	vs	m	
Keto form of β-diketones, —CO—C—CO—	1740–1695	5.75–5.90	vs	m–w	
Enol form of β-diketones, $\overset{\text{OH----O}}{\underset{\text{C=C—C}}{\vert\quad\;\Vert}}$	1640–1580	6.10–6.33	vs	m–w	br, extremely strong band (other bands at ~1500 cm^{-1}, ~1450 cm^{-1}, ~1260 cm^{-1}; O–H str, 3000–2700 cm^{-1})
Cyclic-ketones, enol form	1630–1610	6.14–6.21	s	m	
α-Diketones, –CO–CO–	1730–1705	5.78–5.86	vs	m	
–CO·O·CH$_2$CO–	1745–1725	5.73–5.80	vs	m	ester CO at 1760–1745 cm^{-1}
Cyclopentanone derivatives	1750–1740	5.71–5.75	vs	m	Fermi resonance doublet
Cyclobutanone derivatives	1790–1765	5.59–5.67	vs	m	
Cyclopropenones	1870–1845	5.35–5.42	vs	m	C=C and C=O in-phase and out-of-phase str; as mass of substituents increases, band at ~1475 cm^{-1}, mainly due to C=C str, disappears and strong band at 1655–1620 cm^{-1} appears instead
	1655–1620	6.04–6.17	s	m	
Cyclopropanones	~1820	~5.49	vs	m	Liquid phase (vapour phase, ~1905 cm^{-1})
3,4-Dihydroxy, 3-cyclobutene diones, –(C$_4$O$_4$)–	1820–1785	5.50–5.60	vs	m–w	C=O str
o-Hydroxy-, and o-amino-aryl ketones	1665–1610	6.01–6.21	vs	m	Intramolecular ⎫
β-Diketones, metal chelates	1605–1560	6.23–6.41	s	m	hydrogen bonding ⎪ All four bands due to
	1550–1500	6.45–6.67	m–s	m	occurs position ⎬ C–O and C–C stretching
	1450–1350	6.90–7.41	m–s	m	dependent on complex ⎪ vib
	~1250	~8.00	m	m	stability ⎭

Table 10.2 (*continued*)

Functional Groups	Region		Intensity		Comments
	cm^{-1}	µm	IR	Raman	
Flavones Diagram 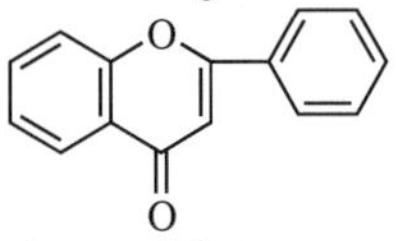	1670–1625	5.99–6.15	vs	m	
Cyclopropyl ketones, 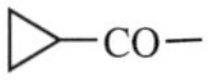 ▷–CO–	1705–1685	5.86–5.94	vs	m	
Aliphatic silyl ketones, 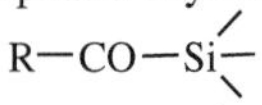 R–CO–Si⟨	1645–1635	6.08–6.12	vs	m	
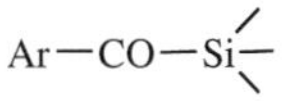 Ar–CO–Si⟨	~1620	~6.17	vs	m	
Benzophenone complexes, –ArCO·AlCl$_3$ and Ar$_2$CO·AlCl$_3$	~1525	~6.56	m–s		

Table 10.3 Ketones: other bands

Functional Groups	Region		Intensity		Comments
	cm^{-1}	µm	IR	Raman	
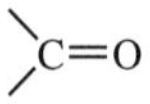 ⟩C=O	3550–3200	2.82–3.13	w	–	C=O stretching vibration overtone
Aliphatic ketones (straight chain)	1170–1100	8.55–9.09	m–w	m–w	C·CO·C asym str
	800–700	12.50–14.29	w	m–s	C·CO·C sym str
	630–620	15.87–16.13	s	s–m	C–CO–C in-plane def vib
α-Branched aliphatic ketones	680–650	14.71–15.38	w	m–s	C·CO·C sym str
	580–565	17.24–17.70	m	s–m	C–CO–C in-plane def vib
Methyl ketones	3045–2965	3.28–3.37	m–w	s–m	asym CH str
	3020–2930	3.31–3.41	m–w	s–m	asym CH str
	2940–2840	3.40–3.52	m	s–m	sym CH str
	1390–1340	7.19–7.46	m–s	m–w	CH$_3$ def. Often 1360–1355 cm^{-1}.
	1170–700	8.55–14.29	m–w	m–w	C·CO·C asym str. Ethyl ketones 1130–1100 cm^{-1}
Aliphatic methyl ketones	600–580	16.67–17.24	s–m	m–s	C–CO–C in-plane def vib
	540–510	18.52–19.61	m	m	C–CO in-plane def vib
Aromatic methyl ketones	600–580	16.67–17.24	s–m	m–s	C–CO–C in-plane def vib
–CH$_2$CO–	1435–1405	6.97–7.12	w	m–s	CH$_2$–CO def vib
Alkyl ketones	1325–1215	7.55–8.23	s	m–w	
	1170–1100	8.55–9.09	m–w	m–w	C·CO·C asym str
	800–700	12.50–14.29	w	m–s	C·CO·C sym str
	490–460	20.41–21.74		vw	Out-of-plane C·CO·C def vib
	430–390	23.26–25.64		m	In-plane C·CO·C def vib

(*continued overleaf*)

Table 10.3 (*continued*)

Functional Groups	Region cm^{-1}	Region µm	Intensity IR	Intensity Raman	Comments
Aryl ketones	1225–1075	8.16–9.30	s	m	Phenyl–carbon str
	~1300	~7.69	m	m	C–C–C bending and C–CO–C
Small-ring cyclic ketones	505–480	19.80–20.83	s	m	C–CO in-plane def vib
3,4-Dihydroxy 3-cyclobutene diones, –(C$_4$O$_4$)–	~1725	~5.80	s	m	
	1590–1560	6.29–6.41	v	s	C=C str
	1485–1405	6.73–7.12	s	m	C–O + C=C str (free acid ~1515 cm^{-1})
	1380–1315	7.25–7.61	m	m–w	
	1285–1220	7.78–8.20	m–s	m–w	
	1200–1140	8.33–8.77	m–s	m–w	
	1090–1020	9.17–9.80	w	w	
	990–800	10.10–12.50	m	w	
	800–635	12.50–15.75	m–w	w	Number of bands

(8.55 µm), the former band being due to CH$_3$ deformation vibrations.[65] Methyl ketones generally (including aromatic methyl ketones) have a strong absorption at 600–580 cm^{-1} (16.67–17.24 µm) which is due to the in-plane deformation vibration of the C–CO–C group. Other aromatic ketones also exhibit this absorption band.

Quinones and

Either one or two carbonyl absorption bands may be observed for *para*-quinones, the range being 1690–1655 cm^{-1} (5.92–6.04 µm), even though only one might be expected from symmetry considerations.[68–72] On the other hand, *ortho*-quinones exhibit only one carbonyl band, which is in the same range although usually at about 1660 cm^{-1} (6.02 µm). The carbonyl absorption frequency of polycyclic quinones increases with the number of fused rings. Quinones with electronegative substituents absorb at the higher end of the frequency range given.

In the absence of hydroxyl and amino- groups, anthraquinones[73] absorb strongly in the region 1680–1650 cm^{-1} (5.95–6.06 µm) due to the carbonyl group. The presence of hydroxyl and amino- groups results in a lowering of this frequency. Charge-transfer complexes of benzoquinone and hydroquinone have been dealt with.[74]

Aldehydes, –CHO

Aldehyde C=O Stretching Vibrations

The C=O stretching vibration is influenced in a similar manner to that observed for ketones[79,122] (see earlier). In non-polar solvents, saturated aliphatic aldehydes absorb strongly in the region 1740–1720 cm^{-1} (5.75–5.82 µm),[75,76] aryl aldehydes at 1715–1685 cm^{-1} (5.83–5.93 µm),[77,78] and α,β-unsaturated aliphatic aldehydes at 1705–1685 cm^{-1} (5.87–5.93 µm), with additional unsaturation lowering the frequency only slightly (approximately 5–10 cm^{-1}). In the solid or liquid phase, the absorption frequencies are lowered by 10–20 cm^{-1} compared with those for dilute solution in non-polar solvents. A study has been made of the temperature dependence of the acetaldehyde C=O stretching vibration.[169]

Aldehydic C–H Vibrations

Two characteristic bands are usually observed due to the stretching vibrations of the aldehydic C–H,[79] both of which are of weak-to-medium intensity, one at about 2820 cm^{-1} (3.55 µm) and the other in the region 2745–2650 cm^{-1} (3.64–3.77 µm). In Raman spectra, the CH stretching band is often of weak intensity, being a shoulder to the band at ~2720 cm^{-1} (3.66 µm) which is normally strong.

Benzaldehydes with bulky *ortho*- substituents such as nitro-, halogen or methoxy groups absorb at 2900–2800 cm^{-1} (3.45–3.57 µm) and

Table 10.4 Quinone C=O stretching vibrations

Functional Groups	Region		Intensity		Comments
	cm^{-1}	μm	IR	Raman	
Quinones	1690–1655	5.92–6.04	vs	m	One or two bands
Polycyclic quinones	1655–1635	6.04–6.03	vs	m	
Anthraquinones (absence of OH and NH$_2$ groups)	1680–1650	5.95–6.06	vs	m	
1-Hydroxyl anthraquinones	1675–1645	6.01–6.08	vs	m	
	1640–1620	6.10–6.17	vs	m	
1,4- or 1,5-dihydroxyl anthraquinones	1645–1605	6.08–6.23	vs	m	
1,8-Dihydroxyl anthraquinones	1680–1660	5.95–6.02	vs	m	
	1625–1615	6.16–6.19	vs	m	
1,4,5-Trihydroxyl anthraquinones	1615–1590	6.19–6.29	vs	m	
1,4,5,8-Tetrahydroxyl anthraquinones	1590–1570	6.29–6.37	vs	m	
Tropones	1600–1575	6.25–6.35	vs	m	
Tropolones,	1620–1590	6.17–6.29	vs	m	Intramolecular bonding to CO group

2790–2720 cm^{-1} (3.58–3.68 μm). Otherwise, aryl aldehydes absorb at 2830–2810 cm^{-1} (3.53–3.56 μm) and 2745–2720 cm^{-1} (3.65–3.68 μm).

The presence of a sharp band at about 2720 cm^{-1} (3.68 μm) and a band due to the carbonyl stretching vibration in the region 1740–1685 cm^{-1} (5.75–5.95 μm) may usually be taken as indicating the presence of an aldehyde.

The CH stretching band, although weak, is useful for characterisation purposes. However, the overtone of the CH in-plane deformations may disturb the position of the CH stretching band or result in some confusion. The presence of two bands in the region 2895–2650 cm^{-1} (3.45–3.77 μm) is due

to an interaction between the C–H stretching vibration and the overtone of the C–H bending vibration near 1390 cm^{-1} (7.19 μm). This involves Fermi resonance since aldehydes for which the latter band is shifted have only one band, this being in the region 2895–2805 cm^{-1} (3.45–3.57 μm).

A weak-to-medium intensity band due to the aldehydic C–H deformation vibration is found in the region 975–780 cm^{-1} (10.26–12.82 μm). However, because of its variable position and intensity, this band may be difficult to identify.

Other Aldehyde Bands

Aliphatic aldehydes absorb weakly in the region 1440–1325 cm^{-1} (6.94–7.55 μm) and aromatic aldehydes absorb weakly at 1415–1350 cm^{-1} (7.07–7.41 μm), 1320–1260 cm^{-1} (7.58–7.94 μm), and 1230–1160 cm^{-1} (8.13–8.62 μm), the last band being due to the C–C stretching vibration. These bands are not normally useful for assignment purposes in infrared spectra. In Raman spectra, the C–C stretching band for *n*-alkyl compounds is of medium-to-strong intensity, occurring at 1120–1090 cm^{-1} (8.23–9.17 μm) with a weak-to-medium intensity band, due to C–C=O in-plane deformation, at 565–520 cm^{-1} (17.70–19.23 μm). For dialkyl aldehydes, this latter band occurs at 665–580 cm^{-1} (15.04–17.24 μm). For aliphatic aldehydes with branching occurring adjacent to the α-carbon atom, a medium-to-strong band

Table 10.5 Quinone C–H out-of-plane deformation vibrations

Functional Groups	Region		Intensity		Comments
	cm^{-1}	μm	IR	Raman	
Monosubstituted *p*-benzonquinones	915–900	10.93–11.11	w–m	w–m	
	865–825	11.56–12.12	m–s	w–m	
2,3-Disubstituted *p*-benzoquinones	860–800	11.63–12.50	s	w–m	
2,5- and 2,6-disubstituted *p*-benzoquinones	920–895	10.87–11.17	s	w–m	

Table 10.6 Aldehyde C=O stretching vibrations

Functional Groups	Region		Intensity		Comments
	cm^{-1}	μm	IR	Raman	
Saturated aliphatic aldehydes	1740–1720	5.75–5.81	vs	w–m	General range for saturated compounds (not aliphatic) 1790–1710 cm^{-1}.
α,β-Unsaturated aliphatic aldehydes	1705–1685	5.87–5.93	vs	w–m	
α,β-γ,δ-Conjugated aliphatic aldehydes	1690–1650	5.91–6.06	vs	w–m	Further conjugation has little effect
Aryl aldehydes	1715–1685	5.83–5.93	vs	v	Most benzaldehydes ~1700 cm^{-1}
o-Hydroxy- and o-amino-aryl aldehydes	1665–1625	6.01–6.16	vs	w–m	Frequency lowered due to hydrogen bonding
α-Keto aldehydes in enol form, $\overset{\|}{-C(OH)}=C-CHO$	1670–1645	6.17–6.25	vs	w–m	Frequency lowered due to hydrogen bonding
α-Di- and trichloroaldehydes	1770–1740	5.65–5.75	vs	v	
$-CF_2CHO$	1790–1755	5.59–5.70	vs	w–m	

Table 10.7 Aldehydes: other bands

Functional Groups	Region		Intensity		Comments
	cm^{-1}	μm	IR	Raman	
Aldehydes, –CHO	2900–2800	3.45–3.57	w–m	w	C–H str
	2745–2650	3.64–3.77	w–m	s–m	C–H str, usually ~2720 cm^{-1}. Fermi resonance with band near 1390 cm^{-1}.
	1450–1325	6.90–7.55	m–s	s–m	In-plane C–H rocking vib. Most aldehydes: 1375–1350 cm^{-1}
	975–780	10.26–12.82	w–m	m	C–H def vib
Aliphatic aldehydes	2870–2800	3.48–3.57	w–m	w	CH str
	2740–2700	3.65–3.70	w–m	s–m	Overtone CH in-plane def vib
	1440–1325	6.94–7.55	m–s	s–m	In-plane C–H rocking vib
	695–635	14.39–15.75	m–s	m–w	C–C–CO in-plane def vib
	565–520	17.70–19.23	m–s	m–w	C–CO in-plane def vib
Aryl aldehydes	2900–2800	3.45–3.57	w–m	w	CH str
	2790–2720	3.58–3.68	w–m	s–m	Overtone CH in-plane def vib
	1415–1350	7.07–7.41	m–w	s–m	In-plane C–H rocking vib
	1320–1260	7.58–7.94	m	m	Due to aromatic ring
	850–720	11.76–13.89	w	m	CH/CO wagging vib, but has been assigned to band ~1000 cm^{-1} for some benzaldehydes
	1230–1160	8.13–8.62	m	m	Possibly ring C–CHO str
	700–580	14.29–17.24	m–s	m	=C–CHO in-plane def vib
α-Branched aliphatic aldehydes	665–635	15.04–15.75	s–m	m–w	C–C–CO in-plane def vib vib
	565–520	17.70–19.23	s–m	m–w	C–CO in-plane def vib

is observed at 800–700 cm^{-1} (12.50–14.29 µm) due to the symmetric skeletal stretching vibration of the quaternary carbon group.

In general, aromatic aldehydes have a strong absorption at 700–580 cm^{-1} (14.29–17.24 µm) due to in-plane deformation vibrations of the C–CHO group.[80]

Aliphatic aldehydes have a medium-to-strong band at 695–635 cm^{-1} (14.39–15.75 µm) and 565–520 cm^{-1} (17.70–19.23 µm) due to C–C–C=O and C–C=O deformations respectively.

Carboxylic Acids, –COOH

Due to the presence of strong intermolecular hydrogen bonding, carboxylic acids normally exist as dimers. Their spectra exhibit a broad band due to the O–H stretching vibration and a strong band due to the C=O stretching vibration. The marked spectral changes which occur when a carboxylic acid is converted to its salt may be used to distinguish it from other C=O containing compounds.

Carboxylic Acid O–H Stretching Vibrations

As a result of the presence of hydrogen bonding, carboxylic acids in the liquid and solid phases exhibit a broad band at 3300–2500 cm^{-1} (3.30–4.00 µm), due to the O–H stretching vibration,[81,82] which sometimes, in the lower half of the frequency range, has two or three weak bands superimposed on it. In the main, it is only chelated O–H groups, e.g. the OH group of the enol form of β-diketones, and carboxylic acids which absorb in the region 2700–2500 cm^{-1} (3.70–4.00 µm), and these two structural groups may be distinguished by their C=O stretching vibrations. Although other groups absorb in the region 3300–2500 cm^{-1} (3.04–4.00 µm), e.g. C–H, P–H, S–H, Si–H, their bands are all sharp. The O–H deformation band may also be useful for distinguishing between groups.

Carboxylic acid monomers have a weak, sharp band at 3580–3500 cm^{-1} (2.79–2.86 µm). Usually, monomers only exist in the vapour phase, but of course some dimeric structure may also be present in this phase too.

Carboxylic Acid C=O Stretching Vibrations

In general, the C=O stretching vibration for carboxylic acids gives rise to a band which is stronger than that for ketones or aldehydes. In the solid or liquid phases, the C=O group of saturated aliphatic carboxylic acids[83,175] absorbs very strongly in the region 1740–1700 cm^{-1} (5.75–5.88 µm). In the Raman spectra of aliphatic compounds, the symmetric C=O stretching band occurs at 1685–1640 cm^{-1} (5.93–6.10 µm).

As mentioned above, most carboxylic acids exist as dimers. However, in very dilute solution in non-polar solvents, or in the vapour phase, when the acid may exist as a monomer, the C=O stretching vibration band is at about 1760 cm^{-1} (5.68 µm). In aqueous solution, polycarboxylic acids exhibit a strong band in their Raman spectra at 1750–1710 cm^{-1} (5.71–5.85 µm).

The frequency of the C=O stretching vibration for saturated n-aliphatic acids usually decreases with increase in chain-length. Electronegative atoms or groups adjacent to carboxylic acid groups have the effect of increasing the C=O stretching vibration frequency, while hydrogen bonding tends to decrease it.[84,85] For example, α-halo-carboxylic acids[7,86] absorb strongly in the region 1740–1715 cm^{-1} (5.75–5.83 µm) and intramolecularly hydrogen-bonded acids absorb at 1680–1650 cm^{-1} (5.95–6.06 µm). Sometimes, α-halo-carboxylic acids exhibit two bands due to the C=O stretching vibration, this being the result of partially restricted rotation.

Aryl and α,β-unsaturated carboxylic acids absorb in the region 1715–1660 cm^{-1} (5.83–6.02 µm). Further conjugation has little effect on the C=O stretching vibration. Aryl carboxylic acids with a hydroxyl group in the ortho- position absorb at about 50 cm^{-1} lower (0.18 µm higher) and with an ortho-amino-group the frequency lowering is about 30 cm^{-1} (0.09 µm). Aryl carboxylic acid monomers absorb at 1755–1735 cm^{-1} (5.70–5.76 µm).

Some saturated dicarboxylic acids have a doublet structure for this C=O band in solid-phase spectra, even though both acid groups are chemically equivalent. This structure may be used to distinguish between optical isomers.

Association of the acid with a solvent such as pyridine, dioxane, etc., generally lowers the C=O stretching vibration frequency.

Other Vibrations of Carboxylic Acids

C–H stretching vibration bands in the region 3100–2800 cm^{-1} (3.23–3.57 µm) sometimes have broad wings due to overlap with the bands due to the O–H stretching vibration. A band at about 1440–1395 cm^{-1} (6.95–7.17 µm), which may be overlooked because of its weak nature, is due to the combination of the C–O stretching and O–H deformation vibrations. A –CH$_2$CO– deformation vibration may further complicate matters since it gives rise to a medium-intensity band at 1410–1405 cm^{-1} (7.09–7.12 µm) which is characteristic of the group.

A medium-to-strong band at 1320–1210 cm^{-1} (7.58–8.28 µm) is observed but this band is not usually much help in identification as other compounds containing the carbonyl group have bands in this region. Carboxylic acid dimers absorb in the narrower range 1320–1280 cm^{-1} (7.58–7.81 µm) and also have a broad, usually asymmetric, band of medium-to-strong intensity in the region 955–915 cm^{-1} (10.47–10.93 µm) due to the out-of-plane deformation of the carboxylic acid OH···O group. This latter band is usually very

Table 10.8 Carboxylic acid C=O stretching vibrations

Functional Groups	Region cm^{-1}	Region μm	Intensity IR	Intensity Raman	Comments
Saturated aliphatic carboxylic acids (hydrogen-bonded or as dimer)	1740–1700	5.75–5.88	vs	w–m	May be found 1785–1685 cm^{-1}. For Raman, C=O sym str occurs at 1685–1640 cm^{-1}
Saturated aliphatic carboxylic acid (as monomer)	1800–1740	5.16–5.75	vs	w–m	In very dilute solution or as vapour
Aryl carboxylic acids (as dimers)	1710–1660	5.85–6.02	vs	w–m	For Raman, the C=O sym str usually occurs at 1710–1625 cm^{-1}
α,β-Unsaturated aliphatic carboxylic acids (as dimer)	1715–1690	5.83–5.92	vs	w–m	Band for triple bond compounds usually at 1690–1680 cm^{-1}
α-Halo-carboxylic acids (as dimer)	1740–1715	5.75–5.83	vs	w–m	(Band for –CF$_2$COOH is at 1785–1750 cm^{-1})
Intramolecular hydrogen-bonded carboxylic acid	1680–1650	5.95–6.06	vs	w–m	Sharp–medium width band. For amino acids see refs: 164–168 and Chapter 23
Saturated dicarboxylic acids	1740–1700	5.75–5.88	vs	w–m	Sometimes broad
α-Unsaturated dicarboxylic acids	1700–1685	5.88–5.94	vs	w–m	Sometimes broad
Peroxy acids, –CO–OOH	1760–1730	5.68–5.77	vs	w–m	
γ-Ketocarboxylic acids	1750–1700	5.71–5.88	vs	w–m	Compounds exist in keto–lactol equilibrium, 2 or 1 band(s)
Thiol acids, –COSH	1700–1690	5.88–5.92	s	w–m	Also see ref. 99 (band due to C–S stretching vibration at 990–945 cm^{-1})
R$_2$N·CH$_2$–COOH	1730–1700	5.78–5.88	vs	w–m	

weak or absent for hydroxy aliphatic acids, but is often more prominent and narrow for aromatic acids.

In the solid phase, the spectra of aliphatic long-chain carboxylic acids exhibit band patterns in the range 1345–1180 cm^{-1} (7.43–8.47 μm). The number of these almost equally-spaced weak bands is related to the length of the aliphatic chain.[87,88] For acids with an even number of carbon atoms, the number of bands observed equals half the number of carbon atoms. For acids with an odd number of carbon atoms, the number of bands is half (the number of carbon atoms plus one). Unfortunately, the band due to the C–O stretching vibration also occurs in this region so that these weak bands may appear as shoulders.

Carboxylic acids have an out-of-plane deformation band in the region 970–875 cm^{-1} (10.42–11.43 μm) which is of medium intensity. Most carboxylic acids have a medium-to-weak band in the region 680–480 cm^{-1} (14.70–20.83 μm) due to the CO out-of-plane deformation.

Normal-aliphatic monocarboxylic acids,[89] except those smaller than n-butyric acid, exhibit, in liquid-phase spectra, three strong bands that are not usually well-resolved in the region 675–590 cm^{-1} (14.81–16.95 μm) due to the in-plane vibration of the O–CO group. In addition, a strong band is found at 495–465 cm^{-1} (20.20–21.51 μm) which is attributed to the in-plane vibration of the C–C=O group. This may be coalesced with

a sharp, strong band observed at about 500 cm^{-1} (20.00 μm). If branching occurs, it affects the position of these bands, as does the physical state of the sample. For example, the in-plane vibrations mentioned occur, in the solid phase, at 680–625 cm^{-1} (14.71–16.00 μm) and 550–525 cm^{-1} (18.18–19.05 μm). α-Branched aliphatic carboxylic acids have three strong bands in the region 665–610 cm^{-1} (15.04–16.39 μm) and a strong band in the region 555–520 cm^{-1} (18.02–19.32 μm). Other branched monocarboxylic acids have three medium-to-strong bands in the region 700–600 cm^{-1} (14.29–16.67 μm).

In the far infrared spectra of acetic acid derivatives,[90–92] a band due to the deformation of the OH···O group is observed at 185–100 cm^{-1} (57.14–100.00 μm) for monosubstituted compounds, at 125–95 cm^{-1} (80.00–105.26 μm) for disubstituted compounds and at 105–80 cm^{-1} (95.24–125.00 μm) for trisubstituted compounds. A study of halogenated acids has been published.[93]

Aromatic acids and esters have a medium-to-strong band at 570–495 cm^{-1} (17.54–20.20 μm) which is due to the rocking vibration of the CO$_2$ group. They also have a band of medium-to-strong intensity which is usually broad for acids and is observed at 370–270 cm^{-1} (27.03–37.04 μm). For *para*-substituted aromatic acids, the bending vibration of the CO$_2$ group results in a band at 620–610 cm^{-1} (16.13–16.39 μm).

Table 10.9 Carboxylic acids: other vibrations

Functional Groups	Region		Intensity		Comments
	cm^{-1}	μm	IR	Raman	
−OH (associated carboxylic acids)	3300−2500	3.00−4.00	m	w	br, −OH str, hydrogen bonding present multiple structure
−OH (free carboxylic acid)	3580−3500	2.79−2.86	w−m	w	sh, as monomer
−OD (deuterated carboxylic acids)	690−650	14.49−15.38	v	w−m	O−D out-of-plane def vib, usually broad
Carboxylic acids, −COOH (dimer)	1440−1395	6.95−7.17	w	w−m	Combination band due to C−O str and O−H def
	1320−1210	7.58−8.26	m−s	w−s	C−O str, sometimes a doublet
	970−875	10.31−11.43	m	w−m	O−H···O out-of-plane def vib, usually broad
	800−630	12.50−15.87	m−w	m−w	CO def vib
	680−480	14.70−20.83	m−w	m−w	Out-of-plane CO def vib
	545−385	18.35−25.97	m−w		Rocking vib
Carboxylic acids, −COOH (monomer)	1380−1280	7.25−7.81	m−s	m−w	O−H def vib
	1190−1075	8.40−9.39	s	w	C−O str
	960−875	10.42−11.43	m	w−m	O−H···O out-of-plane def vib, usually broad
Long-chain aliphatic carboxylic acids	1345−1180	7.43−8.48	w	w−m	CH$_2$ def vib, number of bands determined by aliphatic chain length
Peracids, −CO−OOH	~3280	~3.05	m	w	O−H str
	~950	~10.53	m	w−m	O−H out-of-plane bending vib
	900−700	11.11−14.29	w	s	O−O str
Thiol acids, −CO−SH	2595−2560	3.81−3.91	w	s, p	S−H str
	~950	~10.53	s	s	CS str
	910−825	10.99−12.12	s−m	m−w	In-plane CSH def vib
	750−500	13.33−20.00		s	
	465−430	21.50−23.26		s−m	def vib
n-Aliphatic monocarboxylic acids	675−590	14.81−16.95	s	m−w	O−CO in-plane def vib, three bands usually at ~665, ~630, and ~600 cm^{-1}
	~500	~20−00	s		sh
	495−465	20.20−21.51	s	m−w	C−C=O in-plane def vib
α-Branched aliphatic monocarboxylic acids	665−610	15.04−16.39	s	m−w	O−CO in-plane def vib, three bands usually at ~655, ~635, and ~620 cm^{-1}
	555−520	18.18−19.23	s	m−w	C−CO in-plane def vib
β- and γ-branched aliphatic monocarboxylic acids	700−600	14.29−16.67	s		Three bands
	495−465	20.20−21.51	s		
Aromatic carboxylic acids	1000−900	10.00−11.11	m	m−s	
	820−720	12.20−13.89	m−w	m−w	OH def vib, br
	715−605	13.99−16.53	m−w	m−w	CO$_2$ in-plane def vib
	570−495	17.54−20.20	m−s		CO$_2$ out-of-plane rocking def vib
	370−270	27.03−37.04	m−s		br, esters also absorb in this region but band usually narrower

Table 10.10 Carboxylic acid salts (solid-phase spectra)

Functional Groups	Region		Intensity		Comments
	cm^{-1}	μm	IR	Raman	
Carboxylic acid salts, $-CO_2^-$	1695–1540	5.90–6.49	s	w	asym CO_2^- stretching. Excludes $CX_3CO_2^-$ where X = halogen
	1440–1335	6.94–7.49	m–s	m–s, p	br, sym CO_2^- stretching, usually two or three peaks
	860–615	11.63–16.26	m		CO_2^- scissor vib
	700–450	14.29–22.22	v		CO_2^- wagging vib
	590–350	16.95–28.57	w–m		Rocking vib
Acetate salts	1600–1550	6.25–6.45	s, br	w	asym CO_2^- str
	1440–1400	6.94–7.14	m	m–s, p	sym CO_2^- str
	~1050	~9.52	w	w	
	~1020	~9.80	w	w	
	~925	~10.81	w	s	
$-CF_2CO_2^-$	1695–1615	5.90–6.19	s	w	
	1450–1335	6.90–7.49	m	m–s, p	
Thiol acid salts, $-CO-S^-$	~1525	~6.56	s, br		COS^- str
Monothiol carbonic acid salts, $R-O-CO-S^-$	~1580	~6.33	s		COS^- str, see ref. 100
(Sat)-carboxylic acid salts	1610–1550	6.21–6.45	s, br	w	asym CO_2^- str
	1440–1355	6.94–7.38	m–s	m–s, p	br, sym CO_2^- str
	790–610	12.66–16.39	m		CO_2^- def vib
	625–505	16.00–19.80	v		CO_2^- wagging vib
	490–370	20.41–27.03	w–m		CO_2^- rocking vib
	200–80	50.00–125.00			Torsional vib
α-Halo-carboxylic acid salts	1675–1580	5.97–6.33	s, br	w–m	Fluoro compounds at higher end of frequency range
$-CCl_2CO_2^-$	1680–1640	5.95–6.10	s, br	w	
Aromatic acid salts, $ArCO_2^-$	1605–1525	6.23–6.56	s, br	w	asym CO_2^- str. (α,β-unsat.compds, 1620–1550 cm^{-1})
	1445–1375	6.92–7.27	m–s	m–s	br, sym CO_2^- str.
	860–730	11.63–13.70	m		def vib. (α,β-unsat.compds, 855–625 cm^{-1})
	700–640	14.29–15.63	v		Wagging vib. (α,β-unsat.compds 590–440 cm^{-1})
	580–450	17.24–22.22	w–m		Rocking vib. (α,β-unsat.compds 550–410 cm^{-1})
	245–145	40.82–68.97			Torsional vib
Ammonium carboxylates	1630–1620	6.14–6.17	s, br		Solution
α-Amino carboxylates, $R_2N\cdot CH_2\cdot COO^-$	1595–1575	6.27–6.17	s		Solution. For amino acids, proteins and peptides see refs: 164–168 and 179 respectively plus Chapter 23
3,4-dihydroxy-3-cyclobutene-1,2-dione ion, $C_4O_4^{2-}$	~1530	~6.54	vs	w–m	br. C–O str
	~1090	~9.17	s	m	C–C str
	~660	~15.15	vw		C–O def vib
	~360	~27.78	m		C–O def vib
	~260	~38.46	s		C–O def vib

The C=C stretching vibration band of α,β-unsaturated acids occurs at 1660–1630 cm^{-1} (6.02–6.14 µm), *trans* isomers absorbing 10–20 cm^{-1} higher than *cis* isomers.

Carboxylic Acid Salts

Carboxylic acid salts[94-97] have a very strong, characteristic band in the region 1695–1540 cm^{-1} (5.90–6.49 µm) due to the asymmetric stretching vibration of CO_2^-. The symmetric stretching vibration of this group gives rise to a band in the range 1440–1335 cm^{-1} (6.94–7.49 µm) and is of medium intensity, broad, and generally has two or three peaks. Unfortunately, water, which may be present in the sample, has an absorption at around 1640 cm^{-1} (6.10 µm) and may cause difficulties in identification, as might also the presence of primary or secondary amides due to their amide II band which also occurs in this region. However, Raman spectroscopy does not suffer from these problems.

The asymmetric and symmetric stretching bands for the acetate ion occur at about 1580 cm^{-1} (6.33 µm) and 1425 cm^{-1} (7.02 µm) respectively and weak bands are also observed near 1050 cm^{-1} (9.52 µm), 1020 cm^{-1} (9.80 µm) and 925 cm^{-1} (10.81 µm). Formate salts absorb near 2830 cm^{-1} (3.53 µm), 1600 cm^{-1} (6.25 µm), 1360 cm^{-1} (7.35 µm) and 775 cm^{-1} (12.90 µm).

Table 10.11 Carboxylic acid anhydride C=O stretching vibrations

Functional Groups	Region		Intensity		Comments
	cm^{-1}	µm	IR	Raman	
Saturated aliphatic acid anhydrides	1850–1800	5.41–5.56	vs	m–w	Asymmetric stretching
	1790–1740	5.59–5.75	s	m–w	Symmetric stretching
Aryl and α,β-unsaturated acid anhydrides	1830–1780	5.46–5.62	vs	m–w	
	1755–1710	5.70–5.85	s	m–w	
Saturated five-membered ring acid anhydrides	1870–1820	5.35–5.50	s	m–w	Separation ~70 cm^{-1} except for aromatic compounds for which it is ~50 cm^{-1}
	1800–1775	5.56–5.63	vs	m–w	
α,β-unsaturated five-membered ring acid anhydrides	1860–1850	5.38–5.41	s	m–w	asym C=O str (Raman: strong ring vib band 655–640 cm^{-1})
	1780–1760	5.62–5.68	vs	m–w	sym C=O str
Saturated six-membered ring acid anhydrides	1820–1780	5.49–5.62	s	m–w	Separation ~40 cm^{-1}
	1780–1740	5.62–5.75	vs	m–w	

Table 10.12 Carboxylic acid anhydrides: other bands

Functional Groups	Region		Intensity		Comments
	cm^{-1}	µm	IR	Raman	
Acyclic aliphatic and cyclic six-membered ring acid anhydrides	1135–980	8.81–10.20	s	v	C–O–C str (good negative indicator), often a doublet at ~1050 cm^{-1}
Cyclic five-membered ring acid anhydrides	1310–1210	7.63–8.26	s	m–s	C–O–C str (Raman ring vib at 660–625 cm^{-1})
	955–895	10.47–11.17	s	m	C–O–C str

For acid salts with a strongly electron-withdrawing group, such as CF_3, the asymmetric stretching vibration band may be found outside the normal range quoted and as high as $1690\,cm^{-1}$ ($5.92\,\mu m$). The symmetric vibration band for $CF_3COO^- Na^+$ occurs at about $1450\,cm^{-1}$ ($6.90\,\mu m$), for $CBr_3COO^- Na^+$ at about $1340\,cm^{-1}$ ($7.46\,\mu m$) and $1355\,cm^{-1}$ ($7.38\,\mu m$) (two bands) and for acetic acid salts at about $1425\,cm^{-1}$ ($7.02\,\mu m$).

The asymmetric CO_2^- stretching frequency increases with the electron-withdrawing ability of directly attached groups but is not greatly affected by the mass of the group, whereas the symmetric CO_2^- stretching vibration is affected by mass (increasing the mass results in the frequency decreasing) and is not greatly affected by polar effects. The stretching vibration of the $-CO_2^-$ group depends on both the metal ion and the organic portion of the salt.

Due to the rocking in-plane and out-of-plane deformation vibrations of the carboxylic ion, medium-to-strong bands are observed in the region $760-400\,cm^{-1}$ ($13.16-25.00\,\mu m$).

The salts of complexes of carboxylic acids and their derivatives are reviewed elsewhere.[98]

Carboxylic Acid Anhydrides, $-CO-O-CO-$

Due to the asymmetric and symmetric stretching vibrations of the two $C{=}O$ groups, saturated aliphatic anhydrides[21,31] absorb at $1850-1800\,cm^{-1}$ ($5.41-5.56\,\mu m$) and at $1790-1740\,cm^{-1}$ ($5.59-5.75\,\mu m$) respectively, both bands being sharp and strong. In most cases, these two bands are separated by about $60\,cm^{-1}$ ($0.18\,\mu m$). For acyclic anhydrides, the higher frequency band is usually the more intense.[8] The presence of conjugation results in a shift of about $20\,cm^{-1}$ downward ($0.05\,\mu m$ upward) for both bands. α,β-Unsaturated acid anhydrides and aryl anhydrides absorb at $1830-1780\,cm^{-1}$ ($5.46-5.62\,\mu m$) and at $1755-1710\,cm^{-1}$ ($5.70-5.85\,\mu m$). All these frequencies are increased in strained-ring situations and also by electronegative atoms on the α-carbon atom.

Acid anhydrides also have a strong band in the range $1135-980\,cm^{-1}$ ($8.81-10.20\,\mu m$) due to the $C-O-C$ stretching vibration which appears at $1310-1210\,cm^{-1}$ ($7.63-8.26\,\mu m$) for strained-ring compounds (five-membered ring anhydrides). Straight-chain alkyl anhydrides absorb in the narrow range $1050-1040\,cm^{-1}$ ($9.52-9.62\,\mu m$), the band usually being broad, an exception to this being acetic anhydride which absorbs at about $1135\,cm^{-1}$ ($8.81\,\mu m$). Acyclic anhydrides absorb at about $1050\,cm^{-1}$ ($9.52\,\mu m$), but branching at the α-carbon atom tends to decrease the frequency of this vibration. Cyclic anhydrides[14,30] (five-membered ring) have a strong band at $955-895\,cm^{-1}$ ($10.47-11.17\,\mu m$) and often a weak band near $1060\,cm^{-1}$ ($9.44\,\mu m$) is observed also. Unconjugated cyclic anhydrides absorb strongly at $1130-1000\,cm^{-1}$ ($8.85-10.00\,\mu m$). Aromatic anhydrides absorb in the region $1150-1050\,cm^{-1}$ ($8.70-9.52\,\mu m$). All these bands are believed to involve the stretching vibration of the $C-O-C$ group.

Carboxylic Acid Halides, $-CO-X$

Due to the $C{=}O$ stretching vibration, aliphatic acid chlorides[101,174,182,183] absorb strongly in the region $1830-1770\,cm^{-1}$ ($5.46-5.65\,\mu m$). Acid bromides and iodides absorb in the same region or at very slightly lower wavenumbers than acid chlorides, whereas the fluorides absorb at about $50\,cm^{-1}$ higher ($0.16\,\mu m$ lower). Some α-methyl substituted acid halides exhibit a doublet.

Aryl[46,102,103] and α,β-unsaturated acid halides[11] (of Cl, Br, I) absorb in the range $1795-1735\,cm^{-1}$ ($5.57-5.76\,\mu m$) with fluorides absorbing at higher wavenumbers. In non-polar solvents, a double peak is often observed for aryl acid halides. The second band is probably an overtone band of the strong band which occurs at about $850\,cm^{-1}$ ($11.76\,\mu m$). Fluorides exhibit a single band. The carbonyl stretching vibration frequency for α,β-unsaturated acid halides has been observed to decrease in the order

$$\text{fluoride} > \text{bromide} > \text{chloride}.$$

Compounds with one or more halogen atoms directly bonded to a carbonyl group absorb strongly, due to the carbonyl stretching vibration, in the region $1900-1790\,cm^{-1}$ ($5.26-5.59\,\mu m$), $F_2{\cdot}CO$ absorbing outside this range at about $1930\,cm^{-1}$ ($5.18\,\mu m$). For saturated aliphatic acid chlorides, a strong $C-Cl$ stretching band is observed at $780-560\,cm^{-1}$ ($12.82-17.86\,\mu m$) and the in-plane deformation bands, which are of medium-to-strong intensity, are observed at $490-230\,cm^{-1}$ ($20.41-43.48\,\mu m$). Benzoyl chlorides have a strong band at $900-800\,cm^{-1}$ ($11.11-12.50\,\mu m$) due to the $C-Cl$ and phenyl-C stretching vibrations. Most acid chlorides exhibit a strong band due to the $C-Cl$ stretching vibration at $900-600\,cm^{-1}$ ($11.11-16.67\,\mu m$).

Diacyl Peroxides, $R-CO-O-O-CO-R$, (Acid Peroxides), and Peroxy Acids, $-CO-OO-H$

All acid peroxides[21] have a weak absorption band in the region $900-800\,cm^{-1}$ ($11.11-12.50\,\mu m$) due to the $-O-O-$stretching vibration. Acid peroxides also have strong bands due to their $C{=}O$ stretching vibration. For saturated aliphatics, two bands are usually observed, one at $1820-1810\,cm^{-1}$ ($5.50-5.53\,\mu m$) and the other at $1800-1780\,cm^{-1}$ ($5.56-5.62\,\mu m$), this latter band being more intense. For aryl and α,β-unsaturated acid peroxides, these bands occur at $1805-1780\,cm^{-1}$ ($5.54-5.62\,\mu m$) and $1785-1755\,cm^{-1}$ ($5.60-5.70\,\mu m$). The nature and position of the substituent(s) in the aromatic

Table 10.13 Carboxylic acid halide C=O stretching vibrations

Functional Groups	Region cm⁻¹	Region μm	Intensity IR	Intensity Raman	Comments
	cm⁻¹	μm	IR	Raman	
Saturated aliphatic acid chlorides	1830–1770	5.46–5.65	vs	m–w, p	Mostly 1815–1785 cm⁻¹. Generally, fluorides at higher wavenumbers, bromides and iodides slightly lower
Aryl and α,β-unsaturated acid chlorides	1795–1765	5.57–5.66	vs	m, p	
	1750–1735	5.71–5.76	m	m	Involves overtone of band at 890–850 cm⁻¹
(Sat)·COF	1900–1790	5.26–5.59	s	m	
–CF₂COF	1900–1870	5.26–5.35	s	m–w	See ref. 104
–CF₂COCl	1820–1795	5.50–5.57	s	m–w	
–CO·COF	1900–1850	5.26–5.41	s	m–w	
–CO·COCl	1845–1775	5.42–5.63	s	m–w	
–COBr	1830–1730	5.46–5.78	s	m	
Aliphatic (saturated)·CO·Br	1830–1770	5.46–5.65	s	m	
(α,β-Unsaturated)·CO·Br	1795–1735	5.57–5.76	s	m	See ref. 182
Acetyl chloride complexes, e.g. CH₃COCl·AlCl₃	~1635	~6.12	s		Also strong bands ~2305 and ~2205 cm⁻¹. 2305 cm⁻¹ band possibly due to CH₃COCl⁺ ion, (possibly ⁺C=O contribution)
Acid halide complexes, CH₃CO⁺A⁻, A = BF₄, SbF₆, AsF₆	~1620	~6.17	m–w		Very strong band at 2230–2300 cm⁻¹ due to CH₃CO⁺
	~1555	~6.43	v		
–ArCO⁺A⁻, A = SbF₆, AsF₆	~1540	~6.49	m		Very strong band ~2225 cm⁻¹ due to ArCO⁺

Table 10.14 Carboxylic acid halides: other bands

Functional Groups	Region cm⁻¹	Region μm	Intensity IR	Intensity Raman	Comments
	cm⁻¹	μm	IR	Raman	
Saturated aliphatic acid chlorides	965–920	10.36–10.87	m		C–C=str
	780–560	12.82–17.86	s	m, p	C–Cl str
	670–480	14.93–20.83	w–m	m	CO/CCl def vib
	520–410	19.23–24.39	w–m		CO/CCl def vib
	450–230	22.22–43.48	s–m	s, p	Cl–C=O in-plane def vib
Unsaturated acid chlorides	800–600	12.50–16.67	s	m–s, p	C–Cl str
	760–620	13.16–16.13	s	m–s, p	C–Cl str
Aryl acid chlorides	~1200	~8.33	m	m	C–C str
	930–800	10.75–12.50	s	m–s, p	C–Cl str. Benzoyl chlorides 900–800 cm⁻¹.
	670–570	14.93–17.54	w–m		
	540–420	18.52–23.81	w–m		
	380–280	26.32–35.71			
Acid fluorides	1290–1010	7.75–9.90	s	m–w, p	C–F str
	770–570	12.99–17.54	m	m	CO, CF bending vib
	600–420	16.67–23.81			CO, CF wagging vib
	500–340	20.00–29.41		s, p	CO, CF rocking vib
Acid bromides	850–520	11.76–19.23	s, br	s, p	C–Br str, usually 745–565 cm⁻¹
	680–360	14.71–27.78	w–m		CO/C–Br out-of-plane def vib, usually 560–440 cm⁻¹
	490–310	20.41–32.26	w–m		CO/C–Br in-plane def vib
Saturated acid bromides	700–535	14.29–18.69	s	s, p	C–Br str

Table 10.15 Diacyl peroxide and peroxy acid C=O stretching vibrations

Functional Groups	Region		Intensity		Comments
	cm^{-1}	μm	IR	Raman	
Saturated aliphatic acid peroxides, –CO–O–O–CO–	1820–1810	5.50–5.53	s	m–w	Separation ~25 cm^{-1}, see ref. 21
	1800–1780	5.56–5.62	vs	m–w	
Aryl and α,β-unsaturated acid peroxides	1805–1780	5.54–5.62	s	m–w	
	1785–1755	5.60–5.70	vs	m–w	
Peroxy acids, –CO–OOH	1760–1730	5.68–5.77	vs	m–w	Intramolecular hydrogen bonding

Table 10.16 Diacyl peroxides and peroxy acids: other bands

Functional Groups	Region		Intensity		Comments
	cm^{-1}	μm	IR	Raman	
Peroxides, –O–O–	900–800	11.11–12.50	w	s	All peroxides, O–O str at ~865 cm^{-1} for peroxy acids
Peroxy acids, –CO–OOH	~3280	~3.05	m–s	w	Associated intramolecularly, due to O–H str
	1460–1430	6.85–7.00	m	w–m	O–H bending vib near 1430 cm^{-1} for long-chain linear acids
	1300–1050	7.69–9.52	m–s	w–m	C–O str, often near 1175 cm^{-1}

portion of acid peroxides may significantly influence the position of these bands.

The C–O stretching vibrations are not very useful in the characterization of acid peroxides. They are found in the region 1300–1050 cm^{-1} (7.69–9.52 μm).

Esters, –CO–O–, Carbonates, –O–CO–O–, and Haloformates, –O–CO–X

All esters have two strong characteristic bands,[180] one due to the C=O stretching vibration and the other due to the C–O stretching vibration. The frequency of the C=O stretching vibration for esters is influenced in a very similar way to that observed for ketones, except that the decrease in wavenumber for aliphatic esters due to the presence of α,β-unsaturation[105] is less, being approximately 10–20 cm^{-1}.

Ester C=O Stretching Vibrations

With the exception of formates,[113] which absorb in the region 1730–1715 cm^{-1} (5.78–5.83 μm), saturated aliphatic esters absorb at

1750–1725 cm^{-1} (5.71–5.80 μm).[9,62,114] Electronegative groups or atoms directly bonded to the alcoholic oxygen atom of the ester group tend to increase the frequency of the C=O stretching vibration. Aryl and α,β-unsaturated esters ($\overset{\beta}{C}=\overset{\alpha}{C}–CO–O–$)[112,175,178] absorb at 1740–1705 cm^{-1} (5.75–5.83 μm).

Further conjugation has almost no effect on the C=O stretching vibration frequency. Strongly polar groups substituted on the benzene ring of aryl esters tend to increase the frequency of the C=O stretching vibration.

Esters with electronegative α-substituents ($\overset{\diagdown}{X}C–CO–O–$), e.g. α-halo-esters,[7,106,107] absorb at 1770–1730 cm^{-1} (5.65–5.78 μm), i.e. about 10–20 cm^{-1} higher than for the normal aliphatic ester. α,α'-Dihalo-esters[107] also absorb in the same region but, in general, two closely-spaced bands are observed. Vinyl and phenyl esters (–CO–O–C=C$\diagup$) absorb at about 1770 cm^{-1} (5.65 μm), e.g. vinyl acetate absorbs at 1760 cm^{-1} (5.63 μm).

No change is observed in the position of the band due to the C=O stretching vibration when a carbonyl group is present in the α-position of an ester, –CO–CO–O–, e.g. α-keto-esters and α-diesters both absorb in the range 1760–1740 cm^{-1} (5.68–5.75 μm). In general, for esters of saturated dicarboxylic acids, the C=O band occurs in approximately the same range, 1760–1735 cm^{-1}, (5.68–5.76 μm), as for monoesters and the same influences on the position of this band are observed. If the two ester groups[62,122] are close together in the molecule then a doublet is observed, otherwise a single band is observed. Geminal diesters may absorb at slightly higher wavenumbers than those given above. A study of glycidic esters has been published.[123]

With β-keto-esters, $-CO-\overset{|}{C}H-CO-O-$, keto–enol tautomerism is possible:

$$CO-\underset{|}{C}H-CO-O- \rightleftharpoons -\underset{|}{C}=\overset{OH}{\underset{|}{C}}-CO-O-$$

In this case, a strong band at about 1650 cm^{-1} (6.07 μm) is observed due to the C=O stretching vibration of the hydrogen-bonded C=O group, i.e.

$$-\underset{|}{C}=\overset{\overset{OH----O}{|\quad\quad\|}}{C}-C-O-$$

The C=O stretching vibration frequency is lowered due to the presence of the hydrogen bonding. There is also a band due to the C=C stretching vibration at about 1630 cm^{-1} (6.14 μm) and a sharp band due to the O–H stretching vibration at 3590–3420 cm^{-1} (2.79–2.92 μm). In addition, other bands due to the carbonyl stretching vibration, etc., may be observed, these being due to the keto- form of the ester. The relative intensities of these bands of the β-keto-esters depend on the relative amounts of each tautomer.

Due to intramolecular hydrogen bonding, o-hydroxyl (or o-amino-) benzoates absorb at 1690–1670 cm^{-1} (5.92–5.99 μm).

The effect of converting a methyl ester to a phenyl ester is normally to increase the wavenumber of the band due to the carbonyl stretching vibration by 10–20 cm^{-1} (a decrease of 0.03–0.07 μm).

Intensity correlations for the carbonyl band of esters have been studied extensively.[6,10,15,109]

The carbonyl band for aliphatic chloroformates (–CO·Cl)[53,105,120,121] is observed at higher wavenumbers than that for esters, at about 1780 cm^{-1} (5.62 μm), and for aryl chloroformates at about 1785 cm^{-1} (5.60 μm).

Most noncyclic carbonates[53,108–110] absorb strongly in the region 1790–1740 cm^{-1} (5.59–5.75 μm) whilst five-membered-ring cyclic carbonates[27,32] absorb at 1850–1790 cm^{-1} (5.41–5.59 μm).

The carbonyl band of thiol carbonyl esters –S–CO–[54,111] occurs at lower frequencies than that of normal esters.

A weak band due to the overtone of the C=O stretching vibration of esters occurs at about 3450 cm^{-1} (2.90 μm) and may sometimes be used in confirming the presence of a C=O group.

Ester C–O–C Stretching Vibrations

The bands due to the ester C–O stretching vibration are strong, partly due to an interaction with the C–C vibration, and occur in the range 1300–1100 cm^{-1} (7.69–9.09 μm). Often a series of strong overlapping bands is observed. Caution is required when using these bands in making assignments since the C–O stretching vibrations of alcohols and acids, and possibly ketones also, occur in this region.

The band due to the C–O–C asymmetric stretching vibration for aliphatic esters occurs at 1275–1185 cm^{-1} (7.85–8.44 μm) and that due to the symmetric stretching vibration occurs at 1160–1050 cm^{-1} (8.62–8.70 μm). Both these bands are strong, the former band being usually more intense than that due to the C=O stretching vibration.

Esters of aromatic acids and α,β-unsaturated aliphatic acids have two strong absorption bands, one at 1310–1250 cm^{-1} (7.63–8.00 μm) and the other at 1200–1100 cm^{-1} (8.33–9.09 μm). For esters G–CO·OG′, where G′ is an aromatic or α,β-unsaturated group, a very strong absorption near 1210 cm^{-1} (8.26 μm) is observed. If, in addition, the other group G is aromatic in nature, then the band due to the asymmetric stretching vibration occurs at 1310–1250 cm^{-1} (7.64–8.00 μm) and that due to the symmetric stretching vibration at 1150–1080 cm^{-1} (8.70–9.26 μm). The C–O stretching vibration frequencies do not appear to vary as much as in alcohols, ethers, and acids. Some of the C–O asymmetric stretching vibration band positions are given in Table 10.17. Although it is not possible to distinguish between neighbouring esters in a homologous series, Table 10.17 is still useful in a more general sense.

Table 10.17 Some C–O asymmetric stretching vibration band positions

Ester	Approximate position cm^{-1}	μm	Ester	Approximate position cm^{-1}	μm
Formates	1190	8.40	Acetates	1245	8.03
Propionates	1190	8.40	n-Butyrates	1200	8.33
Isobutyrates	1200	8.33	Isovalerates	1195	8.33
Adipates	1175	8.51	Oleates	1170	8.54
Stearates	1175	8.51	Citrates	1180	8.46
Sebacates	1170	8.53	Laurates	1165	8.59
Benzoates	1280	7.81	Phthalates	1120	8.93
	(sym) 1120	8.91		(sym) 1070	9.35

The position of the band due to the C–O stretching vibration is dependent on the nature of both the acidic and the alcoholic components, although the latter is less important. Alkyl chloroformates have a very strong band due to the asymmetric COC stretching vibration at 1200–1130 cm^{-1} (8.33–8.85 µm), a strong band is also observed at 850–770 cm^{-1} (11.76–2.99 µm).

Methyl esters of long-chain aliphatic acids normally exhibit three bands, the strongest of which is at 1175 cm^{-1} (8.50 µm), the others being near 1250 cm^{-1} (8.00 µm) and 1205 cm^{-1} (8.30 µm).

Acetates of primary alcohols have a medium-intensity band at 1060–1035 cm^{-1} (9.39–9.64 µm) due to the asymmetric stretching of the O–CH$_2$–C group. For acetates of other than primary alcohols, this band is shifted to higher wavenumbers.

Other Ester Bands

Acetates[9] have a medium-to-strong band near 1375 cm^{-1} (7.30 µm), due to the CH$_3$ symmetric deformation, and medium-to-weak bands near 1430 cm^{-1} (6.99 µm) and 2990 cm^{-1} (3.34 µm), due to the asymmetric deformation and stretching vibrations respectively of this group. For other saturated esters containing the –CH$_2$CO–O– group, the CH$_2$ deformation band occurs near 1420 cm^{-1} (7.04 µm).

Most aliphatic esters[66,114] have bands in the regions 645–585 cm^{-1} (15.50–17.09 µm) and 350–300 cm^{-1} (28.57–33.33 µm).

All acetates absorb strongly at 665–635 cm^{-1} (15.04–15.75 µm) due to the bending of the O–C–O group and at 615–580 cm^{-1} (16.29–17.24 µm) due to the out-of-plane deformation vibration of the acetate group. A band at 325–305 cm^{-1} (30.77–32.79 µm) is also often observed. This last band decreases in intensity with increase in molecular weight.

Branched alkyl formates absorb at 520–485 cm^{-1} (19.23–20.62 µm) and 340–285 cm^{-1} (29.41–35.09 µm), whereas n-alkyl formates (ethyl to amyl) have three bands: near 620 cm^{-1} (16.13 µm), in the region 475–460 cm^{-1} (21.05–21.74 µm), and near 340 cm^{-1} (29.41 µm). This last band is always strong and the first, weak. The first two (higher-frequency) bands decrease in intensity as the molecular weight of the formate increases.

Methyl esters[88] have bands near 2960 cm^{-1} (3.38 µm) and 1440 cm^{-1} (6.94 µm) due to the CH$_3$ asymmetric stretching and deformation vibrations and weak bands near 1425 cm^{-1} (7.02 µm) and 1360 cm^{-1} (7.35 µm). In addition, with the exception of the formate and isobutyrate, methyl esters have a band of medium intensity at 450–430 cm^{-1} (22.22–23.26 µm). The characteristic absorptions of methyl and ethyl esters are given in Table 10.18.

For the $\overset{\text{R}}{\underset{\ }{|}}$ $-\text{O}-\text{CH}-\text{CH}_3$ group, a medium intensity band is observed near 1380 cm^{-1} (7.25 µm).

n-Propyl esters have a band near 1390 cm^{-1} (7.19 µm) and bands of variable intensities at 605–585 cm^{-1} (16.53–17.09 µm), near 495 cm^{-1} (20.20 µm) and at 350–340 cm^{-1} (28.57–29.41 µm). The band near 600 cm^{-1} is not present for the formate. Isopropyl esters have bands of variable intensity at 605–585 cm^{-1} (16.53–17.09 µm) and 505–480 cm^{-1} (19.80–20.83 µm) and strong bands near 435 cm^{-1} (22.99 µm) and at

Table 10.18 Characteristic absorptions of formates, acetates, methyl and ethyl esters (excluding C=O stretching vibrations)

Functional Groups	Region cm^{-1}	Region µm	Intensity IR	Intensity Raman	Comments
Formates	2970–2890	3.37–3.46	w–m	w	CH str
	1385–1350	7.22–7.41	w–m	m, p	CH def vib
	1210–1120	8.26–8.93	s	w	CH in-plane def vib
	1070–1010	9.35–9.90	s–m	m	CO–O str
	775–620	12.90–16.13	m	m, p	CH out-of-plane def vib/O–C=O in-plane def vib
	410–230	24.39–43.48		m, p	C–O–R in-plane def vib
	145–65	68.97–153.85			Torsional vib
Acetates	3040–2980	3.29–3.36	w	m–s	asym CH$_3$ str
	3030–2940	3.30–3.40	w	m–s	asym CH$_3$ str
	2960–2860	3.38–3.50	w	m–s	sym CH$_3$ str
	1465–1415	6.83–7.08	m–w	m–w	sym CH$_3$ def vib
	1460–1400	6.85–7.14	m–w	m–w	asym CH$_3$ def vib
	1390–1340	7.19–7.46	m–s	m	sym CH$_3$ def vib
	1265–1205	7.91–8.30	vs	m–s	CO–O str
	1080–1020	9.26–9.80	w–m	w	CH$_3$ rocking vib

Table 10.18 (*continued*)

Functional Groups	Region cm⁻¹	μm	Intensity IR	Raman	Comments
	1025–930	9.76–10.75	w	w	CH₃ rocking vib
	910–810	10.99–12.35	w	m–s	CC str
	665–590	15.04–16.95	v	w	C=O def vib
	620–580	16.13–17.24	v	w	C=O def vib
	465–365	21.50–27.40	w		CCO def vib
	325–230	30.77–43.48	v		COR def vib
	210–110	47.62–90.91			Torsional vib
Methyl esters, (sat.)–CO·OCH₃	3050–2980	3.28–3.36	w–m	m–s	asym CH₃ str
	3030–2950	3.30–3.39	w–m	m–s	asym CH₃ str
	3000–2860	3.33–3.50	m	m–s	sym CH₃ str
	1485–1435	6.73–6.97	m	m–w	asym CH₃ def vib
	1465–1420	6.83–7.04	m–s	m–w	asym CH₃ def vib
	1460–1420	6.85–7.04	w–m	m–w	sym CH₃ def vib
	1220–1150	8.20–8.70	v	w	Rocking vib, generally w–m
	1190–1120	8.40–8.93	v	w	Rocking vib, generally w–m
	290–160	34.48–62.50			
Methyl esters, (unsat.)–CO·OCH₃	3020–3055	3.31–3.27	w–m	m–s	asym CH₃ str
	2975–2925	3.36–3.42	w–m	m–s	asym CH₃ str
	2880–2820	3.47–3.55	w–m	m–s	sym CH₃ str
	1475–1445	6.78–6.92	m–s	m–w	asym CH₃ def vib
	1235–1145	8.10–8.73	w–m	w	Rocking CH₃ vib
	1180–1120	8.47–8.93	w–m	w	Rocking vib
Ethyl esters, –O–CH₂CH₃	2995–2930	3.34–3.41	m	m–s	asym CH₃ & CH₂ str
	2930–2890	3.41–3.46	w	m–s	sym CH₃ str
	2920–2860	3.42–3.50	w	m–s	CH₃ str
	1490–1460	6.71–6.85	m–w	m–w	OCH₂ def vib
	1480–1435	6.76–6.97	m	m–w	asym CH₃ def vib
	1390–1360	7.19–7.35	m–s	w–m	sym CH₃ def vib
	1385–1335	7.22–7.49	m–w	m–w	CH₂ wagging vib
	1325–1340	7.55–7.46	m–w	m–w	CH₂ twisting vib
	1195–1135	8.37–8.81	w	w	CH₃ rocking vib
	1150–1080	8.70–9.26	w	w	CH₃ rocking vib
	940–850	10.64–11.76	w	m	C–C str
	825–775	12.12–12.90	w	w	CH₂ rocking vib
	755–625	13.25–16.00	w		CO in-plane def vib
	700–550	14.29–18.18	w	w	CO out-of-plane def vib
	485–365	20.62–27.40	w–m		CO–O rocking vib
	395–305	25.32–32.79	w–m		C–O–O def vib
	370–250	27.03–40.00	w–m		C–O–O def vib

425–410 cm⁻¹ (23.53–24.39 μm), but isopropyl formate exhibits only the band near 435 cm⁻¹.

n-Butyl esters have medium-to-strong absorptions near 505 cm⁻¹ (19.80 μm) and 435 cm⁻¹ (22.99 μm) and a weak band at 350–335 cm⁻¹ (28.57–29.85 μm). Isobutyl esters have a band of medium intensity near

505 cm⁻¹ (19.80 μm), a strong band near 430 cm⁻¹ (23.26 μm), and a band of variable intensity near 385 cm⁻¹ (25.97 μm), the formate and isobutyrate not exhibiting the band near 505 cm⁻¹.

α,β-Unsaturated esters (e.g. acrylates, methacrylates, fumarates) have a band at 695–645 cm⁻¹ (14.39–15.50 μm) due to the wagging vibration of

Table 10.19   Ester, haloformate, and carbonate C=O stretching vibrations

Functional Groups	Region cm^{-1}	Region μm	Intensity IR	Intensity Raman	Comments
Formates	1730–1715	5.78–5.83	vs	m	Usual range, but may be 1760–1690 cm^{-1}
Acetates	1750–1740	5.70–5.75	vs	m	Usual range, but may be 1770–1730 cm^{-1}
Saturated aliphatic esters	1750–1725	5.71–5.80	vs	m	Except for formates
Aryl and α,β-unsaturated aliphatic esters (esters of aromatic acids, etc.), $\underset{}{\overset{\beta\quad\alpha}{\text{C=C–CO–O–}}}$	1740–1705	5.75–5.87	vs	m, p	Usually at lower end of frequency range in cases of olefinic conjugation
Acrylates, CH$_2$=CHCOOR and methacrylates CH$_2$=CCH$_3$COOR	1725–1710	5.80–5.85	vs	m	C=C str at 1640–1630 cm^{-1}
H–C≡CCOOR	1720–1705	5.81–5.87	vs	m–s	
Dialkyl phthalates	1740–1725	5.75–5.80	vs	s, p	
Vinyl and phenyl esters, R—CO—O—C=C	1800–1750	5.56–5.71	vs	m–s	Phenyl acetates at ~1775 cm^{-1}
α-Halo- and α-cyano-esters	1770–1730	5.65–5.78	vs	m	
CH$_2$Cl·COOR	1750–1735	5.71–5.76	vs	m	
CH$_2$Br·COOR	1740–1730	5.75–5.78	vs	m	
CHCl$_2$·COOR	1760–1745	5.68–5.73	vs	m	
CCl$_3$·COOR	1770–1760	5.65–5.68	vs	m	
α,α-Difluoro esters, –CF$_2$CO·O–	1800–1775	5.56–5.63	vs	m	See ref. 106
α-Keto-esters and α-diesters, –CO·COOR	1760–1740	5.68–5.75	vs	m	
β-Keto-ester, enol form, $\underset{\overset{\|}{\text{OH}}}{-\text{C=C–}}\overset{\overset{\|}{\|}}{\underset{\text{O}}{\text{C}}}\text{–OR}$	1655–1635	6.04–6.12	vs		Sometimes broad, usually ~1650 cm^{-1} (intramolecular hydrogen bonding), strong band near 1630 cm^{-1} due to C=C str
o-Hydroxyl (or o-amino-) benzoates	1690–1670	5.92–5.99	vs	m	sh, intramolecular hydrogen bonding
Esters, CH$_3$COOX (X ≠ carbon atom)	1810–1710	5.53–5.85	vs	m	
Aliphatic chloroformates, R–O–CO–Cl	1780–1775	5.62–5.63	vs	m–w, p	Unsaturation tends to increase frequency, strong band near 690 cm^{-1} due to C–Cl str
Aryl chloroformates	1810–1780	5.52–5.62	vs	m–w, p	
Fluoroformates, –O–CO–F	1900–1790	5.26–5.50	vs	m	(C–F str 1290–1010 cm^{-1}, m–s)
Dialkyl oxalates, R–O–CO–CO–O–R'	~1765	~5.67	vs	m	
	~1740	~5.75	vs	m	Absent for *trans* isomers
Diaryl oxalates, Ar–O–CO–CO–O–Ar	~1795	~5.57	vs	m	See ref. 62
	~1770	~5.65	vs	m	Absent for *trans* isomers
Carbamoyl chlorides, NR$_2$COCl	1745–1735	5.73–5.76	vs	m	C–Cl str at 680–600 cm^{-1}
Alkyl and aryl thiol chloroformates, –S–COCl	1775–1765	5.63–5.67	vs	m	
Thiol fluroformates, –S–COF	1850–1790	5.41–5.59	vs	m	
Peresters, –CO–O–O–	1785–1750	5.60–5.71	vs	m	
Dialkyl thiolesters, R–S–CO–R'	1700–1680	5.88–5.62	vs	m	
Alkyl aryl thiolesters:					
Ar–S–CO–R	1710–1690	5.85–5.92	vs	m	
R–S–CO–Ar	1680–1665	5.88–6.01	vs	m	*Ortho*-halogen-substituted compounds absorb at higher frequencies
Diaryl thiolesters	1700–1650	5.88–6.06	vs	m	*Ortho*-halogen-substituted compounds absorb at higher frequencies
Thiol acetates	1770–1680	5.65–5.62	vs	m	Usually 1710–1680 cm^{-1}.
HCO–S–R	~1675	~5.97	vs	m	
HCO–S–Ar	~1700	~5.88	vs	m	

Table 10.19 (*continued*)

Functional Groups	Region		Intensity		Comments
	cm^{-1}	μm	IR	Raman	
Alkyl carbonates, –O–CO–O–	1760–1740	5.68–5.75	vs	m	
Alkyl aryl carbonates	1790–1755	5.59–5.70	vs	m	
Diaryl carbonates	1820–1775	5.50–5.63	vs	m	See ref. 170
Cyclic carbonates (five-membered ring)	1860–1750	5.38–5.71	vs	m	Halogen substitution of ring increases frequency
Cyclic carbohydrate carbonates	1845–1800	5.42–5.56	vs	m	
Dialkyl thiolcarbonates, R–S–CO–O–R	1720–1700	5.81–5.88	vs	m	See ref. 111 [cyclic compounds (five-membered ring) at 1760–1735 cm^{-1}]
Alkyl aryl thiolcarbonates:					
Ar–S–CO–O–R	1730–1715	5.78–5.83	vs	m	
R–S–CO–O–Ar	1740–1730	5.75–5.78	vs	m	
Dialkyl dithiolcarbonates, R–S–CO–S–R	1655–1640	6.04–6.10	vs	m	
Diaryl dithiolcarbonates, Ar–S–CO–S–Ar	1720–1715	5.81–5.83	vs	m	
	1730–1715	5.78–5.83	vs	m	
R–O–CO–NH–R	1740–1730	5.75–5.78	s	w–m	See ref. 117
R–S–CO–NH–R	~1695	~5.90	vs	m–w	
G–S–CO–NH–Ar	1665–1650	6.01–6.06	vs	m–w	
R–S–CO–NH$_2$	~1700	~5.88	vs	m–w	
Silyl esters, R·CO·SiR$_3$	~1620	~6.17	vs	m	

Table 10.20 Ester, haloformate, and carbonate C–O–C stretching vibrations

Functional Groups	Region		Intensity		Comments
	cm^{-1}	μm	IR	Raman	
R–CO·OR′	1275–1185	7.85–8.44	vs	m–s	asym str
	1160–1050	8.62–8.70	s	w	sym str
Formates, H·CO–OR	1215–1180	8.23–8.47	vs	m–s	Also a strong band at 1165–1150 cm^{-1}
Acetates CH$_3$COOR	1265–1205	7.91–8.30	vs	m–s	Often split
Propionates and higher	1200–1150	8.33–8.70	vs	m–s	Two bands in region 1275–1050 cm^{-1} due to asym and sym C–O–C str. Band at higher wavenumbers (asym) usually the more intense
Esters of aromatic acids (e.g. benzoates, phthalates, etc.)	1310–1250	7.63–8.00	vs	m–s, p	asym C–O–C str
	1150–1100	8.70–9.09	s	w	sym C–O–C str
Unsubstituted benzoates, 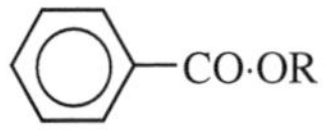 –CO·OR	1280–1270	7.81–7.87	vs	m–s	Weak shoulder at ~1315 cm^{-1}
	~1110	~9.01	s	w	Weak shoulder at ~1175 cm^{-1}
Ortho-substituted benzoates, 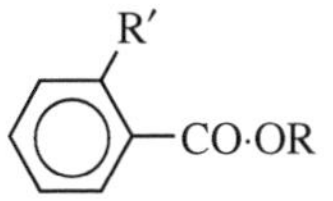 –CO·OR	1265–1250	7.91–8.00	vs	m–s	Shoulder at ~1300 cm^{-1}
	1120–1070	8.93–9.35	s	w	

(*continued overleaf*)

Table 10.20   (*continued*)

Functional Groups	Region		Intensity		Comments
	cm^{-1}	μm	IR	Raman	
Meta-substituted benzoates	1295–1280	7.72–7.81	vs	m–s	Shoulder at ~1305 cm^{-1}
R′					
(◯)–CO·OR					
	1135–1105	8.81–9.05	s	w	
Para-substituted benzoates,	~1310	~7.63	s	m–s	Very strong doublet
R′–(◯)–CO·OR					
	~1275	~7.84	vs	m–s	
	~1180	~8.48	s	w	
	1120–1100	8.93–9.09	s	w	
Dialkyl phthalates	1295–1275	7.72–7.84	vs	s, p	asym COC str
	1170–1115	8.55–8.97	s	w–m, p	sym COC str
α,β-Unsaturated aliphatic esters (e.g. etc. acrylates fumarates	1310–1250	7.63–8.00	vs	m–s	asym C–O–C str
⟩C═Ċ–CO–OR					
	1200–1130	8.33–8.85	s	w	sym C–O–C str
Acrylates CH$_2$═CH–CO–O–R	1290–1280	7.75–7.81	vs	m–s	Shoulder at ~1300 cm^{-1}
	1200–1195	8.33–8.36	s	w	
Methacrylates CH$_2$═C(CH$_3$)CO·OR	1305–1295	7.66–7.72	vs	m–s	Shoulder at ~1330 cm^{-1}
	1180–1165	8.48–8.58	s	w	
Crotonates CH$_3$CH═CH–CO–OR	1290–1275	7.75–7.84	vs	m–s	Usually two shoulders
	1195–1180	8.36–8.48	s	w	
Cinnamates,	1290–1210	7.75–8.00	vs	m–s	Usually two shoulders
(◯)–CH═CH–CO·OR					
	1185–1165	8.44–8.58	s	w	
R–CO·OG′ (G′ vinyl or aromatic)	~1210	~8.26	vs	m–s	asym str. Vinyl C═C str 1690–1650 cm^{-1} of greater intensity than usual
Ar–CO·OAr′	1310–1250	7.64–8.00	vs	m–s	asym str
	1150–1080	8.70–9.26	s	w	sym str
Methyl ester, R–CO·OCH$_3$	~1245	~8.03	s	m–s	O–CH$_3$ str. General range 1315–1195 cm^{-1}
	1175–1155	8.51–8.66	s	w	O–C str. General range 1200–850 cm^{-1} but mostly 1060–900 cm^{-1}, with variable intensity
	530–340	18.87–29.41	w		CO–O rocking vib
	390–250	25.64–40.00			COC def vib
Aliphatic chloroformates	1205–1115	8.30–8.97	vs	m–s	br, asym C–O–C str
	1170–1140	8.55–8.77	s	w	br, sym C–O–C str
	850–770	11.76–12.99	s	w	C–O–C str. (C–O–R in-plane def vib, 300–250 cm^{-1}, weak band)
Aromatic chloroformates	1175–1130	8.51–8.85	s	w	br, asym C–O–C str, usually difficult to identify

Table 10.20 (*continued*)

Functional Groups	Region		Intensity		Comments
	cm^{-1}	μm	IR	Raman	
Dialkyl carbonates, RO·R'O·CO	1290–1240	7.75–8.06	s	m–s	Also weak band at ~1000 cm^{-1} and medium intensity band at ~1160 cm^{-1}
R–O–CO–O–Ar	1250–1210	8.00–8.26	s	m–s	
Diaryl carbonates	1220–1205	8.20–8.30	s	m–s	
Dialkyl thiolcarbonates, R–O–CO–S–R	1165–1140	8.58–8.77	s	w	
Alkyl aryl thiolcarbonates:					
R–O–CO–S–Ar	1140–1125	8.77–8.88	s	w	
Ar–O–CO–S–R	1105–1055	9.05–9.48	s	w	

Table 10.21 Esters, haloformates, and carbonates: other bands

Functional Groups	Region		Intensity		Comments
	cm^{-1}	μm	IR	Raman	
Esters, –CO–O–	~3450	~2.90	w	–	C=O str overtone
Formates	1385–1350	7.22–7.41	m–w	m, p	CH in-plane rocking vib. See Table 10.18
	775–620	12.90–16.13	m	m, p	O–C=O in-plane def vib
n-Alkyl formates (ethyl to amyl)	~620	~16.13	w	m–w	
	475–460	21.05–21.74	v		Frequency increases with molecular weight increase
	~340	~29.41	s		vib
Branched alkyl formates	520–485	19.23–20.62	v		
	340–285	29.41–35.09	v		
Acetates	~2990	~3.34	m–w	m–s	See Table 10.18 and ref. 9
	~1430	~6.99	w–m	m–w	
	~1375	~7.27	m–s	m	CH$_3$ def vib
	1080–1020	9.26–9.80	m–w	w	
	1025–930	9.76–10.75	w	w	
	845–835	11.83–11.93	w	m–s	CH$_3$–C str, usual range
	665–635	15.04–15.75	w–s	w	Weak for tertiary (and sometimes secondary) acetates
	620–580	16.13–17.24	v	m	C=O wagging vib. Intensity may vary from weak to strong.
	325–230	30.77–43.48	v		Absent for isopropyl and sec-butyl acetates
Propionates	1085–1080	9.21–9.26	m	w	OCH$_2$ def vib
	~1020	~9.80	m	w	OCH$_2$ def vib
	~810	~12.35	w	m	
	620–575	16.13–17.39	w–m	m–w	br. Two bands. (Not isoamyl)
Butyrates	~1095	~9.13	m	m	
	1050–1040	9.52–9.62	m	m	
	930–865	10.75–11.56	w	m–w	
	850–830	11.76–12.05	w	m	
	635–625	15.75–16.00	w	m	
	605–580	16.53–17.24	m–s	m	
α,β-Unsaturated aliphatic esters	845–765	11–83–13–07	m	m	Mainly C–O–C def vib
	695–645	14.39–15.50	m	m–w	Mainly C–O–C def vib

(*continued overleaf*)

Table 10.21 (*continued*)

Functional Groups	Region cm^{-1}	Region μm	Intensity IR	Intensity Raman	Comments
Acrylates	~1640	~6.10	s–m	s	C=C str
	~1625	~6.15	m	s	C=C str, less intense than 1640 cm^{-1} band
	~1410	~7.09	m	m	=CH$_2$ def vib
	~1280	~7.81	m	m	=CH rocking vib
	1070–1065	9.35–9.40	m	m	Skeletal vib
	990–980	10.10–10.20	m	w	CH def vib
	970–960	10.30–10.40	s	w	=CH$_2$ wagging vib
	810–800	12.35–12.50	m–s	m–w	=CH$_2$ twisting vib
	675–660	14.81–15.15	m	w	Mainly C–O–C def vib
Methacrylates	~1640	~6.10	m	s	C=C str
	~1410	~7.09	m	m–s	=CH$_2$ def vib
	~1325	~7.55	m	m	=CH rocking vib
	~1300	~7.69	m	m	
	~1010	~9.90	m	m	Skeletal vib
	~1000	~10.00	m	m	Skeletal vib
	950–935	10.53–10.70	s	w	=CH$_2$ wagging vib
	~815	~12.27	m–s	m–s	Skeletal vib
	660–645	15.15–15.50	m	w	Mainly C–O def vib
Crotonates	~1660	~6.02	m	s	C=C str
	~1280	~7.81	m	m	
	1105–1100	9.05–9.09	m	m	Skeletal vib
	970–960	10.31–10.42	s	w	CH=CH twisting vib
	920–900	10.87–11.11	m	m–s	Skeletal vib
	840–830	11.90–12.05	m	m–s	Skeletal vib
	695–675	14.39–14.81	m	w	Mainly C–O–C def vib
Methyl esters	~2960	~3.38	m–w	m–s	asym CH$_3$ str
	~1440	~6.94	m–s	m–w	CH$_3$ def vib. See Table 10.18
	1430–1420	6.99–7.04	w–m	m–w	
	~1360	~7.53	w	m	
	450–430	22.22–23.26	m–s		Not formate or isobutyrate
–CO–O–CH$_2$–	~1475	~6.78	m	m–w	–OCH$_2$ def vib
	~1400	~7.14	m	m–w	OCH$_2$ wagging vib
n-Propyl esters	605–585	16.53–17.09	v		
	~495	~20.20	v		
	350–340	28.57–29.41	v		
Isopropyl esters	605–585	16.53–17.09	v		Not formate
	505–480	19.80–20.83	v		Not formate
	~435	~22.09	s	m	
	435–410	22.99–24.39	s	m	Not formate
n-Butyl esters	~505	~19.08	m–s		
	~435	~22.99	m–s	m	
	350–335	28.57–29.85	w		
Isobutyl esters	~505	~19.08	m		Not formate
	~430	~23.26	s	m	

Table 10.21 (*continued*)

Functional Groups	Region		Intensity		Comments
	cm^{-1}	μm	IR	Raman	
	~385	~25.97	v		
Aromatic esters	650–585	15.38–17.09	v		Rocking vib or in-plane def vib of CO$_2$ group
	370–270	27.03–37.04	m–s		Acids also absorb in this region
Phthalates	3090–3075	3.24–3.25	m	m, p	CH str
	3045–3035	3.28–3.31	w	w, p	CH str
	1610–1600	6.21–6.25	w–m	s, p	Ring str
	1590–1580	6.29–6.33	w–m	m, p	Ring str
	1500–1485	6.67–6.73	m	w	Ring str
	1050–1040	9.52–9.62	w–m	s, p	
	~745	~13.42	s	w	Out-of-plane CH vib. (Raman: ring vib, strong band ~650 cm^{-1})
	410–400	24.39–25.00		m, p	
Benzoates	~710	~14.08	s	–	
Isophthalates	~730	~13.70	s	–	
Teraphthalates	~730	~13.70	s	–	
α-Hydroxy esters	1300–1260	7.69–7.94	s	m–w	br, O–H def vib
Acetylated pyranose sugars	670–625	14.93–16.00	s		See refs 115, and 116
	610–600	16.39–16.67	v		
	405–365	24.69–27.40	v		
Fluroformates, –O·CO·F	1290–1010	7.75–9.90	s	w–m	C–F str
	790–750	12.66–13.33		m	CO, CF def vib
	670–630	14.93–15.87			CO, CF def vib
	570–510	17.54–19.61		s–m	CO, CF rocking vib
Aliphatic chloroformates	850–770	11.76–12.99	s	s, p	sym COC str. Most alkyl compounds
	695–680	14.39–14.71	s	v	CCl str
	490–470	20.41–21.28	m–s	s, p	C–Cl def vib
	~435	~22.99	m–s	s, p	Two bands, COC def vib
–S·CO·F	1100–1040	9.09–9.62	s	m	C–F str
Thiocarbonyl compounds:					
R–CO–S–R	1140–1070	8.77–9.35	w	m	C–C str
	1035–930	9.66–10.75	s	s	C–S str
R–CO–S–Ar	1110–1160	9.01–9.43	w	m	C–C str
	1020–920	9.80–10.87	s	s	C–S str
Ar–CO–S–R	1210–1190	8.26–8.40	w	m	C–C str
	940–905	10.64–11.05	s	s	C–S str
H–CO–S–Ar and H–CO–S–R	780–730	12.82–13.70	s	s	C–S str
R–O–CO–NHR	1250–1210	8.00–8.26	s	s–m	C–N str. See ref. 117
R–S–CO–NHR	1230–1170	8.13–8.55	s	s–m	C–N str
Ar–S–CO–NHAr	1160–1150	8.62–8.70	s	s–m	C–N str

the C=O group. These esters, of course, have a band due to the C=C stretching vibration and also bands due to the =C–H and =CH$_2$ groups, for instance, acrylates and methacrylates have a medium-to-strong band at 820–805 cm^{-1} (12.20–12.42 μm) and a strong band at 970–935 cm^{-1} (10.31–10.70 μm) due to the twisting and wagging respectively of the =CH$_2$ group. For acrylates, the C=C stretching vibration results in a doublet at 1640–1620 cm^{-1} (6.10–6.17 μm) due to the interaction with the overtone of the band near 810 cm^{-1} (12.35 μm). Benzoates with an unsubstituted ring have a strong band near 710 cm^{-1} (14.08 μm) and other bands, due to ring vibrations, of medium intensity near 1070 cm^{-1} (9.35 μm) and 1030 cm^{-1} (9.71 μm). Disubstituted aromatic esters often do not have the usual band pattern expected in the region 880–750 cm^{-1} (11.36–11.33 μm), which may

Table 10.22 Lactone C=O and C–O stretching vibrations

Functional Groups	Region		Intensity		Comments
	cm⁻¹	μm	IR	Raman	
β-Lactones (four-membered ring)	1840–1815	5.44–5.51	s	w–m	C=O str, halogen substitution results in higher frequencies
γ-Lactones (saturated five-membered ring)	1790–1770	5.59–5.65	s	w–m	C=O str
α,β-Unsaturated γ-lactones (unsaturated five-membered ring)	1790–1775	5.59–5.63	s	w–m	C=O str ⎱ Doublet due to
	1765–1740	5.67–5.75	s	w–m	C=O str ⎰ Fermi resonance
β, γ-Unsaturated γ-lactones (unsaturated five-membered ring)	1815–1785	5.51–5.60	s	w–m	C=O str
δ-Lactones and larger					As for open-chain ester
α,β-Unsaturated δ-lactones	1745–1730	5.73–5.78	s	w–m	C=O str ⎱
β-γ, δ-Unsaturated δ-lactones (unsaturated six-membered ring)	1775–1740	5.63–5.75	s	w–m	C=O str ⎰ Doublet due to Fermi
	1740–1715	5.75–5.83	s	w–m	C=O str ⎰ resonance
2-Benzofuranones,	1820–1800	5.50–5.56	s	w–m	C=O str, see ref. 124
Phthalides,	1775–1710	5.63–5.85	s	w–m	C=O str
	1290–1280	7.75–7.81	m–s	m	⎱
	1120–1100	8.93–9.01	m–s	m	Characteristic
	1020–1010	9.80–9.90	w–m	m	phthalide ring
	515–490	19.42–20.41	w–m	m	vibrations
	490–470	20.41–21.28	w–m		
Lactones	1370–1160	7.30–8.62	s	w	C–O str

be due to an interaction with the CO–O group. Because of their centre of symmetry, terephthalates do not have a band near $1600\,\text{cm}^{-1}$ ($6.25\,\mu\text{m}$).

Aromatic acids and esters absorb strongly at $570–545\,\text{cm}^{-1}$ ($17.54–18.35\,\mu\text{m}$) due to the rocking of the CO_2 group and also have a band of medium-to-strong intensity, which is usually broad for acids, at $370–270\,\text{cm}^{-1}$ ($27.03–37.04\,\mu\text{m}$). Aromatic esters have a band of variable intensity in the range $650–585\,\text{cm}^{-1}$ ($15.38–17.09\,\mu\text{m}$) which is due to a deformation vibration of the CO_2 group. A study of phthalides has also been published.[123] Thiol formates have medium-intensity bands at $2835–2825\,\text{cm}^{-1}$ ($3.53–3.54\,\mu\text{m}$) and near $1340\,\text{cm}^{-1}$ ($7.46\,\mu\text{m}$) due to the stretching and deformation vibrations respectively of the CH group, and a weak band at $2680–2660\,\text{cm}^{-1}$ ($3.73–3.76\,\mu\text{m}$) which is an overtone of the CH deformation vibration.

Lactones,

Lactones have bands due to the stretching of the C=O and C–O groups. The C=O stretching vibration for saturated γ-lactones[29,118] (five-membered ring) is at higher frequencies, $1790–1770\,\text{cm}^{-1}$ ($5.59–5.65\,\mu\text{m}$), than for aliphatic esters. Electronegative substituents on the γ-carbon atom tend to increase the frequency. The absorptions of δ-lactones[119] (six-membered ring) and larger lactones are similar to those of open-chain esters. α,β-Unsaturated γ-lactones have two bands due to the carbonyl stretching vibration, at $1790–1775\,\text{cm}^{-1}$ ($5.59–5.63\,\mu\text{m}$) and $1765–1740\,\text{cm}^{-1}$ ($5.67–5.75\,\mu\text{m}$), even though only one carbonyl group is present. This is probably due to Fermi resonance.[14] α,β-γ,δ-Unsaturated δ-lactones similarly have two carbonyl absorption bands which are at $1775–1740\,\text{cm}^{-1}$ ($5.63–5.75\,\mu\text{m}$) and $1740–1715\,\text{cm}^{-1}$ ($5.63–5.83\,\mu\text{m}$).

The band due to the C–O stretching vibration of lactones occurs in the region 1370–1160 cm^{-1} (7.29–8.62 μm), usually being at 1240–1220 cm^{-1} (8.07–8.12 μm) for δ-lactones.

Amides, −CO−N⟨

All amides exhibit a band due to the C=O stretching vibration,[172,175,176] with primary and secondary amides also having bands due to the N–H stretching and deformation vibrations. The positions of the carbonyl band and the N–H bands (if present) are dependent on the amount of hydrogen bonding occurring. The position of the carbonyl band depends also on the substituents on the nitrogen atom. Overtones of the bands due to the N–H stretching vibration for primary and secondary amides occur in the near infrared region.[125]

The absorption bands of even quite small molecules cannot strictly be considered as arising from a single vibration source. In other words, a given absorption band is never due solely to, say, the stretching vibration of the A–B group since in reality the whole molecule is involved. However, because of the complexity of the actual situation, the tendency is to simplify (in some cases, to oversimplify) mainly because it is useful to identify the major cause of any given band. In fact, all statements as to the vibration source of any band should always be interpreted as meaning that the stated type of vibration is the major, not the sole, contribution to that band. In the case of amides, in acknowledgement of the complexity of the situation, the bands observed are given names such as amide I, amide II, etc., rather than C=O stretching, etc.

For example, for primary amides the names given are as follows:

Amide Band	Major Contribution to Vibration
Amide I	C=Z stretching, Z=O, S, Se, etc
Amide II	NH$_2$ deformation
Amide III	C–N stretching
Amide IV	C=Z deformation
Amide V	NH$_2$/CZ wagging
Amide VI	C=Z out-of-plane deformation
Amide VII	NH$_2$/CZ twisting

Amide N–H Stretching Vibrations

For primary amides, two sharp bands of medium intensity are observed due to the asymmetric and symmetric stretching vibrations. In dilute, non-polar solvents, i.e. in the absence of hydrogen bonding, these bands occur at about 3500 cm^{-1} (2.86 μm) and 3400 cm^{-1} (2.94 μm).[126,127] In the solid state and in the presence of hydrogen bonding, these bands are shifted by about 150 cm^{-1} (0.16 μm) to about 3350 cm^{-1} (2.99 μm) and 3200 cm^{-1} (3.13 μm). Both primary and secondary amides may exhibit a number of bands due to different hydrogen-bond states, e.g. dimers, trimers, etc. The bands are concentration- and solvent-dependent. Free (unassociated) secondary amides have a sharp, strong band at 3460–3300 cm^{-1} (2.89–3.03 μm).[128] This band may appear as a doublet due to the presence of cis–trans isomerism.[129,144] In the solid or liquid phases, secondary amides generally exhibit a strong band at about 3270 cm^{-1} (3.06 μm) and a weak band at 3100–3070 cm^{-1} (3.23–3.26 μm).

The cis- and trans- forms of secondary amides may be distinguished by examination of their N–H vibration bands, as indicated in Table 10.23.

Amide C=O Stretching Vibrations: Amide I Band

The amide band due to the C=O stretching vibration is often referred to as the amide I band.[130] Primary amides[6] have a very strong band due to the C=O stretching vibration at 1670–1650 cm^{-1} (5.99–6.06 μm) in the solid phase, the band appearing at 1690–1670 cm^{-1} (5.92–5.99 μm) for a dilute solution using a non-polar solvent. In the solid phase, secondary amides absorb strongly at 1680–1630 cm^{-1} (5.95–6.14 μm), and in dilute solution at 1700–1665 cm^{-1} (5.88–6.01 μm).[131–133,145] The carbonyl absorption band of tertiary amides[134,135] is independent of physical state, since hydrogen bonding to another amide molecule is not possible, and occurs in the region 1670–1630 cm^{-1} (5.99–6.14 μm). If the substituent on the nitrogen is an aromatic for either secondary or tertiary amides then the carbonyl absorption occurs at the higher end of the frequency ranges given,[136–138] whereas aliphatic secondary amides absorb at 1650–1630 cm^{-1} (6.06–6.14 μm). The carbonyl absorption band is obviously greatly influenced by solvents with which hydrogen bonds may be formed.

Primary α-halogenated amides[15] absorb at higher frequencies than the corresponding alkyl compound, up to about 1750 cm^{-1} (5.71 μm), and may, in fact, have two carbonyl bands due to the presence of rotational isomerism. The carbonyl band of N-halogen secondary amides also occurs at higher frequencies than that of the corresponding N-alkyl compound.[131c,132]

In dilute solution in non-polar solvents, acetanilides and benzanilides absorb in the region 1710–1695 cm^{-1} (5.85–5.90 μm).[131c,136,137] Ortho-nitro-substituted anilides, in the solid phase, exhibit two carbonyl bands, one near 1700 cm^{-1} (5.88 μm) and the other at about 1670 cm^{-1} (6.00 μm). Compounds of the type CH$_3$(Ar)NCOCH$_3$ absorb in the region 1685–1650 cm^{-1} (5.93–6.06 μm).

Table 10.23 The N–H vibration bands of secondary amides

Type of secondary amide	Region cm^{-1}	μm	Intensity IR	Raman	Comments
Hydrogen-bonded *trans*- form (solid or liquid phase)	3370–3270	2.97–3.06	m	m–w	N–H str
	3100–3070	3.23–3.26	w	–	Overtone of amide II band
	1570–1515	6.37–6.60	s	m	Amide II band
Hydrogen-bonded *cis*- form (solid or liquid phase) (may be as dimers)	3180–3140	3.15–3.19	m	m–w	N–H str
	~3080	~3.25	w	–	
	1450–1440	6.90–6.94	s	w	N–H def vib
Trans- form (in dilute solution)	3460–3420	2.89–2.92	m	m–w	N–H str
	1550–1510	6.45–6.62	s	m	Amide II band
Cis- form (in dilute solution)	3440–3300	2.91–3.03	m	m–w	N–H str, *cis*- form remains mainly association even in very dilute solution whereas *trans*- form does not

Table 10.24 Amide N–H stretching vibrations (and other bands in same region)

Functional Groups	Region cm^{-1}	μm	Intensity IR	Raman	Comments
(Free) primary amides, –CO–NH$_2$	3540–3480	2.83–2.88	m–s	m–w	asym N–H str
	3420–3380	2.92–2.96	m–s	m–w	asym N–H str
(Associated) primary amides	3375–3320	2.96–3.01	m–s	m–w	asym N–H str
	3205–3155	3.12–3.17	m–s	m–w	sym N–H str
(Free) secondary amides, –CO–NH–	3460–3420	2.89–2.93	m–s	m–w	Doublet if *cis–trans* isomerism present, N–H str
(Associated) secondary amides:					
trans- form	3370–3270	2.97–3.06	m	m–w	N–H str
cis- form	3180–3140	3.15–3.19	m	m–w	N–H str
trans- form	3100–3070	3.23–3.26	w	–	Overtone of amide II band near 1550 cm^{-1}
Hydroxamic acids (solid phase), –CO–NHOH	3300–2800	3.03–3.57	w–m	m–w	Three bands, N–H str and O–H str
–CO·NH·CH$_3$	3360–3270	2.98–3.06	s, br	m–w	Also weak band near 3080 cm^{-1} due to overtone of amide II band
–NH·CO·CH$_3$	3340–3220	2.99–3.10	s, br	m–w	Dilute solutions: 3480–3340 cm^{-1}

The carbonyl stretching vibration frequency of *N*-acetyl and *N*-benzoyl groups in compounds where the nitrogen atom forms part of a heterocyclic ring increases as the resonance energy is increased, e.g. by increasing the number of nitrogen atoms in the ring.[139] For example, in the case of pyrroles, the carbonyl band occurs near 1730 cm^{-1} (5.78 μm) and in the case of tetrazoles, at about 1780 cm^{-1} (5.62 μm).

Oxamides,[146] thioamides,[146–149] amides of *n*-fatty acids,[150] polyamides,[142] phosphonamides,[131b] polyglycines,[141] and numerous other related compounds have been studied.

Amide N–H *Deformation and* C–N *Stretching Vibrations: Amide II Band*

In the solid phase, primary amides have a weak-to-medium intensity band at 1650–1620 cm^{-1} (6.06–6.17 μm) which is generally too close to the strong carbonyl band to be resolved. In dilute solution, this band occurs at 1620–1590 cm^{-1} (6.17–6.31 μm). The position of this band is not greatly influenced by the nature of the primary amide, e.g. aliphatic or aromatic. This band is known as the amide II band and is due to a motion combining both the

N–H bending and the C–N stretching vibrations of the group –CO–NH– in its *trans*- form. The amide II band appears to be mainly due to the N–H bending motion. Secondary amides in the solid phase have a characteristic, strong absorption at 1570–1515 cm^{-1} (6.37–6.60 μm) and in dilute solution, at 1550–1510 cm^{-1} (6.45–6.62 μm). In general, the amide II band of primary amides is more intense than that of secondary amides. In fact, it has been observed that the amide II band is absent in *trans-N*-halogen secondary amides although it is present for *N*-iodo-amides in the solid phase.[131a,b,132,140,143] Secondary aliphatic amides usually have a strong, polarised band in their Raman spectra at 900–800 cm^{-1} (11.11–12.50 μm) due to the symmetrical CNC stretching vibration, the band being of weak intensity in the infrared. For tertiary amides, this band is normally at 870–700 cm^{-1} (11.49–14.29 μm).

Other Amide Bands

Primary amides absorb at 1420–1400 cm^{-1} (7.04–7.14 μm) and secondary amides at 1305–1200 cm^{-1} (7.67–8.33 μm) and at about 700 cm^{-1} (14.3 μm). This last band may not be observed in that position in the spectra of dilute solutions.

In general, all amides have one or more bands of medium-to-strong intensity, which may be broad, in the region 695–550 cm^{-1} (14.39–18.18 μm) which

are probably due to the bending motion of the O=C–N group.[141,142] Primary aliphatic amides absorb at 635–570 cm^{-1} (15.75–17.54 μm), probably due to the out-of-plane bending of the C=O group, whereas α-branched primary amides absorb at 665–580 cm^{-1} (15.04–17.24 μm). Secondary aliphatic amides absorb at 610–590 cm^{-1} (16.39–16.95 μm) and in the case of α-branching, at 670–625 cm^{-1} (14.93–16.00 μm). With the exception of formamides, anilides, and diamides, amides have a medium-to-strong absorption at 520–430 cm^{-1} (19.23–23.26 μm) and, with the exception of *N*-methyl secondary amides, *N*-substituted anilides, lactams, and diamides (also acetamide and propionamide), a band which is usually observed at 390–305 cm^{-1} (25.64–32.79 μm). This last band is sensitive to conformational changes and has been observed as low as 215 cm^{-1} (46.51 μm). Formamide has a strong, broad absorption in the range 700–500 cm^{-1} (14.29–20.00 μm).

Hydroxamic Acids, –CO–NHOH

Hydroxamic acids have a strong carbonyl absorption at about 1640 cm^{-1} (6.10 μm). In the solid phase, three medium-intensity bands are observed at 3300–2800 cm^{-1} (3.03–3.57 μm), a strong amide II band is observed near

Table 10.25 Amide C=O stretching vibrations: amide I bands

Functional Groups	Region		Intensity		Comments
	cm^{-1}	μm	IR	Raman	
Primary amides (solid phase)	1670–1650	5.99–6.06	s	w–m	Usually a doublet involving NH$_2$ def at ~1620 cm^{-1}
Primary amides (dilute solution)	1690–1670	5.92–5.99	s	w–m	
Secondary amides (solid phase)	1680–1630	5.95–6.14	s	w–m	
Secondary amides (dilute solution)	1700–1665	5.88–6.01	s	w–m	Strongly electron-accepting groups on nitrogen increase frequency
Acetylamides, –NH·CO·CH$_3$	1735–1645	5.76–6.08	s	w–m	
Acetanilides (dilute solution), ArNH·CO·CH$_3$	1710–1695	5.85–5.90	s	w–m	
Secondary amides of the type ArCO·NH– (dilute solution)	~1660	~6.02	s	w–m	
Tertiary amides (dilute solution or solid phase)	1670–1630	5.99–6.14	s	m	Strongly electron-accepting groups on nitrogen increase frequency
Amides containing –CO–NH–CO– (diacylamines)	1790–1720	5.59–5.81	s	w	Doublet, separation usually small, but larger for ring amides
	1720–1670	5.81–5.99	s	w–m	
Monosubstituted hydrazides, –CONHNH$_2$	1700–1640	5.88–6.10	s	w–m	Acid hydrazides, see ref. 138
Disubstituted hydrazides, –CONH·NHCO–	1745–1700	5.73–5.88	s	w–m	Doublet, usually marked difference between phase and solution spectra, amide II band for aliphatic compounds at 1500–1480 cm^{-1} (6.67–6.76 μm)

(continued overleaf)

Table 10.25 (*continued*)

Functional Groups	Region		Intensity		Comments
	cm^{-1}	μm	IR	Raman	
Alkyl hydroxamic acids, R–CO·NH·OH, (solid phase)	1710–1680	5.85–5.95	s	w–m	
	~1640	~6.10	s	w–m	
Amides of the type Ar–SO$_2$·NHCOCH$_3$ (solid phase)	1720–1685	5.81–5.93	s	w–m	
Aromatic isocyanates (dimers)	1785–1775	5.60–5.64	s	w–m	
Aliphatic isocyanurates (isocyanate trimers)	1700–1680	5.88–5.95	s	w–m	Shoulder at ~1755 cm^{-1}
Aromatic isocyanurates	1715–1710	5.83–5.85	s	w–m	Shoulder at ~1780 cm^{-1}
–CF$_2$CONH$_2$	1730–1700	5.78–5.88	s	w–m	
CF$_3$CONH–	1740–1695	5.75–5.90	s	w–m	
Methyl carbamoyls, –CO·NH·CH$_3$	1740–1620	5.75–6.17	s	w–m	
Carbamoyl chlorides, Cl·CO·N	~1740	~5.75	s	w–m	See ref. 170
Polypeptides	~1650	~6.06	s	w–m	Mainly C=O str but coupled with C=N also (due to group –CO–NH–)
CHCl$_2$CONH–	1715–1700	5.83–5.88	s	w–m	
CCl$_3$CONH–	~1730	~5.78	s	w–m	
RN(CO)(CO)NR′	1770–1740	5.65–5.75	s	w–m	May be doublet. Also strong band at ~1300 cm^{-1}
R—O—CO—N	1750–1730	5.71–5.75	s	w–m	
R—S—CO—N	1700–1680	5.88–5.95	s	w–m	
R–S–CO–NAr–	1670–1650	5.99–6.06	s	w–m	

Table 10.26 Amide N–H deformation and C–N stretching vibrations: amide II band

Functional Groups	Region		Intensity		Comments
	cm^{-1}	μm	IR	Raman	
Primary amides (solid phase)	1650–1620	6.06–6.17	w–m	w–m	Exception is *o*-cyanobenzamide at 1667 cm^{-1}
Primary amides (dilute solution)	1620–1590	6.17–6.31	w–m	w–m	*n*-alkylamides ~1590 cm^{-1}
Secondary amides (*trans-* form) (solid phase)	1570–1515	6.37–6.60	s	w	
Secondary amides (*trans-* form) (dilute solution)	1550–1510	6.45–6.62	s	w	
–NH·CO·CH$_3$ (*trans-* form)	1600–1480	6.25–6.76	s	w	Most acetylamides absorb in range 1580–1520 cm^{-1}
Aliphatic disubstituted hydrazides, –CONH·NHCO–	1500–1480	6.67–6.76	s	w	
–CF$_2$CONH–	1630–1610	6.14–6.21	m	w	
Hydroxamic acids, R–CO–NHOH	~1550	~6.45	m–s	w	
Hydrazides, –CO–NHNH$_2$	1545–1520	6.47–6.58	m–s	w–m	
Methyl carbamoyls, –CO·NH·CH$_3$	1600–1500	6.25–6.67	m	w	
HCONR$_2$	870–820	11.49–12.20	w–m	w–m	CN str. (Also band ~650 cm^{-1})
R′CONR$_1$R$_2$	750–700	13.33–14.29	w–m	w–m	CN str. (Also band at 620–590 cm^{-1})

Table 10.27 Amides: other bands

Functional Groups	Region		Intensity		Comments
	cm^{-1}	μm	IR	Raman	
Primary amides	1420–1400	7.04–7.14	m	m	C–N str, known as amide III band
	~1150	~8.70	w	w	NH$_2$ in-plane rocking vib, not always seen
	750–600	13.33–16.67	m	w	br, NH$_2$ def vib
	600–550	16.67–18.18	m–s	m	N–C=O def vib
	500–450	20.00–22.22	m–s	m–w	C–C=O def vib
Secondary amides (*trans*- form)	1305–1200	7.67–8.33	w–m	s	Amide III band, usually at ~1260 cm^{-1}
	770–620	13.00–16.13	m	w	br, out-of-plane N–H def vib, for hydrogen-bonded amides usually at ~700 cm^{-1}
Secondary amides (*cis*- form)	1450–1440	6.90–6.94	m	w	N–H bending vib
	1350–1310	7.41–7.63	w–m	s	C–N str (amide III band)
	~800	~12.50	m–s	m–s	br, N–H wagging vib
Methyl carbamoyls, –CO·NH·CH$_3$	1330–1215	7.51–8.23	m–s	s	Amide III band
	860–675	11.63–14.81	m, br	m	Amide V band
	770–525	12.99–19.05	m–s	m–s	Amide IV band
	695–530	14.39–18.87	m	m–s	Amide IV band
	530–350	18.87–28.57	w–m		
Monosubstituted hydrazides	1150–950	8.70–10.53	m–s	m–s	Two bands, NH$_2$ def vib
Primary aliphatic amides, R–CH$_2$–CONH$_2$	635–570	15.75–17.54	s	m	N–C=O def vib
	480–450	20.83–22.22	m–s	m	C–C=O in-plane def vib ⎫ Not formamides or anilides
	360–320	27.78–31.25	s		C–CO–N def vib ⎭
Primary α-branched aliphatic amides, >C–CO–NH$_2$	665–580	15.04–17.24	s	m	
	520–495	19.23–20.20	m–s	m	NCO in-plane bending vib
	320–305	31.25–32.79	s		
n-Aliphatic secondary amides and *N*-methyl aliphatic amides, R$_1$–CH$_2$–CO–NHR$_2$	610–590	16.39–16.95	m–s	m	NCO in-plane bending vib
	480–430	20.83–23.26	s		
	380–330	26.32–30.30	m–s		Absent for *N*-methyl aliphatic amides
α-Branched aliphatic secondary amides, R$_1$R$_2$CH–CO–N(R$_3$)H	670–625	14.93–16.00	m	m	NCO in-plane bending vib
	520–510	19.23–19.61	s	m	
	350–330	28.57–30.30	s		
–CO·NH·CH$_3$	3010–2970	3.32–3.37	w	m	asym CH$_3$ str
	3000–2930	3.33–3.41	w	m	asym CH$_3$ str
	2945–2855	3.40–3.50	w–m	m	sym CH$_3$ str
	1480–1420	6.76–7.04	w–m	m–w	asym CH$_3$ def vib
	1440–1400	6.94–14.29	w–m	m–w	asym CH$_3$ def vib
	1375–1355	7.27–7.38	w	m	sym CH$_3$ def vib
	1330–1220	7.52–8.20	m–s	s	Amide III band
	1130–1050	8.85–9.52	w–m	w	CH$_3$ rocking vib
	1050–980	9.52–10.20	w–m	w	CH$_3$ rocking vib
	975–850	10.26–11.76	w	m–s	CC str
	860–675	11.63–14.81	m, br	m	Amide V band

(*continued overleaf*)

Table 10.27   (*continued*)

Functional Groups	Region		Intensity		Comments
	cm^{-1}	μm	IR	Raman	
	695–530	14.39–18.87	m–s	m–s	Amide IV band
	475–365	21.05–27.40			Skeletal vib
	375–255	26.67–39.22			Torsional vib
	290–160	34.48–62.50			Torsional vib
	265–135	37.74–74.07			CH$_3$ torsional vib
	140–60	71.43–166.67			CO·CH$_3$ torsional vib
Tertiary amides, $\diagdown$C—CON$\diagup$R$_1$ $\diagdown$R$_2$	870–700	11.49–14.29	w	s, p	asym CNC str
	620–570	16.13–17.54	s	m	Absent for formamides, NCO bending vib
	480–440	20.83–22.73	m–s		Absent for formamides
	390–320	25.64–31.25	m		
Tertiary formamides, H–CO–NR$_1$R$_2$	750–700	13.33–14.29	w	s, p	asym CNC str
	700–645	14.29–15.50	s	m	Usually broad
	620–590	16.13–16.95	m–s	m	N–C=O bending vib
	390–340	25.64–29.41	m–s		
N-Substituted anilides, —CO—N(R)—Ar	630–610	15.87–16.39	m	m	
	~445	~22.47	m		
	~405	~24.69	s		
Primary aromatic amides	645–590	15.50–16.95	s	m	N–C=O bending vib
Lactams, $\diagdown$C—(C$\diagup$)$_n$—CO—N—	695–655	14.39–15.27	m–s	m	
	500–470	20.00–21.28	s	m	
Diamides, $\diagup$N—C(=O)—C—C(=O)—N$\diagdown$	675–600	14.81–16.67	s	m	
	595–540	16.81–18.52	m–s	m	
Hydroxamic acids R–CO–NHOH	1440–1360	6.94–7.35	v	m	
	~900	~11.11	s	s	

1550 cm^{-1} (6.45 μm), a strong band at about 900 cm^{-1} (11.11 μm), and a band of variable intensity at 1440–1360 cm^{-1} (6.94–7.35 μm).

Hydrazides, –CO–NH–NH$_2$ and –CO–NH–NH–CO–

Amides of the type –CO–NH–NH$_2$[138] have a number of medium-intensity bands in the region 3350–3180 cm^{-1} (2.99–3.15 μm) due to the NH and NH$_2$ stretching vibrations. The band due to the carbonyl stretching vibration, which is very strong, occurs at 1700–1640 cm^{-1} (5.88–6.10 μm).

A medium-intensity band due to the deformation of the NH$_2$ group occurs at 1635–1600 cm^{-1} (6.12–6.25 μm). The amide II band, which is strong, occurs at 1545–1520 cm^{-1} (6.47–6.58 μm) and a weak-to-medium intensity band, due to the C–N stretching vibration, occurs at 1150–1050 cm^{-1} (8.70–9.52 μm).

For solid-phase spectra, aliphatic amides with the

–CO–NH–NH–CO–

group usually have only one very strong absorption due to the carbonyl groups, at 1625–1580 cm^{-1} (6.16–6.33 μm), whereas aromatic compounds usually

Table 10.28 Hydrazides

Functional Groups	Region cm^{-1}	Region μm	Intensity IR	Intensity Raman	Comments
Amides with –CO–NH–NH$_2$ group	3350–3180	2.99–3.15	m	m–w	N–H str
	1700–1640	5.88–6.10	vs	w	C=O str
	1635–1600	6.12–6.25	m	w	NH$_2$ def vib
	1545–1520	6.47–6.58	s	w	Amide II band
	1150–1050	8.70–9.52	w–m	m–s	C–N str
Aliphatic and aryl amides with –CO–NH–NH–CO–group (in solution)	3330–3280	3.00–3.05	m	m–w	N–H str
	1745–1700	5.73–5.88	vs	w	C=O str
	1710–1680	5.85–5.95	vs	w	C=O str
	1535–1480	6.52–6.64	m	w	Amide II band, aliphatics at 1500–1480 cm^{-1}
Aliphatic amides with –CO–NH–NH–CO–group (solid phase)	3210–3100	3.12–3.23	m	m–w	N–H str
	3060–3020	3.27–3.31	m	m–w	N–H str
	1625–1580	6.15–6.33	vs	w	C=O str
	1505–1480	6.65–6.76	s	w	Amide II band
	1260–1200	7.94–8.33	m	m–w	C–N str
Aromatic amides with –CO–NH–NH–CO–group (solid phase)	3280–2980	3.05–3.36	m	m–w	N–H str
	1730–1670	5.78–5.99	vs	w	C=O str
	1660–1635	6.02–6.12	vs	w	C=O str
	1535–1525	6.52–6.56	s	w	Amide II band
	1285–1245	7.78–8.03	m	m–w	C–N str
Phthalhydrazides,	~3000	~3.33	m	m–w	Very br, N–H str
	1670–1635	5.99–6.12	vs	m–w	C=O str

have two strong bands, one at 1730–1670 cm^{-1} (5.78–5.99 μm) and the other at 1660–1635 cm^{-1} (6.02–6.12 μm). There are usually marked differences in the carbonyl band positions between the solution and solid-phase spectra of hydrazides.

Lactams $-\overset{\text{NH}}{\underset{}{\text{C}}}-(\text{C})_n-\text{CO}$ *(Cyclic Amides)*

The N–H and C=O stretching vibrations of lactams[34] give rise to bands in the same regions as those for secondary amides. Where ring strain occurs, as for β-lactams (four-membered ring) and γ-lactams (five-membered ring), the carbonyl stretching frequency is increased, the band regions being 1760–1730 cm^{-1} (5.68–5.78 μm) and 1720–1700 cm^{-1} (5.81–5.88 μm) respectively. The amide II band is not exhibited by lactams unless the ring

consists of nine or more members, this band being associated with the group –CO–NH–in the *trans*- form. The N–H out-of-plane deformation band occurs at about 700 cm^{-1} (14.3 μm) and is generally broad as for secondary amides.

α,β-Unsaturation results in an increase in the carbonyl stretching vibration frequency by about 15 cm^{-1} (0.05 μm). Fused ring β- and γ-lactams have the frequency of their carbonyl stretching vibration increased by about 20–30 cm^{-1} (0.06–0.07 μm) compared with that of simple β- and γ-lactams. β-Lactams fused to unoxidized thiazolidine rings absorb at 1780–1770 cm^{-1} (5.62–5.65 μm).[149]

Imides, –CO–NH–CO–

Imides[151] may exist in two forms: (a) the two carbonyl groups both *trans*-to the NH group, (b) one carbonyl group being *cis*- and the other *trans*-

Table 10.29 Lactam C=O stretching vibrations: amide I band

Functional Groups	Region		Intensity		Comments
	cm^{-1}	µm	IR	Raman	
β-Lactams (four-membered ring) (dilute solution)	1760–1730	5.68–5.78	s	w	
γ-Lactams (five-membered ring)(dilute solution)	1720–1700	5.81–5.88	s	w	
δ-Lactams (six-membered ring)(dilute solution)	1690–1670	5.92–5.99	s	w	
β-Lactams (ring fused)(dilute solution)	1780–1770	5.62–5.65	s	w	
γ-Lactams (ring fused) (dilute solution)	1750–1700	5.71–5.88	s	w	

Table 10.30 Lactams: other bands

Functional Groups	Region		Intensity		Comments
	cm^{-1}	µm	IR	Raman	
Lactams	1315–1250	7.60–8.00	w	m–s	Amide III band, C–N str
	695–655	14.39–5.27	m–s	m	
	500–470	20.00–21.28	s	m	

Table 10.31 Imides

Functional Groups	Region		Intensity		Comments
	cm^{-1}	µm	IR	Raman	
Imides (solid phase)	3280–3200	3.05–3.13	m	m–w	N–H str
	1740–1670	5.75–5.99	vs	w	C=O str, amide I band
	1510–1500	6.62–6.67	vs	w	br, amide II band
	1235–1165	8.10–8.58	s	m–s	Amide III band
	740–730	13.51–13.70	m–s	w	br, N–H wagging (N–D wagging ~540 cm^{-1})
Cyclic imides (five-membered ring), CO—NH—CO with —C— —C—	~1770	~5.65	s	w	Unsaturation results in an increase in wavenumber of about 15 cm^{-1}
	~1700	~5.88	s	w	
Cyclic imides (six-membered ring), CO—NH—CO with —C—C—C	~1710	~5.85	s	w	
	~1700	~5.88	s	w	Usually of greater intensity than the other band
Maleimides,	1805–1745	5.54–5.73	s	w	C=O str ⎫ Doublet, see refs 152, 153
	1730–1685	5.78–5.93	s	w	C=O str ⎭
	1550–1450	6.45–6.90	s	s	C=C str
	1365–1340	7.33–7.46	s	m–s	C–N str
	1080–1040	9.26–9.62	m	m	
	780–730	12.82–13.70	s	m–w	

Table 10.31   (*continued*)

Functional Groups	Region		Intensity		Comments
	cm^{-1}	μm	IR	Raman	
Phthalimides,	1790–1735	5.59–5.76	s	w–m	See ref. 154
	1745–1670	5.73–5.99	s	w–m	
	1235–1165	8.10–8.58	w–m	m–s	Amide III band
α-Pyridones,	1690–1650	5.92–6.06	s	w–m	C=O str
γ-Pyridones,	1650–1630	6.06–6.14	s	w–m	C=O str

to the NH group. The *trans–trans* type has, in its solid-phase spectra, a medium intensity absorption at 3280–3200 cm^{-1} (3.05–3.13 μm), due to the N–H stretching vibration, and strong bands at 1740–1730 cm^{-1} (5.75–5.78 μm) (the carbonyl band), at 1510–1500 cm^{-1} (6.62–6.67 μm) (the amide II band), at 1235–1165 cm^{-1} (8.10–8.58 μm) (the amide III band), and at 740–730 cm^{-1} (13.51–13.70 μm) (due to the N–H wagging vibration). The spectra of the *cis–trans* forms differ from the above in that the carbonyl band occurs near 1700 cm^{-1} (5.88μm) with weaker bands near 1630 cm^{-1} (6.14 μm) and 1650 cm^{-1} (6.06 μm). The band due to the N–H stretching vibration occurs at 3250 cm^{-1} (3.08 μm) with weak bands on either side and the band due to the N–H wagging vibration occurs at 835–815 cm^{-1} (11.98–12.27 μm).

The carbonyl band of cyclic imides is shifted to higher frequencies if the ring is strained. Cyclic imides do not have an amide II band near 1510 cm^{-1} (6.62 μm).

In general, acyclic imides exhibit two amide I bands and weak amide IV bands have also been observed near 610 cm^{-1} (16.39 μm) and 560 cm^{-1} (17.86 μm).

Ureas, N–CO–N (Carbamides)

The band due to the stretching vibration of the carbonyl group of ureas[29,155–157] occurs at 1705–1635 cm^{-1} (5.82–6.12 μm). The presence of

ring strain tends to increase the frequency of this vibration. Strongly electron-accepting groups on the nitrogen also raise this frequency (amides behave in a similar manner).

In dilute carbon tetrachloride solution, *N*-monoalkyl ureas have three bands due to the N–H stretching vibrations, the bands due to the NH$_2$ asymmetric and symmetric vibrations are at about 3515 cm^{-1} (2.85 μm) and 3415 cm^{-1} (2.93 μm) respectively and that due to the N–H stretching vibration, which varies according to the alkyl substituent, occurs at 3465–3440 cm^{-1} (2.89–2.91 μm). The carbonyl stretching vibration gives rise to a band near 1705 cm^{-1} (5.86 μm).

In dilute solution (non-polar solvent), sym-*N*,*N'*-dialkylureas have, essentially, a single band due to the N–H stretching vibration in the region 3465–3435 cm^{-1} (2.89–2.91 μm) and a strong band due to the C=O stretching vibration at about 1695 cm^{-1} (5.90 μm). Also in dilute solution (non-polar solvent), unsym-*N*,*N'*-dialkylureas may exhibit one or two bands due to the N–H stretching vibration.

The amide II band of ureas is usually found at 1560–1515 cm^{-1} (6.41–6.60 μm), for *N*,*N'*-dialkyl substituted ureas two weak bands are observed near 1585 cm^{-1} (6.31 μm) and 1535 cm^{-1} (6.51 μm). For associated (hydrogen-bonded) ureas, the band due to the N–H stretching vibration occurs at 3400–3360 cm^{-1} (2.94–2.98 μm) and that due to the C=O stretching vibration at about 1635 cm^{-1} (6.11 μm). For the monomer, this last band is round at about 1690 cm^{-1} (5.92 μm).

Ureas have a strong, characteristic band at 1360–1300 cm^{-1} (7.35–7.69 μm) due to the asymmetric stretching vibration of the N–C–N group, the band due to the symmetric vibration being of medium intensity and occurring at 1190–1140 cm^{-1} (8.40–8.77 μm).

Table 10.32 Urea C=O stretching vibrations: amide I band

Functional Groups	Region cm^{-1}	Region μm	Intensity IR	Intensity Raman	Comments
Ureas (solid phase)	1680–1635	6.33–6.12	s	w–m	br, primary ureas, i.e. with NH$_2$ group
Ureas (in solution)	1705–1660	5.86–6.02	s	w–m	
Cyclic ureas (five-membered ring) (in solution)	1735–1685	5.76–5.93	s	w–m	Ketone groups in ring increase frequency
–HNCONH$_2$ and $\diagdown$NCONH$_2$ (solid phase)	1680–1635	6.33–6.12	s	w–m	br
–HNCONH–(solid phase)	1670–1615	5.99–6.19	s	w–m	
$\diagdown$NCON$\diagdown$ (solid phase)	1660–1625	6.02–6.15	s	w–m	
Cyclic ureas (solid phase) HN—NH / CO	1690–1660	5.92–6.02	s	w–m	
Diaryl ureas, ArNH–CO–NHAr (solid phase)	~1640	~6.10	s	w–m	
N-Chloro diaryl ureas, ArNCl–CO–NClAr (solid phase)	1735–1710	5.76–5.85	s	w–m	

Table 10.33 Ureas: other bands

Functional Groups	Region cm^{-1}	Region μm	Intensity IR	Intensity Raman	Comments
Ureas	3440–3200	2.91–3.13	m	m–w	NH str
	1605–1515	6.23.6.60	m	w	Amide II band
	1360–1300	7.35–7.69	s–m	m	asym N–C–N str
	1190–1140	8.40–8.77	m	m	sym N–C–N str
–NHCONH$_2$	3440–3400	2.91–2.94	m	m–w	asym NH$_2$ str
	3360–3320	2.98–3.01	m	m–w	NH str
	3240–3200	3.09–3.13	m	m–w	sym NH$_2$ str
	1605–1515	6.23–6.60	s	w	NH$_2$ def vib
	1360–1300	7.35–7.69	s–m	m	asym N–C–N str
	1190–1140	8.40–8.77	m	m	sym N–C–N str
	620–530	16.13–18.87	v		NH$_2$ def vib
$\diagdown$NCONH$_2$	3440–3400	2.91–2.94	m	m–w	asym NH$_2$ str
	3240–3200	3.09–3.13	m	m–w	sym NH$_2$ str
	1605–1515	6.23–6.60	s	w	NH$_2$ def vib
	1360–1300	7.35–7.69	s–m	m	asym N–C–N str
	1190–1140	8.40–8.77	m	m	sym N–C–N str
	620–530	16.13–18.87	v		NH$_2$ def vib
–NHCONH–	3360–3320	2.98–3.01	m	m–w	NH str
	1585–1515	6.31–6.60	v	w	NH def vib
	1360–1300	7.35–7.69	s–m	m	asym N–C–N str
	1190–1140	8.40–8.77	m	m	sym N–C–N str

Table 10.33 (*continued*)

Functional Groups	Region cm^{-1}	Region μm	Intensity IR	Intensity Raman	Comments
NCON	1360–1300	7.35–7.69	s–m	m	asym N–C–N str
	1190–1140	8.40–8.77	m	m	sym N–C–N str
Cyclic ureas HN NH CO	3315–3200	3.02–3.13	m	m–w	
	1450–1440	6.90–9.94	v	m	
	1275–1250	7.84–8.00	m	m	

Table 10.34 Urethane N–H stretching vibrations

Functional Groups	Region cm^{-1}	Region μm	Intensity IR	Intensity Raman	Comments
Primary urethanes, H_2N–CO–O–	3450–3400	2.90–2.94	m	m–w	NH_2 asym str
	3240–3200	3.09–3.13	m	m–w	NH_2 sym str
(Associated) secondary urethanes, –HN–CO–O–	3340–3250	2.99–3.08	m	m–w	Hydrogen- bonded
(Unassociated) secondary urethanes	3410–3390	2.92–2.95	m	m–w	
N-Aryl urethanes (associated), Ar–NH–CO·OR	3460–3295	2.89–3.03	m	m–w	
N-Aryl urethanes (unassociated)	3460–3410	2.89–2.93	m	m–w	

Table 10.35 Urethane C=O stretching vibrations: amide I band

Functional Groups	Region cm^{-1}	Region μm	Intensity IR	Intensity Raman	Comments
Alkyl urethanes, NCO–O–	1740–1680	5.75–5.95	s	w–m	
NH_2CO·OR	1695–1680	5.90–5.95	s	w–m	
RO·CO·NHR	1740–1730	5.75–5.78	s	w–m	
N-Aryl urethanes, Ar–NH–CO·OR (associated) (solid phase)	1735–1705	5.75–5.87	vs	w–m	One or two peaks, strong hydrogen bonding may result in band as low as 1690 cm^{-1}
N-Aryl urethanes (unassociated)	1760–1730	5.68–5.78	vs	w–m	See ref. 163
Alkyl thiocarbamates, –NH–CO–S–	~1695	~5.90	s	w–m	See ref. 53
N-Aryl thiocarbamates	1700–1660	5.88–6.02	s	w–m	
Alkyl carbamoyl chlorides, NR_2·COCl	1745–1735	5.73–5.75	vs	w–m	
Cyclic urethanes (five-membered ring), CO N O C–C	1785–1745	5.60–5.73	vs	w–m	Ring carbonyl groups increase frequency (N-acetyloxazolidones have bands: at ~1795 cm^{-1} and ~1710 cm^{-1}

Table 10.36 Urethane combination N–H deformation and C–N stretching vibrations (amide II band) and other bands

Functional Groups	Region cm^{-1}	Region µm	Intensity IR	Intensity Raman	Comments
Primary urethanes	1630–1610	6.13–6.21	m–s	m	NH$_2$ def
Secondary urethanes (dilute solution)	1530–1500	6.54–6.67	s	m	Absorption due to CHN group 1540–1530 cm^{-1}
Secondary urethanes (associated or in solid phase)	1600–1500	6.25–6.67	s	m	
Urethanes	1265–1200	7.90–8.33	m	m–s	Amide IV band (coupled C–N and C–O stretching vibrations)
N-Aryl urethanes (unassociated)	1550–1500	6.45–6.67	s	m	In-plane N–H bending vib
	1285–1235	7.78–8.10	m	m–s	Ar-N str, does not alter significantly on phase change
	1225–1195	8.14–8.36	vs	m	Amide V band, stronger than C=O band in solution spectra, in solid-phase band occurs at 1260–1200 cm^{-1} (7.94–8.33 µm)
	1090–1040	9.17–9.62	m–s	m–w	C–O str
	570–500	17.54–20.00	w–m		Out-of-plane N–H def vib, in solid phase band occurs at 680–625 cm^{-1} (14.71–16.00 µm) *ortho*-halogen – substituted compounds absorb in range 570–550 cm^{-1} (17.54–18.18 µm)

Urethanes, $\diagdown$N–CO–O– (Carbamates)

The band due to the carbonyl stretching vibration (amide I band) of urethanes[158–163,175,177] occurs in the region 1740–1680 cm^{-1} (5.75–5.95 µm). Primary urethanes have a number of absorptions in the region 3450–3200 cm^{-1} (2.90–3.13 µm) due to the N–H stretching vibration. Secondary urethanes absorb near 3300 cm^{-1} (3.03 µm) if hydrogen bonding occurs and at 3450–3390 cm^{-1} (2.92–2.95 µm) if it is absent. For alkyl primary urethanes in chloroform solution, the amide I band is observed at 1730–1720 cm^{-1} (5.78–5.81 µm), for secondary urethanes (N-monosubstituted) at 1720–1705 cm^{-1} (5.81–5.87 µm), and for tertiary urethanes (N,N-disubstituted) at 1690–1680 cm^{-1} (5.92–5.95 µm). These ranges may be slightly lower in frequency than for other solvents. In the solid phase, primary urethanes may have very broad C=O bands and absorb as low as 1690 cm^{-1} (5.92 µm), otherwise the same general absorption pattern is observed.

Primary urethanes have a medium-to-strong band near 1620 cm^{-1} (6.17 µm) due to the deformation vibrations of the NH$_2$ group. Associated secondary urethanes absorb strongly at 1540–1530 cm^{-1} (6.49–6.54 µm) due to the CHN group vibration (similar to that of secondary amides) and in dilute solution this band is found at 1530–1510 cm^{-1} (6.54–6.62 µm).

References

1. P. Combelas *et al.*, *Ann. Chim. France*, 1970, **5**, 315.
2. L. J. Bellamy and R. J. Pace, *Spectrochim. Acta*, 1963, **19**, 1831.
3. W. A. Seth-Paul, *Spectrochim. Acta*, 1974, **30A**, 1817.
4. V. G. Boitsov and Y. Y. Gotlib, *Opt. Spectrosc.*, 1961, **11**, 372.
5. C. N. R. Rao and R. Venkataraghavan, *Can. J. Chem.*, 1961, **39**, 1757.
6. M. St. Flett, *Spectrochim. Acta*, 1962, **18**, 1537.
7. J. Bellantano and J. R. Baroello, *Spectrochim. Acta*, 1960, **16**, 1333.
8. C. Fayat and A. Foucaud, *Compt. Rend.*, 1966, **263B**, 860.
9. T. L. Brown, *Spectrochim. Acta*, 1962, **18**, 1617.
10. J. L. Mateos *et al.*, *J. Org. Chem.*, 1961, **26**, 2494.
11. H. N. Al-Jallo and M. G. Jalhoom, *Spectrochim. Acta*, 1975, **31A**, 265.
12. J. Dabrowski, *Spectrochim. Acta*, 1963, **19**, 475.
13. K. L. Toack, *Spectrochim. Acta*, 1962, **18**, 1625.
14. R. V. Jones *et al.*, *Can. J. Chem.*, 1959, **37**, 2007.
15. T. L. Brown *et al.*, *J. Phys. Chem.*, 1959, **63**, 1324.
16. R L. Erskine and E, S. Waight, *J. Chem. Soc.*, 1960, 3425.
17. R. Mecke and K. Noack, *Chem. Ber.*, 1960, **93**, 210.
18. T. Gramstad and W. J. Fuglevik, *Spectrochim. Acta*, 1965, **21**, 343.
19. J. A. Pullin and R. L. Werner. *Spectrochim. Acta*, 1965, **21**, 1257,
20. A. R. Katrilzky and R. A. Jones, *Spectrochim. Acta*, 1961, **17**, 64.
21. L. J. Bellamy *et al.*, *Z. Elektrochim.*, 1960, **64**, 563.
22. L. J. Bellamy and P. E. Rogash, *Spectrochim. Acta*, 1960, **16**, 30.
23. L. H. Bellamy and R. L. Williams, *Trans. Faraday Soc.*, 1959, **55**, 14.

24. K. B. Whetsel and R. E. Kagarise, *Spectrochim. Acta*, 1962, **18**, 315, 329 and 341.
25. J. Dabrowski and K. Kamiensa-Trela, *Spectrochim. Acta*, 1966, **22**, 211.
26. H. P. Figeys and J. Nasielski, *Spectrochim. Acta*, 1967, **23A**, 465.
27. R. A. Pethrick and A. D. Wilson, *Spectrochim. Acta*, 1974, **30A**, 1073.
28. P. Bassignana *et al.*, *Spectrochim. Acta*, 1965, **21**, 677.
29. H. K. Hall and R. Zbinden *J. Am. Chem. Soc.*, 1958, **80**, 6428.
30. W. G. Dauben and W. W. Epstein, *J. Org. Chem.*, 1959, **24**, 1595.
31. F. Marquarat, *J. Chem. Soc. B*, 1966, 1242.
32. L. Hough and J. E. Priddle, *J. Chem. Soc.*, 1961, 3178.
33. B. F. Kucherov *et al.*, *Izvest. Acad. Nauk. USSR Otel. Khim. Nauk.*, 1958, 186.
34. H. Zahn and J. Kunde, *Chem. Bar.*, 1961, **94**, 2470.
35. G. L. Caldow and W. H. Thompsom, *Proc. R, Soc. London*, 1960, **245A**, 1.
36. A. D. Buckingham, *Trans. Faraday Soc.*, 1960, **56**, 753.
37. L. J. Bellamy, in *Spectroscopy, Rept Conf. Organ. Hydrocarbon Res. Group, Inst. Petroleum*, 1962, p. 205.
38. H. H. Freedman, *J. Am. Chem. Soc.*, 1960, **82**, 2454.
39. D. Peltier *et al.*, *Compt. Rend.*, 1959, **248**, 1148.
40. C. J. W. Brooks, *J. Chem. Soc.*, 1961, 106.
41. K. Shimzu *et al.*, *Spectrochim. Acta*, 1966, **22**, 1528
42. P. J. Kruger, *Can. J. Chem.*, 1973, **51**, 1363.
43. W. A. Seth-Paul and A. van Duyse, *Spectrochim. Acta*, 1972, **28A**, 211.
44. F. A. Long and R. Bakule, *J. Am. Chem. Soc.*, 1963, **85**, 2313.
45. R. Bauke and F. A. Long, *J. Am. Chem. Soc.*, 1963, **85**, 2309.
46. C. N. R. Rao and R. Venkataraghavan, *Spectrochim. Acta*, 1962, **18**, 273.
47. R. Cataliotti and R. N. Jones, *Spectrochim. Acta*, 1971, **27A**, 2011.
48. C. I. Angell *et al.*, *Spectrochim. Acta*, 1959, **15**, 926.
49. G. Allen *et al.*, *J. Chem. Soc.*, **1960**, 1909.
50. R. N. Jones *et al.*, *Can. J. Chem.*, 1959, **37**, 2007.
51. R. N. Jones and B. S. Gallagher, *J. Am. Chem. Soc.*, 1959, **81**, 5242.
52. L. J. Bellamy and R. L. Williams, *Trans. Faraday Soc.*, 1959, **55**, 14.
53. R. A. Nyquist and W. J. Potts, *Spectrochim. Acta*, 1961, **17**, 679.
54. R. A. Nyquist and W. J. Potts, *Spectrochim. Acta*, 1959, **15**, 514.
55. B. Subrahmanyam *et al.*, *Curr. Sci.*, 1964, **33**, 304.
56. N. L. Allinger *et al.*, *J. Am. Chem. Soc.*, 1960, **82**, 5876.
57. J. Cantacuzene, *J. Chem. Phys.*, 1962, **59**, 186.
58. E. M. Marek *et al.*, *Zh. Prinkl. Spektrosk.*, 1973, **19**, 130.
59. C. E. Griffin, *Spectrochim. Acta*, 1960, **16**, 1464.
60. R. N. Jones and E. Spinner, *Can. J. Chem.*, 1958, **36**, 1020.
61. R. D. Campbell and H. M. Gilow, *J. Am. Chem. Soc.*, 1962, **84**, 1440.
62. R. A. Abramovitch, *Can. J. Chem.*, 1959, **37**, 361 and 1146.
63. R. Mecke and E. Funck, *Z. Electkrochem.*, 1956, **60**, 1124.
64. A. N. Hambly and B. V. O'Grady, *Austral. J. Chem.*, 1963, **16**, 459.
65. M. A. Gianturco and R. G. Pitcher, *Appl. Spectrosc.*. 1965, **19**, 109.
66. J. K. Katon and F. F. Bentley, *Spectrochim. Acta*, 1963, **19**, 639.
67. E. I. Matrosov and M. I. Kabachnik, *Spectrochim. Acta*, 1972, **28A**, 191.
68. R. L. Edwards, *J. Appl. Chem.*, 1960, **10**, 246.
69. J. Y. Savoie and P. Brassard, *Can. J. Chem.*, 1966, **44**, 2867.
70. J. F. Baghi, *J. Am. Chem. Soc.*, 1962, **84**, 177.
71. T. L. Brown, *Spectrochim. Acta*, 1962, **18**, 1065.
72. E. D. Becker *et al.*, *Spectrochim. Acta*, 1963, **19**, 1871.
73. H. Bloom *et al.*, *J. Chem. Soc.*, **1959**, 178.

74. M. A. Slifkin, *Spectrochim. Acta*, 1973, **29A**, 835.
75. A. Marco *et al.*, *Compt. Rend.*, 1972, **274B**, 400.
76. E. Sanicki and T. R. Hauser, *Anal. Chem.*, 1959, **31**, 523.
77. M. R. Padhye and B. G. Vilader, *J. Sci. Instrum. Res.*, 1960, **19B**, 45.
78. W. F. Forbes, *Can. J. Chem.*, 1962, **40**, 1891.
79. E. L. Saier *et al.*, *Anal. Chem.*, 1962, **34**, 824.
80. J. V. Pustinger *et al.*, *Appl. Spectrosc.*, 1964, **18**, 36.
81. R. Blinc *et al.*, *Z. Elecktrochem.*, 1960, **64**, 567.
82. M. Oki and M. Hirota, *Bull. Chem. Soc. Jpn*, 1963, **36**, 290.
83. M. Josien *et al.*, *Compt. Rend.*, 1960, **250**, 4146.
84. C. J. W. Brooks *et al.*, *J. Chem. Soc.*, **1961**, 661.
85. A. J. Collins and K. J. Morgan, *J. Chem. Soc.*, **1963**, 3437.
86. J. R. Barcello and C. Otero, *Spectrosc. Acta*, 1962, **18**, 1231.
87. H. Susi, *Anal. Chem.*, 1959, **31**, 910.
88. R. N. Jones, *Can. J. Chem.*, 1962, **40**, 301.
89. R. J. Jakobsen *et al.*, *Appl. Spectrosc.*, 1968, **22**, 641.
90. J. W. Brasch *et al.*, *Appl. Spectrosc. Rev.*, 1968, **1**, 187.
91. R. J. Jakobsen *et al*, *Appl. Spectrosc.*, 1968, **22**, 641,
92. F. F. Bentley *et al.*, *Spectrosc. Acta*, 1964, **20**, 685.
93. R. J. Jakobsen and J. E. Katon, *Spectrosc. Acta*, 1973, **29A**, 1953.
94. E. Spinner, *J. Chem. Soc.*, **1964**, 4217.
95. M. K. Hargreaves and E. A. Stevenson, *Spectrosc. Acta*, 1965, **21**, 1681.
96. F. Vratny *et al.*, *Anal. Chem.*, 1961, **33**, 1455.
97. J. H. S. Green *et al.*, *Spectrosc. Acta*, 1961, **17**, 486.
98. L. L. Shevchenko, *Russ. Chem. Rev.*, 1963, **32**, 201.
99. A. S. N. Murthy *et al.*, *Trans. Faraday Soc.*, 1962, **58**, 855.
100. F. G. Pearson and R. B. Stasiak, *Appl. Spectrosc.*, 1958, **12**, 116.
101. L. J. Bellamy and R. L. Williams, *J. Chem. Soc.*, **1958**, 3465.
102. S. Pinchas *et al.*, *J. Chem. Soc.*, **1961**, 2382.
103. H. N. Al-Jallo and M. G. Jalhoon, *Spectrosc. Acta*, 1972, **28A**, 1655.
104. K. R. Loos and R. G. Lord, *Spectrosc. Acta*, 1965, **21**, 119.
105. A. R. Katritzky *et al.*, *Spectrosc. Acta*, 1960, **16**, 964.
106. J. Radell and L. A. Harrah, *J. Chem. Phys.*, 1962, **36**, 1571.
107. R. N. Jones and E. Spinner, *Can. J. Chem.*, 1958, **36**, 1020.
108. B. J. Hales *et al.*, *J. Chem. Soc.*, **1957**, 618.
109. H. W. Thompson and D. A. Jameson, *Spectrosc. Acta*, 1958, **13**, 236.
110. J. R. Durig *et al.*, *J. Mol. Struct.*, 1970, **5**, 67.
111. A. W. Baker and G. H. Harris, *J. Am. Chem. Soc.*, 1960, **82**, 1923.
112. F. Dalton *et al.*, *J. Chem. Soc.*, **1960**, 3681.
113. J. K. Wilmshurst, *J. Mol. Spectrosc.*, 1957, **1**, 201.
114. J. L. Lucier and F. F. Bentley, *Spectrosc. Acta*, 1964, **20**, 1.
115. R. S. Tipson and H. S. Isbell, *J. Res. NBS*, 1960, **64A**, 405.
116. R. S. Tipson and H. S. Isbell, *J. Res. NBS*, 1961, **65A**, 249.
117. J. P. Freeman, *J. Am. Chem. Soc.*, 1958, **80**, 5954.
118. B. F. Kucherov *et al.*, *Izvest. Acad. Nauk USSR Otdel. Khim. Nauk*, **1958**, 186.
119. F. Korte *et al.*, *Angew. Chem.*, 1959, **71**, 523.
120. R. A. Nyquist, *Spectrosc. Acta*, 1972, **28A**, 285.
121. H. A. Ory, *Spectrosc. Acta*, 1960, **16**, 1488.
122. E. P. Blanchard and G. Buechi, *J. Am. Chem. Soc.*, 1963, **85**, 955.
123. R. J. Jakobsen and R. E. Wyant, *Appl. Spectrosc.*, 1960, **14**, 61.
124. W. H. Washburn, *Appl. Spectrosc.*, 1964, **18**, 61.
125. S. E. Krikorian and M. Mahpour, *Spectrosc. Acta*, 1973, **29A**, 1233.

126. A. R. Katritzky and R. A. Jones, *J. Chem. Soc.*, **1959**, 2067.
127. A. R. Katritzky and R. A. Jones, *J. Chem. Soc.*, **1960**, 679.
128. R. L. Jones, *J. Mol. Spectrosc.*, 1963, **11**, 411.
129. R. A. Russell and H. W. Thompsom, *Spectrosc. Acta*, 1956, **8**, 138.
130. J. Jakes and S. Krimm, *Spectrosc. Acta*, 1971, **27A**, 35.
131. R. A. Nyquist, *Spectrosc. Acta*, 1963, **19**, 509, 713 and 1595.
132. R. D. Mclachlan and R. A. Nyquist, *Spectrosc. Acta*, 1964, **20**, 1397.
133. M. Bear *et al.*, *J. Chem. Phys.*, 1958, **29**, 1097.
134. C. D. Schmulbach and R. S. Drago, *J. Phys. Chem.*, 1960, **64**, 1956.
135. A. J. Speziale and R. C. Freeman, *J. Am. Chem. Soc.*, 1960, **82**, 903.
136. H. H. Freedman, *J. Am. Chem. Soc.*, 1960, **82**, 2454.
137. E. J. Forbes *et al.*, *J. Chem. Soc.*, **1963**, 835.
138. M. Mashima, *Bull. Chem. Soc. Jpn*, 1962, **35**, 332 and 1862.
139. N. Ogata, *Bull. Chem. Soc. Jpn*, 1961, **34**, 245 and 249.
140. J. E, Devia and J. C. Carter, *Spectrosc. Acta*, 1973, **29A**, 613.
141. T. Miyazawa, *Bull. Chem. Soc. Jpn*, 1961, **34**, 691.
142. C. G. Cannon, *Spectrosc. Acta*, 1960, 16, 302.
143. W. J. Klein and A. R. Plesman, *Spectrosc. Acta*, 1972, **28A**, 673.
144. T. L. Brown, *J. Phys. Chem.*, 1959, **63**, 1324.
145. Y. Kuroda *et al.*, *Spectrosc. Acta*, 1973, **29A**, 411.
146. B. Milligan *et al.*, *J. Chem. Soc.*, **1961**, 1919.
147. H. O. Desseyn and M. A. Herman, *Spectrosc. Acta*, 1967, **23A**, 2457.
148. H. O. Desseyn, *Spectrosc. Acta*, 1974, **30A**, 503.
149. P. J. F. Griffiths and G. D. Morgan, *Spectrosc. Acta*, 1972, **28A**, 1899.
150. K. Machida *et al.*, *Spectrosc. Acta*, 1972, **28A**, 235.
151. T. Uno and K. Machida, *Bull. Chem. Soc. Jpn*, 1961, **34**, 545 and 551.
152. R. H. Wiley and S. C. Slaymaker, *J. Am. Chem. Soc.*, 1958, **80**, 1385.
153. D. E. Ames and T. F. Gray, *J. Chem. Soc.*, **1955**, 631.
154. N. A. Borisevitch and N. N. Khoratovitch, *Opt. Spectrosc.*, 1961, **10**, 309.
155. E. Spinner, *Spectrosc. Acta*, 1959, **15**, 95.
156. Y. Mido, *Spectrosc. Acta*, 1973, **29A**, 1 and 431.
157. D. F. Kutepov and S. S. Dubov, *J. Gen. Chem. Moscow*, 1960, **30(92)**, 3416.
158. S. Pinchas and D. Ben Ishai, *J. Am. Chem. Soc.*, 1957, **79**, 4099.
159. M. Sato, *J. Org. Chem.*, 1961, **26**, 770.
160. A. R. Katritzky and R. A. Jones, *J. Chem. Soc.*, **1960**, 676.
161. J. C. Carter and J. E. Devia, *Spectrosc. Acta*, 1973, **29A**, 623.
162. H. S. Randhawa *et al.*, *Spectrosc. Acta*, 1974, **30A**, 1915.
163. R. A. Nyquist, *Spectrosc. Acta*, 1973, **29A**, 1635.
164. R. J. Koegel *et al.*, *Ann. NY Acad. Sci.*, 1957, **69**, 94.
165. G. B. B. M. Sulherland, *Adv. Protein Chem.*, 1952, **7**, 291.
166. R. J. Koegel *et al.*, *J. Amer. Chem. Soc.*, 1955, **77**, 5708.
167. M. Tsuboi *et al.*, *Spectrosc. Acta*, 1963, **19**, 271.
168. E. Steger *et al.*, *Spectrosc. Acta*, 1963, **19**, 293.
169. J. Weinman *et al.*, *Chim. Phys.-Chim. Biol.*, 1976, **73**, 331.
170. G. Guiheneut and C. Laurence, *Spectrosc. Acta*, 1978, **34A**, 15.
171. R. J. Jakobsen, F. M. Wasacz, *Appl. Spectrosc.*, 1990, **44**, 1478.
172. R. A. Nyquist *et al.*, *Appl. Spectrosc.*, 1990, **44**, 426.
173. R. A. Nyquist *et al.*, *Appl. Spectrosc.*, 1991, **45**, 92, 860, 1075.
174. J. Durig *et al.*, *Int. Rev. Phys. Chem.*, 1990, **9(4)**, 349.
175. R. A. Nyquist, *The Interpretation of Vapour-Phase Spectra*, Sadtler, 1985.
176. G. Eaton *et al.*, *J. Chem. Soc., Faraday Trans.*, 1989, **85**, 3257.
177. P. A. Lang and J. E. Katon, *J. Mol. Struct.*, 1988, **172**, 113.
178. N. A. Shimanko and M. V. Shishkina, *Infrared and U. V. Absorption Spectra Aromatic Esters*, Nauka, Moscow, 1987.
179. W. K. Sutewiz *et al.*, *Biochemistry*, 1993, **32(2)**, 389.
180. D. Steele and A. Muller, *J. Phys. Chem.* 1991, **95**, 6163.
181. T. K. Ha *et al.*, *Spectroschim. Acta*, 1992, 48, 1083.
182. J. R. Durig *et al.*, *J. Mol. Struct.*, 1989, **212**, 169.
183. J. R. Durig *et al.*, *J. Mol. Struct.*, 1989, **212**, 187.

11   Aromatic Compounds

For simplicity and convenience, the modes of vibration of aromatic compounds are considered as separate C–H or ring C=C vibrations. However, as with any 'complex' molecule, vibrational interactions occur and these labels really only indicate the predominant vibration. Substituted benzenes have a number of substituent sensitive bands, that is, bands whose position is significantly affected by the mass and electronic properties (inductive or mesomeric properties) of the substituents. These bands are sometimes referred to as X-sensitive bands. For example, mono-substituted benzenes have six X-sensitive bands, where X represents a substituent. Obviously, the region in which an X-sensitive band may be found is quite large.

In their infrared spectra, the strongest absorptions for aromatic compounds[1,2] occur in the region 900–650 cm^{-1} (11.11–15.27 μm) and are due to the C–H vibrations out of the plane of the aromatic ring (Figure 11.1). These bands are generally weak in Raman spectra. In infrared spectra, most mononuclear and polynuclear aromatic compounds have three or four peaks in the region 3080–3010 cm^{-1} (3.25–3.32 μm), these being due to the stretching vibrations of the ring CH bonds. In Raman spectra, these bands may be strong, but skeletal vibration bands may be even stronger. Ring carbon–carbon stretching vibrations occur in the region 1625–1430 cm^{-1} (6.16–6.99 μm). A number of weak combination and overtone bands occur in the region 2000–1650 cm^{-1} (5.00–6.06 μm). These bands are highly characteristic (Figure 11.2) and can be very useful in the evaluation of the number of substituents on the aromatic ring. Unfortunately, part of this region may be overlapped by strong absorptions due to carbonyl or alkene groups.

As mentioned earlier, some vibrational modes of the aromatic carbon–X bond are affected by the mass of the substituent and so these vibrations are known as X-sensitive modes. The X-sensitive bands normally occur in the regions 1300–1050 cm^{-1} (7.69–9.52 μm), 850–620 cm^{-1} (11.76–16.13 μm), and 580–200 cm^{-1} (17.24–50.00 μm). These vibrational modes are due to (a) the ring carbons 1, 3, and 5 moving radially in phase while the substituent on carbon 1 moves radially out of phase, (b) the in-plane bending of a quadrant of the ring in which the C–X bond length increases as the distance between the carbons 1 and 4 decreases, and (c) the distance between carbons 1 and 4

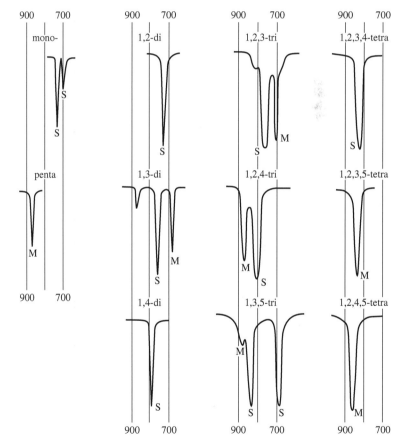

Figure 11.1 In infrared spectra, band intensities and positions in the region 900–600 cm^{-1} (11.11–15.39 μm) are very characteristic of the number of adjacent hydrogen atoms on the aromatic ring. The patterns above are typical of what is observed. These are averages of a great many spectra and so should only be used as a guide to what might be seen

and the C–X bond-length both increasing simultaneously. In general, due to their large range, X-sensitive bands are not very useful in characterisation.

For compounds of the type aryl–metal, a band at 1120–1050 cm^{-1} (8.93–9.52 μm) is observed whose position is dependent on the nature of the metal. References specific to aromatic compounds are: benzyl compounds,[22,23] bridged aromatic compounds,[24] dibenzene oxacyclanes,[30] and phenyl derivatives of Bi, Sb, Si, Ge, Sn, Pb, and P.[25]

Aromatic C–H Stretching Vibrations

As already mentioned, these bands occur in the region 3080–3010 cm^{-1} (3.25–3.32 μm)[3] and are of strong-to-medium intensity. A band with up to five peaks may be observed in this region. As might be expected, monosubstituted benzenes usually exhibit more peaks than di- or trisubstituted benzenes. Alkene C–H stretching vibrations also result in bands in this region, as do both O–H

Chart 11.1 Infrared – substituted benzenes, absorption ranges and intensities in region 1000–300 cm^{-1}

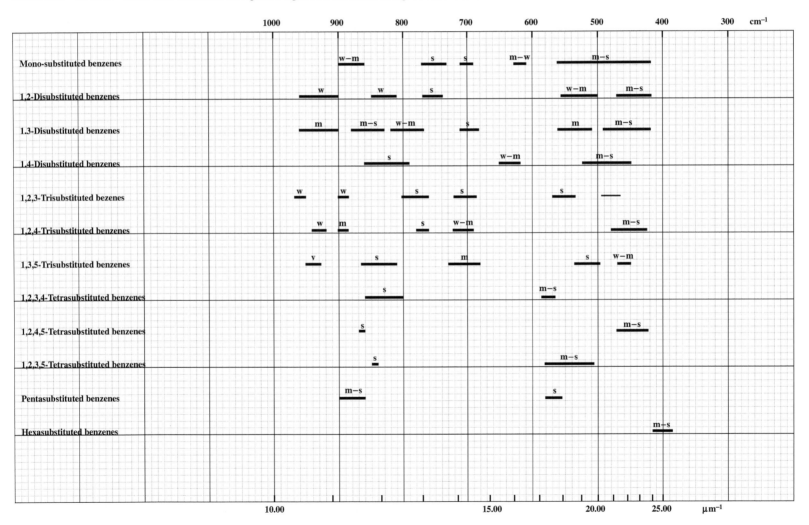

Chart 11.2 Characteristic bands observed in the Raman spectra of substituted benzenes in the region $4000-200\,\text{cm}^{-1}$

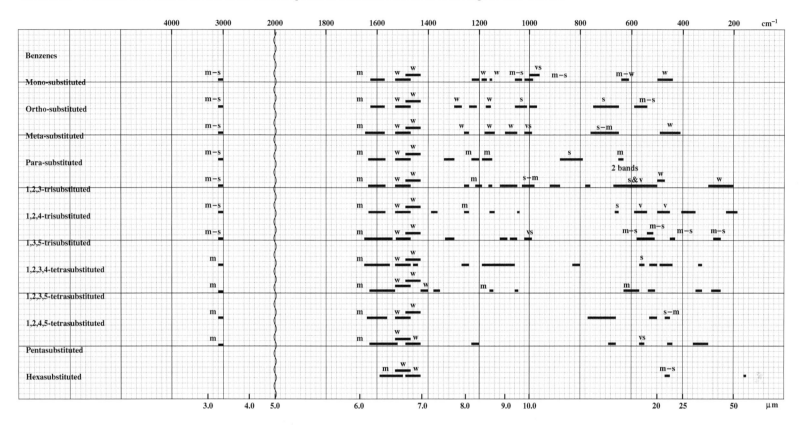

Aromatic In-plane C–H Deformation Vibrations

In Raman spectra, the bands due to the C–H in-plane deformation vibrations, which occur in the region $1290-990\,\text{cm}^{-1}$ ($7.75-10.10\,\mu\text{m}$), are very useful for characterisation purposes and may be very strong indeed. For example, a very strong band in the Raman spectra of mono-, 1,3 di-, and 1,3,5 trisubstituted benzenes is observed near $1000\,\text{cm}^{-1}$ ($10.00\,\mu\text{m}$) which may be the strongest band in the spectrum. In the infrared, a number of C–H in-plane deformation bands (up to six) occur in the region $1290-900\,\text{cm}^{-1}$

($7.75-11.11\,\mu\text{m}$). Although these bands are usually sharp, they are of weak-to-medium intensity. In infrared, these bands are not normally of importance for interpretation purposes although they can be used. In fact, a number of interactions are possible, thus necessitating great care in the interpretation of bands in this region. Polar ring substituents may result in an increase in the intensity of these bands. Additional difficulties may also arise due to the presence of other bands in this region, e.g. due to C–C, C–O stretching vibrations.

Aromatic Out-of-plane C–H Deformation Vibrations and Ring Out-of-plane Vibrations in the Region $900-650\,\text{cm}^{-1}$

The frequencies of the C–H out-of-plane deformation vibrations are mainly determined by the number of adjacent hydrogen atoms on the ring and not very

and N–H stretching vibrations, although the latter bands are much broader than those due to the aromatic C–H stretching vibration. In general, a strong band due to C–H stretching vibrations is observed in the Raman spectra of benzenes at $3070-3030\,\text{cm}^{-1}$ ($3.26-3.30\,\mu\text{m}$).

much affected by the nature of the substituent(s),[4,5] although strongly electron-attracting substituent groups, such as nitro-, can result in an increase of about $30\,cm^{-1}$ in the frequency of the vibration. In infrared spectra, these bands give an important means for determining the type of aromatic substitution. Although normally strong, they are often not the only strong bands in the region since, for example, the carbon–halogen bond vibration may also give rise to absorptions in this region. As always, any interpretation should, if possible, be supported by the presence of more than one band. The patterns observed in infrared spectra are given in Figure 11.1 and may be used as a guide to the absorptions in this region.

The C–H out-of-plane deformation bands are as follows:

(a) Monosubstituted benzenes[4] have two strong absorptions, one at $820-720\,cm^{-1}$ $(12.20-13.89\,\mu m)$ and the other at $710-670\,cm^{-1}$ $(14.08-14.93\,\mu m)$. The second of these bands is usually not as intense as the first.

(b) *Ortho*-disubstituted benzenes[6] have a strong absorption at $790-720\,cm^{-1}$ $(12.66-13.89\,\mu m)$.

(c) *Meta*-disubstituted benzenes have two medium-intensity absorptions, one at $960-900\,cm^{-1}$ $(10.42-11.11\,\mu m)$, the other at $880-830\,cm^{-1}$ $(11.36-12.05\,\mu m)$, a weak band at $820-765\,cm^{-1}$ $(12.20-13.07\,\mu m)$, and a medium-to-strong band at $710-680\,cm^{-1}$ $(14.08-14.71\,\mu m)$.

(d) *Para*-disubstituted benzenes[6] absorb strongly at $860-780\,cm^{-1}$ $(11.63-12.82\,\mu m)$.

(e) 1,2,3-Trisubstituted benzenes[7,8] absorb strongly at $800-750\,cm^{-1}$ $(12.50-13.33\,\mu m)$ and at $740-685\,cm^{-1}$ $(13.51-14.60\,\mu m)$, the first band often not being as intense as the second.

(f) 1,2,4-Trisubstituted benzenes[7] have a medium absorption at $940-840\,cm^{-1}$ $(10.64-11.90\,\mu m)$ and a strong band at $780-760\,cm^{-1}$ $(12.82-13.16\,\mu m)$.

(g) 1,3,5-Trisubstituted benzenes[7] have a strong absorption at $865-810\,cm^{-1}$ $(11.56-12.35\,\mu m)$ and a band of lesser intensity at $730-660\,cm^{-1}$ $(13.70-15.15\,\mu m)$.

(h) 1,2,3,4-Tetrasubstituted benzenes absorb strongly at $860-800\,cm^{-1}$ $(11.63-12.50\,\mu m)$.

(i) 1,2,3,5-Tetrasubstituted benzenes[9,34] absorb strongly at $850-840\,cm^{-1}$ $(11.76-11.90\,\mu m)$.

(j) 1,2,4,5-Tetrasubstituted benzenes absorb strongly at $870-860\,cm^{-1}$ $(11.49-11.63\,\mu m)$.

(k) Pentasubstituted benzenes have a band of medium-to-strong intensity at $900-860\,cm^{-1}$ $(11.11-11.63\,\mu m)$.

Mono-, 1,3-di-, and 1,3,5-trisubstituted benzenes have a strong band in the region $730-660\,cm^{-1}$ $(13.70-15.15\,\mu m)$. In the same region, 1,2- and 1,4-disubstituted benzenes absorb weakly or not at all, depending on whether the two substituent groups are different or not. When the substituents are identical, symmetry results in this vibration being infrared inactive. Trisubstituted 1,2,3- and 1,2,4-benzenes also absorb in this range.

It is both useful and convenient to summarize the C–H out-of-plane vibrations in terms of the number of adjacent hydrogen atoms:

1. Six adjacent hydrogen atoms (e.g. benzene), band at $671\,cm^{-1}$
2. Five adjacent hydrogen atoms (e.g. monosubstituted aromatics), band at $820-720\,cm^{-1}$.
3. Four adjacent hydrogen atoms (e.g. *ortho*-substituted aromatics), band at $790-720\,cm^{-1}$.
4. Three adjacent hydrogen atoms (*meta*- and 1,2,3-trisubstituted aromatics), band at $830-750\,cm^{-1}$.
5. Two adjacent hydrogen atoms (e.g. *para*- and 1,2,3,4-tetrasubstituted aromatics), band at $880-780\,cm^{-1}$.
6. An isolated hydrogen atom (e.g. *meta*-, 1,2,3,5-tetra-, 1,2,4,5-tetra-, and pentasubstituted aromatics), band at $935-810\,cm^{-1}$.

An additional band is observed at $745-690\,cm^{-1}$ $(13.42-14.49\,\mu m)$ in the spectra of monosubstituted, 1,3-disubstituted, compounds.

A coupling between adjacent hydrogen atoms is also observed for naphthalenes, phenanthrenes, pyridines, quinolines (the nitrogen atom being treated as a substituted carbon atom of a benzene ring), and other aromatic compounds.

Nitro-substituted benzenes have a band, in addition to that expected, near $700\,cm^{-1}$ which is believed to involve an NO_2 out-of-plane bending vibration.[32]

In Raman spectra, the out-of-plane deformation bands are usually weak.

Aromatic C=C Stretching Vibrations

The ring carbon–carbon stretching vibrations occur in the region $1625-1430\,cm^{-1}$ $(6.16-6.99\,\mu m)$.[10–14] For aromatic six-membered rings, e.g. benzenes and pyridines, there are two or three bands in this region due to skeletal vibrations, the strongest usually being at about $1500\,cm^{-1}$ $(6.67\,\mu m)$. In the case where the ring is conjugated further, a band at about $1580\,cm^{-1}$ $(6.33\,\mu m)$ is also observed.

In general, the bands are of variable intensity and are observed at $1625-1590\,cm^{-1}$ $(6.15-6.29\,\mu m)$, $1590-1575\,cm^{-1}$ $(6.29-6.35\,\mu m)$, $1525-1470\,cm^{-1}$ $(6.56-6.80\,\mu m)$, and $1465-1430\,cm^{-1}$ $(6.83-6.99\,\mu m)$. In Raman spectra, the band near $1600\,cm^{-1}$ $(6.25\,\mu m)$ is sharp and strong. A band at $1380-1250\,cm^{-1}$ $(7.25-8.00\,\mu m)$ may also be observed but this band is often overlapped by the CH deformation vibrations of alkyl groups. A weak band near $1000\,cm^{-1}$ may also be observed.

For substituted benzenes with identical atoms or groups on all *para*- pairs of ring carbon atoms, the vibrations causing the band at 1625–1590 cm^{-1} (6.15–6.29 µm) (and also the band at 730–680 cm^{-1} (13.70–14.71 µm) – see above) are infrared inactive due to symmetry considerations, the compounds having a centre of symmetry at the ring centre. Hence, benzenes with a centre of symmetry i.e. 1,4 di-, 1,2,4,5 tetra- and hexasubstituted benzenes have no infrared bands near 1600 cm^{-1} (6.25 µm) and 1580 cm^{-1} (6.33 µm). If the groups on a *para*- pair of carbon atoms are different then there is no centre of symmetry and the vibration(s) are infrared active. With heavy substituents, the bands near 1600, 1580, 1490 and 1440 cm^{-1} shift to lower wavenumbers. They also become broader with increase in the number of substituents. If there is no ring conjugation, the band near 1600 cm^{-1} is stronger than that near 1580 cm^{-1}. For alkyl substituents, the band near 1580 cm^{-1} appears as a shoulder on that near 1600 cm^{-1}.

When a substituent is C=O, C=C, C≡N, or NO$_2$ and is directly conjugated to the ring, or is a heavy element such as Cl, Br, I, S, P, or Si, a doublet is observed at 1625–1575 cm^{-1} (6.15–6.35 µm).[36,37] Substituents resulting in conjugation, such as C=C and C=O, increase the intensity of this doublet.[35]

For monosubstituted benzenes with strong electron acceptor or donor groups, the bands at 1625–1590 cm^{-1} (6.15–6.29 µm) and 1590–1575 cm^{-1} (6.29–6.35 µm) are of medium intensity, the second band being the weaker, but for weakly-interacting groups these bands are both weak.

For meta-disubstituted benzenes, the intensity of the band at about 1600 cm^{-1} (6.25 µm) is directly dependent on the sum of the electronic effects of the substituents whereas for *para*-disubstituted benzenes it is dependent on the difference of the electronic effects of the substituents. For example, due to the large dipole changes possible for *para*-disubstituted compounds in which one group is *ortho–para*-directing and the other is *meta*-directing, the band at 1625–1590 cm^{-1} (6.15–6.29 µm) is quite intense. In general, mono-, *meta*-, di-, and 1,3,5-trisubstituted benzenes have strong bands at 1625–1590 cm^{-1} (6.15–6.29 µm) and at 730–680 cm^{-1} (13.70–14.71 µm).

A fairly weak band is observed in the region 1465–1430 cm^{-1} (6.83–6.99 µm) for aromatic compounds, except *para*-disubstituted benzenes for which the range is 1420–1400 cm^{-1} (7.04–7.14 µm). A band in the range 1510–1470 cm^{-1} (6.62–6.80 µm) is observed for monosubstituted, *ortho*- and *meta*-disubstituted, and 1,2,3-trisubstituted benzenes, whereas for *para*-disubstituted and 1,2,4-trisubstituted compounds this band occurs at 1525–1480 cm^{-1} (6.56–6.76 µm). (The differences noted for *para*- compounds are useful in isomer studies.) This last band (at ∼1500 cm^{-1}) is relatively strong for electron donor groups but is otherwise weak or absent, e.g. for the carbonyl group it is very weak. The bands at 1500–1400 cm^{-1} (6.67–7.14 µm) cannot be misinterpreted as due to olefinic C=C stretching vibrations since the latter lie outside this range. However, the band near

1450 cm^{-1} (6.90 µm) may be obscured by the band due to the aliphatic C–H deformation vibration.

Overtone and Combination Bands

Overtone and combination bands due to the C–H out-of-plane deformation vibrations occur in the region 2000–1660 cm^{-1} (5.00–6.02 µm).[15] The absorption patterns observed are characteristic of different benzene ring substitutions (see Figure 11.2, which gives a guide as to what may be observed for a given compound). These bands are weak and it may, therefore, be necessary in some cases to use cells of longer path length or to use a more concentrated sample. Interference in this region from olefinic C=C and carbonyl C=O absorptions may also occur.

Aromatic Ring Deformation Vibrations Below 700 cm^{-1}

Some bands in this region are quite sensitive to changes in the nature and position of substituents,[16–23] although other bands (due to certain vibrations of

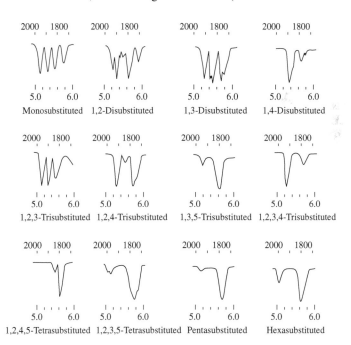

Figure 11.2

aromatic rings) depend mainly on the distribution and number of substituents rather than on their chemical nature or mass, so that these latter vibrations, together with the out-of-plane vibrations of the ring hydrogen atoms, are extremely useful in determining the positions of substituents.

Two bands usually observed are those due to the in-plane and out-of-plane ring deformation vibrations. The in-plane deformation vibration is at higher frequencies than the out-of-plane vibration and is generally weak for mono- and *para*-substituted benzenes, often also being masked by other stronger absorptions which may occur due to the substituent group.

For monosubstituted aromatics, the band due to the out-of-plane ring deformation vibration occurs as follows for the stated substituents:

(a) $\diagup C = C \diagdown$, $-C \equiv C-$, or $-C \equiv N$: near 550 cm^{-1} (18.18 µm);

(b) an electron donor such as $-OH$ or $-NH_2$: near 500 cm^{-1} (20.00 µm);

(c) a halogen or alkyl groups: in the range 500–440 cm^{-1} (20.00–22.73 µm);

(d) an electron acceptor such as NO_2 or $COOH$: below 450 cm^{-1} (above 22.22 µm).

For *meta*-disubstituted compounds, this band occurs in the region 460–415 cm^{-1} (21.74–24.10 µm) except when the substituents are electron-accepting groups in which case the range is 490–460 cm^{-1} (20.41–21.74 µm). The band for *para*-disubstituted benzenes with electron-donating substituents occurs at 520–490 cm^{-1} (19.23–20.41 µm), exceptions being cyano- compounds which absorb at about 545 cm^{-1} (18.35 µm). Phthalides have bands at 520–490 cm^{-1} (19.23–20.41 µm) and 490–470 cm^{-1} (20.41–21.28 µm).

Alkyl-substituted diphenyl compounds exhibit three bands of medium-to-strong intensity, due to ring deformation vibrations, at 620–605 cm^{-1} (16.13–16.53 µm), 490–455 cm^{-1} (20.41–21.98 µm), and 410–400 cm^{-1} (24.39–25.00 µm). A number of 1,2-dialkyl-substituted diphenyls have a band

Table 11.1 Aromatic =C–H and ring C=C stretching vibrations

Functional Groups	Region		Intensity		Comments
	cm^{-1}	µm	IR	Raman	
=C–H	3105–3000	3.22–3.33	m	s	A number of peaks, decreasing in number with increase in substitution.
–C=C–	1625–1590	6.16–6.29	v	m–s, dp	Usually ~1600 cm^{-1}
	1590–1575	6.29–6.35	v	m	Strongest band if conjugated, usually ~1580 cm^{-1}
	1525–1470	6.56–6.80	v	w	Usually ~1470 cm^{-1} for acceptors and ~1510 cm^{-1} for electron donors
	1470–1430	6.80–6.99	v	w	

Table 11.2 Aromatic =C–H out-of-plane deformation vibrations and other bands in region 900–675 cm^{-1}

Functional Groups	Region		Intensity		Comments
	cm^{-1}	µm	IR	Raman	
Monosubstituted benzenes	900–860	11.11–11.63	w–m	w	Out-of-plane def vib (5H)
	820–720	12.20–13.89	s	w	Out-of-plane def vib (5H)
	710–670	14.08–14.93	s	w	Ring out-of-plane def vib
1,2-Disubstituted benzenes	960–900	10.42–11.05	w	w	Out-of-plane def vib (4H)
	850–810	11.76–12.35	w	w	Out-of-plane def vib (4H)
	790–720	12.66–13.89	s	w	Out-of-plane def vib (4H)
1,3-Disubstituted benzenes	960–900	10.42–11.11	m	w	Out-of-plane def vib (1H)
	880–830	11.36–12.05	m	w	Out-of-plane def vib (3H)
	820–765	12.20–13.07	m–s	w	Out-of-plane def vib (3H)
	710–680	14.08–14.71	s	w	Ring out-of-plane def vib
	650–630	15.38–15.87	–	m	

Table 11.2 (*continued*)

Functional Groups	Region cm⁻¹	Region μm	Intensity IR	Intensity Raman	Comments
1,4-Disubstituted benzenes	860–780	11.63–12.82	s	w	Out-of-plane def vib (2H)
	710–680	14.08–14.71	w–m	w	
1,2,3-Trisubstituted benzenes	965–950	10.36–10.53	w	w	Out-of-plane def vib (3H)
	900–885	11.11–11.30	w	w	Out-of-plane def vib (3H)
	830–760	12.05–13.10	s	w	Out-of-plane def vib (3H)
	740–685	13.51–14.60	s	w	Out-of-plane def vib (3H)
1,2,4-Trisubstituted benzenes	940–885	10.64–11.30	m–s	w	Out-of-plane def vib (1H)
	860–840	11.63–11.90	m–s	w	Out-of-plane def vib (2H)
	780–760	12.82–13,16	s	w	Out-of-plane def vib (2H)
	740–690	13.51–14.49	w–m	w	Out-of-plane def vib (2H)
1,3,5-Trisubstituted benzenes	890–830	11.24–12.05	w–m	w	Out-of-plane def vib (1H)
	865–810	11.56–12.35	s–m	w	Out-of-plane def vib (1H)
	730–660	13.70–15.15	m–s	w	Ring out-of-plane def vib
1,2,3,4-Tetrasubstituted benzenes	860–780	11.63–12.82	s	w	Out-of-plane def vib (2H)
1,2,4,5-Tetrasubstituted benzenes	870–860	11.49–11.63	s	w	Out-of-plane def vib (1H)
	820–790	12.20–12.66	w–m	w	
1,2,3,5-Tetrasubstituted benzenes	850–840	11.76–11.90	s	w	Out-of-plane def vib (1H), see ref. 34
Pentasubstituted benzenes	900–860	11.11–11.63	m–s	w	Out-of-plane def vib (1H)

Table 11.3 Aromatic ring deformation vibrations

Functional Groups	Region cm⁻¹	Region μm	Intensity IR	Intensity Raman	Comments
Monosubstituted benzenes	630–605	15.87–16.53	m–w	m, dp	In-plane ring def vib
	560–415	17.86–24.10	m–s	m–w	Out-of-plane ring def vib
1,2-Disubstituted benzenes	555–495	18.02–20.20	w–m	m	In-plane ring def vib
	470–415	21.28–24.10	m–s	w	Out-of-plane ring def vib
1,3-Disubstituted benzenes	560–505	17.86–19.80	m	m	In-plane ring def vib. Medium intensity Raman band at 765–645 cm⁻¹
	490–415	20.41–24.10	m–s	w	Out-of-plane ring def vib
1,4-Disubstituted benzenes	650–615	15.38–16.26	w–m	m, p	In-plane ring def vib
	520–445	19.23–22.47	m–s	w	Out-of-plane ring def vib (except for CN-substituted benzenes)
1,2,3-Trisubstituted benzenes	670–500	14.93–20.00		s	
	570–535	17.54–18.69	s	v	Out-of-plane ring def vib
	~485	~20.62		w	
	300–200	33.33–50.00		w	Two bands
1,2,4-Trisubstituted benzenes	580–540	17.24–18.52		v	
	475–425	21.05–23.53	m–s	w	Out-of-plane ring def vib
1,3,5-Trisubstituted benzenes	580–510	17.24–19.61	m–s		

(*continued overleaf*)

Table 11.3 (*continued*)

Functional Groups	Region		Intensity		Comments
	cm^{-1}	μm	IR	Raman	
	535–495	18.69–20.20	s	m–s	Out-of-plane ring def vib
	470–450	21.28–22.22	w–m		
	280–250	35.71–40.00		m–s	
1,2,3,4-Tetrasubstituted benzenes	585–565	17.09–17.70	m–s	s	
1,2,3,5-Tetrasubstituted benzenes	580–505	17.24–19.80	m–s	v	Out-of-plane ring def vib
1,2,4,5-Tetrasubstituted benzenes	470–420	21.28–23.81	m–s	s–m	
Pentasubstituted benzenes	580–555	17.24–18.02	s	vs	
Hexasubstituted benzenes	415–385	24.10–25.97	m–s	m–s	
Alkyl-substituted diphenyls	620–605	16.13–16.53	m–s		
	490–455	20.41–21.98	m–s		
	410–400	24.39–25.00	m–s		

Table 11.4   Aromatic =C–H in-plane deformation vibrations

Functional Groups	Region		Intensity		Comments
	cm^{-1}	μm	IR	Raman	
Monosubstituted benzenes	1250–1230	8.00–8.13	w		
	1195–1165	8.37–8.58	w–m	w	
	1175–1130	8.51–8.85	w	w	
	1085–1050	9.22–9.52	m	w	
	1040–1000	9.62–10.00	w–m	m–s, p	
	1010–990	9.90–10.10	w	s, p	sh
1,2-Disubstituted benzenes	1290–1250	7.75–8.00	w	w	
	1230–1215	8.13–8.23	–	m	Alkyl benzenes
	1170–1150	8.55–8.70	w–m	w	
	1150–1110	8.70–9.01	w–m	v	
	1055–1020	9.48–9.80	m	s	
1,3-Disubstituted benzenes	1300–1240	7.69–8.06	w	w	
	1170–1150	8.55–8.70	w–m	w	
	1105–1085	9.05–9.22	w	w	
	1085–1065	9.22–9.39	v	w	
	1010–990	9.90–10.10	w	s, p	sh
1,4-Disubstituted benzenes	1270–1250	7.87–8.00	w–m	w	
	1230–1215	8.13–8.23	–	m	Alkyl benzenes
	1185–1165	8.44–8.58	v	m	
	1130–1110	8.85–9.01	v	v	
	1025–1005	9.76–9.95	w–m	w	
	995–975	10.05–10.26	w		
1,2,3-Trisubstituted benzenes	1170–1150	8.55–8.70	w	w	
	1085–1065	9.22–9.39	m	w	

Table 11.4 (*continued*)

Functional Groups	Region		Intensity		Comments
	cm^{-1}	μm	IR	Raman	
1,2,4-Trisubstituted benzenes	1030–1010	9.71–9.90	m–w	s–m	
	1220–1200	8.20–8.33	w	w	
	1160–1140	8.62–8.77	m–w	w	
1,3,5-Trisubstituted benzenes	1040–1020	9.62–9.80	m–w	w	
	1275–1255	7.84–7.97	m–w	w	
	1180–1160	8.47–8.62	m–w	w	
1,2,4,5-Tetrasubstituted benzenes	1040–995	9.62–10.05	w	vs	
	1280–1260	7.81–7.94	w	w	
	1205–1185	8.30–8.44	w	w	

at 560–545 cm^{-1} (17.86–18.35 μm) and 1,3-dialkyl-substituted diphenyls have a band near 530 cm^{-1} (18.87 μm).

A band due to ring breathing coupled with C–X stretching occurs in the region 540–490 cm^{-1} (18.52–20.41 μm), where X = CH$_3$, CD$_3$, OH, NO$_2$, NH$_2$, F, CN, CHO. Out-of-plane deformations of the benzene ring occur in the region 550–440 cm^{-1} (18.18–22.73 μm) C–X (X as above). Ring deformations also occur in the region 240–140 cm^{-1} (41.67–71.43 μm).

Polynuclear Aromatic Compounds

Polynuclear, aromatic, condensed-ring compounds absorb in the same general regions as benzene derivatives[26-31] and therefore the previous section should be noted carefully. (A study of pyrenes has been published.[31])

Naphthalenes,

Naphthalenes[26-28] have a band of medium intensity in the region 1620–1580 cm^{-1} (6.17–6.33 μm) and a band near 1515 cm^{-1} (6.56 μm) and 1395 cm^{-1} (7.17 μm). As a result of C–H out-of-plane deformation vibrations, 1-substituted naphthalenes absorb at 810–775 cm^{-1} (12.35–12.90 μm) due to the presence of three adjacent hydrogen atoms on a ring, and at 780–760 cm^{-1} (12.82–13.16 μm) due to four adjacent hydrogen atoms. 2-Substituted naphthalenes absorb at 760–735 cm^{-1} (13.16–13.61 μm) due to four adjacent hydrogen atoms, at 835–800 cm^{-1} (11.98–12.50 μm) due to two adjacent hydrogen atoms, and at 895–825 cm^{-1} (11.17–12.12 μm) due to a single atom. Table 11.2 correlating the C–H out-of-plane bending vibrations

to the number of adjacent hydrogen atoms on the aromatic ring is, of course, applicable here. There are also a number of bands in the region 1400–1000 cm^{-1} (7.14–10.00 μm).

Mono- and dialkyl-substituted naphthalenes have a strong band at 645–615 cm^{-1} (15.50–16.26 μm) and a band of variable intensity at 490–465 cm^{-1} (20.41–21.51 μm). Both naphthalenes and anthracenes have a band at about 475 cm^{-1} (21.05 μm) due to the out-of-plane ring vibrations.

As a result of the C–H out-of-plane vibrations of adjacent aromatic hydrogen atoms, tetrahydronaphthalenes (tetralins), and polynuclear aromatic compounds in general, have absorption bands as follows:

four adjacent hydrogen	770–740 cm^{-1}	(12.99–13.51 μm);
three adjacent aromatic hydrogen atoms	815–775 cm^{-1} 760–730 cm^{-1}	(12.27–12.90 μm); (13.10–13.70 μm);
two adjacent aromatic hydrogen atoms	850–800 cm^{-1}	(11.76–12.50 μm);
one isolated aromatic hydrogen atom,	900–825 cm^{-1}	(11.11–12.12 μm).

Anthracenes, *, and Phenanthrenes*

Anthracenes absorb near 1630 cm^{-1} (6.14 μm) and near 1550 cm^{-1} (6.45 μm) whilst phenanthrenes have two bands near 1600 cm^{-1} (6.25 μm) and one band near 1500 cm^{-1} (6.67 μm). Anthracenes also have one or two strong bands in the region 900–650 cm^{-1} (11.11–15.38 μm). The higher frequency band near 900 cm^{-1} (11.11 μm) is associated with the 9,10 hydrogen atoms and

Table 11.5 Polynuclear aromatic compounds

Functional Groups	Region		Intensity		Comments
	cm^{-1}	μm	IR	Raman	
Naphthalenes	1620–1580	6.17–6.33	m	s–m	C=C str, often a doublet
	1550–1505	6.45–6.65	m	m	C=C str
	1400–1370	7.14–7.19	m	s–m	A strong band is observed for mono- and di-substituted naphthalenes
	645–620	15.50–16.13	m–s	m	In-plane ring vib
	485–465	20.62–21.51	v	w–m	Out-of-plane vib
1-Monosubstituted naphthalenes	810–775	12.35–12.90	s	w–m	C–H out-of-plane def vib (3H)
	780–760	12.82–13.16	s	w–m	C–H out-of-plane def vib (4H)
2-Monosubstituted naphthalenes	895–825	11.17–12.12	s	w–m	Out-of-plane def vib (1H)
	835–800	11.98–12.50	s	w–m	Out-of-plane def vib (2H)
	760–735	13.16–13.61	m–s	w–m	Out-of-plane def vib (4H)
Mono- and dialkyl-substituted naphthalenes	645–615	15.50–16.25	m–s	m	
	490–465	20.41–21.51	v	w–m	Out-of-plane ring def vib
Phenanthrenes	~1600	~6.25	m	m–s	Two bands
	~1500	~6.67	m	m	
	750–730	13.33–13.70	s	w	
Anthracenes	1640–1620	6.10–6.17	m	m	
	~1550	~6.45	m	m	Not present with 9,10 substitution
	1415–1380	7.07–7.25	w	vs	ring str
	900–650	11.11–15.38	s	m–w	One or two bands
	430–390	23.26–25.64		s	
Mono- and dimethyl 1,2-benzanthracenes	~680	~14.71			Possibly skeletal vib

Table 11.6 Substituted naphthalenes: characteristic C–H vibrations

Hydrogen atom positions on one ring[†]	Region		Intensity		Comments
	cm^{-1}	μm	IR	Raman	
1,2,3,4	800–760	12.50–13.10	s	w–m	Four adjacent hydrogen atoms, out-of-plane vib
	770–725	12.99–13.79	s–vs	w–m	Four adjacent hydrogen atoms, out-of-plane vib
1,2,3	820–775	12.20–12.90	s	w–m	Three adjacent hydrogen atoms, out-of-plane vib
	775–730	12.90–13.70	s	w–m	Three adjacent hydrogen atoms, out-of-plane vib
1,2,4	925–885	10.81–11.30	m	w–m	
	900–835	11.11–11.98	m–s	w–m	Isolated hydrogen atom, out-of-plane vib
	850–805	11.76–12.42	vs	w–m	Two adjacent hydrogen atoms, out-of-plane vib

Table 11.6 (*continued*)

Hydrogen atom positions on one ring[†]	Region cm⁻¹	Region μm	Intensity IR	Intensity Raman	Comments
1,2	835–800	11.98–12.50	s	w–m	Two adjacent hydrogen atoms, out-of-plane vib
	755–720	13.25–13.89	m–s	w–m	
	775–765	12.90–13.07		m–s	
	525–515	19.05–19.42		w	
2,3	835–810	11.98–12.35	s–vs	w–m	Two adjacent hydrogen atoms, out-of-plane vib
1,3	905–865	11.05–11.56	m–s	w–m	Isolated hydrogen atom, out-of-plane vib
	875–840	11.43–11.90	s	w–m	Isolated hydrogen atom, out-of-plane vib
1,4	890–870	11.24–11.49	s	w–m	Isolated hydrogen atom, out-of-plane vib
1 or 2	900–855	11.11–11.70	m	w–m	Isolated hydrogen atom, out-of-plane vib
	720–650	13.89–15.38		m–s	
	535–510	18.69–19.61		w	

† The numbers refer to the substitution pattern of hydrogen atoms so chosen as to give the lowest possible numbering.

disappears if these are substituted. Both anthracenes and naphthalenes have a band at about 475 cm⁻¹ (21.05 μm) due to out-of-plane ring vibrations. Spectra of anthracene and acridene derivatives are available elsewhere,[33] although the complete normal infrared range is not covered.

References

1. G. Varsányi, *Vibrational Spectra of Benzene Derivatives*, Academic Press, New York, 1969.
2. T. F. Ardyukova *et al.*, *Atlas of Spectra of Aromatic and Heterocyclic Compounds*, Nauka Sib. Otd., Novosibirsk, 1973.
3. S. E. Wiberly *et al.*, *Anal. Chem.*, 1960, **32**, 217.
4. S. Higuchi *et al.*, *Spectrochim. Acta*, 1974, **30A**, 463.
5. D. H. Wiffen, *Spectrochim. Acta*, 1955, **7**, 253.
6. A. Stojikjkovic and D. H. Whiffen, *Spectrochim. Acta*, 1958, **12**, 47 and 57.
7. J. H. S. Green *et al.*, *Spectrochim. Acta*, 1971, **27A**, 793 and 807.
8. V. P. Fedorov *et al.*, *Vorp. Mol. Spectrosc.*, 1971, 41.
9. G. Varsányi and P. Soyar, *Acta Chim. Budapest*, 1973, **76**, 243.
10. A. R. Katritzky, *J. Chem. Soc.*, 1958, 4162.
11. A. R. Katritzky and P. Simmons, *J. Chem. Soc.*, 1959, 2058.
12. A. R. Katritzky and J. M. Lagowski, *J. Chem. Soc.*, 1958, 4155.
13. A. R. Katritzky and R. A. Jones, *J. Chem. Soc.*, 1959, 3670.
14. A. R. Katritzky, *J. Chem. Soc.*, 1959, 2051.
15. C. W. Young *et al.*, *Anal. Chem.*, 1951, **23**, 709.
16. R. J. Jakobsen, *Wright–Patterson Air Force Base Tech. Report*, 1962, Documentary Report No. ASD-TDR-62-895 Oct.
18. W. S. Wilcox *et al.*, *WADD Tech. Report*, 1960, 60–333.
19. R. J. Jakobsen, *WADD Tech. Report*, 1960, 60–204.
20. R. J. Jakobsen and F. F. Bentley, *Appl. Spectrosc.*, 1964, **18**, 88.
21. A. Mansingh, *J. Chem. Phys.*, 1970, **52**, 5896.
22. L. Verdonck *et al.*, *Spectrochim. Acta*, 1973, **29A**, 813.
23. L. Verdonck and G. P. van der Kelen, *Spectrochim. Acta*, 1972, **28A**, 51 and 55.
24. B. H. Smith, *Bridged Aromatic Compounds*, Academic Press, New York, 1964, pp. 385–391.
25. L. A. Harrah *et al.*, *Spectrochim. Acta*, 1962, **18**, 21.
26. S. E. Wiberley and R. D. Gonzalez, *Appl. Spectrosc.*, 1961, **15**, 174.
27. J. G. Hawkins *et al.*, *Spectrochim. Acta*, 1957, **10**, 105.
28. B. W. Cox *et al.*, *Spectrochim. Acta*, 1965, **21**, 1633.
29. S. Califano, *J. Chem. Phys.*, 1962, **36**, 903.
30. G. Karogounis and J. Agathokli, *Pract. Acad. Athenon, Greece*, 1970, **44**, 388.
31. R. Mecke and W. E. Klee, *Z. Elektrochem.*, 1961, **65**, 327.
32. J. H. S. Green and D. J. Harrison, *Spectrochim. Acta*, 1970, **26A**, 1925.
33. V. A. Koptyug (ed.), *Atlas of Spectra of Aromatic and Heteroaromatic Compounds*, No. 7. Nauka Sib. Otd., Novosibirsk, 1974.
34. G. Varsányi *et al.*, *Acta Chim. Acad. Sci. Hungary*, 1977, **93**, 315.
35. N. A. Shimanko and M. V. Shishkina, *Infrared and U. V. Absorption Spectra of Aromatic Esters*, Nauka, Moscow, 1987.
36. R. Shanker *et al.*, *J. Raman Spectrosc.*, 1992, **23**, 141.
37. V. Suryanarayana *et al.*, *Spectrochim. Acta*, 1992, **48A**, 1481.

12 Six-membered Ring Heterocyclic Compounds

Pyridine Derivatives,

The spectra of pyridine compounds[1-10] have many of the features of the spectra of homonuclear aromatic compounds, such as bands due to the aromatic C–H stretching vibration, overtones in the region 2080–1750 cm^{-1} (4.81–5.88 µm) etc., with the nitrogen atom behaving in a similar fashion to that observed for a substituted carbon atom. Therefore, the contents of the previous chapter should be noted.

Aromatic C–H Stretching Vibrations

The aromatic C–H stretching vibration of nitrogen heterocyclic aromatic compounds gives rise to a band at 3100–3010 cm^{-1} (3.23–3.32 µm). This band is in the same region as that expected for benzene derivatives and is also similar in that the band is of medium-to-strong intensity and consists of a number of peaks.

Overtone and Combination Bands

As with benzene derivatives, weak overtone and combination bands are observed in the region 2080–1750 cm^{-1} (4.81–5.88 µm), these being characteristic of the position of the substitution (see Figure 12.1). These patterns are intended to serve as a guide as to what might be observed.

C=C and C=N Stretching Vibrations

Interactions between ring C=C and C=N stretching vibrations result in two strong-to-medium intensity absorptions about 100 cm^{-1} (0.4 µm)

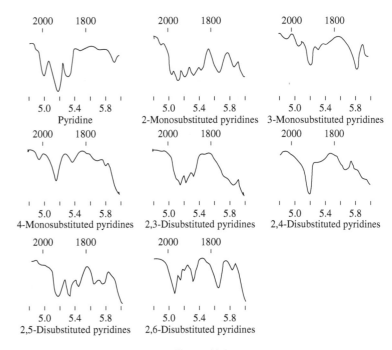

Figure 12.1

apart. These absorptions occur at 1615–1575 cm^{-1} (6.19–6.35 µm) and 1520–1465 cm^{-1} (6.58–6.83 µm), the higher-frequency band often having another medium-intensity band on its low-frequency side which is found at 1590–1555 cm^{-1} (6.29–6.43 µm). A strong band is usually observed in the region 1000–985 cm^{-1} (10.00–10.15 µm), but this band may be very weak or undetectable for 3-substituted pyridines.

Ring C–H *Deformation Vibrations*

Bands of variable intensity are observed in the regions 1300–1180 cm^{-1} (7.69–8.48 µm) and 1100–1000 cm^{-1} (9.09–10.00 µm) due to in-plane deformations vibrations. Strong bands are observed in the region 850–690 cm^{-1} (11.76–14.49 µm) which are characteristic of the position of the substitution, these bands being due to C–H out-of-plane deformation vibrations. See Table 11.2 for the correlation C–H out-of-plane vibrations with the number of adjacent hydrogen atoms on aromatic ring.

Other Bands

Monosubstituted pyridines,[7,24] with the exception of 4-substituted pyridines,[24] have a medium-to-strong band at 635–600 cm^{-1} (15.75–16.67 µm) and a strong band at 420–385 cm^{-1} (23.81–25.97 µm). 4-Monosubstituted pyridines

appear to have bands below 650 cm^{-1} (15.38 µm) similar to those for the corresponding monosubstituted benzenes. (Studies of di- and trisubstituted pyridines have been published.[4,8])

Pyridine N-Oxides

Pyridine *N*-oxides have similar absorptions to those of pyridines.

A particular feature of pyridine *N*-oxides[11] is a strong band in the region 1310–1220 cm^{-1} (7.64–8.20 µm) due to the N–O stretching vibration.

Other Comments

Pyridine may form charge transfer complexes.[12] Studies on picolines,[13] bipyridines,[14] and pyrazine *N*-oxides[15] have also been published.

Table 12.1 Pyridine ring and C–H stretching vibrations

Functional Groups	Region		Intensity		Comments
	cm^{-1}	µm	IR	Raman	
Pyridines	3100–3010	3.23–3.32	m–s	m–s	CH str, number of peaks
	1615–1570	6.19–6.37	m–s	m–s	
	1590–1575	6.29–6.43	m–s	m	
	1520–1465	6.58–6.83	m–s	m	C=C and C=N in plane vib (ring str vib) general ranges
	1450–1410	6.90–7.09	m–s	m–w	
	1000–985	10.00–10.15	w–m	s	
2-Monosubstituted pyridines	1615–1575	6.19–6.35	m–s	m–s	Ring str
	1575–1570	6.35–6.37	m–s	m	Ring str
	1480–1450	6.76–6.90	m–s	m	Ring str
	1440–1425	6.94–7.02	m–s	w–m	Ring str
	1050–1040	9.52–9.62	m	m–s	Ring str
	1000–985	10.00–10.15	m	vs	Ring str
3-Monosubstituted pyridines	1600–1590	6.25–6.29	m–s	m–s	Ring str
	1585–1560	6.31–6.41	m–s	m	Ring str
	1485–1465	6.73–6.83	m–s	m	Ring str
	1430–1410	9.99–7.09	m–s	w	Ring str
	1030–1010	9.71–9.90	m	vs	Ring str
4-Monosubstituted pyridines	1610–1565	6.21–6.40	m–s	m–s	Ring str
	1570–1550	6.27–6.45	m	m	Ring str
	1520–1480	6.58–6.76	m	m	Ring str
	1420–1410	7.04–7.09	m–s	w–m	Ring str
	1000–985	10.00–10.15	m	vs	Ring str
Polysubstituted pyridines	1610–1595	7.09–6.27	m	s	Ring str
	1590–1565	6.29–6.39	m	v	Ring str
	1555–1490	6.43–6.71	m	m–w	Ring str
	1035–900	9.66–11.11	w–m	vs	Ring str

Table 12.2 Pyridine C–H deformation vibrations

Functional Groups	Region		Intensity		Comments
	cm⁻¹	μm	IR	Raman	
2-Monosubstituted pyridines	1305–1265	7.66–7.90	w	w	Also *N*-oxides
	~1150	~8.70	w	m–s	Also *N*-oxides
	1115–1085	8.97–9.22	w	w–m	Also *N*-oxides
	890–800	11.24–12.50	w	m–s	
	780–740	12.82–13.51	s	–	Four adjacent hydrogen atoms
	740–720	13.51–13.89	m	w	
	635–600	15.75–16.67	w	m–s	In-plane ring def vib
	420–385	23.81–25.97		w	Out-of-plane ring bending vib, often absent
3-Monosubstituted pyridines	1320–1230	7.58–8.13	w	w–m	
	1200–1180	8.33–8.48	v	m–s	
	~1125	~8.89	w		
	1105–1095	9.05–9.13	w	w–m	
	1045–1030	9.57–9.71	m	s–vs	
	920–890	10.87–11.24	w	w	
	810–750	12.35–13.33	w	m	
	820–770	12.20–12.29	m–s	–	Also *N*-oxides, three adjacent hydrogen atoms
	730–690	13.70–14.49	m–s	w	Ring bending vib
	630–600	15.87–16.67	w	m–s	In-plane ring def vib
	420–385	23.81–25.97	v	m–w	Out-of-plane bending of ring. Often absent
4-Monosubstituted pyridines	1320–1280	7.58–7.81	w	w–m	
	1230–1210	8.13–8.26	v	m–s	
	1100–1075	9.09–9.30	s–m	w–m	
	~1070	~9.35	s	s–vs	
	850–790	11.76–12.66	s	–	Also *N*-oxides, two adjacent hydrogen atoms
	805–780	12.42–12.82	w	m–s	
	730–720	13.70–13,89	m	w	
	670–660	14.93–15.15	–	m	
	420–385	23.81–25.97	w	w	Out-of-plane bending of ring. Often absent
2,3-Disubstituted pyridines	815–785	12.27–12.74	m		(For 2-fluoropyridine methyl derivatives see ref. 26.)
	740–690	13.51–14.49	m–s	w	
2,5-Disubstituted pyridines	825–810	12.12–12.35	m–s		
	735–725	13.60–13.75	m–s		
2,6-Disubstituted pyridines	815–770	12.27–12.99	m–s		See ref. 10
	750–720	13.33–13.89	m–s	w	
3,4-Disubstituted pyridines	860–840	11.63–11.90	m		
Trisubstituted pyridines	~725	~13.79	s	w	See refs 4, 8

Table 12.3 Pyridinium salts

Functional Groups	Region		Intensity		Comments
	cm^{-1}	μm	IR	Raman	
Pyridinium salts (free)	3340–3210	2.99–3.12	v	m–w	N$^+$-H str, a number or bands
Pyridinium salts (hydrogen-bonded)	3300–2370	3.03–4.22	v	m–w	N$^+$-H str, a number of bands
	1250–1240	8.00–8.06	m–w	m	NH def vib
Pyridinium salts	1640–1600	6.10–6.25	v	s–vs	Ring vib
	1620–1585	6.17–6.31	v	m	Ring vib
	1550–1505	6.45–6.64	v	m	Ring vib
	1335–1280	7.49–7.81	m–w	w–m	CH def vib
	1270–1220	7.87–8.20	m–w	m	CH def vib
	1220–1185	8.20–8.44	m–w	m	CH def vib
	1110–1075	9.01–9.30	m–w	m	CH def vib
	1030–1005	9.71–9.95	w	vs	Ring vib
	~1010	~9.90	w	vs	Ring vib
	940–880	10.64–11.36		v	Out-of-plane NH def vib
	655–620	15.27–16.13		m	Ring vib

Table 12.4 Pyridine N-oxide C–H and ring stretching vibrations

Functional Groups	Region		Intensity		Comments
	cm^{-1}	μm	IR	Raman	
Pyridine N-oxides	3100–3010	3.23–3.32	m–s	m–s	C–H str
	1645–1600	6.08–6.25	v	m–s	
	~1570	~6.37	v	m	
	1540–1475	6.49–6.78	v	m	C=C and C=N in-plane vibs general ranges
	1450–1425	6.90–7.02	v	m	
	1310–1220	7.64–8.20	s	m	N$^+$–O$^-$ str
	880–835	11.36–11.98	m–s	s	
	~540	~18.52		w	
2-Monosubstituted pyridine N-oxides	1640–1600	6.10–6.25	v	m–s	
	1580–1555	6.33–6.43	v	m	
	1540–1480	6.49–6.76	v	m	
	1445–1425	6.92–7.02	v	m	
3-Monosubstituted pyridine N-oxides	~1605	~6.23	v	s	See ref. 11
	1564–1560	6.39–6.41	v	m	
	1490–1475	6.71–6.78	v	m	
	~1435	~6.97	v	m	
	~1015	~9.86	s–m	s	Ring vib
4-Monosubstituted pyridine N-oxides	1645–1610	6.08–6.21	v	s	
	1490–1475	6.71–6.78	v	m	
	1450–1435	6.90–6.97	v	m	

Table 12.5 Pyridine N-oxide C–H deformation vibrations

Functional Groups	Region		Intensity		Comments
	cm⁻¹	μm	IR	Raman	
2-Monosubstituted pyridine N-oxides	~1150	~8.70	w	m–s	Also pyridines
	1115–1090	8.97–9.17	w	w–m	Also pyridines
	1055–1040	9.48–9.61	w	m–s	Also pyridines
	990–960	10.10–10.42	m	s	
	790–750	12.66–13.33	m–s		Four adjacent hydrogen atoms
3-Monosubstituted pyridine N-oxides	~1160	~8.62	v	m–w	
	1120–1080	8.93–9.26	w–m		
	980–930	10.20–10.75	s		
	820–770	12.20–12.29	m–s	w	Also pyridines, three adjacent hydrogen atoms
	680–660	14.71–15.15	m		Ring bending vib
4-Monosubstituted N-oxides	~1170	~8.55	s		
	1110–1095	9.01–9.13	w		
	~1035	~9.66	m		
	850–820	11.76–12.20	s	w	Also pyridines, two adjacent hydrogen atoms
3,4-Disubstituted N-oxides	890–860	11.24–11.63	s		
	825–310	12.12–12.35	s		

Table 12.6 2-Pyridols , and 4-pyridols

Functional Groups	Region		Intensity		Comments
	cm⁻¹	μm	IR	Raman	
2-Pyridols	1670–1655	5.99–6.04	vs	w–m	C=O str
	1630–1590	6.14–6.29	vs	m–s	
	1570–1535	6.37–6.52	s	m–s	
	1500–1470	6.67–6.80	m	m–s	
	1445–1415	6.92–7.06	m–s	m	
4-Pyridols	1660–1620	6.02–6.17	vs	m–s	
	1580–1550	6.33–6.45	vs	w–m	C=O str
	1515–1485	6.60–6.74	w–m	m	
	1470–1400	6.80–7.14	m–s	m	
2-Pyridthiones	1145–1100	8.73–9.09	m–s	s	C=S str
4-Pyridthiones	1120–1105	8.93–9.05	vs	s	C=S str

Quinolines, , and Isoquinolines,

Quinolines and isoquinolines[1,16] have three bands near 1600 cm^{-1} (6.25 μm) and five in the range 1500–1300 cm^{-1} (6.67–7.69 μm). Disubstituted methyl quinolines have four bands in the region 1600–1500 cm^{-1} (6.25–6.67 μm). The aromatic C–H out-of-plane deformation vibrations are similar to those observed for naphthalenes. (Reviews of the infrared spectra of acridines have been published.[17,18])

Pyrimidines,

Pyrimidines[19,25] absorb strongly at 1600–1500 cm^{-1} (6.25–6.67 μm) due to the C=C and C=N ring stretching vibrations. Absorptions are also observed at 1640–1620 cm^{-1} (6.10–6.17 μm), 1580–1520 cm^{-1} (6.33–6.58 μm), 1000–960 cm^{-1} (10.00–10.42 μm), and 825–775 cm^{-1} (12.12–12.90 μm). 2-Monosubstituted pyrimidines have three medium-to-strong absorption bands at 650–630 cm^{-1} (15.38–15.87 μm), 580–475 cm^{-1} (17.24–21.05 μm), and 515–440 cm^{-1} (19.42–22.73 μm).

4-Monosubstituted pyrimidines have a band of variable intensity at 685–660 cm^{-1} (14.60–15.15 μm) which is usually at 680 cm^{-1} (14.71 μm), a medium-to-strong band at 555–500 cm^{-1} (18.02–20.00 μm), and a strong band at 500–430 cm^{-1} (20.00–23.26 μm). Due to tautomerism, pyrimidines substituted with hydroxyl groups are generally in the keto form and therefore have a band due to the carbonyl group. In their Raman spectra, pyrimidines with substituents on the 2- and/or 4-positions have a strong band at 1005–960 cm^{-1} (9.95–10.42 μm) and 5-substituted pyrimidines have a strong band at ~1050 cm^{-1} (9.52 μm).

Quinazolines,

Due to aromatic ring vibrations, quinazolines[20] absorb strongly at 1635–1610 cm^{-1} (6.13–6.21 μm), 1580–1565 cm^{-1} (6.33–6.39 μm), and 1520–1475 cm^{-1} (6.58–6.78 μm), with six bands of variable intensity usually being observed at 1500–1300 cm^{-1} (6.67–7.69 μm). In the region 1000–700 cm^{-1} (10.00–14.29 μm), bands of variable intensity are observed due to the C–H out-of-plane deformation vibrations. These bands are useful for assignment purposes since different types of monosubstitution may be recognized. Bands of variable intensity, usually weak, due to C–H in-plane deformation vibrations, are observed at 1290–1010 cm^{-1} (7.75–9.90 μm), six bands often being observed.

Purines,

Purines[21] are not, in general, easily distinguished from pyrimidines.

All purines have a characteristic, strong band at about 640 cm^{-1} (15.63 μm). 2-Monosubstituted purines have two bands of medium-to-strong intensity at 650–610 cm^{-1} (15.38–16.39 μm) and 630–585 cm^{-1} (15.87–17.09 μm) and one of variable intensity at 495–375 cm^{-1}

Table 12.7 Acridines

Functional Groups	Region		Intensity		Comments
	cm^{-1}	μm	IR	Raman	
Acridines	3100–3010	3.23–3.32	m–s	m–s	See refs 17, 18
	1630–1360	6.13–7.35	m–s	m–s	Ring vib, 7–9 bands
	~1000	~10.00	w–m	m–s	Ring vib, 2 bands
9-Monosubstituted acridines	~1630	~6.13	m–s	m–s	
	1610–1595	6.21–6.27	m–s	m–s	
	~1545	~6.47	m–s	m–s	
	~1520	~6.58	m–s	m–s	
	~1460	~6.85	m–s	m–s	
	~1435	~6.97	m–s	m–s	
	~1400	~7.14	m–s	m–s	

Table 12.8 Pyrimidines

Functional Groups	Region cm⁻¹	μm	Intensity IR	Raman	Comments
Pyrimidines	3120–3010	3.21–3.32	m	m–s	=C–H str
	1590–1520	6.29–6.58	m–s	m–s	C=C, C=N str
	1480–1400	6.76–7.15	v	m	C=C, C=N str
	1410–1375	7.09–7.28	v	m	C=C, C=N str
	1350–1250	7.41–8.00	v	m	C=C, C=N str
	1005–960	9.95–10.42	m–s	m–s	C=C, C=N str
	825–775	12.12–12.90	m–s	m–s	C=C, C=N str
2-Pyrimidines	1005–960	9.95–10.42		s	
	650–630	15.38–15.87	m–s	m	
	580–475	17.24–21.05	m–s		
	515–440	19.42–22.73	m–s		
4-Pyrimidines	1000–960	10.00–10.42		s	
	685–660	14.60–15.15	v	m	Usually at ~680 cm⁻¹
	555–500	18.02–20.00	m–s		
	500–430	20.00–23.26	s		
Pyrimidine N-oxides	1300–1240	7.69–8.07	s–vs	m	N–O str, often ~1280 cm⁻¹

Table 12.9 Quinazoline aromatic ring stretching vibrations

Functional Groups	Region cm⁻¹	μm	Intensity IR	Raman	Comments
Quinazolines	1635–1565	6.13–6.39	s	m–s	Two or three bands
2-Monosubstituted quinazolines	1630–1620	6.14–6.17	s	m–s	
	1600–1580	6.25–6.33	m–s	m–s	
	1585–1570	6.31–6.37	s	m–s	
	1495–1480	6.69–6.76	s	m–s	
	1475–1445	6.78–6.92	m–s	m–s	
	1415–1395	7.07–7.17	m–s	m–s	
	1390–1355	7.19–7.38	s	m–s	
	1335–1325	7.49–7.55	w–m	m–s	
4-Monosubstituted quinazolines	1620–1615	6.17–6.19	m–s	m–s	
	1575–1565	6.35–6.39	s	m–s	
	1505–1485	6.65–6.73	s	m–s	
	1470–1455	6.80–6.87	w	m–s	
	1410–1365	7.09–7.33	m–s	m–s	
	1360–1340	7.35–7.46	m–s	m–s	
5-Monosubstituted quinazolines	1630–1615	6.13–6.19	m–s	m–s	
	1585–1575	6.31–6.35	s	m–s	
	1580–1560	6.33–6.41	s	m–s	
	1490–1480	6.71–6.76	s	m–s	
	1470–1445	6.80–6.92	w	m–s	
	1420–1415	7.04–7.07	v	m–s	
	1400–1395	7.14–7.17	w–m	m–s	
	1385–1360	7.22–7.35	s	m–s	

Table 12.9 (*continued*)

Functional Groups	Region cm^{-1}	Region μm	Intensity IR	Intensity Raman	Comments
	1315–1305	7.61–7.67	w–m	m–s	
6-Monosubstituted quinazolines	1630–1620	6.14–6.17	m–s	m–s	
	1605–1595	6.23–6.27	v	m–s	
	1580–1565	6.33–6.39	s	m–s	
	1505–1490	6.65–6.71	s	m–s	
	1475–1430	6.78–7.00	w–m	m–s	
	1425–1405	7.02–7.12	v	m–s	
	1390–1380	7.19–7.25	s	m–s	
	1375–1360	7.27–7.35	s	m–s	
	1325–1310	7.55–7.63	v	m–s	
7-Monosubstituted quinazolines	1630–1615	6.14–6.19	m–s	m–s	
	1595–1575	6.27–6.35	m–s	m–s	
	1575–1545	6.35–6.47	m	m–s	
	1495–1475	6.69–6.78	v	m–s	
	1475–1445	6.78–6.92	w	m–s	
	1425–1410	7.02–7.07	v	m–s	
	1390–1380	7.19–7.25	s	m–s	
	1375–1360	7.27–7.35	s	m–s	
	1325–1305	7.55–7.66	w–m	m–s	
8-Monosubstituted quinazolines	1635–1615	6.12–6.19	m–s	m–s	
	1585–1580	6.31–6.33	s	m–s	
	1575–1560	6.35–6.41	m–s	m–s	
	1490–1475	6.71–6.78	m–s	m–s	
	1470–1460	6.80–6.85	w	m–s	
	1450–1445	6.90–6.92	w	m–s	
	1410–1390	6.09–6.19	v	m–s	
	1390–1380	7.19–7.25	s	m–s	
	1310–1300	7.63–7.69	w–m	m–s	

Table 12.10 Purines

Functional Groups	Region cm^{-1}	Region μm	Intensity IR	Intensity Raman	Comments
2-Monosubstituted purines 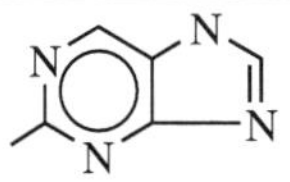	650–610	15.38–16.39	m–s		
	630–585	15.87–17.09	m–s	w	C–H out-of-plane bending vib
	495–375	20.20–26.67	v		Out-of-plane pyrimidine ring def vib
6-Monosubstituted purines, 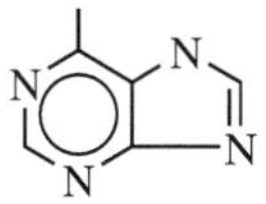	690–645	14.49–15.50	v		

(*continued overleaf*)

Table 12.10 (*continued*)

Functional Groups	Region cm^{-1}	Region μm	Intensity IR	Intensity Raman	Comments
8-Monosubstituted purines,	650–625	15.38–16.00	s		
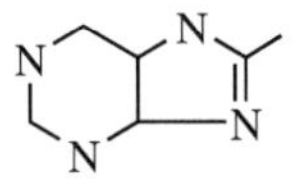	660–640	15.15–15.63	v		
2,6-Disubstitituted purines	630–610	15.87–16.39	s		
	650–630	15.38–15.87	s		
	575–535	17.39–18.69	v		Not observed for di(methyl-amino) purine or 6-amino-2-methylamino purine

Table 12.11 Pyrazines, and pyrazine *N*-oxides

Functional Groups	Region cm^{-1}	Region μm	Intensity IR	Intensity Raman	Comments
Pyrazines and pyrazine *N*-oxides	1600–1575	6.25–6.35	v	m–w	See ref. 15
	1570–1520	6.37–6.58	w–m	m	
	1500–1465	6.67–6.83	m–s	m–w	
	1420–1370	7.04–7.30	m–s	w	
Monosubstituted pyrazines	1025–1000	9.76–10.00		vs	
	840–785	11.90–12.74		s	
	660–615	15.15–16.26		w	
2,3-Disubstituted pyrazines	1100–1080	9.09–9.26		s	
	760–685	13.16–14.60		vs	
2,6-Disubstituted pyrazines	1025–1020	9.76–9.80		m	
	710–705	14.08–14.18		vs	
2,5-Disubstituted pyrazines	865–835	11.56–11.98		vs	
	650–640	15.38–15.63		m	
Trisubstituted pyrazines	955–915	10.47–10.93		m	
	750–725	13.33–13.79		m	
	710–695	14.08–14.39		m	
Tetrasubstituted pyrazines	720–710	13.89–14.08		s	
Pyrazine *N*-oxides	1350–1260	7.41–7.94	s	m	N–O str, pyrazine mono-*N*-oxide ~1320 cm^{-1}

(20.20–26.67 μm). 6-Monosubstituted purines have a strong band at 650–625 cm^{-1} (15.38–16.00 μm) and a band of variable intensity at 690–645 cm^{-1} (14.49–15.50 μm), with the exception of 6-cyanopurine for which this last band is not observed. 8-Monosubstituted purines have a strong band at 630–610 cm^{-1} (15.87–16.39 μm) and one of variable intensity at 660–640 cm^{-1} (15.15–15.63 μm). Most 2,6-disubstituted purines have a strong band at 650–630 cm^{-1} (15.38–15.87 μm) and a band of variable intensity at 575–535 cm^{-1} (17.39–18.69 μm). For some 2,6-compounds, this last band is not observed, e.g. some methyl aminopurines.

Phenazines,[22]

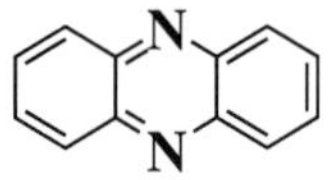

Bands due to the stretching vibrations of N–H and C–H are observed at 3500–3150 cm^{-1} (2.86–3.18 μm) and 3070–3050 cm^{-1} (3.26–3.28 μm) respectively.

A number of strong bands due to the C–H out-of-plane deformation vibrations are observed in the region 900–680 cm^{-1} (11.11–13.79 μm). As with aromatic hydrocarbons, the position of these bands correlates with the number of adjacent hydrogen atoms on the rings:

one hydrogen atom	900–850 cm^{-1}	(11.11–11.76 μm)	s
two adjacent hydrogen atoms	860–800 cm^{-1}	(11.63–12.50 μm)	s
three adjacent hydrogen atoms	810–750 cm^{-1}	(12.35–13.33 μm)	s
	725–680 cm^{-1}	(13.79–14,71 μm)	m
four adjacent hydrogen atoms	770–735 cm^{-1}	(12.99–13.61 μm)	s

Sym-triazines,

Alkyl and aryl sym-triazines[23] have at least one strong band at 1580–1520 cm^{-1} (6.33–6.58 μm) which may be a doublet, at least one band at 1450–1350 cm^{-1} (6.90–7.41 μm), and at least one weak band at 860–775 cm^{-1} (11.63–12.90 μm). This last band is due to an out-of-plane deformation of the ring, the others being due to in-plane stretching vibrations. 'Hydroxyl'-substituted triazines have a strong band at 1775–1675 cm^{-1} (5.63–5.97 μm) due to the C=O stretching vibrations of the keto form and a sharp, medium–intensity band at 795–750 cm^{-1} (12.58–13.25 μm) due to the iso form (see melamines). The normal triazine ring out-of-plane bending vibration band is found at 825–795 cm^{-1} (12.12–12.58 μm).

Since the trisodium salt of cyanuric acid is in the enol form, it has the band normally observed for triazines near 820 cm^{-1} (12.20 μm) (see melamine), as do trialkyl cyanurates.

Ammelide (6-amino-sym–triazine-2,4-diol) and ammeline (4,6-di-amino-sym–triazine-2-ol) have a broad absorption near 2650 cm^{-1} (3.77 μm) resulting from the ring NH group which is intramolecularly bonded to the C=O group, e.g. one form of ammeline is

Table 12.12 Sym-triazines

Functional Groups	Region		Intensity		Comments
	cm^{-1}	μm	IR	Raman	
Sym-triazines	3100–3000	3.23–3.33	m	m, p	C–H str
	1580–1520	6.33–6.58	vs	m–w	Ring str, at least one band
	1450–1350	6–90-7.41	v	w, d	Ring str, at least one band
	1000–980	10.00–10.20	w	s, p	Ring str
	860–775	11.63–12.90	w–m		out-of-plane bending vib, at least one band
Amino-substituted triazines	1680–1640	5.95–6.10	m–s		NH$_2$ def vib
Trialkyl cyaurates	1600–1540	6.25–6.49	s	s	Ring str
	1380–1320	7.25–7.58	v		
	1160–1110	8.62–9.01	m	m–w	OCH$_2$ str
	820–805	12.20–12.42	m		Triazine out-of-plane bending vib
Ammelines and ammelides	~2650	~3.77	w–m		br, ring NH···O=C vib
Thioammelines	2900–2800	3.45–3.57	w–m		br, ring NH···S=C vib
	~1200	~8.33	s	s	C=S str
	~775	~12.90	m		Iso form, ring out-of-plane bending vib

Table 12.13 Melamines

Functional Groups	Region		Intensity		Comments
	cm^{-1}	μm	IR	Raman	
Melamines	3500–3100	2.86–3.23	v	m–w	NH$_2$ str
	1680–1640	5.95–6.10	m	w	NH$_2$ def
	1600–1500	6.25–6.67	s	m–s	Ring str
	1450–1350	6.90–7.41	v	m	sh, number of bands
	825–800	12.12–12.50	m		} Only one of the two is present
	795–750	12.58–13.25	m		

Table 12.14 Sym-tetrazines,

Functional Groups	Region		Intensity		Comments
	cm^{-1}	μm	IR	Raman	
Sym–tetrazines	1600–1500	6.25–6.67	m–s	s	Ring str, absent for molecules with centre of symmetry
	1495–1320	6.69–7.58	m	s	Ring str
	970–880	10.31–11.36	m		

Table 12.15 α-Pyrones, , and γ-pyrones,

Functional Groups	Region		Intensity		Comments
	cm^{-1}	μm	IR	Raman	
α-Pyrones	1740–1720	5.75–5.81	s	w–m	C=O str often split
	1650–1635	6.06–6.12	m	s	C=C str
	~1565	~6.39	s	s	C=C str
γ-Pyrones	1570–1540	6.37–6.49	vs	m	} Combination of C=O and C=C str vib
	1535–1525	6.12–6.56	vs		
	1465–1445	6.83–6.92	m–s		
	1420–1400	7.04–7.14	m		
γ-Thiopyrones,	~1610	~6.21	s	m	br, C=O, C=C overlap
γ-Pyrthiones	~1100	~9.09	s	m–s	C=S str

Table 12.16 Pyrylium compounds,

Functional Groups	Region		Intensity		Comments
	cm^{-1}	μm	IR	Raman	
Pyrylium derivatives	3100–3010	3.23–3.32	w–m	m–s	=C–H str, a number of bands
	1650–1615	6.06–6.19	vs	s	Ring in-plane vib
	1560–1520	6.41–6.58	vs	s	Ring in-plane vib
	1520–1465	6.58–6.83	m	s	Ring in-plane vib
	1450–1400	6.90–7.14	v	s	Ring in-plane vib
	1000–970	10.00–10.31	v	s	Ring in-plane vib
Unsubstituted pyrylium salts	~960	~10.42	s	m–s	C–H out-of-plane vib
	~775	~12.92	m		
2,6-Disubstituted pyrylium compounds	~935	~10.70	m	m–s	C–H out-of-plane vib
	~800	~12.50	s		
2,4,6-Trisubstituted pyrylium compounds	960–900	10.42–11.11	v	m–s	C–H out-of-plane vib, two bands
2,3,4,6-Tetrasubstituted pyrylium compounds	~920	~10.87	w		Out-of-plane vib
	890–870	11.24–11.49	m	m–s	
2,3,5,6-Tetrasubstituted pyrylium compounds	900–880	11.11–11.36	w	m–s	C–H out-of-plane vib
	~705	~14.18	m		

Melamines,

Melamine may exist in tautomeric forms, e.g.

Melamines have an absorption of variable intensity at 3500–3100 cm^{-1} (2.86–3.23 μm) due to the NH$_2$ stretching vibrations and a band of medium intensity at 1680–1640 cm^{-1} (5.95–6.10 μm) due to NH$_2$ deformations. A strong band in the region 1600–1500 cm^{-1} (6.25–6.67 μm), usually at 1550 cm^{-1} (6.45 μm), and a number of absorptions at 1450–1350 cm^{-1} (6.90–7.41 μm) are also observed. A sharp, medium-intensity band is usually found at 825–800 cm^{-1} (12.12–12.50 μm) although this band may be at 795–750 cm^{-1} (12.58–13.25 μm) when the triazine ring is in the iso form in which at least one double bond is external to the ring. The ring N-alkyl iso-melamines and hydrohalide melamine salts also absorb in this region.

References

1. A. F. Ardyukova *et al.*, *Atlas of Spectra of Aromatic and Heterocyclic Compounds*, Nauka Sib. Otd., Novosibirsk, 1973.
2. G. L. Cook and F. M. Church, *J. Phys. Chem.*, 1957, **61**, 458.
3. A. R. Katritzky, *Quart. Rev.*, 1959, **13**, 353.
4. J. H. S. Green and D. J. Harrison, *Spectrochim. Acta*, 1973, **29A**, 293.
5. D. Hement *et al.*, *J. Am. Chem. Soc.*, 1959, **81**, 3933.
6. J. K. Wilmshurst and H. J. Bernstein, Can. *J. Chem.*, 1957, **35**, 1183.
7. R. Isaac *et al.*, *Appl. Specrtosc.*, 1963, **17**, 90.
8. J. H. S. Green *et al.*, *Spectrochim. Acta*, 1973, **29A**, 1177.
9. J. H. S. Green *et al.*, *Spectrochim. Acta*, 1963, **19**, 549.
10. R. Tripathi, *Indian J. Pure Appl. Phys.*, 1973, **11**, 277.
11. A. R. Katritzky *et al.*, *J. Chem. Soc.*, **1959**, 3680.
12. J. Yarwood, *Spectrochim. Acta*, 1970, **26A**, 2077.
13. G. Varsányi *et al.*, *Acta Chim. Hungary*, 1965, **43**, 205.
14. J. S. Strukl and J. L. Walter, *Spectrochim. Acta*, 1971, **27A**, 209.
15. H. Shindo, *Chem. Pharm. Bull. Jpn*, 1960, **8**, 33.
16. A. Leifer *et al.*, *Spectrochim. Acta*, 1964, **20**, 909.
17. R. Acheson, *Chem. Heterocyclic Compds*, 1973, **9**, 665.

18. J. Reisch *et al.*, *Pharmazie*, 1972, **27**, 208.
19. A. F. Ardyukova *et al.*, *Atlas of Spectra of Aromatic and Hetrocyclic Compounds, No. 4, Infrared Spectra of Pyrimidine Series*, Nauka Sib. Otd., Novosibirsk, 1974.
20. A. R. Katritzky *et al.*, *Spectrochim. Acta*, 1964, **20**, 593.
21. J. H. Lister, *Chem. Heterocyclic Compds*, 1971, **7**, 496.
22. C. Stammer and A. Taurins, *Spectrochim. Acta*, 1963, **19**, 1625.
23. W. A. Heckle *et al.*, *Spectrochim. Acta*, 1961, **17**, 600.
24. O. P. Shkurko and I. K. Korobeinicheva, Zh. Prinkl. *Spectrosk.*, 1975, **23**, 860.
25. F. Billes *et al.*, *Theoretical Chem.*, 1998, **423(3)**, 225.
26. A. Puszko and H. Ciurla, *Chem. Heterocycl. Compd.*, 1999, **35(6)**, 677.

13 Five-membered Ring Heterocyclic Compounds

Heteroaromatic compounds of the type

generally have three bands due to C=C in-plane vibrations at about 1580 cm^{-1} (6.33 μm), 1490 cm^{-1} (6.71 μm), and 1400 cm^{-1} (7.14 μm). In addition, those with a CH=CH group have a strong band in the region 800–700 cm^{-1} (12.50–14.29 μm) due to an out-of-plane deformation vibration.

Pyrroles, and Indoles,

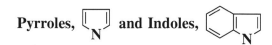

In dilute solution, the band due to the N–H stretching vibration occurs at 3500–3400 cm^{-1} (2.86–2.94 μm). In the presence of hydrogen bonding, a broad absorption occurs at 3400–3000 cm^{-1} (2.94–3.33 μm). The bands due to the C=C and C=N stretching vibrations occur in the region 1580–1390 cm^{-1} (6.33–7.19 μm).[1]

Pyrroles have one or two bands in the region 1580–1545 cm^{-1} (6.33–6.47 μm) depending on whether or not there is substitution on the nitrogen atom. A very strong band is observed at 1430–1390 cm^{-1} (6.99–7.19 μm) and a weak band near 1470 cm^{-1} (6.80 μm). 1-Alkyl pyrroles have a strong band in infrared spectra (which is usually strong in Raman spectra) due to the N–C stretching vibration near 1285 cm^{-1} (7.78 μm). 1,2-Disubstituted pyrroles have a medium-intensity band at 1500–1475 cm^{-1} (6.67–6.78 μm) and a weak band at 1530 cm^{-1} (6.54 μm). This latter band is also observed for 1,2,5- and 1,3,4-trisubstituted pyrroles.

Indoles[2,3] absorb at 3480–3020 cm^{-1} (2.87–3.31 μm) and near 1460 cm^{-1} (6.85 μm), 1420 cm^{-1} (7.04 μm), and 1350 cm^{-1} (7.41 μm).

Pyrrolines,

Pyrrolines[4] have a medium-to-strong band at 1660–1560 cm^{-1} (6.02–6.41 μm) due to the C=N stretching vibration. The other forms, such as

are normally unstable and therefore it is usual for no band due to the N–H stretching vibration to be observed.

Furans,

Bands due to the C–H stretching vibration for furans occurs above 3000 cm^{-1} (3.33 μm) in the region 3180–3000 cm^{-1} (3.14–3.33 μm). Furan derivatives[8,13,14] have medium-to-strong bands at 1610–1560 cm^{-1} (6.21–6.41 μm), 1520–1470 cm^{-1} (6.58–6.80 μm), and 1400–1390 cm^{-1} (7.14–7.19 μm) which are due to the C=C ring stretching vibrations. Furans with electronegative substituents usually have strong bands in these regions.

For 2-substituted furans,[13] the out-of-plane deformation vibrations of the C–H group give bands at 935–915 cm^{-1} (10.70–10.93 μm), 885–880 cm^{-1} (11.29–11.34 μm), and 835–780 cm^{-1} (11.98–12.82 μm).

All furans have a strong absorption near 595 cm^{-1} (16.81 μm) which is probably due to a ring deformation vibration.

Tetrahydrofurans[11,12] have a strong band at 1100–1075 cm^{-1} (9.09–9.30 μm) due to the C–O stretching vibration and another band near 915 cm^{-1} (10.93 μm).

Table 13.1   Pyrroles (and similar five-membered ring compounds): N–H, C–H, and ring stretching

Functional Groups	Region cm^{-1}	Region μm	Intensity IR	Intensity Raman	Comments
Pyrroles	3500–3400	2.86–2.94	v	m–w	N–H str, free pyrroles
	3400–3000	2.94–3.33	s	m–w	br, N–H str, hydrogen-bonded pyrroles
	3100–3010	3.23–3.32	m	m–s	=C–H str, multiple peaks
	1580–1540	6.33–6.49	w–m	w	Two bands for 1-substituted pyrroles, C=C and C=N in-plane vib
	1510–1460	6.62–6.85	w–m	vs	C=C and C=N in-plane vib
	1430–1380	6.99–7.25	vs	s	C=C and C=N in-plane vib
	~480	~20.83	m–s		Ring def vib, not greatly affected by substitution
1-Alkyl pyrroles	1510–1490	6.62–6.71		vs	Ring vib
	1400–1380	7.14–7.25	m–s	s	Ring vib
	1290–1280	7.75–7.81	s	vs	N–C ring str
	1095–1080	9.13–9.26	m	s, p	CH in-plane def vib
	1065–1055	9.39–9.48	w–m	s	CH in-plane def vib
	620–605	16.13–16.53		m	In-plane ring def vib
2-Alkyl pyrroles	1605–1590	6.23–6.29	m–w	s	C=C str
	1570–1560	6.37–6.41	v	m	Ring vib
	1515–1490	6.60–6.71	m	vs	C=C str
	1475–1460	6.78–6.85	m	vs	Ring vib
	1420–1400	7.04–7.14	m–s	s	Ring vib
	1120–1100	8.93–9.09	w–m	w	NH def vib
	1090–1080	9.17–9.26	w	s	CH in-plane def vib
3-Alkyl pyrroles	1570–1360	6.37–7.35	v	m	Ring vib
	1490–1480	6.71–6.76	m	s	Ring vib
	1430–1420	6.99–7.04	m–s	m	Ring vib
	1080–1060	9.26–9.43	w	s	CH in-plane def vib
1,2-Disubstituted pyrroles	~1530	~6.54	w	s	C=C in-plane vib
	1500–1475	6.67–6.78	m–s	s	C=C in-plane vib
1,2,5- and 1,3,4-trisubstituted pyrroles,	~1530	~6.54	w	s	C=C in-plane vib
	1585–1560	6.31–6.41	w–m	s	C=C str
	1480–1460	6.76–6.85	m	vs	C=C str
	1165–1130	8.58–8.85	m		
	1040–1010	9.62–9.90	m		
	825–795	12.12–12.58	w		
Indoles,	1630–1615	6.14–6.19	m	s–m	Ring vib
	1600–1575	6.25–6.35	m	s–m	Ring vib
	1565–1540	6.39–6.49	v		Sometimes absent, ring vib
	1520–1470	6.58–6.80	m		Ring vib
Pyrrolines,	1660–1560	6.02–6.41	m	s	C=N str, see ref. 4

Table 13.1 (*continued*)

Functional Groups	Region cm^{-1}	Region μm	Intensity IR	Intensity Raman	Comments
Oxazoles, 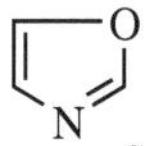	1585–1555	6.31–6.43	m	s	C=N str
Thiazoles, 	1550–1505	6.45–6.64	m	s	C=N str, see refs 6, 7. Monosubstituted: ring str vib gives strong Raman bands at 1550–1485, 1410–1380, 1320–1295 and ~870 cm^{-1}, also bands of variable intensity at ~750 and ~600 cm^{-1}
Imidazoles, 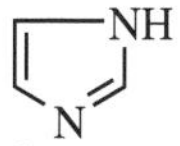	1560–1520	6.41–6.58	m	s	C=N str, see ref. 5
Benzimidazoles, 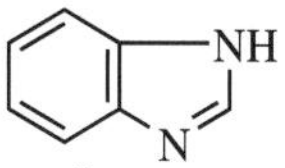	1560–1520	6.41–6.58	m	s	C=N str
Oxazolines, 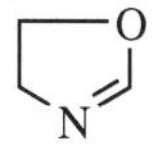	1695–1645	5.90–6.08	s	s	C=N str
Oxadiazoles	~3150	~3.17	m	s	CH str
	1550–1420	6.45–7.04	m–w	vs	Ring str
	1500–1310	6.67–7.63	m–w	vs	Ring str
	1275–1035	7.84–9.66	w	m–s	CH def vib
	1100–990	9.09–10.10	w	m	In-plane ring def vib
	955–860	10.47–11.63	w	m	In-plane ring def vib
	945–820	10.58–12.20	w	w	CH def vib
	655–620	15.27–16.13		w–m	Ring def vib
1,2,4-Oxadiazoles, 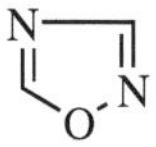	1590–1560	6.29–6.41	s	vs	Ring vib
	1390–1360	7.19–7.35	s	vs	Ring vib
1,2,5-Oxadiazoles, 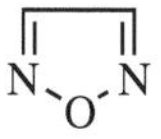	1625–1560	6.15–6.41	s	vs	

Thiophenes,

Thiophenes[15,16] absorb at 3120–3000 cm^{-1} (3.21–3.33 μm) due to the C–H stretching vibration and also have four bands of variable intensity in the region 1555–1200 cm^{-1} (6.43–8.33 μm) due to in-plane ring vibrations.

All monosubstituted thiophenes have two bands of variable, often medium-to-strong, intensity, one at 745–695 cm^{-1} (13.42–14.39 μm) and the other at 700–660 cm^{-1} (14.29–15.15 μm), possibly due to the out-of-plane bending of the =C–H group.

2-Monosubstituted thiophenes usually have two bands of variable intensity, one at 570–490 cm^{-1} (17.54–20.41 μm) and the other at 470–430 cm^{-1} (21.28–23.26 μm). For esters of thiophene-2-carboxylic acid, the former band is usually near 565 cm^{-1} (17.70 μm).[16]

3-Monosubstituted thiophenes have a band of medium intensity at 540–515 cm^{-1} (18.52–19.42 μm) and a band of variable intensity at 500–465 cm^{-1} (20.00–21.51 μm). Sometimes only one band is observed.

Table 13.2 Substituted pyrroles: N–H and C–H deformation vibrations

Functional Groups	Region		Intensity		Comments
	cm^{-1}	μm	IR	Raman	
1-Substituted pyrroles	~1080	~9.26	s–m		Four adjacent hydrogen atoms
	1035–1015	9.66–9.85	m		Four adjacent hydrogen atoms
	~925	~10.81	m		Four adjacent hydrogen atoms
	~725	~13.79	vs		Four adjacent hydrogen atoms
2-Substituted pyrroles	~1115	~8.97	w–m		Three adjacent hydrogen atoms
	1105–1070	9.05–8.55	m–s		Three adjacent hydrogen atoms
	~1030	~9.71	m–s		Three adjacent hydrogen atoms
	~925	~10.81	w		Three adjacent hydrogen atoms
	~880	~11.36	w–m		Three adjacent hydrogen atoms
1,2-Disubstituted pyrroles	~1090	~9.17	m		Three adjacent hydrogen atoms
	1065–1050	9.39–9.52	v		Three adjacent hydrogen atoms
1,2,5-Trisubstituted pyrroles	~1035	~9.66	m		Two adjacent hydrogen atoms
	980–965	10.20–10.36	w		Two adjacent hydrogen atoms
	~755	~13.25	vs		Two adjacent hydrogen atoms
1,3,4-Trisubstituted pyrroles	~1055	~9.48	s		One hydrogen atom
	~930	~10.75	m		One hydrogen atom
	~770	~12.99	vs		One hydrogen atom

Table 13.3 Furans

Functional Groups	Region		Intensity		Comments
	cm^{-1}	μm	IR	Raman	
Furan derivatives	3180–3000	3.14–3.33	m	m–s	=C–H str
	1610–1560	6.21–6.41	v	v	C=C str, usually m–s
	1520–1460	6.58–6.85	m–s	vs	C=C str
	1400–1390	7.14–7.19	m–s	s	C=C str
	1025–1000	9.76–10.00	m–s	m	
	595–515	16.81–19.42	s	w	Ring def vib
2-Monosubstituted furans	1610–1590	6.21–6.29	v	s	Ring vib
	1585–1560	6.31–6.41	v	s	Ring vib
	1515–1490	6.60–6.71	m	vs	Ring vib
	1480–1460	6.76–6.85	m	vs	Ring vib
	1240–1200	8.07–8.33	v	m	C–H def vib, see ref. 14
	1175–1145	8.51–8.73	m–s	m–w	C–H def vib
	1085–1070	9.22–9.35	m	m–w	C–H def vib
	1020–990	9.80–10.10		m–s	Ring vib
	935–915	10.70–10.93	w–m	m–w	Out-of-plane C–H def vib
	885–880	11.29–11.34	w–m	m–w	Out-of-plane C–H def vib
	835–780	11.98–12.82	w–m	w	Out-of-plane C–H def vib
	595–515	16.81–19.42	s	w	Ring def vib
3-Monosubstituted furans	1170–1150	8.55–8.70	s	m	C–H def vib, see ref. 14
	1080–1050	9.26–9.52	m–s	m	C–H def vib
	1025–1000	9.76–10.00	vs	m–w	C–H def vib
	~920	~10.87	v	m–w	C–H def vib

Table 13.3 (*continued*)

Functional Groups	Region cm^{-1}	Region µm	Intensity IR	Intensity Raman	Comments
	~875	~11.43	s	m–w	C–H def vib
	790–720	12.66–13.89	s		Usually two bands
2,5-Disubstituted furans	~1620	~6.17	v	w	Ring vib
	1600–1570	6.25–6.37	v	vs	Ring vib
	1530–1500	6.54–6.67	m	v	Ring vib
	1255–1225	7.97–8.17	w–m		see ref. 13
	1165–1140	8.58–8.77	w–m		
	~1020	~9.90	m	s	Ring def vib
	990–960	10.10–10.42	m	vs	
	930–915	10.75–10.93	w–m	m–w	C–H out-of-plane def vib
	835–780	11.98–12.82	w–m	m–w	C–H out-of-plane def vib
Polysubstituted furans	~1560	~6.41	m–s	s	C=C str
	~1510	~6.62	m–s	s	C=C str
Oxazoles, 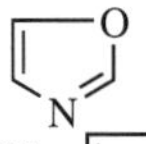	1585–1555	6.31–6.43	m	s	C=N str
Iso-oxazoles, 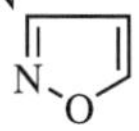	~1600	~6.25	m–s		
	~1460	~6.85	m–s		
	~1380	~7.25	m–s		
1,2,4-Oxadiazoles, 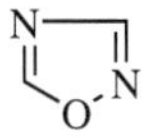	1590–1560	6.29–6.41	m–s	s	Ring str, see ref. 9
	1470–1430	6.80–6.99	m–s	s	Ring str
	1390–1360	7.19–7.35	m–s	s	Ring str
	1070–1050	9.35–9.52	m		
	915–885	10.93–11.30	m–s		
	750–710	13.33–14.08	m–s		
1,2,5-Oxadiazoles (furazanes), 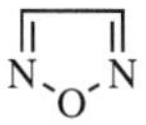	1625–1560	6.15–6.41	m–s	s	Ring str
	~1570	~6.37	m–s		
	~1425	~6.78	m–s		
	1395–1370	7.17–7.30	m–s	s	Ring str
1,2,5-Oxadiazole oxides, 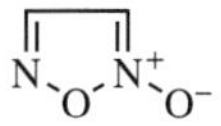	1635–1600	6.12–6.25	m–s	s	Ring str, see ref. 10
	1530–1515	6.54–6.60	m–s	s	Ring str
	1475–1410	6.78–7.09	m–s	s	Ring str
1,3-Dioxolanes, 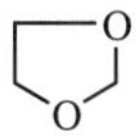	1170–1145	8.55–8.73	s	w–m	Ring vib
	1100–1050	9.09–9.52	s	w–m	
	1055–1025	9.48–9.76	m	w	
	~940	~10.64	vs	vs	Ring vib, may be absent
Oxalolidines, 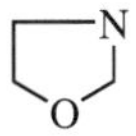	1190–1050	8.40–9.52	m		Ring def vib, at least three bands

(*continued overleaf*)

Table 13.3 (*continued*)

Functional Groups	Region		Intensity		Comments
	cm^{-1}	μm	IR	Raman	
Tetrahydrofurans, 	2980–2700	3.36–3.70	s	m–s	Several bands, see refs 11, 12
	1500–1450	6.67–6.90	v	m	CH$_2$ def vib
	1325–1275	7.27–7.84	v	m	CH$_2$ def vib
	1260–1175	7.94–8.51	v	m	CH$_2$ def vib
	1100–1075	9.09–9.30	s	m–s	C–O str
	~915	~10.93	w	s	
	860–760	11.63–13.16	v	w	CH$_2$ def vib

Table 13.4 Thiophenes

Functional Groups	Region		Intensity		Comments
	cm^{-1}	μm	IR	Raman	
Thiophenes	3120–3000	3.21–3.33	m	m–s	=C–H str
	1585–1480	6.31–6.56	v	v	C=C in-plane vib
	1445–1390	6.92–7.19	v	vs	C=C in-plane vib
	1375–1340	7.33–7.46	v	s	C=C in-plane vib
	1240–1195	8.07–8.37	v	m	C=C in-plane vib
	530–450	18.87–22.22	v	m	Ring def
Monosubstituted thiophenes	745–695	13.42–14.39	v	m–w	=C–H out-of-plane def vib
	700–660	14.29–15.15	v	m–w	=C–H out-of-plane def vib
2-Monosubstituted thiophenes	1540–1490	6.49–6.71	v	v	C=C in-plane vib, see ref. 17
	1455–1430	6.87–6.99	m-s	vs	C=C in-plane vib
	1365–1345	7.33–7.44	m-s	s	C=C in-plane vib
	570–490	17.54–20.41	v		Esters at ~565 cm^{-1}
	470–430	21.28–23.26	v	s	
2-Alkyl thiophenes	1240–1215	8.06–8.23	m–w	m	CH in-plane def vib
	1160–1140	8.62–8.77	w	m	CH def vib
	1085–1060	9.22–9.43	w	m	C–H def vib
	1055–1030	9.48–9.71	w–m	m	
	940–905	10.64–11.05	m	w	Out-of-plane CH def vib
	870–840	11.49–11.90	m–s	s	Out-of-plane CH def vib
	855–800	11.70–12.50	m	w	Out-of-plane CH def vib
	770–735	12.99–13.60		m	Ring def vib
	725–670	13.80–14.93	m	w	Out-of-plane CH def vib
3-Monosubstituted thiophenes	1540–1490	6.49–6.71	v	v	C=C in plane vib
	1410–1380	7.09–7.25	m	vs	
	1380–1360	7.25–7.35	m–s	s–m	Ring vib
	935–880	10.70–11.36	w	s–m	C–S asym str
	850–825	11.76–12.12	w	vs–s	C–S sym str
	540–515	18.52–19.42	m		Sometimes only one present
	500–465	20.00–21.51	v		

Table 13.4   (*continued*)

Functional Groups	Region cm^{-1}	Region µm	Intensity IR	Intensity Raman	Comments
3-Alkyl-substituted thiophenes	~1530	~6.54	v	v	C=C in-plane vib
	~1410	~7.09	v	vs	C=C in-plane vib
	~1370	~7.30	v	s	C=C in-plane vib
	~1155	~8.66	w	w–m	C–H def vib
	1100–1070	9.09–9.35	w	w	C–H def vib
	895–850	11.17–11.76	m		
	795–745	12.58–13.42	s	w	Out-of-plane C–H def vib
2,3-Disubstituted thiophenes	715–690	14.01–14.49	m	w	Out-of-plane C–H def vib
2,4-Disubstituted thiophenes	825–805	12.11–12.41	m	w	Out-of-plane C–H def vib
2,5-Diaklyl thiophenes	1600–1570	6.25–6.37		vs	Ring vib
	1530–1500	6.54–6.67		v	Ring vib
	~795	~12.58	m–s	w	C–H def vib
2-Nitro-5-substituted thiophenes	555–525	18.02–19.05	v		
	490–445	20.41–22.47	v		
	~430	~23.26	w		One or two bands
3,4-Disubstituted thiophenes	925–910	10.80–11.00	m	w	Out-of-plane C–H def vib
	860–835	11.63–11.98	m	w	Out-of-plane C–H def vib
	780–775	12.82–12.90	m	w	Out-of-plane C–H def vib
Tetrahydrothiophenes	~685	~14.60	m	s, p	C–S str
Selenophenes, mono- and dimethyl substituted	440–405	22.73–24.69			See ref. 18
2-Monosubstituted selenophenes	1550–1530	6.45–6.54	v	m–w	Ring vib
	1460–1430	6.85–6.99	m–s	s	Ring vib
	1345–1325	7.43–7.55	v	m	Ring vib
	1100–1075	9.09–9.30	w	m	CH in-plane def vib
	1040–1015	9.62–9.85	w	m	CH in-plane def vib
	810–765	12.35–13.07	v	m	Ring vib
	635–615	15.75–16.26	v	m-w	In-plane def vib
Thiazoles,	~1610	~6.21	v	v	See ref. 6
	1550–1505	6.45–6.64	m	v	
	~1380	~7.25	v	s	

Most 2-nitro-5-substituted thiophenes have bands of variable intensity at 555–525 cm^{-1} (18.02–19.05 µm) and 490–445 cm^{-1} (20.41–22.47 µm), and usually one or two weak bands near 430 cm^{-1} (23.26 µm).

In general, mono-, di-, tri-, and tetrasubstituted thiophenes all have bands in the region 530–450 cm^{-1} (18.87–22.22 µm) due to the out-of-plane ring deformation.

Thiophenes have a band near 675 cm^{-1} (14.81 µm) due to the C–S stretching vibration. This band is usually of medium intensity in the infrared and of strong intensity (also polarised) in Raman spectra.

Imidazoles,

In general, azoles have three or four bands in the region 1670–1320 cm^{-1} (5.99–7.58 µm) due to C=C and C=N stretching vibrations. The intensities of these bands depend on the nature and positions of the substituent and on the position and nature of the ring heteroatoms.

Table 13.5   Imidazoles

Functional Groups	Region cm^{-1}	Region μm	Intensity IR	Intensity Raman	Comments
Imidazoles, 	1660–1610	6.02–6.21	v	m–s	Imidazole I
	1605–1585	6.23–6.31	w–m	m–s	band
	1560–1520	6.41–6.58	s	s	Ring C=C str N=C–N str
4-Monosubstituted, 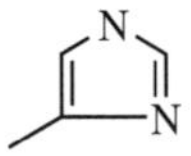	670–625	14.93–16.00	s		
	630–605	15.87–16.53	s		
	445–355	22.47–28.17	m		
	360–325	27.78–30.77	m		
4,5-Disubstituted, 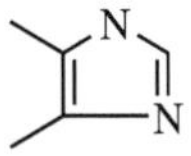	665–650	15.04–15.38	m–s		
	645–610	15.50–16.39	m–s		
1,4,5-Trisubstituted, 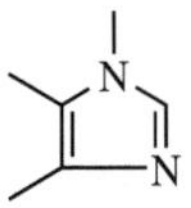	660–640	15.15–15.63	m–s		
	420–390	23.81–25.64	w–m		

Table 13.6   Pyrazoles

Functional Groups	Region cm^{-1}	Region μm	Intensity IR	Intensity Raman	Comments
N-Alkyl-substituted pyrazoles	3125–3095	3.20–3.23	m	w–m	CH str
	~1520	~6.58	v	m–s	Ring vib
	~1400	~7.14	v	m–s	Ring vib
	1090–1060	9.17–9.43	m–w	w	See ref. 20. CH def vib
	1040–1030	9.62–9.71	m	s	Ring vib
	970–950	10.31–10.53	m	s	Ring def vib
	~755	~13.28	m–s		
3-Alkyl-substituted pyrazoles	~3175	~3.15	m	m–w	N–H str. Hydrogen bonded: br, 3175–3155 cm^{-1}
	3125–3095	3.20–3.23	m	w–m	CH str
	~1580	~6.33	v	m–s	Ring vib
	~1470	~6.80	v	m–s	Ring vib
	~1050	~9.52	w	w	CH def vib
	1020–1010	9.80–9.90	m		Ring vib
	~935	~10.69	m	s	Ring def vib
	~770	~12.99	s		br
4-Alkyl substituted pyrazoles	~1575	~6.35	v	m–s	Ring vib
	~1490	~6.71	v	m–s	Ring vib
	1060–1040	9.43–9.62	m	m–w	Ring vib
	1010–990	9.90–10.10	m	s	Ring def vib
	~950	~10.53	s		
	~860	~11.63	s		
	~805	~12.42	s		

Table 13.6 (*continued*)

Functional Groups	Region		Intensity		Comments
	cm^{-1}	μm	IR	Raman	
3- 2,3-, 3,4-, 1,3,4-, and 2,3,4- substituted pyrazol-5-ones	3000–2200	3.33–4.55	m	m–w	br, O–H and N–H str
	1670–1450	5.99–6.90	w–m	m–s	Three or four bands due to C=C and C=N str
1,2,3-Trisubstituted pyrazol-5-ones	1675–1655	5.97–6.04	s	m–w	C=O str
3,4,4-Trisubstituted pyrazol-5-ones	~3150	~3.18	m	m–w	br, N–H str
	1760–1675	5.87–5.97	s	w–m	C=O str

In the solid phase, five-membered heteroatomic compounds with two or more nitrogen atoms in the ring have a broad absorption at 2800–2600 cm^{-1} (3.57–3.85 μm) due to the NH$\cdots$N bond.

Imidazoles[5] have several bands of variable intensity in the range 1660–1450 cm^{-1} (6.02–6.90 μm) due to C=N and C=C stretching vibrations. Most 4-monosubstituted imidazoles have two strong bands, at 670–625 cm^{-1} (14.93–16.00 μm) and 630–605 cm^{-1} (15.87–16.53 μm). They also have two bands of medium intensity, at 445–355 cm^{-1} (22.47–28.17 μm) and 360–325 cm^{-1} (27.78–30.77 μm), although this last band is absent for some imidazoles. The first of these two bands is probably due to out-of-plane bending of the –N–H group.

4,5-Disubstituted imidazoles have two medium-to-strong bands, at 665–650 cm^{-1} (15.04–15.38 μm) and 645–610 cm^{-1} (15.50–16.39 μm). 1,4,5-Tri-substituted imidazoles have a medium-to-strong absorption at 660–650 cm^{-1} (15.15–15.63 μm) and a weak-to-medium band at 420–390 cm^{-1} (23.81–25.64 μm). A study of metal complexes with imidazole ligands can be found elsewhere.[19]

Some pyrazol-5-one derivatives[22] exist as a form in which the carbonyl group is no longer present, and indeed two such forms may exist:

In the case of 4,4- and 1,2-disubstituted pyrazol-5-ones, the carbonyl group[24] is present and hence for these compounds an absorption band due to the carbonyl stretching vibration is observed.

5-Aminopyrazoles have a band of medium intensity near 1595 cm^{-1} (6.27 μm) and weaker bands near 1660 cm^{-1} (6.02 μm) and 1550 cm^{-1} (6.45 μm). All three bands have been attributed to ring vibrations.

For bonded pyrazoles, the N–H stretching vibration is weak and occurs at 3175–3155 cm^{-1} (3.15–3.17 μm).

Pyrazoles,

Due to tautomerism,[25] positions 3 and 5 of pyrazoles[20,21] are equivalent:

References

1. R. A. Jones and A. G. Moritz, *Spectrochim. Acta*, 1965, **21**, 295.
2. F. Millich and E. I. Becker, *J. Org. Chem.*, 1958, **23**, 1096.
3. A. R. Katritzky and A. P. Ambler, *Physical Methods in Heterocyclic Chemistry*, Academic Press, New York, 1963, p. 161.
4. A. I. Meyers, *J. Org. Chem.*, 1959, **24**, 1233.
5. P. Bassignana *et al.*, *Spectrochim. Acta*, 1965, **21**, 605.
6. J. Chouteau *et al.*, *Bull. Soc. Chim. Fr.*, 1962, **18**, 1794.
7. M. P. V. Mijovic and J. Walker, *J. Chem. Soc.*, **1961**, 3381.
8. W. H. Washburn, *Appl. Spectrosc.*, 1964, **18**, 61.
9. J. Baran, *Compt. Rend.*, 1959, **249**, 1096.
10. J. H. Bayer *et al.*, *J. Am. Chem. Soc.*, 1957, **79**, 1748.

11. N. Baggett *et al.*, *J. Chem. Soc.*, **1960**, 4565.
12. P. Grünager and F. Pozzi, *Gazz. Chim. Ital.*, 1959, **89**, 897.
13. P. Grünager *et al.*, *Gazz. Chim. Ital.*, 1959, **89**, 913.
14. A. R. Katritzky and J. M. Lagowski, *J. Chem. Soc.*, **1959**, 657.
15. M. Rico *et al.*, *Spectrochim. Acta*, 1965, **21**, 689.
16. A. Hidalgo, *J. Phys. Rad.*, 1955, **16**, 366.
17. A. R. Katritzky and A. J. Boulton, *J. Chem. Soc.*, **1959**, 3500.
18. N. A. Chumaevskii *et al.*, *Opt. Spectrosc.*, 1959, **6(1)**, 25.

19. M. T. Forel *et al.*, *Colloq. Int. Cent. Nat. Rech. Sci.*, 1970, **191**, 167.
20. G. Zerbi and C. Alberti, *Spectrochim. Acta*, 1962, **18**, 407.
21. A. A. Novikova and F. M. Shemyakin, *Khim.-Farm. Zh.*, 1968, **2(10)**, 45.
22. S. Refn, *Spectrochim. Acta*, 1961, **17**, 40.
23. J. H. Lister, *Chemistry of Heterocyclic Compounds*, Vol. 24, Wiley, New York, 1971, p. 496.
24. W. Freyer, *J. Prakt. Chem.*, 1977, **319**, 911.
25. J. M. Orza *et al.*, *Spectrochim. Acta Part A*, 2000, **56(8)**, 1469.

14 Organic Nitrogen Compounds

Nitro Compounds, $-NO_2$[1,2]

Saturated primary and secondary aliphatic nitro compounds,[3-8,36] $-CH_2NO_2$ and $\diagdown CHNO_2$, have very strong bands at about $1550\,cm^{-1}$ $(6.45\,\mu m)$ and $1390-1360\,cm^{-1}$ $(7.19-7.35\,\mu m)$ which are due to the asymmetric and symmetric stretching vibrations respectively of the NO_2 group. In Raman spectra, these bands generally have medium-to-strong intensities. Electron-withdrawing substituents adjacent to the nitro-group tend to increase the frequency of the asymmetric vibration and decrease that of the symmetric vibration.[7,32,33] For saturated nitro compounds, the asymmetric stretching band may be found in the region $1660-1500\,cm^{-1}$ $(6.02-6.67\,\mu m)$. For molecules with an NO_2 group or a halogen atom on the α-carbon atom, the NO_2 asymmetric stretching vibration band range is $1625-1540\,cm^{-1}$ $(6.15-6.49\,\mu m)$ and that of the symmetric stretching vibration is $1400-1360\,cm^{-1}$ $(7.14-7.35\,\mu m)$.

The band due to the C–N stretching vibration is of weak-to-medium intensity and occurs at $920-850\,cm^{-1}$ $(10.87-11.76\,\mu m)$. Other groups have strong absorptions in this region which may obscure this band. In general, organic nitro compounds have a very strong band at $655-605\,cm^{-1}$ $(15.27-16.53\,\mu m)$ due to the deformation vibration of the NO_2 group. Primary nitro compounds[36] have a weak-to-medium absorption in the region $615-525\,cm^{-1}$ $(16.26-19.05\,\mu m)$ due to the NO_2 wagging vibration, whereas secondary and tertiary nitro compounds have a weak-to-medium absorption in the region $650-570\,cm^{-1}$ $(15.38-17.54\,\mu m)$ and α-unsaturated and aromatic compounds[36] have a medium-to-strong band at $790-690\,cm^{-1}$ $(12.66-14.49\,\mu m)$. Primary aliphatic straight-chain nitro compounds absorb strongly at $620-600\,cm^{-1}$ $(16.13-16.67\,\mu m)$ and also have a medium-to-strong band at $490-465\,cm^{-1}$ $(20.41-21.51\,\mu m)$, both bands being due to the NO_2 deformation vibration. Secondary nitroalkanes absorb at $630-610\,cm^{-1}$ $(15.87-16.39\,\mu m)$ and $550-515\,cm^{-1}$ $(18.18-19.42\,\mu m)$. For saturated nitro compounds, the NO_2 in-plane deformation band is of weak-to-medium intensity and occurs in the region $775-605\,cm^{-1}$ $(12.90-16.53\,\mu m)$ but, for most saturated halogen- or NO_2-substituted nitro compounds, this band appears at $695-605\,cm^{-1}$ $(14.39-16.53\,\mu m)$ whereas for conjugated or aromatic compounds this band is observed at $910-790\,cm^{-1}$ $(10.99-12.66\,\mu m)$.

α,β-Unsaturated nitroalkenes absorb strongly at $1565-1505\,cm^{-1}$ $(6.39-6.64\,\mu m)$ and $1360-1335\,cm^{-1}$ $(7.35-7.49\,\mu m)$ due to the $-NO_2$ asymmetric and symmetric stretching vibrations. These bands are almost of equal intensity. The nitro group does not appear to affect the position of the characteristic alkene C=C and C–H bands. However, the relative intensities of the bands due to the =C–H stretching and wagging vibrations are increased when the nitro group is bonded to the same olefinic carbon as the hydrogen atom, the intensity of the band due the C=C stretching vibration $1650-1600\,cm^{-1}$ $(6.06-6.25\,\mu m)$ also being increased.

Aromatic nitro compounds[9-12] have strong absorptions due to the asymmetric and symmetric stretching vibrations of the NO_2 group at $1570-1485\,cm^{-1}$ $(6.37-6.73\,\mu m)$ and $1370-1320\,cm^{-1}$ $(7.30-7.58\,\mu m)$ respectively. The intensity of this latter band is increased for electron-donating ring substituents. The former band is usually found in the range $1540-1515\,cm^{-1}$ $(6.49-6.60\,\mu m)$. *Ortho*-substituted nitrocompounds whose substituent is a strongly electron-donating atom or group absorb at $1515-1485\,cm^{-1}$ $(6.60-6.73\,\mu m)$, whereas those with electron-accepting groups absorb at $1570-1540\,cm^{-1}$ $(6.37-6.49\,\mu m)$. The asymmetric NO_2 stretching vibration of most singly-substituted aromatic *para*-nitro compounds gives a band in the range $1535-1510\,cm^{-1}$ $(6.52-6.62\,\mu m)$, exceptions to this being *p*-dinitrobenzene and some *p*-aminonitrobenzenes. Singly-substituted aromatic *meta*-nitro compounds absorb in the range $1540-1525\,cm^{-1}$ $(6.49-6.59\,\mu m)$ and nitro compounds with small substituents in the *ortho* position absorb at $1540-1515\,cm^{-1}$ $(6.49-6.60\,\mu m)$. The band due to the asymmetric stretching vibration for nitro groups forced out of the plane of the ring by bulky substituents in the *ortho* positions is at $1565-1540\,cm^{-1}$ $(6.39-6.49\,\mu m)$. Hydrogen bonding has little effect on the NO_2 asymmetric stretching vibration.[12]

The symmetric vibration of the NO_2 group for aromatic *para*-substituted nitro compounds occurs at $1355-1335\,cm^{-1}$ $(7.38-7.49\,\mu m)$ whereas for *meta*

Table 14.1   Nitro compounds

Functional Groups	Region cm^{-1}	Region μm	Intensity IR	Intensity Raman	Comments
Saturated primary and secondary aliphatic nitro compounds, CH$_2$–NO$_2$ and $>$CH—NO$_2$	1555–1545	6.43–6.47	vs	m–w	asym NO$_2$ str, see ref: 13, stronger than sym str
	1385–1360	7.22–7.35	vs	s	sym NO$_2$ str (CH$_2$ def vib also occurs in this region)
	1000–915	10.00–10.93	m–w	m–s, p	C–N str, *trans-* form
	920–850	10.87–11.76	m–w	m–s, p	br, C–N str, *gauche-* form
	655–605	15.27–16.53	vs	m, p	NO$_2$ def vib. Two weak bands 670–605 cm^{-1} in IR and Raman
	560–470	14.86–21.28	m–s	v	NO$_2$ rocking vib
Straight-chain primary nitroalkanes	620–600	16.13–16.67	m–w	m–w	sym NO$_2$ def vib (except nitromethane at ~649 cm^{-1})
	490–465	20.41–21.51	m–s	m–w	NO$_2$ rocking vib (except nitromethane at ~602 cm^{-1})
Secondary nitroalkanes	630–610	15.87–16.39	m–w	m	sym NO$_2$ def vib
	550–515	18.18–19.42	m–s	v	NO$_2$ rocking vib
Saturated tertiary aliphatic nitro compounds, $-$$\overset{|}{\underset{|}{C}}NO_2$	1555–1530	6.43–6.54	s	m–w	asym NO$_2$ str
	1375–1340	7.27–7.46	s	s	sym NO$_2$ str
Dinitroalkanes, $>$C(NO$_2$)$_2$	1590–1570	6.29–6.37	s	m–w	asym NO$_2$ str usual range (but may be found in 1610–1540 cm^{-1}). Medium intensity bands due to NO$_2$ in-plane def vib 690–630 cm^{-1} and wagging vib 650–510 cm^{-1}
	1340–1325	7.46–7.55	s	s	sym NO$_2$ str usual range, may be split (but may be found in 1405–1285 cm^{-1})
–C(NO$_2$)$_3$	1605–1595	6.23–6.27	vs	m–w	
	1310–1295	7.63–7.72	s	s	
α,β-Unsaturated nitro compounds	1565–1505	6.39–6.64	s	m–s	asym NO$_2$ str
	1360–1335	7.35–7.49	s	m–s	sym NO$_2$ str
α-Halo-nitro compounds	1580–1555	6.33–6.43	s	m–w	asym NO$_2$ str. General range 1625–1555 cm^{-1}
	1370–1340	7.30–7.46	s	s	sym NO$_2$ str. General range 1375–1305 cm^{-1}
α,α'-Dihalo nitro compounds	1600–1570	6.25–6.37	s	m–w	asym NO$_2$ str
	1340–1320	7.46–7.58	s	s	sym NO$_2$ str
Aromatic nitro compounds	1580–1485	6.33–6.73	s	m–w	asym NO$_2$ str, stronger str. For *o-* or *p-* strong electron donors at lower end of range
	1370–1315	7.30–7.60	s	s	sym NO$_2$ str. For *o-* or *p-* strong electron donors at 1375–1285 cm^{-1}
	1180–865	8.47–11.56	m	m–s	CN str
	865–830	11.56–12.05	s–m	m–w	NO$_2$ def vib
	790–690	12.66–14.49	s	m	Not always present
	590–500	16.95–20.00	v	w	In-plane bending vib of –NO$_2$ group
o-Aminonitro-aromatic compounds	1260–1215	7.94–8.26	s	s	sym NO$_2$ str
Nitroamines, $>$N—NO$_2$	1630–1530	6.14–6.54	s	–	asym NO$_2$ str, solids may be as low as 1500 cm^{-1}
	1315–1260	7.61–7.94	s	v, p	sym NO$_2$ str, solids may be as low as 1250 cm^{-1}
	1030–980	9.71–10.20	m	s, p	N–N str
	775–755	12.90–13.25	w–m	m	NO$_2$ def vib
	730–590	13.70–16.95	w–m	m	NO$_2$ wagging vib

Table 14.1 (*continued*)

Functional Groups	Region		Intensity		Comments
	cm⁻¹	μm	IR	Raman	
Nitrates, –O–NO₂	620–560	16.13–17.86		v	NO₂ rocking vib
	1660–1615	6.02–6.19	s	–	asym NO₂ str, see ref. 35. Not observed in Raman
	1300–1270	7.69–7.87	s	s, p	sym NO₂ str
	870–840	11.49–11.90	m	m	NO str
	765–745	13.07–13.42	w–m	m	NO₂ wagging vib
	720–680	13.89–14.71	w–m	m	NO₂ def vib
	570–500	17.54–20.00	m	v	NO₂ rocking vib
Carbonitrates, $\diagdown \!\!\! \diagup \! C\!=\!NO_2^-$	1605–1575	6.23–6.35	s	s	C=N str, see ref. 15, low, due to resonance
	1315–1205	7.60–8.30	s	m–w	asym NO₂ str
	1175–1040	8.51–9.62	s	s	sym NO₂ str
	735–700	13.61–14.29	m–s	m	NO₂ def vib
Nitrocycloalkanes (three-membered ring and larger)	1550–1535	6.45–6.51	s	m–w	asym NO₂ str
	1380–1355	7.25–7.38	s	s	sym NO₂ str

compounds, and also *ortho* compounds with small substituents, the range is 1355–1345 cm⁻¹ (7.38–7.44 μm). In the case of bulky *ortho* substituents, this band may be found as high as 1380 cm⁻¹ (7.25 μm). In cases where strong hydrogen bonding occurs, this band may be found at about 1320 cm⁻¹ (7.58 μm), an example being *o*-nitrophenol.

Aromatic nitro compounds have a band of weak-to-medium intensity in the region 590–500 cm⁻¹ (16.95–20.00 μm) which is due to the in-plane deformation of the –NO₂ group.[30,31] A strong band observed at 865–835 cm⁻¹ (11.56–11.98 μm) and a band is also sometimes observed at about 750 cm⁻¹ (13.33 μm).

Due to the deformation vibration of the adjacent methylene group, primary nitroalkanes[36] have a band of medium intensity near 1430 cm⁻¹ (6.99 μm). In general, the band due to the symmetric deformation vibration of the methyl group is overlapped by that due to the NO₂ symmetric stretching vibration. However, in compounds where both the methyl and nitro groups are attached to the same carbon atom, two well-separated bands are observed – one near 1385 cm⁻¹ (7.22 μm) and the other near 1370 cm⁻¹ (7.30 μm).

For molecules with an NO₂ group or a halogen atom on the α-carbon atom, the rocking vibration occurs at 530–430 cm⁻¹ (18.17–23.26 μm), with secondary nitro compounds absorbing at 530–470 cm⁻¹ (18.17–21.28 μm) and tertiary nitro compounds at 500–430 cm⁻¹ (20.00–23.26 μm).

Alkali metal nitroparaffins[14] have a very strong absorption at 1605–1575 cm⁻¹ (6.23–6.35 μm) due to the C=N stretching vibration, and a weak band near 1660 cm⁻¹ (6.06 μm).

Nitroso Compounds, –N=O,[16–19,34] (and Oximes, $\diagdown \!\!\! \diagup \! C\!=\!N\!-\!OH$)

In the solid and liquid phases, organic nitroso compounds normally exist as dimers and may have *cis*- or *trans*-forms.

The fact that primary and secondary nitroso compounds readily form oximes may present difficulties:

$$\diagdown \!\!\! \diagup \! CH\!-\!N\!=\!O \longrightarrow \diagdown \!\!\! \diagup \! C\!=\!N\!-\!OH$$

This reaction of nitroso compounds, which in some cases occurs very easily due to either heat or light, may be used to identify bands associated with the nitroso group by observing their disappearance from the spectrum. This conversion can easily be detected since nitroso compounds are highly coloured and oximes are not.

Aliphatic nitroso compounds in the solid phase have two strong absorptions when in the *cis*- form, one at 1425–1330 cm⁻¹ (7.02–7.52 μm) and the other at 1345–1320 cm⁻¹ (7.43–7.58 μm), whereas in the *trans*- form they have a band at 1290–1175 cm⁻¹ (7.75–8.50 μm).

Aromatic nitroso compounds, as dimers in the *cis*- form, absorb strongly at 1400–1390 cm⁻¹ (7.14–7.19 μm) and at about 1410 cm⁻¹ (7.10 μm) whereas, in the *trans*- form, a band at 1300–1250 cm⁻¹ (7.69–8.00 μm) is observed.

As monomers,[20] which only occur in the gas phase and in dilute solution, aromatic nitroso compounds absorb strongly at 1515–1480 cm^{-1} (6.06–6.75 μm) and aliphatic nitroso compounds at 1590–1540 cm^{-1} (6.29–6.49 μm) due to the –N=O stretching vibration.

α-Halogenated nitroso compounds absorb near 1620 cm^{-1} (6.17 μm). The position of the band due to the N=O stretching vibration is affected by substituent groups in a very similar manner to that of the carbonyl band.

Nitroso compounds usually have a band at 1180–1000 cm^{-1} (8.48–10.00 μm) and another at 865–750 cm^{-1} (11.56–13.33 μm), these being due to strong coupling of the C–N stretching vibration and the vibration of the carbon skeleton. The presence of chlorine atoms increases the intensity of these bands.

The C–N=O bending vibration results in a band of medium intensity near 575 cm^{-1} (17.39 μm). Free oximes have a characteristic absorption at

Table 14.2 Organic nitroso compound N–O stretching vibrations

Functional Groups	Region		Intensity		Comments
	cm^{-1}	μm	IR	Raman	
Cis-dimers $\diagdown$ N=N $\diagup$ / O / \ O					
Aliphatic compounds	1425–1330	7.02–7.52	m–s	s	
	1345–1320	7.43–7.58	vs		
Aromatic compounds	~1410	~7.10	vs	s	
	1400–1390	7.14–7.19	vs		
Trans-dimers $\diagdown$ N=N $\nearrow$ O / \ O					
Aliphatic compounds	1290–1175	7.75–8.50	s	s	
Aromatic compounds	1300–1250	7.69–8.00	s	s	Raman: very strong band at 1480–1450 cm^{-1} due to N=N which is infrared inactive
Monomers					
Aliphatic compounds	1625–1540	6.15–6.49	s	s	N=O str, usually at ~1550 cm^{-1}
α-Halogenated compounds	1620–1565	6.17–6.39	s	s	N=O str
Aromatic compounds	1525–1485	6.66–6.73	s	s	N=O str, usually at ~1500 cm^{-1}
Halogen-substituted compounds	1510–1485	6.62–6.73	s	s	N=O str

Table 14.3 Nitrosamines, $\diagdown$ N–N=O $\diagup$

Functional Groups	Region		Intensity		Comments
	cm^{-1}	μm	IR	Raman	
Nitrosamines (vapour phase)	1500–1480	6.67–6.76	s	s	N=O str, monomer
Nitrosamines (dilute solution), see refs 21–23	~3200	~3.13	w		Overtone
	1460–1435	6.85–6.97	s	s	N=O str (aromatics 1500–1450 cm^{-1})
	1150–1025	8.70–9.76	s	s–m	br, N–N str (aromatics 1030–925 cm^{-1})
	1030–980	9.71–10.20	w	s	CN str (aromatics 1200–1160 cm^{-1})
	~660	~15.15	m–s		N–N=O def vib
Nitrosamides, –N(NO)CO–	1535–1515	6.52–6.60	s	s	N=O str, see ref. 24
Alkyl thionitrites, R–S–N=O	~1535	~6.52	s	s	N=O str, multiple peaks

Table 14.4 Nitroamines $\diagdown$N·NO$_2$, and nitroguanidines, $-$N$=$C(N$-$NO$_2$)·N$\diagup$

Functional Groups	Region		Intensity		Comments
	cm^{-1}	μm	IR	Raman	
Nitroamines	1315–1260	7.60–7.94	s	m–s	sym NO$_2$ (see Table 14.1)
Saturated aliphatic nitroamines	790–770	12.66–12.99	m		
	1585–1530	6.31–6.54	s	m–s	asym NO$_2$ str
Alkyl nitroguanidines	1640–1605	6.10–6.23	s	m–s	asym NO$_2$ str
Aryl nitroguanidines	1590–1575	6.29–6.35	s	m–s	asym NO$_2$ str
Aryl nitroureas	1590–1575	6.29–6.35	s	m–s	asym NO$_2$ str

3650–3500 cm^{-1} (2.74–2.86 μm) due to the O–H stretching vibration whose frequency is reduced, of course, in the presence of hydrogen bonding. The band is then broad and found in the region 3300–3150 cm^{-1} (3.03–3.17 μm). A band which is weak, except for conjugated compounds, is observed at 1690–1650 cm^{-1} (5.92–6.06 μm) due to the C$=$N stretching vibration, the frequency of the band being increased in ring-strained situations. The band due to the N–O stretching vibration occurs at 960–930 cm^{-1} (10.42–10.75 μm).

In quinone mono-oximes the N–O stretching vibration appears at 1075–975 cm^{-1} (9.30–10.26 μm).

Covalent Nitrates, –ONO$_2$

Organic nitrates[25,35] have strong absorptions due to the asymmetric and symmetric stretching vibrations of the NO$_2$ group which occur at 1660–1615 cm^{-1} (6.02–6.19 μm) and 1285–1270 cm^{-1} (7.78–7.87 μm) respectively. The symmetric NO$_2$ stretching vibration band of secondary

alkyl nitrates and monocyclic nitrates consists of a doublet. The N–O stretching vibration also results in a very strong band, at 870–855 cm^{-1} (11.49–11.70 μm). Bands of weak-to-medium intensity are observed due to the NO$_2$ deformation vibrations at 760–755 cm^{-1} (13.10–13.25 μm) and 710–695 cm^{-1} (14.08–14.39 μm).

Inorganic nitrate salts[26] have a characteristic, sharp, weak-to-medium band in the region 860–710 cm^{-1} (11.63–14.08 μm) due to the bending vibration of the NO group.

Nitrato-metal complexes[27] absorb in the regions 1530–1480 cm^{-1} (6.54–6.76 μm) and 1290–1250 cm^{-1} (7.75–8.00 μm) due to the asymmetric and symmetric vibrations respectively of the NO$_2$ group.

Nitrites, –O–N$=$O

Nitrites[28,29] have very strong bands at 1680–1650 cm^{-1} (5.95–6.06 μm) and 1625–1610 cm^{-1} (6.16–6.21 μm) due to the N$=$O stretching vibration of the

Table 14.5 Organic (covalent) nitrates

Functional Groups	Region		Intensity		Comments
	cm^{-1}	μm	IR	Raman	
Nitrates, –ONO$_2$	1660–1615	6.02–6.19	vs	s–m	asym NO$_2$ str
	1300–1250	7.79–8.00	vs	s, p	sym NO$_2$ str
	870–840	11.49–11.90	vs	m	br, N–O str
	765–745	13.07–13.42	w–m	m	NO$_2$ out-of-plane def vib
	720–680	13.89–14.70	w–m	m	NO$_2$ def vib
	610–560	16.39–17.86		s	NO$_2$ in-plane def vib
Inorganic nitrate salts	1410–1350	7.09–7.41	vs	m	br, asym NO$_3$ str
	860–800	11.63–12.50	m	m–s	sh
	730–710	13.70–14.08	m–w	m–w	
	315–190	31.75–52.63	m		

Table 14.6 Organic nitrites, $-O-N{=}O$

Functional Groups	Region cm^{-1}	Region µm	Intensity IR	Intensity Raman	Comments
Nitrite compounds	3360–3220	2.98–3.11	w–m		Overtones of N=O str
Nitrites, *cis-* form	1625–1610	6.16–6.21	vs	s	N=O str. Secondary ~1615 cm^{-1}, tertiary ~1610 cm^{-1}
	850–810	11.76–12.35	s	m	N–O str
	690–615	14.49–16.26	s		O–N=O def vib
Nitrites, *trans-* form	1680–1650	5.95–6.06	vs	s	N=O str. Primary ~1675 cm^{-1}, secondary 1665 cm^{-1} and tertiary ~1625 cm^{-1}
	815–750	12.27–13.33	vs	m	N–O str
	625–565	16.00–17.00	s		O–N=O def vib
Alkyl thionitrites, $-S-N{=}O$	~1535	~6.52	s	s	N=O str, multiple peaks
Inorganic nitrite salts	1275–1235	7.84–8.10	s	m–s	asym NO$_2$ str
	835–800	11.98–12.50	m	m	sh

Table 14.7 Amine oxides, $-N^{+}-O^{-}$

Functional Groups	Region cm^{-1}	Region µm	Intensity IR	Intensity Raman	Comments
Aliphatic *N*-oxides, $-N^{+}-O^{-}$	970–950	10.31–10.53	s	m	N–O str
Pyridine and pyrimidine *N*-oxides (non-polar solution)	1320–1230	7.58–8.13	m–s	m	N–O str, hydrogen bonding lowers frequency by 10–20 cm^{-1}, band position affected by ring substituents
	895–840	11.17–11.90	m	s	N–O def vib
Pyridine *N*-oxides	1190–1150	8.40–8.70	m–s		
Pyrazine *N*-oxides	1380–1280	7.25–7.81	m–s	m	N–O str, band position affected by ring substituents
	1040–990	9.62–10.10	m–s		
	~850	~11.76	m		N–O def vib
Nitrile oxides	1380–1290	7.25–7.75	s		N–O def vib
Oximes, $C{=}N-OH$	960–930	10.42–10.75	s	m	N–O str

Table 14.8 Azoxy compounds $-N{=}N^{+}-O^{-}-$

Functional Groups	Region cm^{-1}	Region µm	Intensity IR	Intensity Raman	Comments
Aliphatic azoxy compounds	1530–1495	6.54–6.69	m–s	vs	N=N str
	1345–1285	7.43–7.78	m–s	m	NO str
Aromatic azoxy compounds	1480–1450	6.76–6.90	m–s	m–s	asym N=N–O str
	1335–1315	7.49–7.60	m–s	m–s	sym N=N–O str

trans- and *cis-* forms respectively. The overtone band is of weak-to-medium intensity and occurs at 3360–3220 cm^{-1} (2.98–3.11 μm). Halogen substitution tends to increase these frequencies.

A strong absorption due to the N-O stretching vibration is observed at 815–750 cm^{-1} (12.27–13.33 μm) for the *trans-* form and at 850–810 cm^{-1} (11.76–12.35 μm) for the *cis-* form. Strong bands also occur at 690–615 cm^{-1} (14.49–16.26 μm) and 625–565 cm^{-1} (16.00–17.70 μm) for the *cis-* and *trans-* forms respectively, due to the deformation vibrations of the O–N=O group.

References

1. T. Y. Paperno and Y. V. Perekalin, *Spectra of Nitro Compounds*, Gas. Pedayag, Leningrad, 1974.
2. M. Colette, *Ann. Sci. Univ. Besancon Chim.*, 1972, **9**, 3.
3. R. N. Hazeldine, *J. Chem. Soc.*, 1953, 2525.
4. J. F. Brown, *J. Am. Chem. Soc.*, 1955, **77**, 6341.
5. Z. Eckstein *et al.*, *J. Chem. Soc.*, 1961, 1370.
6. F. Borek, *Naturwiss.*, 1963, **50**, 471.
7. W. H. Lunn, *Spectrochim. Acta*, 1960, **16**, 1088.
8. N. Jonathan, *J. Mol. Spectrosc.*, 1961, **7**, 105.
9. C. J. W. Brooks and J. F. Morman, *J. Chem. Soc.*, **1961**, 3372.
10. T. Kinugasa and R. Nakushina, Nippon Kaguku Zasshi, 1963, **84**, 365.
11. C. P. Conduit, *J. Chem. Soc.*, **1959**, 3273.
12. W. F. Baitinger *et al.*, Tetrahedron, 1964, **20**, 1635.
13. A. S. Wexler, *Appl. Spectrosc. Rev.*, 1968, **1**, 29.
14. A. G. Lee, *Spectrochim. Acta*, 1972, **28A**, 133.
15. H. Feurer *et al.*, *Spectrochim. Acta*, 1963, **19**, 431.
16. M. Colette, *Ann. Sci. Univ. Besancon Chim.*, 1971, **8**, 80.
17. B. C. Gowenlock and W. Lüttke, *Quart. Rev.*, 1958, **12**, 321.
18. L. J. Bellamy and R. L. Williams, *J. Chem. Soc.*, **1957**, 863.
19. W. Lüttke, *Z. Elektrochem.*, 1957, **61**, 976.
20. W. Lüttke, *Z. Elektrochem.*, 1957, **61**, 302.
21. R. L. Williams *et al.*, *Spectrochim. Acta*, 1964, **20**, 225.
22. P. Tarte, *Bull. Soc. Chim. Belges*, 1954, **63**, 525.
23. C. E. Looney *et al.*, *J. Am. Chem. Soc.*, 1957, **79**, 6136.
24. E. H. White, *J. Am. Chem. Soc.*, 1955, **77**, 6008.
25. R. A. G. Carrington, *Spectrochim. Acta*, 1960, **16**, 1279.
26. F. A. Miller and C. H. Wilkins, *Anal. Chem.*, 1952, **24**, 1253.
27. E. Bannister and F. A. Cotton, *J. Chem. Soc.*, 1960, 2276.
28. P. Tarte, *J. Chem. Phys.*, 1952, **20**, 1570.
29. R. N. Hazeldine and B. J. H. Mattinson, *J. Chem. Soc.*, **1955**, 4172.
30. J. H. S. Green and D. J. Harrison, *Spectrochim. Acta*, 1970, **26A**, 1925.
31. J. H. S. Green and H. A. Lauwers, *Spectrochim. Acta*, 1971, **27A**, 817.
32. J. Durig *et al.*, *J. Mol. Struct.*, 1983, **99**, 45.
33. N. S. Sundra, *Spectrochim. Acta*, 1985, **41A**, 905,
34. R. P. Müller *et al.*, *J. Mol. Spectrosc.*, 1984, **104**, 209.
35. J. R. Durig and N. E. Lindsay, *Spectrochim. Acta*, 1990, **46A**, 112.
36. J. R. Hill *et al.*, *J. Phys. Chem.*, 1991, **95**, 3037

15 Organic Halogen Compounds

Organic Halogen Compounds, $-\!\!\!\!\!\diagdown\!\!\!C-X$ (where X=F, Cl, Br, I)

Strong characteristic absorptions due to the C–X stretching vibration are observed, the position of the band being influenced by neighbouring atoms or groups – the smaller the halide atom, the greater the influence of the neighbour. Different rotational isomers may often be identified since, in general, the *trans*- form absorbs at higher frequencies than the *gauche*- form. Bands of weak-to-medium intensity are also observed due to the overtones of the C–X stretching vibration. In Raman spectra, the C–X stretching vibrations result in strong bands for X=Cl, Br and I, but for fluorine the bands are weaker, the intensity increasing from F to I.

Monohaloalkanes (excluding fluorine as the atom is too small) often exhibit more than one C–X stretching band due to the different possible rotational isomeric configurations available. The population of a given isomer is, obviously, determined by energy considerations and this has a bearing on the intensity of the C–X stretching bands observed. In other words, the more stable the isomer, the greater the intensity of the C–X stretching band associated with it.

Organic Fluorine Compounds

The band due to the C–F stretching vibration may be found over a wide frequency range, 1360–1000 cm^{-1} (7.35–10.00 μm),[1–6,25,26,29,30] since the vibration is easily influenced by adjacent atoms or groups. Monofluorinated compounds have a strong band at 1110–1000 cm^{-1} (9.01–10.00 μm) due to the C–F stretching vibration. With further fluorine substitution, two bands are observed due to the asymmetric and symmetric stretching vibrations, these occurring at higher frequencies.[19–21,29,30]

Due to the strong coupling of the C–F and C–C stretching vibration, polyfluorinated compounds[2–4] have a series of very intense bands in the region 1360–1090 cm^{-1} (7.36–9.18 μm). A –CF$_3$ group[2,25,30,31] attached to an alkyl group absorbs strongly near 1290 cm^{-1} (7.75 μm), 1280 cm^{-1} (7.81 μm), 1265 cm^{-1} (7.91 μm), 1230 cm^{-1} (8.13 μm), and 1135 cm^{-1} (8.81 μm). Compounds with the group CF$_3$CF$_2$– have a medium-intensity absorption in the region 1365–1325 cm^{-1} (7.33–7.55 μm) and a strong band at 745–730 cm^{-1} (13.42–13.70 μm) due to deformation vibrations. Compounds with –CF$_3$ on an aromatic ring have very strong bands near 1320 cm^{-1} (7.58 μm), 1180 cm^{-1} (8.47 μm), and 1140 cm^{-1} (8.77 μm).

The C–H stretching vibration of aliphatic groups with fluorine bonded to the carbon atom, such as –CF$_2$H and $\diagdown\!\!\!C\!\!\!\diagup$FH, gives a band near 3000 cm^{-1} (3.33 μm).

Fluorine atoms directly attached to carbon double bonds have the effect of shifting the C=C stretching vibration to higher frequencies. For example, –CF=CF$_2$ at absorbs at 1800–1780 cm^{-1} (5.56–5.62 μm) and $\diagdown\!\!\!C$=CF$_2$ 1755–1735 cm^{-1} (5.70–5.76 μm).[6,7]

In general, C–F deformation vibrations give bands in the region 830–520 cm^{-1} (12.05–19.23 μm).

Aromatic fluoro compounds have a band of variable intensity in the region 420–375 cm^{-1} (23.81–26.67 μm) due to an in-plane deformation.

The difluoride hydrogen ion FHF$^-$ has a very broad absorption in the region 1700–1400 cm^{-1} (5.88–7.14 μm) due to its asymmetric stretching vibrations and a band in the region 1260–1200 cm^{-1} (7.94–8.33 μm) due to its deformation vibrations.

Organic Chlorine Compounds

The C–Cl stretching vibrations[6,8–11,19,20,22,23,25,26] give generally strong bands in the region 760–505 cm^{-1} (13.10–19.80 μm). Compounds with more than one chlorine atom exhibit very strong bands due to the asymmetric and symmetric stretching modes. Vibrational coupling with other groups may

Table 15.1 Organic fluorine compounds

Functional Groups	Region		Intensity		Comments
	cm^{-1}	μm	IR	Raman	
C–F	1400–1000	7.14–10.00	s	w–m	C–F str, general range
	830–520	12.05–19.23	s	m–s	C–F def, general range
Aliphatic monofluorinated compounds	1110–1000	9.01–10.00	vs	w–m	C–F str
	780–680	12.81–14.71	s	s	C–F def vib
Aliphatic difluorinated compounds	1250–1050	8.00–9.52	vs	w–m	Two bands, C–F str
Polyfluorinated alkanes	1360–1090	7.36–9.18	vs	m	A number of bands
CF$_3$–CF$_2$–	1365–1325	7.33–7.55	m–s	m	C–F str
	745–730	13.42–13.70	s	s	C–F def vib
–CF$_3$	1420–1205	7.04–8.30	s–m	m	CF str. ArCF$_3$ 1345–1265 cm^{-1}, α-unsatCF$_3$ 1390–1105 cm^{-1}
	1350–1120	7.41–8.93	s–m	m	CF str. A number of bands. ArCF$_3$ 1190–1130 and 1165–1105 cm^{-1}, α-unsatCF$_3$ 1215–1175 and 1215–1045 cm^{-1}
	780–680	12.82–14.71	m–w	s	CF def vib, may be as high as 810 cm^{-1}. ArCF$_3$ 720–580 cm^{-1}, α-unsatCF$_3$ 760–610 cm^{-1}
	680–590	14.71–16.95	m–w		asym CF$_3$ def vib. ArCF$_3$ 645–535 cm^{-1}, α-unsatCF$_3$ 640–515 cm^{-1}
	610–440	16.39–22.73	m–w		sym CF$_3$ def vib. May be absent for α-unsaturated compounds. ArCF$_3$ 610–460 cm^{-1}, α-unsatCF$_3$ 570–440 cm^{-1}
	500–220	20.00–45.45	m–w		CF$_3$ rocking vib. ArCF$_3$ 470–340 cm^{-1}, α-unsatCF$_3$ 500–310 cm^{-1}
	390–165	25.64–60.60	w–m		CF$_3$ rocking vib. ArCF$_3$ 360–260 cm^{-1}, α-unsatCF$_3$ 360–280 cm^{-1}
(Sat)–CF$_3$	1420–1210	7.04–8.26	v		C–F str, usually medium intensity in range 1340–1250 cm^{-1}
	1350–1150	7.41–8.69	v		C–F str, usually medium intensity in range 1290–1170 cm^{-1}
	1270–1050	7.87–9.52	v		C–F str, usually medium intensity in range 1225–1090 cm^{-1}
	810–600	12.35–16.67	w–m		CF def vib, usually 780–610 cm^{-1}
	720–520	13.89–19.23	w–m		CF def vib, usually 650–530 cm^{-1}
	595–485	16.81–20.62	w–m		CF def vib, usually 590–500 cm^{-1}
	485–220	20.62–45.45	w–m		Rocking vib, usually 390–260 cm^{-1}
	390–160	25.64–62.50	w–m		Rocking vib, usually 310–220 cm^{-1}
CF$_3$CO·O–	1375–1205	7.27–8.30	v	w–m	C–F str, usually medium intensity in range 1350–1230 cm^{-1}
	1260–1190	7.94–8.40	v	w–m	C–F str, usually medium intensity in range 1250–1160 cm^{-1}
	1220–1110	8.20–9.01	v	w–m	C–F str, usually medium intensity in range 1205–1145 cm^{-1}
	785–615	12.74–16.26	w–m	s	CF def vib, usually 780–690 cm^{-1}
	670–510	14.93–19.61	w–m		CF def vib, usually 590–550 cm^{-1}
	535–495	18.69–20.20	w–m		CF def vib, usually 530–500 cm^{-1}
	485–225	20.62–44.44	w–m		Rocking vib, usually 415–360 cm^{-1}
	270–190	37.04–52.63	w–m		Rocking vib, usually 250–205 cm^{-1}
CF$_3$– (unsat)	1390–1180	7.19–8.47	v	w–m	C–F str, usually medium intensity in range 1345–1245 cm^{-1}
	1215–1175	8.23–8.51	v	w–m	C–F str, usually medium intensity in range 1215–1175 cm^{-1}
	1215–1045	8.51–8.57	v	m	C–F str, usually medium intensity in range 1185–1135 cm^{-1}

(continued overleaf)

Table 15.1 (*continued*)

Functional Groups	Region cm^{-1}	Region μm	Intensity IR	Intensity Raman	Comments
	760–610	13.16–16.39	w–m	s	CF def vib, usually 725–625 cm^{-1}
	640–510	15.63–19.62	w–m	s	CF def vib, usually 640–570 cm^{-1}
	570–440	17.54–22.73	w–m		CF def vib, usually 550–480 cm^{-1}
	500–310	20.00–32.26	w–m		Rocking vib, usually 470–370 cm^{-1}
	360–280	27.78–35.71	w–m		Rocking vib, usually 360–280 cm^{-1}
CF$_3$–Ar	1345–1265	7.43–7.91	v	w–m	C–F str, usually medium intensity in range 1340–1290 cm^{-1}
	1190–1130	8.40–8.85	v	w–m	C–F str, usually medium intensity in range 1190–1150 cm^{-1}
	1165–1105	8.58–9.05	v	w–m	C–F str, usually medium intensity in range 1155–1115 cm^{-1}
	720–570	13.89–17.54	w–m	s	CF def vib, usually 690–630 cm^{-1}
	645–535	15.50–18.69	w–m		CF def vib, usually 640–580 cm^{-1}
	610–440	16.39–22.73	w–m		CF def vib, usually 590–490 cm^{-1}
	470–340	21.28–29.41	w–m		Rocking vib, usually 450–350 cm^{-1}
	360–260	27.78–38.46	w–m		Rocking vib, usually 350–260 cm^{-1}
>CF$_2$	1300–1100	7.69–9.09	s	m–w	asym CF str, Usually found 1275–1175 cm^{-1}.
	1200–1060	8.33–9.43	s	m	sym C–F str. Usually found 1200–1100 cm^{-1}.
	675–375	14.81–26.67	m–s		CF scissor vib. Often 580–440 cm^{-1}
	515–300	19.42–33.33	w		CF$_2$ wagging vib
	470–360	21.28–27.78	w		CF$_2$ rocking vib
	360–130	27.78–76.92	w		Torsional vib
Cyclic –CF$_2$– (four- or five-membered ring)	1350–1140	7.41–8.77	s	m	CF str
–CHF$_2$	1205–1105	8.30–9.05	s	m–w	asym CF str. Medium-to-strong bands 3005–2975, 1445–1345 and 1345–1205 cm^{-1} due to CH str, CH def vib and CH def vib
	1125–1055	8.89–9.48	s	m–w	sym CF str
	780–540	12.82–18.52	m–s		CF$_2$ wagging vib. Usual range 660–600 cm^{-1} but may be shifted by 100 cm^{-1} or more due to isomerism.
	575–475	17.39–21.05	m–s		CF$_2$ twisting def vib
	320–200	31.25–50.00	w		Skeletal vib
–CH$_2$F	3095–2950	3.23–3.39	m–w	m–s	asym CH$_2$ str, usually 3015–2975 cm^{-1}
	2995–2935	3.34–3.41	m–w	m–s	sym str
	1510–1400	6.62–7.14	m	m–w	CH$_2$ def vib, usually 1480–1430 cm^{-1}
	1435–1275	6.97–7.84	m–w	m	CH$_2$ wagging vib, usually 1395–1335 cm^{-1}
	1295–1115	7.72–8.97	m–w	m–w	CH$_2$ twisting vib, usually 1275–1190 cm^{-1}
	1110–990	9.01–10.10	vs	m–w	C–F str, usually 1080–1020 cm^{-1}
	990–800	10.10–12.50	w	w	CH$_2$ rocking vib, usually 970–870 cm^{-1}
	570–270	17.54–37.04	s		C–F def vib, usually 515–330 cm^{-1}
	250–110	40.00–90.91			Torsional vib
>C=CF$_2$	1755–1735	5.70–5.76	vs	s	C=C str
	1340–1300	7.46–7.69	s	m–w	CF str

Table 15.1 (*continued*)

Functional Groups	Region cm^{-1}	Region μm	Intensity IR	Intensity Raman	Comments
	580–560	17.24–17.86		s	CF$_2$ wagging vib
	525–505	19.05–19.80	m–s		Bending vib
	515–335	19.42–28.99	s		
	455–345	21.98–28.99	m–s		Rocking vib
–CF=CF$_2$	1800–1780	5.55–5.62	s	s	C=C str
	1340–1300	7.46–7.69	vs	m–w	C–F str
Ar–F	1270–1100	7.87–9.09			Ring and C–F str
	420–375	23.81–26.67	v		In-plane C–F def vib
Cyclobutylfluoride	~1100	~9.09	s	m–w	C–F str
(Sat)–CO·F	1235–1075	8.10–9.30	m–s	m	C–F str
	770–570	12.99–17.54		m	CO/CF def vib (range too wide to be useful)
	600–420	16.67–23.81		m	CO/CF def vib (range too wide to be useful)
	500–340	20.00–29.41		s	CO/CF rocking vib
(Unsat)–CO·F	1225–1085	8.16–9.22	m–s	m	C–F str
	730–580	13.70–17.24			CO/CF def vib (range too wide to be useful)
–O–CO·F	1140–1010	8.77–9.90	m–s	m	C–F str
	790–750	12.65–13.33		m	CO/CF def vib (range too wide to be useful)
	670–630	14.93–15.87		m	CO/CF def vib (range too wide to be useful)
	570–510	17.54–19.61		s	CO/CF rocking vib

result in a shift in the absorption to as high as 840 cm^{-1} (11.90 μm). For simple organic chlorine compounds, the C–Cl absorptions are in the region 750–700 cm^{-1} (13.33–14.29 μm) whereas for the *trans-* and *gauche-* forms they are near 650 cm^{-1} (15.38 μm),[8] the *trans-* form generally absorbing at higher frequencies.

In the liquid phase, since primary chloroalkanes exist as two or three isomers, two or three bands may be observed due to their C–Cl stretching vibrations. Primary chloro *n*-alkanes and α,ω-dichloro *n*-alkanes absorb strongly at 730–720 cm^{-1} (13.70–13.89 μm) and 655–645 cm^{-1} (15.27–15.50 μm), exceptions being the ethane and propane derivatives. In general, secondary chloroalkanes have a number of rotational isomers which therefore complicate the observed spectrum. For 2-chloroalkanes, strong bands are observed at 680–670 cm^{-1} (14.71–14.93 μm) and 615–610 cm^{-1} (16.26–16.39 μm), the latter band sometimes obscuring a further band which may be observed at about 625 cm^{-1} (16.00 μm).

Most mono- and disubstituted aromatic chloro compounds have a band of strong-to-medium intensity in the region 385–265 cm^{-1} (25.97–37.74 μm) due to C–Cl in-plane deformation.

Overtone bands of medium intensity resulting from the C–Cl stretching vibration are observed in the region 1510–1450 cm^{-1} (6.62–6.90 μm).

Organic Bromine Compounds

Bromine compounds[12] absorb strongly in the region 650–485 cm^{-1} (15.38–20.62 μm) due to the C–Br stretching vibrations, although when there is more than one bromine atom on the same carbon atom, two bands may be observed at higher frequencies. The CH$_2$ wagging vibration of –CH$_2$Br, 1315–1200 cm^{-1} (7.60–8.33 μm), is affected by conformation, so the difference between *trans-* and *gauche-* may be as much as 50 cm^{-1}.

Primary bromoalkanes of *n*-paraffins absorb strongly in the regions 645–635 cm^{-1} (15.50–15.75 μm) and 565–555 cm^{-1} (17.70–18.02 μm) due to the stretching vibration of the C–Br bond of the group C–CH$_2$–CH$_2$Br. Also, for *n*-bromoalkanes a band of variable intensity is observed at 440–430 cm^{-1} (22.73–23.26 μm), exceptions to this being the bromo derivatives of ethane, propane, and *n*-tridecane. With the exception of small molecules, α,ω-dibromoalkanes have similar absorption regions to the monobromo *n*-alkanes except for the lower-frequency region where weak-to-medium intensity bands are observed at 490–480 cm^{-1} (20.41–20.83 μm) and 445–425 cm^{-1} (22.47–23.53 μm).

The spectra of *n*-alkyl bromides exhibit a similar dependence on conformation to those of the chlorides. It has been found that for the compounds

Table 15.2 Organic chlorine compounds

Functional Groups	Region		Intensity		Comments
	cm⁻¹	μm	IR	Raman	
C–Cl	760–505	13.10–19.80	s	s	C–Cl str, general range
	450–250	22.22–40.00	s	s	C–Cl def vib, general range
\CCl₂ /	855–650	11.70–15.38	s	s–m	C–Cl str, ref. 28
	790–545	12.66–18.35	m–s	vs	CCl₂ sym str, usually 690–500 cm⁻¹
	420–340	23.81–35.71	w–m	s	CCl₂ wagging vib
	380–280	26.32–35.71	m–w		CCl₂ rocking vib
	340–260	29.41–38.46		s	Twisting vib
	290–210	34.48–47.62			def vib
–CCl₃	900–710	11.11–14.08	s	s	CCl str, usually 870–760 cm⁻¹
	815–645	12.27–15.50	s	s	CCl str, usually 800–670 cm⁻¹
	680–435	14.71–22.99	s	m	CCl str, usually 630–450 cm⁻¹
	435–295	22.99–33.90	w–m		def vib, usually 415–315 cm⁻¹
	385–265	25.97–37.74	w–m		def vib, usually 375–280 cm⁻¹
	355–225	28.17–44.44	w–m	vs	def vib, usually 340–240 cm⁻¹
	260–190	38.46–52.63			Rocking vib, usually 250–200 cm⁻¹
	230–70	43.48–142.86			Rocking vib, usually 200–115 cm⁻¹
	150–50	66.67–200.00			
\CHCl /	710–590	14.08–16.95	s	s	CCl str (CH str 2980–2900 cm⁻¹, m, CH out-of-plane def vib 1380–1280 cm⁻¹, w, CH in-plane def vib 1290–1200 cm⁻¹, m–s
	400–290	25.00–34.48	w–m		CCl def vib
	330–230	30.30–43.48	w	s	CCl def vib
–CH₂Cl	3035–2985	3.29–3.50	w–m	m–s	asym CH₂ str, ref. 27
	2985–2940	3.50–3.40	w–m	m–s	sym CH₂ str
	1460–1410	6.85–7.09	m	m–w	CH₂ def vib
	1315–1215	7.60–8.23	m–s	m–w	CH₂ wagging vib. (Unsat. compounds 1280–1250 cm⁻¹)
	1280–1145	7.81–8.73	m	m–w	CH₂ twisting vib. (Unsat. compounds 1225–1155 cm⁻¹)
	990–780	10.10–12.82	m–w	w	CH₂ rocking vib. (Unsat. compounds 955–845 cm⁻¹ and aromatic compounds 765–725 cm⁻¹)
	770–630	12.99–15.87	s	s	C–Cl str (Unsat. compounds 740–655 cm⁻¹)
	365–205	27.40–48.78	m	s	C–Cl def. (Unsat. compounds 450–230 cm⁻¹)
	205–85	48.78–117.65			Torsional vib
R–(CH₂)₂Cl and Cl–(CH₂)ₙ₌₃Cl	730–710	13.70–14.08	s	s	–CH₂Cl has a strong band at 1300–1240 cm⁻¹ due to CH₂ wagging vib
R(CH₂)₂CH(CH₃)₂Cl	655–645	15.27–15.50	s	s	
	680–670	14.71–14.93	s	s–m	
	~625	~16.00	w–m	s	Easily overlooked
	615–610	16.26–16.39	s	s	
R(CH₂)₂CR′(CH₃)Cl (R′=Me or Et)	630–610	15.87–16.39	m–s	s	
	580–560	17.24–17.86	m–s	s	

Table 15.2   (*continued*)

Functional Groups	Region cm^{-1}	Region μm	Intensity IR	Intensity Raman	Comments
$-OCH_2Cl$, $-NCH_2Cl$, $-SCH_2Cl$,	3070–3000	3.26–3.33	w–m	m	asym CH_2 str
	3005–2945	3.33–3.40	w–m	m	asym str. ($-SCH_2Cl$ 2970–2930 cm^{-1})
	1465–1415	6.83–7.07	m	m–w	CH_2 def vib
	1350–1280	7.41–7.81	m–s	m–w	CH_2 wagging vib
	1275–1205	7.84–8.30	m	m–w	CH_2 twisting vib ($-SCH_2Cl$ 1160–1120 cm^{-1})
	1020–900	9.80–11.11	m	w	CH_2 rocking vib ($-SCH_2Cl$ 985–840 cm^{-1})
	755–630	13.25–15.87	s	s	C–Cl str
	370–250	27.03–40.00	m		C–Cl def vib
	200–100	40.00–100			Torsional vib
(Sat)–$CHCl_2$	3020–2975	3.31–3.36	m	m	CH str
	1310–1200	7.63–8.33	m	m	CH def vib
	1250–1180	8.00–8.47	m	m	CH wagging vib
	830–660	12.05–15.15	m–s	s	CCl_2 asym str ($-CO-CHCl_2$ 840–710 cm^{-1})
	780–600	12.82–16.67	m–s	s	CCl_2 sym str
	550–320	18.18–31.25		s	CCl_2 def vib ($-CO-CHCl_2$ 420–360 cm^{-1})
	335–235	29.85–42.55			CCl_2 def vib ($-CO-CHCl_2$ 275–175 cm^{-1})
	285–165	35.09–60.60			CCl_2 def vib
Ar–$CHCl_2$	~3005	~3.33	m	m	C–H str
	1300–1250	7.69–8.00	m	m	CH def vib
	1220–1200	8.20–8.33	m	m	CH wagging vib
	770–680	12.99–14.71	m–s	s	asym CCl_2 str
	630–580	15.87–17.24	m–s	s	sym CCl_2 str
	410–360	24.39–27.78		s	CCl_2 def vib
Polychlorinated compounds	800–700	12.50–14.29	vs	s	
$\diagdown$C=CCl_2 $\diagup$	500–320	20.00–31.25	m		Bending vib (C=C str, ~1615 cm^{-1})
	265–235	37.74–42.55	w		
	260–180	38.46–55.56	s		Rocking vib
Chloroformates, RO–CO·Cl	~690	~14.49	s	s	C–Cl str
	485–470	20.62–21.28	s		C–Cl def vib
RS–CO·Cl	~580	~17.24	s	s	C–Cl str
	350–340	28.57–29.41	s		C–Cl def vib
$\diagdown$N—Cl $\diagup$	805–690	12.42–14.49			See ref. 18
Ar–Cl	1100–1090	9.09–9.17			*Para*-substituted ⎱
	1080–1070	9.26–9.35			*Meta*-substituted ⎰ Combined ring and
	1060–1030	9.43–9.71			*Ortho*-substituted ⎰ C–Cl strs
Rotational configurations: chloroalkanes					
Primary chloroalkanes	730–720	13.70–13.89	s	s	Cl atom *trans* to C atom
	660–650	15.15–15.38	s	s	Cl atom *trans* to H atom
	695–680	14.39–14.71	s	s	Cl atom *trans* to H atom in branched alkane

(*continued overleaf*)

Table 15.2   *(continued)*

Functional Groups	Region cm^{-1}	μm	Intensity IR	Raman	Comments
Secondary chloroalkanes	760–740	13.10–13.51	m–w	m	Cl atom *trans* to two C atoms
	675–655	14.81–15.27	m–s	m	Cl atom *trans* to C and H atoms
	640–625	15.63–16.00	m–s	m–w	Cl atom *trans* to two H atoms in bent molecule
	625–605	16.00–16.53	s	s	Cl atom *trans* to two H atoms
Tertiary chloroalkanes	580–540	17.24–18.52	m–s	m–s	Cl atom *trans* to three H atoms
	635–610	15.75–16.39	m–s	m–s	Cl atom *trans* to one C and two H atoms
	385–265	25.97–37.74	m–s		In-plane C–Cl def vib
Cyclobutylchlorides	~620	~16.13	m–w	w	Equatorial
	~530	~18.87	m	m	Axial
Cyclopentylchlorides	~625	~16.00	m	w	Equatorial
	~590	~16.95	m	w	Axial
Cyclohexylchlorides	780–740	12.80–13.51	v	s	Equatorial C–Cl
Cyclohexylchlorides	730–580	13.70–17.25	s–m	m	Axial C–Cl

Table 15.3   Organic bromine compounds

Functional Groups	Region cm^{-1}	μm	Intensity IR	Raman	Comments
C–Br	750–485	13.33–20.62	s	s	C–Br str, general range
	400–140	25.00–71.43	m	s	C–Br def, general range
–CHBr$_2$	730–580	13.70–17.24	s	s–m	asym CBr$_2$ str
	625–480	16.00–20.83	s	s	sym CBr$_2$ str
	400–340	25.00–29.41		s	CBr$_2$ wagging vib
	350–290	28.57–34.48			CBr$_2$ rocking vib
	290–210	34.48–47.62		s	CBr$_2$ twisting vib
	210–150	47.62–66.67		s	CBr$_2$ def vib
–CH$_2$Br	3050–2990	3.28–3.34	m–w	m	asym CH$_2$ str
	2990–2900	3.34–3.44	m–w	m	sym CH$_2$ str
	1450–1410	6.90–7.09	m	m–w	CH$_2$ def vib
	1315–1200	7.60–8.33	m–s	m–w	CH$_2$ wagging vib, (affected by conformation difference by ~50 cm^{-1})
	1245–1105	8.03–9.05	m	m–w	CH$_2$ twisting vib
	945–715	10.58–13.99	w–m	w	CH$_2$ rocking vib
	730–550	13.70–18.18	s–m	s	C–Br str
	355–175	28.17–57.14	m–w		C–Br def vib
	190–70	52.63–142.86			Torsional vib
R–(CH$_2$)$_2$Br	645–615	15.50–16.26	s	s	C–Br str of C–(CH$_2$)$_2$Br–CH$_2$Br has strong band near 1230 cm^{-1} due to CH$_2$ wagging vib
	565–555	17.70–18.02	s	s	
	440–430	22.73–23.26	v	s	

Table 15.3 (*continued*)

Functional Groups	Region cm^{-1}	Region μm	Intensity IR	Intensity Raman	Comments
Br(CH$_2$)$_{n>3}$Br	660–625	15.15–16.00	s	s	C–Br str
	565–555	17.70–18.02	s	s	C–Br str
	490–480	20.41–20.83	w–m		
	445–425	22.47–23.53	w–m		
R–CH$_2$CHR'CH$_2$Br	650–645	15.38–15.50	s	s	C–Br str, *trans*- form
(R'=Me or Et)					
	625–610	16.00–16.39	s	s	C–Br str, *gauche*- form
R–(CH$_2$)$_2$CH(CH$_3$)Br	620–605	16.13–16.53	s	s	
	590–575	16.95–17.39	m–w	s	
	540–530	18.52–18.87	s	s	
R–(CH$_2$)$_2$C(CH$_3$)$_2$Br	600–580	16.67–17.24	m–s	s	
	525–505	19.05–19.80	s	v	
$\diagdown$CBr$_2$ $\diagup$	720–580	13.89–17.24	s	s–m	asym CBr$_2$ str
	580–480	17.24–20.83	s	s	sym CBr$_2$ str
	400–340	25.00–29.41			CBr$_2$ wagging vib
	350–290	28.57–34.48			CBr$_2$ rocking vib
	290–210	34.48–47.62		s	CBr$_2$ twisting vib
	210–150	47.62–66.67		s	CBr$_2$ def vib
$\diagdown$C=CBr$_2$ $\diagup$	310–250	32.26–40.00	s		Bending vib
	185–135	54.05–74.07	m		
	160–120	62.50–83.33	s		Rocking vib
Ar–Br	1075–1065	9.30–9.39	m		*Meta*- and *para*-substituted aromatic compounds ring and C–Br str combinations
	1045–1025	9.57–9.76	m		*Ortho*-substituted aromatic ring and C–Br str combination
	325–175	30.77–51.14	s–m		In-plane and out-of-plane C–Br def vib (2 bands)
Rotational configurations: *Bromoalkanes*					
Primary bromoalkanes	650–635	15.38–15.75	vs	s	Br atom *trans* to C atom
	565–555	17.70–18.02	vs	s	Br atom *trans* to H atom
	625–610	16.00–16.39	s	s	Br atom *trans* to H atom in branched alkane
	590–575	16.95–17.39	m	m–w	Br atom *trans* to two H atoms in bent molecule
	540–530	18.52–18.87	s	s	Br atom *trans* to two H atoms
Tertiary bromoalkanes	520–510	19.23–19.61	vs	v	Br atom *trans* to three H atoms
	590–580	16.95–17.24	m	s	Br atom *trans* to one C and two H atoms
Cyclohexylbromides	750–685	13.33–14.60	s	s	Equatorial C–Br
	690–550	14.50–18.20	s	s–m	Axial C–Br

in the series ethyl to *n*-decyl bromide, the C–Br stretching vibration gives a band at 645–635 cm^{-1} (15.50–15.75 μm) when the bromine atom is *trans*- to a carbon atom and at 565–555 cm^{-1} (17.70–18.02 μm) when *trans*- to a hydrogen atom.

Organic Iodine Compounds

Due to the large mass of the iodine atom, the C–I stretching vibration is coupled with skeletal vibrations. Also, a number of rotational isomers may

Table 15.4 Organic iodine compounds

Functional Groups	Region		Intensity		Comments
	cm^{-1}	μm	IR	Raman	
C–I	610–200	16.39–50.00	s	vs	C–I str, general range may be up to 660 cm^{-1}
	300–50	33.33–200.00	v	s	C–I def vib, general range
–CH$_2$I	1275–1050	7.84–9.52	m–s	m–w	CH$_2$ wagging vib. (Rotational isomerism gives up to 80 cm^{-1} band separation)
	620–490	16.13–20.41	m–s	vs	C–I str
	320–120	31.25–83.33		s	C–I def vib
R(CH$_2$)$_2$I	600–585	16.67–17.09	s	s	C–I str, –CCH$_2$I have strong band ~1170 cm^{-1} due to CH$_2$ wagging vib
	515–500	19.42–20.00	s	vs	C–I str
I(CH$_2$)$_{n>3}$I	615–575	16.26–17.39	s	s	C–I str
	~500	~20.00	s	vs	C–I str
Secondary iodoalkanes	590–575	16.95–17.39	s	s	
	550–520	18.18–19.23	s	vs	
	490–480	20.41–20.83	s		
Tertiary iodides	580–570	17.24–17.54	s	s	C–I str
	510–485	19.61–20.62	m	s	C–I str
	490–465	20.41–21.51	s	s	
$>$C=CI$_2$	~200	~50.00			Bending vib
	~100	~100.00			
	~50	~200.00			Rocking vib
Rotational configurations: *Iodoalkanes*					
Primary iodoalkanes	~600	~16.67	vs	s	I atom *trans* to C atom
	~510	~19.61	vs	s	I atom *trans* to H atom
	590–580	16.95–17.25	s	s	I atom *trans* to H atom in branched alkane
Secondary iodoalkanes	~580	~17.25	m	s	I atom *trans* to C and H atoms
	590–520	16.95–19.23	m	w	I atom *trans* to two H atoms in bent molecule
	490–480	20.41–20.83	s	s	I atom *trans* to two H atoms
Tertiary iodooalkanes	~490	~20.41	s	s	I atom *trans* to three H atoms
	580–570	17.25–17.54	m	s	I atom *trans* to one C and two H atoms
Cyclohexyliodides	~635	~15.27	s	s	Liquid phase. Equatorial C–I
	~640	~15.63	s	s	Liquid phase. Axial C–I

Table 15.5 Aromatic halogen compounds

Functional Groups	Region		Intensity		Comments
	cm^{-1}	μm	IR	Raman	
Aromatic halogen compounds (X =Cl, Br, I)	~1050	~9.52	m		X–sensitive band
Aromatic fluorine compounds	1270–1100	7.87–9.09	m		Approximate range, X–sensitive band
	680–520	14.71–19.23	m–s		Aromatic C–F str and ring def vib

Table 15.5 (*continued*)

Functional Groups	Region cm⁻¹	Region μm	Intensity IR	Intensity Raman	Comments
	420–375	23.81–26.67	v		In-plane aromatic C–F bending vib
	340–240	29.41–41.67	s		Out-of-plane aromatic C–F bending vib
Aromatic chlorine compounds	1060–1030	9.43–9.71	m	s	*Ortho*-substituted benzenes ⎫
	1080–1070	9.26–9.35	m	w	*Meta*-substituted benzenes ⎬ X–sensitive bands
	1100–1090	9.09–9.17	m	v	*Para*-substituted benzenes ⎭
	760–395	13.10–25.32	s		Not always present
	500–370	20.00–27.03	m–s		Aromatic C–Cl str and ring def vib
	390–165	25.64–60.61	m–s		Out-of-plane vib ⎫
	330–230	30.30–43.48	m–s		In-plane aromatic C–Cl bending vib ⎬ Not always present
Aromatic bromine compounds	1045–1025	9.57–9.76	m	s	*Ortho*-substituted benzenes ⎫
	1075–1065	9.30–9.39	m	w	*Meta*- and *para*-benzenes ⎬ X–sensitive bands
	400–260	25.00–38.46	s		Aromatic C–Br str and ring def vib
	325–175	30.77–57.14	m–s		Out-of-plane aromatic C–Br def vib
	290–225	34.48–44.44	m–s		In-plane aromatic C–Br bending vib
Aromatic iodine compounds	1060–1055	9.43–9.48	m–s	w	X–sensitive band for *para*-substituted benzenes
	310–160	32.26–62.50	s		Out-of-plane aromatic C–I bending vib
	265–185	37.74–54.05			Aromatic C–ring def vib
	~200	~50.00			In-plane aromatic C–I def vib

exist thus affecting the position of the C–I band, which is found in the region 600–200 cm⁻¹ (16.67–50.00 μm).[12,24,29,30] In general, primary iodo *n*-alkanes have strong absorptions at 610–585 cm⁻¹ (16.39–17.09 μm) and 515–500 cm⁻¹ (19.42–20.00 μm). It has been suggested that the former of these C–I stretching vibration bands is the result of the iodine atom being *trans* to a carbon atom and the latter the result of it being *trans* to a hydrogen atom. α,ω-Diiodoalkanes absorb in the same regions, strong bands usually being observed near 595 cm⁻¹ (16.81 μm) and 500 cm⁻¹ (20.00 μm).

Aromatic Halogen Compounds

Unlike aliphatic compounds, there appears to be no pure C–X stretching vibration band for aromatic halogen compounds.[5,13–16,19,20] However, several X-sensitive bands[17] are observed, one of which occurs at about 1050 cm⁻¹ (9.52 μm).

Aromatic fluoro compounds[5] have medium-intensity bands in the region 1270–1100 cm⁻¹ (7.87–9.09 μm), those with only one fluorine atom on the ring tending to absorb at about 1230 cm⁻¹ (8.13 μm). Bands due to the C–H out-of-plane vibrations and other aromatic ring vibrations are also observed.

Due mainly to the bending of the ring–halogen bond, aromatic fluorocompounds have a band of variable intensity at 420–375 cm⁻¹ (23.81–26.77 μm), aromatic chloro compounds have a band also of variable intensity (often medium-to-strong) at 390–270 cm⁻¹ (25.64–37.04 μm), and aromatic bromo compounds absorb strongly at 320–255 cm⁻¹ (31.25–39.22 μm). These bands as well as being observed for mono- and disubstituted benzenes, may also sometimes be observed, with different intensities, in polysubstituted aromatic compounds.

Most aromatic chloro and bromo compounds have strong absorptions at 760–395 cm⁻¹ (13.10–25.32 μm) and 650–395 cm⁻¹ (15.38–25.32 μm) respectively, which is due to a combination of vibrational modes. Monosubstituted benzenes, dihalogen-substituted benzenes, and compounds with electron-donor or methyl substituents in the *para* position of halobenzenes all exhibit the former band.

References

1. J. Murto *et al.*, *Spectrochim. Acta*, 1973, **29A**, 1121.
2. O. Risgin and R. C. Taylor, *Spectrochim. Acta*, 1959, **15**, 1036.
3. J. H. Simons (ed.), *Fluorocarbons and Related Compounds – Fluorocarbon Chemistry*, Vol. II. Academic Press, New York, 1954, p. 449.

4. C. Y. Liang and S. Krimm, *J. Chem. Phys.*, 1956, **25**, 563.
5. D. A. Long and D. Stecke, *Spectrochim. Acta*, 1963, **19**, 1947.
6. N. C. Craig and D. A. Evans, *J. Am. Chem. Soc.*, 1965, **87**, 4223.
7. D. E. Mann *et al.*, *J. Chem. Phys.*, 1957, **27**, 51.
8. J. J. Shipman *et al.*, *Spectrochim. Acta*, 1962, **18**, 1603.
9. A. R. Katritzky, *Spectrochim. Acta*, 1960, **16**, 964.
10. G. W. Chantry *et al.*, *Spectrochim. Acta*, 1966, **22**, 125.
11. M. A. Ory, *Spectrochim. Acta*, 1960, **16**, 1488.
12. F. F. Bentley *et al.*, *Spectrochim. Acta*, 1964, **20**, 105.
13. G. Varsanyi *et al.*, *Spectrochim. Acta*, 1963, **19**, 669.
14. T. R. Nanney *et al.*, *Spectrochim. Acta*, 1965, **21**, 1495.
15. T. R. Nanney *et al.*, *Spectrochim. Acta*, 1966, **22**, 737.
16. H. E. Shurvell *et al.*, *Spectrochim. Acta*, 1966, **22**, 333.
17. E. F. Mooney, *Spectrochim. Acta*, 1964, **20**, 1021.
18. R. C. Petterson *et al.*, *J. Org. Chem.*, 1960, **25**, 1595.
19. R. A. Yadav and I. S. Singh, *Spectrochim. Acta*, 1985, **41A**, 191.
20. R. A. Nyquist, *The Interpretation of Vapour-Phase Spectra*, Sadtler, 1985.
21. G. A. Growder and J. M. Lightfoot, *J. Mol. Struct.*, 1983, **99**, 77.
22. M. S. Soliman, *Spectrochim. Acta*, 1993, **49A**, 189.
23. H. G. M. Edwards, *J. Mol. Struct.*, 1991, **263**, 11.
24. J. R. Durig *et al.*, *Struct. Chem.*, 1993, **4**, 103.
25. J. R. Durig *et al.*, *J. Phys Chem.*, 1991, **95**, 4664.
26. M. Monnier *et al.*, *J. Mol. Struct.*, 1991, **243**, 13.
27. J. R. Durig *et al.*, *Spectrochim. Acta*, 1991, **47A**, 105.
28. S. H. Ghough and S. Krimm, *Spectrochim. Acta*, 1990, **46A**, 1419.
29. E. K. Murthy and G. R. Rao, *J. Raman Spectrosc.*, 1989, **20**, 409.
30. E. K. Murthy and G. R. Rao, *J. Raman Spectrosc.*, 1988, **19**, 359 & 439.
31. P. Stoppa and A. Gambi, *J. Mol. Struct.*, 2000, 517–518, 209–216.

16 Sulphur and Selenium Compounds

Mercaptans, –SH

In Raman spectra, the S–H stretching vibration generally gives strong, polarised bands. In the infrared, the band due to the S–H stretching vibration[1-7] is weak (sometimes very weak) and may be missed in dilute solutions. It occurs in the region $2600-2540\,\mathrm{cm^{-1}}$ ($3.85-3.94\,\mu\mathrm{m}$) and is easily recognized since this is a region relatively free of other absorption bands. The N–H stretching vibrations of organic nitrogen compounds in the solid phase give a complex pattern of bands in this region whereas a single band is observed due to the S–H stretching vibration. Carboxylic acids also have bands in this region, forming a broad complex pattern due to the O–H stretching vibration. Aldehydes also may have weak, sharp bands in this region due to the aldehydic C–H stretching vibration, but usually a doublet is observed.

Hydrogen-bonding effects[2,7] are much smaller for the –S–H group than they are for the –O–H and $\diagdown$N—H groups. If dimers and monomers coexist, two S–H bands due to the S–H stretching vibration may be observed.

The C–S stretching vibration gives a weak band in the region $720-570\,\mathrm{cm^{-1}}$ ($13.89-17.54\,\mu\mathrm{m}$) (see the section dealing with sulphides). This vibration results in a strong, polarised band in Raman spectra.

Monothiocarboxylic acids,[5,6] –CO–SH, are a mixture of two forms:

$$-\mathrm{CO-SH} \rightleftharpoons -\mathrm{CS-OH}$$

and therefore exhibit bands due to S–H, O–H, C=O, and C=S vibrations.

C–S and S–S Vibrations: Organic Sulphides, $\diagdown$S, Mercaptans, –SH, Disulphides, –S–S–, and Polysulphides, $-(-\mathrm{S-S-})_n-$

In general, the assignment of the band due to the C–S stretching vibration in different compounds is difficult in the infrared since the band is of variable intensity and may be found over the wide region $1035-245\,\mathrm{cm^{-1}}$ ($9.66-40.82\,\mu\mathrm{m}$), whereas, in general, C–S stretching vibrations result in strong bands in Raman spectra which are normally easy to identify.

Both aliphatic[1] and aromatic[9] sulphides have a weak-to-medium band due to the C–S stretching vibration in the region $750-570\,\mathrm{cm^{-1}}$ ($13.33-17.54\,\mu\mathrm{m}$), primary sulphides absorbing at the higher-frequency end of the range and tertiary sulphides at the lower end. In the Raman spectra of alkyl disulphides,[63] the C–S stretching vibration band may result in one or more strong polarised bands in the region $750-570\,\mathrm{cm^{-1}}$ ($13.33-17.54\,\mu\mathrm{m}$), depending on the rotational isomerism of the compound. For the $-\mathrm{CH_2-S-S-}$group with a hydrogen atom in the *trans* position to sulphur, the C–S band is in the range $670-630\,\mathrm{cm^{-1}}$ ($14.93-15.87\,\mu\mathrm{m}$) and, with the carbon atom in the trans position, the band is at $750-700\,\mathrm{cm^{-1}}$ ($13.33-14.29\,\mu\mathrm{m}$).

Double-bond conjugation with the C–S bond, e.g. either vinyl or phenyl =C–S–, lowers the C–S stretching vibration frequency to about $590\,\mathrm{cm^{-1}}$ ($16.95\,\mu\mathrm{m}$) and increases its intensity significantly. For compounds in which the C–S group is adjacent to a C=O group, the C–S band is normally above $710\,\mathrm{cm^{-1}}$ (below $14.08\,\mu\mathrm{m}$). The band due to the C–Cl stretching vibration also occurs in this region and may, in some cases, make interpretation more difficult. Thioethers absorb in the region $695-655\,\mathrm{cm^{-1}}$ ($14.39-15.27\,\mu\mathrm{m}$) due to the C–S–C stretching vibration.

Chart 16.1 The positions and intensities of bands observed in the infrared spectra of sulphur compounds

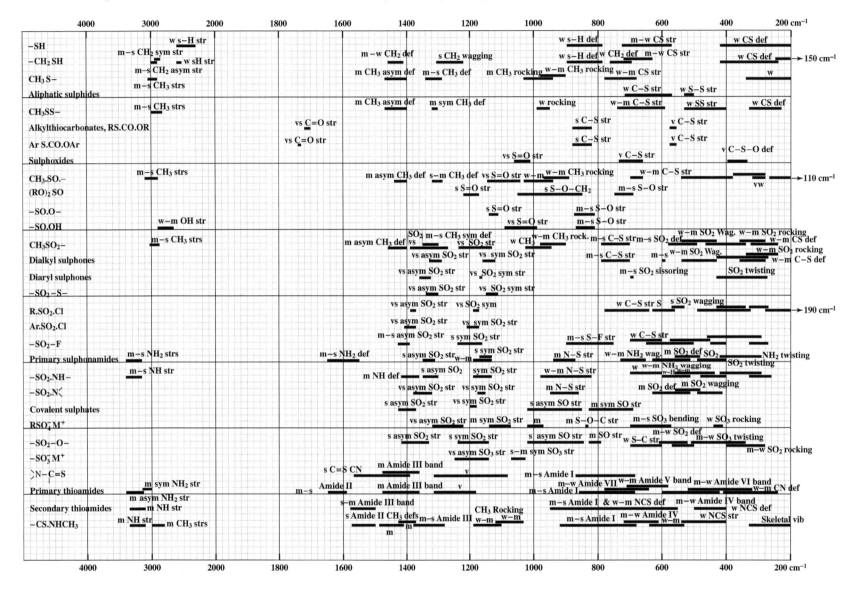

The band due to the C–S–C bending vibration has been observed for a few sulphides and occurs at about 250 cm⁻¹ (40.00 μm), the C–C–S band being near 325 cm⁻¹ (30.77 μm).

In Raman spectra, the S–S stretching vibration gives rise to a strong polarised band whereas, in infrared spectra, because of the symmetry of the S–S group, aliphatic disulphides have two weak bands. These bands occur at

Table 16.1 Mercaptan S–H stretching and deformation vibrations

Functional Groups	Region		Intensity		Comments
	cm^{-1}	μm	IR	Raman	
Mercaptans, aliphatic thiols, and thiophenols (free)	2600–2540	3.85–3.94	w	s, p	S–H str, see ref. 1. May be very weak. For n-alkyl compounds, strong Raman band due to C–S str 660–650 cm^{-1}, general range for C–SH str 740–585 cm^{-1}.
	895–785	11.73–12.74	w		
–CH$_2$SH	2600–2535	3.85–3.95	w	s, p	S–H str, often at 2565 cm^{-1}
	895–785	11.73–12.74	w		S–H def vib
Aryl mercaptans	2600–2450	3.85–4.08	w	s, p	S–H str, see refs 2–4, 7
Dithioacids (hydrogen-bonded)	2500–2400	4.00–4.17	w	s, p	br, S–H str
Dithioacids, –CS–SH (free)	2600–2500	3.85–4.00	w	s, p	S–H str, sometimes a doublet, see ref. 6
Dithioacids, –CS–SH	~860	~11.63	s		br, S–H in-plane def vib
Compounds with –CO–SH (free)	2595–2560	3.85–3.91	w	s, p	S–H str
Trithiocarbonic acids (free)	2560–2550	3.91–3.92	w	s, p	S–H str
Organic compounds containing SeH (free)	2330–2280	4.29–4.39	w	s, p	Se–H str, see ref. 10
Monothioacids, –CO–SH	840–830	11.90–12.05	m		S–H in-plane def vib
R$_2$(P=S)SH	2420–2300	4.13–4.35	m	s	br
(RO)$_2$(P=S)SH	2480–2440	4.03–4.10	m	s, p	br, S–H str, dilute solution 2590–2550 cm^{-1}
	865–835	11.56–11.98	m		

530–500 cm^{-1} (18.87–20.00 μm) and 515–500 cm^{-1} (19.42–20.00 μm). Aryl disulphides absorb at 540–500 cm^{-1} (18.52–20.00 μm) and 505–430 cm^{-1} (19.80–23.36 μm).

Compounds containing S=O: Organic Sulphoxides, >S=O, and Sulphites, –O–SO–O–

In a non-polar solvent such as carbon tetrachloride or n–hexane, sulphoxides[11,14–19,96] have a strong-absorption at 1070–1035 cm^{-1} (9.35–9.66 μm) due to the stretching vibrations of the S=O group, while for solvents in which hydrogen bonding is possible, and for chloroform, the range is 1055–1010 cm^{-1} (9.48–9.90 μm). In the case of strong intramolecular hydrogen bonding, the band due to the S=O stretching vibration of sulphoxides has been observed at about 995 cm^{-1} (10.05 μm) with a very much weaker band being observed in the normal region.[20,22]

In the solid phase, the S=O band appears 10–20 cm^{-1} lower than as given above for the inert solvent and is broad, sometimes consisting of a number of peaks.[20–22] Conjugation has only a small effect on the position of the band.

Dialkyl sulphites have a strong band due to this vibration at 1220–1170 cm^{-1} (8.20–8.55 μm).

The position of the S=O band is dependent on the electronegativity of the attached group, Electronegative substituents tend to raise the frequency since they tend to stabilize the form S=O rather than S$^+$–O$^-$. Hence, the frequency of the S=O stretching vibration increases in the following order:

sulphoxides < sulphinic acids < sulphinic acid esters ~ sulphinyl chlorides

< sulphites

–S=O < –SO–OH < –SO–O– ~ –SO–Cl < O–SO–O–

(For sulphites, there are two electronegative atoms adjacent to the S=O group).

In general, organic compounds of the type >SO may be distinguished from those of the type –(SO$_2$)–which are not ionic in nature, i.e. G–SO$_2$–G or G–SO$_3$–G, since the group >SO has only one strong absorption in the region 1360–1100 cm^{-1} (7.35–9.09 μm) whereas sulphones, etc., have two (see section on sulphones).

Sulphoxides absorb in the region 730–660 cm^{-1} (13.70–15.15 μm) probably due to the stretching vibration of the C–S bond. A band of variable

Table 16.2 CH$_3$ and CH$_2$ vibration bands of organic sulphur compounds CH$_3$–S–and –CH$_2$S–groups

Functional Groups	Region		Intensity		Comments
	cm^{-1}	μm	IR	Raman	
CH$_3$–S–	3040–2980	3.29–3.36	m–w	m–s	asym CH$_3$ str. Sat. compounds 3000–2980 cm^{-1}, unsat. and Ar 3020–2990 cm^{-1}
	3030–2935	3.30–3.41	m	m–s	asym CH$_3$ str. Sat. compounds 3000–2960 cm^{-1}, unsat. and Ar 3015–2965 cm^{-1}
	3000–2840	3.33–3.52	m–s	m–s	sym CH$_3$ str. Sat. compounds 2935–2905 cm^{-1}, unsat. and Ar 2945–2915 cm^{-1}
	1470–1420	6.80–7.06	m	m	asym CH$_3$ def vib. Sat. compounds 1455–1425 cm^{-1}, unsat. and Ar 1460–1430 cm^{-1}
	1460–1400	6.85–7.14	m	m	asym CH$_3$ def vib. Sat. compounds 1440–1400 cm^{-1}, unsat. and Ar 1460–1420 cm^{-1}
	1340–1290	7.46–7.76	m–s	m–w	sym CH$_3$ def vib. Sat. compounds 1340–1300 cm^{-1}, unsat. and Ar 1330–1310 cm^{-1}
	1030–945	9.71–10.58	m	w	CH$_3$ rocking vib (but CH$_3$SH ~1065 cm^{-1}). Sat. compounds 1035–965 cm^{-1}, unsat. and Ar 1025–965 cm^{-1}
	980–900	10.20–11.11	w–m	w	CH$_3$ rocking vib. Sat. compounds 975–905 cm^{-1}, unsat. and Ar 970–950 cm^{-1}
–CH$_2$–S–	2985–2920	3.35–3.43	m	m	asym CH$_2$ str
	2945–2845	3.40–3.51	m	m	sym CH$_2$ str
	1435–1410	6.97–7.09	m	m	CH$_2$ def vib
	1305–1215	7.66–8.203	s	m	CH$_2$ wagging vib
CH$_3$CH$_2$–S–	2995–2965	3.34–3.37	m	m–s	asym CH$_3$ str
	2975–2955	3.36–3.38	m	m–s	asym CH$_3$ str
	2960–2920	3.38–3.42	m	m–s	asym CH$_2$ str
	2945–2895	3.40–3.45	m	m–s	sym CH$_3$ str
	2910–2850	3.45–3.51	m	m–s	sym CH$_2$ str
	1480–1450	6.76–6.90	w	m–w	asym CH$_3$ def vib
	1460–1440	6.85–6.94	w	m–w	asym CH$_3$ def vib
	1445–1415	6.92–7.07	m–w	m–w	CH$_2$ def vib
	1380–1370	7.25–7.30	m–w	m–w	sym CH$_3$ def vib
	1310–1250	7.63–8.00	m–s	m–w	CH$_2$ wagging vib. Usually 1285–1265 cm^{-1}
	1270–1230	7.87–8.13	w	m	CH$_2$ twisting vib
	1105–1045	9.05–9.57	w–m	w	CH$_3$ rocking vib
	1060–1010	9.43–9.90	w–m	w	CH$_3$ rocking vib
	1000–950	10.00–10.53	v	m–s	CC str
	800–730	12.50–13.70	w–m	w	CH$_2$ rocking vib
	280–210	35.71–47.62			CH$_3$ torsional vib
	215–155	46.51–64.52			CH$_3$CH$_2$ torsional vib
	105–45	95.24–222.22			SCH$_3$CH$_2$ torsional vib
–CH$_2$SH	2985–2935	3.35–3.41	m–s	m–s	asym CH$_2$ str
	2945–2855	3.40–3.50	m–s	m	sym CH$_2$ str
	1460–1410	6.85–7.09	m–w	m–w	CH$_2$ def vib
	1305–1215	7.66–8.23	s	m–w	CH$_2$ wagging vib
	765–695	13.07–14.39	w	w	CH$_2$ rocking vib
	250–150	40.00–66.67			Torsional vib

Table 16.2 (*continued*)

Functional Groups	Region		Intensity		Comments
	cm^{-1}	μm	IR	Raman	
	175–85	57.14–117.65		s, p	Torsional vib
$-S-CH{=}CH_2$	~1590	~6.29	m	s, p	C=C str
	~965	~10.36	s	w	C–H out-of-plane def vib
	~860	~11.63	s	w	CH_2 out-of-plane def vib
$-S-SCH_3$	1320–1300	7.58–7.69	m	m–w	sym CH_3 def vib
	985–955	10.15–10.47	w	w	Rocking CH_3 vib

Table 16.3 Organic sulphides, mercaptans, disulphides, and polysulphides: C–S and S–S stretching vibrations

Functional Groups	Region		Intensity		Comments
	cm^{-1}	μm	IR	Raman	
CH_3-S-	775–675	12.90–14.81	w–m	s–m, p	C–S str, occasionally strong. (C–S def vib gives weak band at 340–200 cm^{-1} which is m–w in Raman spectra. For sat. compounds: 290–210 cm^{-1}; for unsat.and aromatic compounds: 325–265 cm^{-1})
CH_3-S-CH_2-R	730–685	13.70–14.60	w	s–m, p	asym C–S str
CH_3CH_2-S-	705–635	14.18–15.75	w–m	s–m, p	~50 cm^{-1} lower than MeS. Affected by conformational changes. Increase in length of the alkyl group(s) decreases the frequency
$R-CH_2-S-$	660–630	15.15–15.87	w–m	s, p	C–S str
$RR'CH-S-$	630–600	15.87–16.67	w–m	s, p	C–S str
$R_1R_2R_3C-S-$	600–570	16.67–17.54	w–m	s, p	C–S str
CH_3CH_2S-	705–635	14.18–15.75	w–m	s	50 cm^{-1} lower than Me–S–(sat.). Affected by conformational changes. (See few lines above)
	390–310	25.64–32.26		m–w	SCC def vib
	305–165	32.79–60.60		s, p	CSC def vib
CH_3SCH_2-	775–675	12.90–14.81	w	s–m	asym CSC str
	725–635	13.79–15.75	w	s	sym CSC str
$-CH_2SH$	720–630	13.89–15.87	m–s	s	CS str. May be as low as 585 cm^{-1}.
	420–240	28.81–41.67	w	s	CS def vib, usually 400–300 cm^{-1}
$-CH_2-S-CH_2-$	695–655	14.39–15.27	w–m	s–m	C–S–C str
Cyclohexyl sulphides	710–685	14.08–14.60	w–m	s	C–S str
Phenyl sulphides	715–670	13.99–14.93	w–m	s, p	C–S str
α,β-Unsaturated sulphides	740–690	13.51–14.49	v	s	C–S str
–S–Ar	1110–1030	9.01–9.37	m	s, p	ring vib with C–S interaction, X–sensitive band
Aliphatic disulphides	715–570	13.99–17.54	w	s, p	C–S str. IR inactive for symmetrical compounds
	530–500	18.87–20.00	w	s–m, p	S–S str. Often two bands due to rotational isomerism
–SSCH–	640–590	15.63–16.95	w	s	Two *trans* hydrogens to sulphur
$-SSCH_3$	530–400	18.87–25.00	w	vs–m, p	S–S str.
	740–690	13.51–14.49	w–m	s, p	C–S str.
	330–230	30.30–43.48	w	s	C–S def vib.
Aromatic disulphides	540–400	18.52–25.00	w–m	vs–m, p	S–S str. Two bands due to rotational isomerism
Polysulphides	510–450	19.61–22.22	w–m	vs–m, p	S–S str

(*continued overleaf*)

Table 16.3 (*continued*)

Functional Groups	Region cm^{-1}	Region μm	Intensity IR	Intensity Raman	Comments
Mono- and disulphonyl chlorides	775–650	12.90–15.38	w–m	s	C–S str
Dithiolcarbonic acid esters, (RS)$_2$C=O	880–825	11.36–12.20	s	s	asym C–S str
	570–560	17.54–17.86	v	s	asym C–S str, review of thiol esters, see ref. 11
(RS)(ArS)C=O	~565	~17.70	v	s	C–S str. CO str at 1715–1660 cm^{-1}
(ArS)$_2$C=O	~560	~17.86	s	s	C–S str
Thiolchloroformates, (RS)ClC=O	850–815	11.76–12.30	s	s	asym C–S str, often strongest band in spectrum
	~580	~17.24	m–s	s	asym C–S str
	~345	~28.99	s	s	C–Cl def vib
(ArS)ClC=O	~820	~12.20	s	s	asym C–S str, often strongest band in spectrum
	~595	~16.81	s	s	asym C–S str
Monothiol esters, $-\overset{\underset{\|}{O}}{C}-S-$	1035–930	9.66–10.75	s	s	C–S str, see ref. 11; Has been suggested C–S str for thiol acids and esters be assigned to band ~625 cm^{-1}, see ref. 12
Monothiol acids, $-\overset{\underset{\|}{O}}{C}-S-H$	~950	~10.53	s	s	C–S str, see ref. 11
Thioketals,	800–245	12.50–40.82	m–s	s	C–S str, a number of bands due to coupling
Xanthates, $-O-\overset{\underset{\|}{S}}{C}-S-$	965–860	10.35–11.65	w–m	s	C–S str
Dithioacids, $-\overset{\underset{\|}{S}}{C}-SH$	~580	~17.25	s	s	C–S str
Dixanthogens $-O-\overset{\underset{\|}{S}}{C}-S-S-\overset{\underset{\|}{S}}{C}-O-$	965–860	10.35–11.65	w–m	s	C–S str
Thionitrites, –S–N=O	730–685	13.70–14.60	m–s	s	C–S str, see ref. 13
Thioacetals and trithiocarbonates, =C(SR)(SR)	900–800	11.11–12.50	m–s	s	asym S–C–S str
Ionic dithiolates, =C(S)(S$^-$)	1050–900	9.52–11.11	m–s	s	asym S–C–S str
Ionic 1,1-dithiolates, =C(S$^-$)(S$^-$)	980–850	10.20–11.76	m–s	s	asym S–C–S str
M=P, As	675–660	14.81–15.15	w	s–m, p	
	655–640	15.27–15.63	m–w	s, p	
Trialkyl arsine sulphides	~480	~20.83	vs		As–S str, band position dependent on size of alkyl groups
R$_3$Ge–S–GeR$_3$	~415	~24.10	s		Ge–S–Ge str
R$_3$Sn–S–SnR$_3$	~375	~26.67	s		Sn–S–Sn str
R$_3$Pb–S–Pb	~335	~29.85	s		Pb–S–Pb str

intensity at 395–335 cm^{-1} (25.32–29.85 μm) is also observed and has been assigned to the C–S=O deformation.

Sulphoxides may act as electron donors to either metals[25–27] or other molecules.[20–22] If coordination to a metal atom occurs through the oxygen atom, the SO stretching frequency decreases when compared with that of the free ligand. For example, for dimethyl sulphoxide complexes the SO frequency occurs in the region 1100–1050 cm^{-1} (9.09–9.52 μm). When coordination occurs through the sulphur atom, there may be an increase in the SO stretching frequency, 1160–1115 cm^{-1} (8.62–8.97 μm). For oxygen bonded complexes the band in the region 1025–985 cm^{-1} (9.76–10.15 μm) is found to be metal sensitive.

For cyclic (six-membered ring) sulphoxides, the S=O group in the equatorial position absorbs at ~20 cm^{-1} higher than when in an axial position.

Cyclic sulphites (five- to seven-membered rings) absorb at 1225–1200 cm^{-1} (8.16–8.33 μm).

Organic Sulphones, $\diagdown$SO$_2$

In dilute solution in non-polar solvents, all organic sulphones[17,28–33,97] have two very strong bands[34] due to the asymmetric and symmetric stretching vibrations[29] of the SO$_2$ group, at 1360–1290 cm^{-1} (7.41–7.75 μm) and 1170–1120 cm^{-1} (8.55–8.93 μm) respectively. In the solid phase, the band due to the asymmetric stretching vibration occurs 10–20 cm^{-1} lower than in dilute solution and usually appears to have a number of peaks whereas the

Table 16.4 Organic sulphoxides, $\diagdown$S=O

Functional Groups	Region cm^{-1}	μm	Intensity IR	Raman	Comments
Sulphoxidies, $\diagdown$S=O (in dilute CCl$_4$ solution)	1070–1030	9.35–9.70	vs	m–w	S=O str, halogen or oxygen atom bonded to S atom increases frequency. Hydrogen bonding decreases frequency
Sulphoxides	730–660	13.70–15.15	v	m–s	C–S str
	395–335	25.32–29.85	v	m, p	sym C–S–O def vib
Cyclic sulphoxldes (six- and seven-membered rings) (in CCl$_4$ solution)	~1060	~9.43	s	m	S=O str, see ref. 17
Cyclic sulphoxides (four-membered rings) (in CCl$_4$ solution)	~1090	~9.17	s	m	S=O str
Methyl sulphoxides –SO·CH$_3$	1145–1045	8.73–9.57	s		Usually 1075–1045 cm^{-1}. Affected by different conformations and solvent
	700–660	14.29–15.15	w–m	s	C–S str
	540–380	18.52–26.32			S=O def vib
	375–330	26.66–30.30			S=O wagging vib
	320–280	31.25–35.71	vw	w–m	C–S–def vib
Dialkyl sulphoxides	1045–1035	9.57–9.66	s	m–w	
Aryl sulphoxides	1060–1040	9.43–9.62	s	m–w	See refs 18, 19; CHCl$_3$ solution spectra quite different
Methyl aryl sulphoxides Ar–SO–CH$_3$	535–495	18.69–20.20	s		C–S=O in-plane def vib
Thiosulphoxides, G$_1$–S·SO·G$_2$					
G$_1$, G$_2$ = CH$_3$ and/or Ar	1110–1095	9.01–9.13	s	m	See ref. 24
G$_1$, G$_2$ = R and Ar	1090–1075	9.17–9.30	s	m	See ref. 24
G$_1$, G$_2$ = Ar	1115–1100	8.97–9.09	s	m	See ref. 24
Compounds of the type R–S–O–S–R′	345–255	28.99–39.22			sym S–O–S str, see ref. 15
	160–125	62.50–80.00			S–O–S bending vib
Dialkyl sulphites, (RO)$_2$SO	1220–1170	8.20–8.55	s	s, p	S=O str

(continued overleaf)

Table 16.4 (*continued*)

Functional Groups	Region		Intensity		Comments
	cm^{-1}	μm	IR	Raman	
	1050–850	9.52–11.76	s	m–w	Due to S–O–CH$_2$ group
	750–690	13.33–14.49	m–s	m	S–O str, two bands
Chloroalkyl sulphites	1225–1210	8.16–8.26	s	m–s	S=O str
Sulphinic acid esters, –SO–O–	1140–1125	8.77–8.89	s	m	S=O str
Sulphinic acids, –SO–OH	2790–2340	3.58–4.27	w–m	w	O–H str (solid phase value)
	1090–990	9.17–10.10	vs	m–s	S=O str
	870–810	11.49–12.35	m–s		S–O str
Aryl sulphinic acids, Ar–SO–OH	~1100	~9.09	s	m	S=O str
Sulphinic anhydrides, R–SO$_2$–SO–R	~1100	~9.09	s	m	S=O str
Sulphinic acid salts, RSO$_2^-$M$^+$	~1030	~9.71	s		asym S⸱⸱O, str, stronger of the two bands
	~980	~10.29	s		sym S⸱⸱O str
Alkyl sulphinyl chlorides, R–SO–Cl	~1135	~8.81	s	m	S=O str
RO–SO–Cl	1225–1210	8.16–8.26	s		
Thionylamines, –N=S=O	1300–1230	7.69–8.13	v	s, p	asym N=S=O str, see ref. 23
	1180–1110	8.48–9.01	v	s	sym N=S=O str
Cyclic sulphites	1260–1230	7.94–8.13	s	s, p	Equatorial S=O str
	1230–1205	8.13–8.30			Twisting vib
	1215–1170	8.23–8.55	s	s, p	Axial S=O str
Cyclic sulphites (five-, six-, and seven-membered rings)	1220–1210	8.20–8.26	s	s, p	S=O str

band due to the symmetric stretching vibration usually consists of a single peak at 1180–1145 cm^{-1} (8.48–8.73 μm).

A number of sulphones have three components of the band due to the asymmetric SO$_2$ stretching vibration when in non-polar solvents such as carbon tetrachloride. In order of decreasing intensity, these bands occur at 1335–1315 cm^{-1} (7.49–7.61 μm), 1315–1305 cm^{-1} (7.61–7.66 μm), and 1305–1285 cm^{-1} (7.66–7.78 μm).

Conjugation does not alter the positions of the bands due to the SO$_2$ stretching vibration.

All sulphones have a characteristic medium–to-strong band at 590–500 cm^{-1} (16.95–20.00 μm) which is due to the scissor vibration of the –SO$_2$ group and a band usually strong, is observed at 555–435 cm^{-1} (18.02–22.99 μm).

Saturated aliphatic sulphones have a medium–intensity band at 525–495 cm^{-1} (19.05–20.20 μm) due to the wagging motion of the –SO$_2$ group.

Sulphonyl Halides, SO$_2$–X

The frequencies of the SO$_2$ stretching vibrations of sulphonyl fluorides and chlorides[28,38–43] are higher than those of the sulphones due to the presence of the electronegative halogen atom.

Aliphatic sulphonyl chlorides[38,39] absorb strongly at 1385–1360 cm^{-1} (7.22–7.35 μm) and 1190–1160 cm^{-1} (8.40–8.62 μm) due to the SO$_2$ asymmetric and symmetric stretching vibrations respectively. For aromatic sulphonyl chlorides,[40,41] these ranges are extended to higher frequencies but the main difference observed is that the band due to the symmetric vibration forms a doublet.

Sulphonyl halides have a medium-to-strong band in the region 600–530 cm^{-1} (16.67–18.87 μm) due to the deformation vibrations of the SO$_2$ group.

Sulphonamides, –SO$_2$–N⟨

In the solid phase, sulphonamides[28,44–47] have strong bands due to their N–H stretching vibrations in the region 3390–3245 cm^{-1} (2.95–3.08 μm), (see sections dealing with amines and amides). In the unassociated state, these bands occur in the same region as for amines.

Also in the solid phase, sulphonamides have a very strong, broad absorption band at 1360–1315 cm^{-1} (7.35–7.61 μm) which generally consists of a number of peaks and is due to the asymmetric stretching vibration

Table 16.5 Organic sulphone SO_2 stretching vibrations

Functional Groups	Region		Intensity		Comments
	cm^{-1}	μm	IR	Raman	
Sulphones (dilute solution)	1360–1290	7.41–7.75	vs	v	asym SO_2 str, usual range, but may be found at 1390–1270 cm^{-1}. Raman band often absent
	1170–1120	8.55–8.93	vs	s, p	sym SO_2 str, usual range, but may be found at 1225–1135 cm^{-1}.
	785–735	12.74–13.61	m–s	s	C–S str, usual range, but may be found at 790–700 cm^{-1}.
	600–480	16.67–20.83	m–s	v	SO_2 def vib. Most compounds found in range 590–510 cm^{-1}.
	555–435	18.02–22.99	w–m	m–s	SO_2 wagging vib
	470–340	21.27–29.41			SO_2 twisting vib
	360–280	27.78–35.71	w–m		SO_2 rocking vib
	335–225	29.85–44.44			CS def
Methyl sulphones, $CH_3 \cdot SO_2 -$	3050–2920	3.28–3.42	m–s	m–s	CH_3 str
	1460–1300	6.85–7.69	m	m–w	CH_3 def. Most sym. def (1340–1310 cm^{-1}) obscured by strong SO_2 str
	1390–1270	7.19–7.87	vs	v	asym SO_2 str. Most compounds 1360–1300 cm^{-1}
	1225–1135	8.16–8.81	vs	s	sym SO_2 str. Most compounds 1180–1140 cm^{-1}
	790–700	12.66–14.29	m–s	s	C–S str. Most compounds 785–735 cm^{-1}
	575–495	17.39–20.20	m–s	v	SO_2 def
	535–435	18.69–22.99	w–m	m–s	SO_2 wagging vib
	470–340	21.28–29.41			SO_2 twisting vib
	360–280	27.78–35.71	w–m		SO_2 rocking vib
	335–245	19.85–40.82			CS def
Dialkyl sulphones	1330–1295	7.52–7.72	vs	w	⎫ Straight-chain alkyl sulphones absorb at slightly higher
	1155–1125	8.66–8.89	vs	s	⎰ frequencies than branched compounds
Alkyl-aryl sulphones	1335–1325	7.49–7.54	vs	w	(For methylvinylsulphones, see ref. 94)
	1160–1150	8.62–8.70	vs	s	
Diaryl sulphones	1360–1335	7.35–7.49	vs	w	See ref. 30
	1170–1160	8.55–8.62	vs	w	
GSO_2CH_2COG	~1330	~7.52	vs	w	
	~1160	~8.62	vs	s	
Disulphones, $-SO_2-SO_2-$	1360–1280	7.35–7.78	vs	w	asym SO_2 str, see refs 9, 37
	1170–1120	8.55–8.93	vs	s	sym SO_2 str,
Thiolsulphonates, $-SO_2-S-$	1340–1305	7.46–7.66	vs	w	asym SO_2 str, see refs 28, 36, 37
	1150–1125	8.70–8.89	vs	s	sym SO_2 str,
Sulphinic acid anhydrides (sulphonyl sulphones), $-SO_2-SO-$	~1340	~7.46	vs	w	asym SO_2 str, see refs 35 and 37
	~1140	~8.77	vs	s	sym SO_2 str,
	~1100	~9.09	s	s	S=O str
Sulphones	600–590	16.67–16.94	m–s	v	SO_2 scissoring vib
	430–275	23.26–36.36			
Saturated aliphatic sulphones	555–435	18.02–22.30	m–s	m–s	SO_2 wagging vib
CH_3SO_2-Ar	575–440	17.39–22.73	m–s	m–s	SO_2 wagging vib

Table 16.6 Sulphonyl halides

Functional Groups	Region cm^{-1}	Region μm	Intensity IR	Intensity Raman	Comments
Aliphatic sulphonyl chlorides	1390–1360	7.19–7.35	vs	m–s	asym SO$_2$ str
	1190–1160	8.40–8.62	vs	s	sym SO$_2$ str
	775–640	12.90–15.63	w	s	C–S str
	610–565	16.39–17.70	s		in-plane SO$_2$ def vib. (usually 590–530 cm^{-1})
	570–530	17.54–18.87	s	m–s	SO$_2$ wagging vib
	490–330	20.41–30.30			SO$_2$ twisting vib
	430–360	23.26–27.78			Cl–S–str
	330–270	30.30–37.04			SO$_2$ rocking vib
	280–190	35.71–52.63			Cl–S–def vib
Aromatic sulphonyl chlorides	1420–1360	7.04–7.22	m–s	m–w	asym SO$_2$ str
	1205–1170	8.30–8.54	s	s	sym SO$_2$ str, doublet
	~1090	~9.17	w		
Sulphonyl fluorides, –SO$_2$–F	1415–1395	7.07–7.17	m–s	m–s	asym SO$_2$ str, but may have range 1505–1385 cm^{-1}
	1240–1165	8.06–8.58	s	s	sym SO$_2$ str, but may occur as high as 1270 cm^{-1}
	900–750	11.11–13.33	m–s	m–w	S–F str
	635–485	15.75–20.62		v	SO$_2$ def vib. Usually 560–490 cm^{-1}
	700–600	14.29–16.67	w	s	C–S str
	570–450	17.54–22.22		m–s	SO$_2$ wagging vib. Usually 515–445 cm^{-1}
	540–400	18.52–25.00			SO$_2$ twisting vib
	460–290	21.74–34.48			SO$_2$ rocking vib
	330–270	30.30–37.04			–SO$_2$F skeletal vib
RO·SO$_2$Cl	1455–1405	6.87–7.12	s	m–w	asym SO$_2$ str
	1225–1205	8.17–8.30	s	s	sym SO$_2$ str
RO·SO$_2$F	1510–1445	6.62–6.92	s	m–w	asym SO$_2$ str, usual range, but may be 1505–1385 cm^{-1}.
	1260–1230	7.94–8.13	s	s	sym SO$_2$ str
Aromatic sulphonyl fluorides	1425–1405	7.02–7.12	vs	m–w	asym SO$_2$ str
	1240–1190	8.06–8.40	vs	s	sym SO$_2$ str

of the SO$_2$ group. In solution, this band is about 10–20 cm^{-1} higher and occurs at 1380–1325 cm^{-1} (7.25–7.55 μm). A very strong band due to the symmetric stretching vibrations of the SO$_2$ group occurs at 1180–1140 cm^{-1} (8.47–8.77 μm) when in the solid phase and at 1170–1150 cm^{-1} (8.55–8.70 μm) when in dilute solution (i.e. there is very little difference in the band position for this vibration)(the NH$_2$ rocking/twisting vibration band, which is of weak-to-medium intensity, occurs at 1190–1130 cm^{-1}). Due to the influence of the electronegative nitrogen atom, the frequencies of the SO$_2$ asymmetric and symmetric stretching vibrations are higher for sulphonamides than for sulphones. The positions of these bands are little influenced by molecular structure, i.e. whether the sulphonamides are aliphatic or aromatic.

A band of medium intensity is observed in the region 950–860 cm^{-1} (10.53–11.63 μm). There is also a band at 710–650 cm^{-1} (14.08–15.38 μm)

due to the NH$_2$ wagging vibration and a torsional vibration has been reported at 420–290 cm^{-1}.

Covalent Sulphonates, R–SO$_2$–OR′[28,37]

Aliphatic sulphonates have a strong band in the region 1420–1330 cm^{-1} (7.04–7.52 μm) which may appear as a doublet and another strong band in the region 1200–1145 cm^{-1} (8.33–8.73 μm) which is usually found near 1175 cm^{-1} (8.51 μm). Aromatic esters of sulphonic acids have strong absorptions at 1380–1350 cm^{-1} (7.25–7.41 μm) and 1200–1190 cm^{-1} (8.33–8.40 μm).

Sulphonates have a weak band (often two) at 600–515 cm^{-1} (16.67–19.42 μm) due to the SO$_2$ deformation.

Table 16.7 Sulphonamides

Functional Groups	Region		Intensity		Comments
	cm^{-1}	μm	IR	Raman	
Primary sulphonamides, $-SO_2NH_2$ (hydrogen bonded or solid phase)	3390–3245	2.95–3.08	s	m–w	Two bands due to asym and sym N–H str
	1650–1550	6.02–6.45	m–s	w	NH$_2$ def vib
	1360–1310	7.35–7.63	s	w–m	asym SO$_2$ str. Ar SO$_2$NH$_2$ 1340–1310 cm^{-1}
	1190–1130	8.40–8.85		w–m	NH$_2$ rocking vib
	1165–1135	8.58–8.81		s	sym SO$_2$ str
	935–875	10.70–11.43	m		N–S str
	730–650	13.70–15.38	w–m	w–m	NH$_2$ wagging vib, br
	630–510	15.87–19.61		v	SO$_2$ def vib
	560–480	17.86–20.83		m–s	SO$_2$ wagging vib
	490–400	20.41–25.00			SO$_2$ twisting vib
	415–290	24.10–34.48			NH$_2$ twisting vib
N-Mono-substituted sulphonamides, $-SO_2NH-$ (hydrogen bonded or solid phase)	3335–3205	3.00–3.12	s	m–w	N–H str, one band only. Dilute solutions 3400–3380 cm^{-1}
	1420–1370	7.04–7.30	m	w	NH def vib
	1360–1300	7.35–7.69	s	w	asym SO$_2$ str.
	1190–1130	8.40–8.85	s	s	sym SO$_2$ str.
	975–835	10.26–11.98	w–m		N–S str
	700–600	14.29–16.67	w, br	w–m	NH def vib
	600–520	16.67–19.23	w–m	v	SO$_2$ def vib
	555–445	18.01–22.47	w–m	m–s	SO$_2$ wagging vib
	480–400	20.83–25.00			SO$_2$ twisting vib
	~350	~28.57			SO$_2$ rocking vib
	~280	~35.71			CNS def vib
Sulphonamides, $-SO_2-N\big\langle$ (dilute solution)	1380–1325	7.25–7.55	vs	m	asym SO$_2$ str (10–20 cm^{-1} lower in solids)
	1170–1150	8.55–8.70	vs	s	sym SO$_2$
	950–860	10.53–11.63	m		N–S str
	630–510	15.87–19.61	m	v	SO$_2$ def vib
	560–480	17.86–20.83	m	m–s	SO$_2$ wagging vib
	490–400	20.41–25.00			SO$_2$ twisting vib
Sulphondiamides, $\big\rangle N\cdot SO_2\cdot N\big\langle$	1340–1320	7.46–7.58	vs	m–w	asym SO$_2$ str, see ref. 47
	1145–1140	8.73–8.77	vs	s	sym SO$_2$ str

Organic Sulphates, $-O-SO_2-O-$

Organic, covalent sulphates[28,48,49,97] have two strong bands, one at 1415–1370 cm^{-1} (7.07–7.30 μm) and the other at 1200–1185 cm^{-1} (8.33–8.44 μm), both of which are due to the stretching vibrations of the SO$_2$ group. As might be expected, electronegative substituents tend to increase the frequencies of the SO$_2$ stretching vibration. Studies of diaryl and alkylaryl sulphates have been published.[95]

Primary alkyl sulphate salts, ROSO$_2$O$^-$M$^+$,[50] have a very strong band at 1315–1220 cm^{-1} (7.61–8.20 μm) and a less intense band at 1140–1075 cm^{-1} (8.85–9.30 μm) due to the asymmetric and symmetric stretching vibrations respectively of the SO$_2$ group, whereas secondary alkyl sulphate salts have a very strong doublet at about 1270–1210 cm^{-1} (7.87–8.26 μm) and a strong band at $1075-1050\,\text{cm}^{-1}$ ($9.30-9.52\,\mu\text{m}$). The positions of these bands are influenced far more by different metal ions than by the nature of the alkyl group.

The asymmetric and symmetric S–O–C stretching vibration bands occur at about $875\,\text{cm}^{-1}$ ($11.43\,\mu\text{m}$) and $750\,\text{cm}^{-1}$ ($13.33\,\mu\text{m}$) respectively, the first band being of medium intensity and the second weak. These bands occur in a region where alkyl bands occur and may therefore be difficult to identify.

Sulphonic Acids, –SO₃H, and Salts, SO₃⁻M⁺

Small traces of water result in ionization of sulphonic acids, therefore extra care must be exercised if one is to observe covalent (non-ionized) sulphonic

acids rather than the ionic (hydrated) form, $-SO_3{}^-H_3{}^+O$. The bands observed due to the SO_3 stretching vibration for both the anhydrous and hydrated form are strong and usually broad. In general, these two bands together form a broad absorption with two maxima and may thus be distinguished from the acid salts which have two separate bands.

The band due to the O–H stretching vibration of hydrated sulphonic acids is very broad and usually has several maxima, occurring in the region $2800-1650\,\text{cm}^{-1}$ ($3.60-6.06\,\mu\text{m}$). Sulphonic acid salts, of course, have no corresponding band. The broadness of the band due to the O–H stretching vibration may be used to distinguish between the hydrated and anhydrous forms of sulphonic acids.

The band due to the SO_3 asymmetric stretching vibration of sulphonic acid salts occurs at $1250-1140\,\text{cm}^{-1}$ ($8.00-8.77\,\mu\text{m}$), the position of the band being mainly dependent on the nature of the metal ion, not on whether the

Table 16.8 Compounds with SO_2

Functional Groups	Region cm⁻¹	Region μm	Intensity IR	Intensity Raman	Comments
Covalent sulphates (RO)₂SO₂	1415–1370	7.07–7.30	s	s–m	asym SO₂ str
	1200–1185	8.33–8.44	vs	s	sym SO₂ str
	1020–850	9.80–11.76	s		SO asym str
	830–690	12.05–14.49	m		SO sym str
Primary alkyl sulphate salts, RSO₄⁻M⁺ (solid phase)	1315–1220	7.61–8.20	vs	s–m	asym SO₂ str, often doublet ~1250 and ~1220 cm⁻¹ (aromatic compounds in same range)
	1140–1075	8.77–9.30	m	s	sym SO₂ str, aromatic compounds, ~1040 cm⁻¹
	~1000	~10.00	m		Often split
	840–835	11.90–11.98	m		S–O–C str
	700–570	14.29–17.54	m–s		SO₃ bending vib, two bands
	440–410	22.73–24.39	w		SO₃ rocking vib
Secondary alkyl sulphate salts, R₁R₂CHSO₄⁻M⁺ (solid phase)	1270–1210	7.87–8.26	vs	s–m	asym SO₂ str, often doublet
	1075–1050	9.30–9.52	s	s	sym SO₂ str
	945–925	10.60–10.81	m		
	700–570	14.29–17.54	m–s		SO₃ bending vib, two bands
	440–410	22.73–24.39	w		SO₃ rocking vib
Covalent sulphonates, R–SO₂–OR	1420–1330	7.04–7.52	s	s–m	asym SO₂ str
	1235–1145	8.10–8.73	s	s	sym SO₂ str
	1020–850	9.80–11.76	s		SO asym str
	830–690	12.05–14.49	m		SO sym str
	700–600	14.29–16.67	w	s	S–C str
	610–500	16.39–20.00	m–w	v	SO₂ def vib, usually two bands
Alkyl sulphonates, RO–SO₂–R	1360–1350	7.35–7.40	m–s	m	
	1175–1165	8.51–8.58	vs	s	
Alkyl aryl sulphonates, Ar–SO₂–OR	1365–1335	7.32–7.49	m–s	m	
	1200–1185	8.33–8.44	vs	s	

Table 16.8 (*continued*)

Functional Groups	Region		Intensity		Comments
	cm^{-1}	μm	IR	Raman	
Alkyl thiosulphonates, RSO$_2$SR	1335–1305	7.49–7.67	s–m	s–m	asym SO$_2$ str
	1130–1125	8.85–8.89	s	s	sym SO$_2$ str
	560–550	17.86–18.18	s–m	v	SO$_2$ def vib
Alkyl sulphonic acids (anhydrous), RSO$_2$·OH	3000–2800	3.33–3.57	s	w	br, O–H str
	2500–2300	4.00–4.35	w–m	w	br, O–H str
	1355–1340	7.38–7.46	s	s–m	asym SO$_2$ str
	1200–1100	8.33–9.10	s	s	sym SO$_2$ str
	1165–1150	8.59–8.70	s		br, S–O str
	1080–1040	9.26–9.62	w	s	
	910–890	10.99–11.24	s		S–O str
	700–600	14.29–16.67	w	s	S–C str
Alkyl sulphonic acids (hydrated), RSO$_3^-$H$_3$O$^+$	~2600	~3.85	m–w	w	Very broad band with three maxima, O–H str
	~2250	~4.45	m–w	w	
	~1680	~6.00	m–w		
	1230–1120	8.13–8.93	s		asym SO$_3$ str
	1120–1025	8.93–9.76	s		sym SO$_3$ str
Aryl sulphonic acids (solid phase)	~2760	~3.60	m–s	w	Broad band with shoulders, O–H str
	~2350	~4.25	m–s	w	
	~1345	~7.44	s		
	~1160	~8.62	s	s	
Aryl sulphonic acids (in inert solvent)	~3700	~2.70	v	w	sh, O–H str
	~2900	~3.45	v	w	
	~2500	~4.00	v		
Alkyl sulphonic acid sodium salts	1195–1175	8.37–8.51	vs		asym SO$_3$ str
	1065–1050	9.39–9.52	s		sym SO$_3$ str
Sulphonic acid salts, SO$_3^-$M$^+$	1250–1140	8.00–8.77	vs		asym SO$_3$ str
	1070–1030	9.35–9.70	s–m		sym SO$_3$ str
Sultones,	1385–1345	7.22–7.44	s	s–m	asym SO$_2$ str, often split
	1175–1165	8.51–8.58	s	s	sym SO$_2$ str, see ref. 28
Sulphate ion, SO$_4^{2-}$	1200–1140	8.33–8.77	m	m–s	
	1130–1080	8.85–9.26	vs	m–s	br, with shoulders, SO$_4$ str
	1065–955	9.39–10.47	w	s	sh, not always present
	680–580	14.71–17.24	m	m–s	Several bands
	530–405	18.87–24.69		m–s	
Sulphite ion, SO$_3^{2-}$	~1215	~8.23	w		
	~1135	~8.81	w		
	1010–900	9.90–11.11	v	s	Often strong
	660–615	15.15–16.26	m	m	
Bisulphate ion, HSO$_4^-$	1190–1160	8.40–8.62	s–m	s	asym SO$_3$ str
	1080–1000	9.26–10.00	s		sym SO$_3$ str
	880–840	11.36–11.90	m	m	Probably S–OH str

compound is alkyl or aryl. The band is usually broad with shoulders. The band due to the symmetric stretching vibration is sharper, also has shoulders and occurs at 1070–1030 cm^{-1} (9.35–9.70 μm). Ionic sulphates, which are a common impurity, have a very strong band in the region 1130–1080 cm^{-1} (8.85–9.26 μm). Substituted benzene and naphthalene sulphonic acid salts also have a band in this region which is not observed for alkyl acid salts.

Thiocarbonyl Compounds, $\diagdown$C=S

The thiocarbonyl[51–59] absorption is not as strong as that due to the carbonyl C=O group, as might be expected since the sulphur atom is less electronegative than the oxygen atom and therefore the C=S group is less polar than the C=O group. In the case of compounds where the thiocarbonyl group is directly attached to a nitrogen atom, i.e. N–C=S,[53,56,60,61,64] the stretching vibration of the C=S portion is strongly coupled to that of the C–N part as a direct consequence of which several bands may, at least partly, be associated with the C=S stretching vibration. Hence thioamides,[53,60,61] thioureas, thiosemicarbazones, thiazoles, and dithio-oxamides have three absorption bands, in the regions 1570–1395 cm^{-1} (6.37–7.17 μm), 1420–1260 cm^{-1} (7.04–7.94 μm), and 1140–940 cm^{-1} (8.77–10.64 μm) which are in part due to the C=S stretching vibration.

The C=S stretching vibration for compounds where the thiocarbonyl group is not directly bonded to nitrogen gives rise to a band which is generally strong, often sharp, and occurs in the region 1230–1030 cm^{-1} (8.13–9.17 μm).

In general, the C=S band behaves in a similar manner to the carbonyl band. When a chlorine atom is directly bonded to the carbon of the C=S group, the band is observed at 1235–1225 cm^{-1} (8.10–8.16 μm).

Carbon disulphide, which is used as a solvent, absorbs strongly near 1510 cm^{-1} (6.62 μm) and 395 cm^{-1} (25.32 μm).

Table 16.9   Organic sulphur compounds containing C=S group

Functional Groups	Region		Intensity		Comments
	cm^{-1}	μm	IR	Raman	
Dialkyl thioketones, R–CS–R′	~1150	~8.70	s	s	C=S str, see ref. 54; normally dimerisation makes the assignment of this band difficult, range has also been reported (see ref. 52) as 1270–1245 cm^{-1}
Diaryl thioketones, Ar–CS–Ar′	1225–1205	8.16–8.30	w–m	s	C=S str, see ref. 55
α,β-Unsaturaled thioketones	1155–1140	8.66–8.77	s	s	C=S str
Dialkyl trithiocarbonates, (RS)$_2$C=S	1075–1050	9.30–9.52	s	s	C=S str, see ref. 62
	~850	~11.76	s	s	asym S–C–S str, C=S mixing
	~700	~14.29	w–m		two bands for small alkyl groups
	~500	~20.00	w–m		two bands, sym S–C–S str and C–S out-of-plane def vib
Thioncarbonates, (RO)$_2$C=S	1235–1210	8.10–8.26	s	s	C=S str, strong bands near 1200 cm^{-1} and 1100 cm^{-1}
Dithioacids, R–CS–SH	~1220	~8.20	s	s	C=S str
Dithioesters, R–CS–SR	1225–1185	8.16–8.44	s	s	C=S str, see ref. 53
	~870	~11.49	m–s	m–s, p	asym CS–S str
Thioacid fluorides, R–CS–F	1125–1075	8.90–9.30		s	C=S str
Thioacid chlorides	1235–1225	8.10–8.16	s	s	C=S str
	1100–1065	9.09–9.39			
RO–CSCH$_2$COOH and RS–CSCH$_2$COOH	~1050	~9.52	s	s	C=S str, see ref. 51
Xanthates	1250–1100	8.00–9.09	m–s	s	At least two bands
	1065–1040	9.39–9.62	vs	s	C=S str
Dixanthates, –O–C=S–S–	1250–1190	8.00–8.40	vs	w	asym C–O–C str, see refs 65–67
	1120–1100	8.93–9.09	m–s	s–m	sym C–O–C str
	1060–1000	9.43–10.00	vs	s	C=S str
Xanthate salts, R–O–CS–S$^-$	1100–1000	9.10–10.00	s		Have three strong bands in region 1250–1030 cm^{-1} (see above)

Table 16.9 (*continued*)

Functional Groups	Region		Intensity		Comments
	cm^{-1}	μm	IR	Raman	
	680–650	14.71–15.38		s	
	480–445	20.83–22.47			
Zn, Cu xanthates	1250–1200	8.00–8.33	s	m–s	
	1140–1110	8.77–9.01	m		
	1070–1020	9.35–9.80	s	s	Out-of-plane CS$_2$ str
Na, K xanthates	1190–1175	8.40–8.51	s	s	Out-of-plane COC str
	1065–1020	9.39–9.80	s	s	Out-of-plane CS$_2$ str
	680–650	14.71–15.38		s	CS$_2$COCH *trans*
	630–600	15.87–16.67		s	CS$_2$COCH *trans*
	480–445	20.83–22.47		m–s	COC def vib
Oxyxanthates	~1580	~6.33	vs		br
	1115–1090	8.97–9.17	s		The only single strong band in region 1200–1000 cm^{-1}
	1050–1000	9.54–10.00	w		
	~695	~14.39	m–s		sh
Pyridthiones	1150–1100	8.70–9.09	s	s	C=S str, see refs 68, 69
Thioamides etc 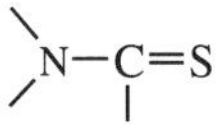 $\overset{\textstyle\diagdown}{\underset{\textstyle\diagup}{N}}-\overset{\textstyle\vert}{C}=S$	1570–1395	6.37–7.71	s	m	br $\big\}$ Due to strong coupling between C=S and C–N vibs
	1480–1360	6.76–7.35	v	m–s	amide III) $\big\}$
	1140–940	8.77–10.64	v	v	Usually strong in Raman
	860–680	11.63–14.70	v	s	C=S str, amide III band
Primary thioamides	3400–3150	2.94–3.17	m	m–w	Several bands NH$_2$ str
	1650–1590	6.06–6.29	m–s	w	NH$_2$ scissoring vib.(–CS·CS·NH$_2$ 1610–1590 cm^{-1}). Amide II band
	1480–1360	6.76–7.35	m	m–s	Amide III band
	1305–1085	7.66–9.22	v	s–m	Most compounds have a band due to rocking vib at 1170–1085 cm^{-1}. These bands are of variable intensity in both infrared and Raman spectra.
	860–680	11.63–14.71	m–s	m–s	C–S str, amide I band
	770–640	12.99–15.63	m–w	s	NH$_2$ twisting/wagging vib, amide VII band
	710–580	14.08–17.24	w–m		NH$_2$ wagging vib, amide V band
	600–420	16.67–23.81	m–w	s	CS def vib, amide IV band
	520–320	19.23–31.25	m–w	s	Out-of-plane NCS def vib, amide VI band
	410–240	24.39–41.67	w		CN def vib
Secondary thioamides	3280–3100	3.05–3.23	m	m	N–H str
	1550–1500	6.45–6.67	m	m–s	Amide III band
	~1350	~7.41	m		
	950–800	10.53–12.50	m	m–s	C–S str
	700–550	14.29–18.18	w–m	vs	NCS def vib
	500–400	20.00–25.00	w	s	NCS def vib
–CS·NH·CH$_3$	3320–3180	3.01–3.14	m–s	m	NH str
	3000–2920	3.33–3.42	m	m	CH$_3$ asym str
	2920–2820	3.42–3.55	m	m–s	CH$_3$ sym str
	1570–1500	6.37–6.67	s	m	Amide II
	1475–1410	6.78–7.09	m	m	CH$_3$ asym def vib
	1425–1375	7.02–7.55	m	m–w	CH$_3$ sym def vib

(*continued overleaf*)

Table 16.9   (*continued*)

Functional Groups	Region		Intensity		Comments
	cm^{-1}	μm	IR	Raman	
	1375–1280	7.55–7.81	m–s	m	Amide III
	1190–1100	8.40–9.09	w–m	w	CH$_3$ rocking vib
	1115–1035	8.97–9.66	w–m	w	CH$_3$ rocking vib
	905–685	11.05–14.60	m	m–s	Amide I (with exception of H–, CH$_3$O–, CH$_3$–and a few other compounds, range is 840–720 cm^{-1})
	720–610	13.89–16.39	m–s	m–s	br, amide V
	640–530	15.63–18.87	w–m	s	Amide IV
	540–400	18.52–25.00	w–m	s	Amide VI
	450–340	22.22–29.41		m–s	Skeletal vib
Tertiary thioamides	1565–1500	6.39–6.67	m–s	m	C–N str
	1285–1210	7.78–8.26	s	w	A number of bands in the region 1000–700 cm^{-1}
	630–500	15.87–20.00	m	vs	NCS def vib
	450–335	22.22–29.85	m–w	m	NCS def vib
Derivatives of H$_2$C⟨CH$_2$–C=S / CH$_2$–NH⟩ and CH$_2$–CH$_2$–C=S / CH$_2$–CH$_2$–NH	~1115	~8.70	s	s	C=S str (also seven-membered rings)
Dithiocarbamates, NH$_2$·CS·SR	~970	~10.30	s	s	C=S str
Cyclic thioureas (five- to seven-membered rings)	~1205	~8.30	s		Solid-phase spectra, also strong band at 1505 cm^{-1}. A strong band is observed in both IR and Raman near 450 cm^{-1} due to C=S def vib
P=S (solid phase)	865–655	11.56–15.27	m–s		
	750–530	13.33–18.87	v	v	P=S str
R$_3$P=S	770–685	12.99–14.60	m–s		
	595–530	16.81–18.87	v		
R$_3$As=S	490–470	20.41–21.28		m–s	As=S str
Methyl dithiocarbazic acids, salts and Ni^{II} and Cr^{III} coordination compounds	1040–960	9.62–10.42	s		asym CS$_2$, see ref. 64
	690–590	14.49–16.95	s		sym CS$_2$ str

Reviews

A review of the infrared spectra of sulphur compounds has been given by Billing.[92,93] A review of the infrared spectra of gaseous diatomic sulphides is given by Barrow and Cousins.[8] The spectra of selenol and thiol esters have been reviewed by Ciurdaru and Denes.[11]

Organic Selenium Compounds

The infrared spectra of selenium compounds[10,51,60,62,70–83] exhibit a great similarity to the corresponding sulphur analogues. This is hardly surprising – the change in mass, the (normally) weaker bonds formed by the selenium atom, and the slight variation in the bond angles account for the spectral differences.

Selenocarbonyls, ⟩C=Se, absorb at 1305–800 cm^{-1} (7.66–12.50 μm).

A review of the spectra of selenium compounds has been published.[70]

Selenoamides, ⟩N–CSe–

Selenoamides[60] do not have a band due solely to the C=Se stretching vibration because of the strong coupling of this vibration with the stretching vibration of the C–N bond. This type of behaviour has also been mentioned for thioamides.

Table 16.10 Other sulphur-containing compounds

Functional Groups	Region		Intensity		Comments
	cm^{-1}	μm	IR	Raman	
S–F	815–755	12.27–13.25	s		S–F str
S–O–CH$_2$–	1020–850	9.80–11.76	s	m	asym S–O–C str
	830–690	12.05–14.49	m	m	sym S–O–C str
Dialkyl thiolesters, R–CO–SR	1700–1690	5.88–5.59	vs	w–m	C=O str
Ar–CO–SR	1670–1665	5.99–6.00	vs	w–m	C=O str
Alkyl thiocarbonates, RS–CO–OR	1720–1700	5.81–5.88	vs	w–m	C=O str
ArS–CO–OAr	1740–1730	5.75–5.78	vs	w–m	C=O str
$\diagdown$C=C–S—	~1590	~6.29	m	s	C=C str, lower than expected, =C–H def vib band in normal position, see ref. 94
Thiooximes	~1620	~6.17	w–m	s	C=N str

Table 16.11 Organic selenium compounds

Functional Groups	Region		Intensity		Comments
	cm^{-1}	μm	IR	Raman	
Selenols, R–Se–H, and other compounds with Se–H	2330–2280	4.29–4.39	w	s	Se–H str, see refs 10, 78, 85
R–Se–D	~1680	~5.95	w	s	Se–D str
Selenides, R–Se–R	610–550	16.39–18.18	w–m	s	C–Se–C str, see ref. 79, 80
	585–505	17.09–19.80	w–m	s	C–Se–C str
Diselenides, R–Se–Se–R′	580–505	17.24–19.80	w–m	s	C–Se str, see ref. 81
	295–285	33.90–35.09	w	s	Se–Se str
Selenoxides, R–SeO–R and R–SeO$_2$–R	625–530	16.00–18.87	s	s	C–Se str, band weak for selenious acids and ions
Selenosemicarbazones, $\diagdown$N—CSe—N—N=C$\diagup$	800–780	12.50–12.82	m	s	C=Se str
Coordinated carbon diselenide, e.g. (phosphine)$_2$PtCSe,	~995	~10.05	s	s	C=Se str, carbon diselenide in CCl$_4$ absorbs ~1270 cm^{-1} and at ~310 cm^{-1}
Diselenocarbonates of the and types RO–CSeSeCH$_2$COOH and RS–CSeSeCH$_2$COOH	~940	~10.64	s	s	C=Se str, see ref. 51, 62
Dialkyl triselenocarbonates	~800	~12.50	s	s	C=Se str, doubling of bands (see refs 71, 86) may be observed due to presence of different conformations
	~750	~13.33	s	m	asym Se–C–Se str
	~600	~16.67	w–m		Two bands
Phosphoniodiselenoformates, R$_3$P$^+$CSeSe$^-$	~900	~11.11	s	m	asym CSe$_2$ str, see ref. 87
Methyl diselenocarbazic acid, salts, and Ni^{II} and Cr^{III} coordination complexes	930–860	10.75–11.63	s	m	asym CSe$_2$ str, see ref. 88
	615–490	16.26–20.41	s	m	sym CSe$_2$ str
Selenoacetals, =C$\diagup$$^{SeR}_{SeR}$	800–700	12.50–14.29	s	m	asym CSe$_2$ str

(*continued overleaf*)

Table 16.11 (*continued*)

Functional Groups	Region		Intensity		Comments
	cm^{-1}	μm	IR	Raman	
Ionic 1,1-ethylene diselenolates, $=C\begin{smallmatrix}Se^-\\Se^-\end{smallmatrix}$	870–750	11.49–11.33	s	m	asym =CSe$_2$ str
Ionic diselenocarbamates, $=C\begin{smallmatrix}Se\\Se^-\end{smallmatrix}$	950–800	10.53–12.50	s	m	asym =CSe$_2$ str
Cyanimidodiselenocarbonate alkali metal salts, CN–N=C(SeH)$_2$	~870	~11.49	s	m	asym CSe$_2$ str
Methyl aromatic selenothioesters, ArCSe·SMe	~980	~10.20	s	s	C=Se str
Aliphatic diselenocarboxylic acid esters, R–CSe·SeR	~780	~12.82	s	m	asym CSe·Se str
Selenoamides, ureas, and hydrazides	700–600	14.29–16.67	s	s	C=Se str, strongly coupled as with thioamides
Selenoamides	1500–1400	6.67–7.14	s	s	C=Se and C–N str, strongly coupled as with thioamides
	1200–1000	8.33–10.00	m		
	700–600	14.29–16.67	s		
Derivatives of $H_2C\begin{smallmatrix}CH_2-C=Se\\CH_2-NH\end{smallmatrix}$ and $\begin{smallmatrix}CH_2-CH_2-C=Se\\CH_2-CH_2-NH\end{smallmatrix}$	~1085	~9.21	s	s	C=Se str
Selenazoles, benzoselenazoles, and selenazolines	1570–1535	6.37–6.52	s	m–s	N=C–Se str, see refs 89, 90
Selenazolines	1680–1650	5.95–6.06	w–m	s	C=N str
Benzoselenazoles	1610–1590	6.21–6.29	w–m	s	C=N str
Selenates, (RO)$_2$SeO$_2$	1040–1010	9.71–9.90	vs		asym O–Se–O str, see ref. 91
	960–930	10.42–10.75	vs		sym O–Se–O str
	700–600	14.29–16.67	m		Se–O–C str
Diselenates, (RO)$_2$SeO$_5$	700–600	14.29–16.67	vs		asym Se–O–Se str
	560–500	17.86–20.00	vs		sym Se–O–Se str
	~230	~43.48			Se–O–Se def
Selenones, R$_2$SeO$_2$	920–910	10.87–10.99	vs		asym O–Se–O str
	890–880	11.24–11.36	vs		sym O–Se–O str
	420–390	23.81–25.64			O–Se–O def vib
Selenites	~930	~10.75	s	m–s	Se=O str
Seleninic acids, R·SeO·OH	900–850	11.11–11.76	s	m–s	Se=O str
	700–680	14.29–14.71	m–s		Se–OH str
Seleninic acid anhydrides, R·SeO–O–SeO·R	900–850	11.11–11.76	s	m–s	Se=O str
	700–600	14.29–16.67	m–s		asym Se–O–Se str
	560–500	17.86–20.00	m–s		sym Se–O–Se str
	230–170	43.48–58.82			Se–O–Se def vib
Seleninic acid esters	900–850	11.11–11.76	s	m–s	Se=O str
Seleninyl halides	1005–930	9.96–10.75	s	m–s	Se=O str
Selenoxides, R$_2$SeO	840–800	11.90–12.50	s	m–s	Se=O str

Table 16.11 (*continued*)

Functional Groups	Region cm^{-1}	Region μm	Intensity IR	Intensity Raman	Comments
Selenious amides	890–880	11.24–11.36	s	m–s	Se=O str
–Se–Se–	370–265	27.03–37.74	w	s	
Dialkyl amino compounds of the type (R$_2$N)$_2$Se, (R$_2$N)$_2$SeR$_2$, (R$_2$N)$_2$SeO, R$_2$N·SeO·Cl	590–540	16.95–18.52			N–Se–N str, asym and sym str coincide
Triaryl phosphine selenides, Ar$_3$PSe	~560	~17.86	s	v	P=Se str, for metal complexes band occurs at 540–530 cm^{-1}
Trialkyl phosphine selenides, R$_3$PSe	~425	~23.53	s	v	P=Se str
(EtO)$_2$·P=Se·SR	~590	~16.95	s	v	P=Se str
Trialkyl arsine selenides, R$_3$As=Se	360–330	27.78–30.30			As=Se str, a doublet
Trialkyl stibine selenides, R$_3$Sb=Se	300–270	33.33–37.04	s		Sb=Se str

Two strong bands observed at 1500–1400 cm^{-1} (6.67–7.14 μm) and 700–600 cm^{-1} (14.29–16.67 μm) both have a contribution from the C=Se stretching vibration.

For metal complexes in which the metal ion is directly bound to the selenium atom, the former band position tends to higher frequencies whilst the latter tends to lower frequencies. Selenoamides have a medium-intensity band at 1200–1000 cm^{-1} (8.33–10.00 μm) which also has a contribution from the C=Se bond vibration. The spectra of other compounds[64,82] with the N–C=Se group are similar to those of selenoamides.

The Se=O Stretching Vibration

The band due to the stretching vibration of the group Se=O is found over a wide range: 1000–800 cm^{-1} (10.00–12.50 μm).[75] As might be expected, the band is lower in frequency for metal complexes[76] than for the corresponding free selenoxide.

The P=Se Stretching Vibration

The band due to the P=Se stretching vibration[72–74] is of medium-to-strong intensity and occurs over the wide range 600–420 cm^{-1} (16.67–23.81 μm), more than one band often being observed.

The band generally occurs at higher frequencies for triaryl phosphine selenides, being at about 560 cm^{-1} (17.86 μm), than for the aliphatic compounds of this type, for which it occurs at about 425 cm^{-1} (23.53 μm). A similar difference is observed for the corresponding sulphur compounds.

For metal complexes[84] of triaryl phosphine selenides, this absorption occurs at about 535 cm^{-1} (18.69 μm).

In the case of amide, ester, salt, and acid chloride derivatives of selenophoric acid, selenophosphonic acid, and selenophosphinic acid, the band is strong and in the range 600–500 cm^{-1} (16.67–20.00 μm).

References

1. C. S. Hsu, *Spectrosc. Lett.*, 1974, **7**, 439.
2. J. G. David and H. E. Hallam, *Spectrochim. Acta*, 1965, **21**, 841.
3. A. R. Cole et al., *Spectrochim. Acta*, 1965, **21**, 1169.
4. S. I. Miller and G. S. Krishnamurthy, *J. Org. Chem.*, 1962, **27**, 645.
5. R. A. Nyquist and W. J. Potts, *Spectrochim. Acta*, 1959, **15**, 514.
6. P. A. Tice and D. R. Powell, *Spectrochim. Acta*, 1965, **21**, 837.
7. J. G. David and H. E. Hallam, *Trans. Faraday Soc.*, 1964, **60**, 2013.
8. R. F. Barrow and C. Cousins, *Adv. High Temp. Chem.*, 1971, **4**, 161.
9. J. Cymerman and J. B. Willis, *J. Chem. Soc.*, 1951, 1332.
10. N. Sharghi and I. Lalezari, *Spectrochim. Acta*, 1964, **20**, 237.
11. G. Ciurdaru and V. I. Denes, *Stud. Cercet. Chim.*, 1971, **19**, 1029.
12. G. A. Crowder, *Appl. Spectrosc.*, 1972, **26**, 486.
13. R. J. Philippe and H. Moore, *Spectrochim. Acta*, 1961, **17**, 1004.
14. T. Cairns et al., *Spectrochim. Acta*, 1964, **20**, 31.
15. T. Cairns et al., *Spectrochim. Acta*, 1964, **20**, 159.
16. D. Barnard et al., *J. Chem. Soc.*, 1949, 2442.
17. W. Otting and F. A. Neugebauer, *Chem. Ber.*, 1962, **95**, 540.
18. S. Pinchas et al., *J. Chem. Soc.*, 1962, 3968.
19. G. Kresze et al., *Spectrochim. Acta*, 1965, **21**, 1633.
20. T. Granstad, *Spectrochim. Acta*, 1963, **19**, 829.
21. R. H. Figueroa et al., *Spectrochim. Acta*, 1966, **22**, 1563.
22. R. H. Figueroa, *Spectrochim. Acta*, 1966, **22**, 1109.

23. W. K. Glass and R. D. E. Pullin, *Trans. Faraday Soc.*, 1961, **57**, 546.
24. S. Ghersetti and G. Modena, *Spectrochim. Acta*, 1963, **19**, 1809.
25. M. F. Lappert and J. K. Smith, *J. Chem. Soc.*, 1961, 3224.
26. R. S. Drago and D. Meek, *J. Phys. Chem.*, 1961, **65**, 1446.
27. R. Francis and F. A. Cotton, *J. Chem. Soc.*, 1961, 2078.
28. E. A. Robinson, *Can. J. Chem.*, 1961, **39**, 247.
29. L. J. Bellamy and R. L. Williams, *J. Chem. Soc.*, 1957, 863.
30. N. Marziano *et al.*, *Ann. Chim. Rome*, 1962, **52**, 121.
31. S. Ghersetti and C. Zauli, *Ann. Chim. Rome*, 1963, **53**, 710.
32. W. R. Fearirheller and J. E. Katon, *Spectrochim. Acta*, 1964, **20**, 1099.
33. P. M. G. Bavin *et al.*, *Spectrochim. Acta*, 1960, **16**, 1312.
34. A. S. Wexler, *Appl. Spectrosc. Rev.*, 1968, **1**, 29.
35. R. J. Gillespie and E. A. Robinson, *Spectrochim. Acta*, 1963, **19**, 741.
36. A. Simon and D. Kunath, *Z. Anorg. Chem.*, 1961, **308**, 21.
37. M. Bredereck *et al.*, *Chem. Ber.*, 1960, **93**, 2736.
38. G. Von Geiseler and K. O. Bindernagel, *Z. Elektrochem.*, 1959, **63**, 1140.
39. G. Von Geiseler and K. O. Bindernagel, *Z. Elektrochem.*, 1960, **64**, 421.
40. G. Malewski and H. J. Weigmann, *Spectrochim. Acta*, 1962, **18**, 725.
41. G. Malewski and H. J. Weigmann, *Z. Chem.*, 1964, **4**, 389.
42. D. A. Long and R. T. Bailey, *Trans. Faraday Soc.*, 1963, **59**, 792.
43. R. J. Gillespie and E. A. Robinson, *Can. J. Chem.*, 1961, **38**, 2171.
44. J. N. Baxter *et al.*, *J. Chem. Soc.*, 1955, 669.
45. A. R. Katritzky and R. A. Jones, *J. Chem. Soc.*, 1960, 4497.
46. E. von Merian, *Helv. Chim. Acta*, 1960, **49**, 1122.
47. A. Vandi *et al.*, *J. Org. Chem.*, 1961, **26**, 1136,
48. A. Simon *et al.*, *Chem. Ber.*, 1956, **89**, 1883, **2378**, and 2384.
49. S. Detoni and D. Hadzi, *Coll. Spectrosc, Int. VI*, 1956, 601.
50. G. Chihara, *Chem. Pharm. Bull.*, 1960, **8**, 988.
51. K. A. Jensen and U. Anthoni, *Acta Chem. Scand.*, 1970, **24**, 2055.
52. C. Andrieu and Y. Mollier, *Spectrochim. Acta*, 1972, **28A**, 785.
53. L. J. Bellamy and P. E. Rogash, *J. Chem. Soc.*, 1960, 2218.
54. R. Mayer *et al.*, *Angew. Chem.*, 1964, **76**, 157.
55. N. Lornch and G. Guillouzo, *Bull. Soc. Chim. France*, 1957, 1221.
56. C. N. R. Rao and R. Venkataraghavan, *Spectrochim. Acta*, 1962, **18**, 541.
57. C. N. R. Rao and R. Venkataraghavan, *Can. J. Chem.*, 1961, **39**, 1757.
58. H. E. Hallam and C. M. Jones, *Spectrochim. Acta*, 1969, **24A**, 1791.
59. G. Keresztury and M. P. Marzocchi, *Spectrochim. Acta*, 1975, **31A**, 271.
60. K. A. Jensen and P. M. Nielsen, *Acta Chem. Scand.*, 1966, **20**, 597.
61. H. O. Desseyn *et al.*, *Spectrochim. Acta*, 1974, **30A**, S03.
62. G. Borch *et al.*, *Spectrochim. Acta*, 1973, **29A**, 1 109.
63. K. Herzog *et al.*, *J. Mol. Struct.*, 1969, **3**, 339.
64. U. Anthoni *et al.*, *Acta Chem. Scand.*, 1970, **29**, 959.
65. F. G. Pearson and R. B. Stasiak, *Appl. Spectrosc.*, 1958, **12**, 116.
66. L. M. Little *et al.*, *Can. J. Chem.*, 1961, **39**, 745.
67. M. L. Shankaranaryana and C. C. Patel, *Can. J. Chem.*, 1961, **39**, 1633.
68. A. R. Katritzky and R. A. Jones, *Spectrochim. Acta*, 1961, **17**, 64.
69. A. R. Katritzky and R. A. Jones, *J. Chem. Soc.*, 1960, 2947.
70. K. A. Jensen *et al.*, in *Organic Selenium Compounds – Their Chemistry and Biology* (D. L. Klayman and W. H. H. Günther, eds), Wiley, New York, 1973.
71. L. Hendriksen *et al.*, *Spectrochim. Acta*, 1975, **31A**, 191.
72. R. A. Zingaro, *Inorg. Chem.*, 1963, **2**, 191
73. K. A. Jensen and P. H. Nielsen, *Acta Chem. Scand.*, 1963, **17**, 1875.
74. J. R. Durig *et al.*, *J. Mol. Spectrosc.*, 1968, **28**, 444.
75. R. von Paetzold, *Z. Chem.*, 1964, **4**, 321.
76. R. von Paetzold and G. Bochmann, *Z. Anorg. Allg. Chem.*, 1969, **368**, 202
77. V. Horn and R. von Paetzold, *Spectrochim. Acta*, 1974, **30A**, 1489.
78. A. B. Harvey and M. K. Wilson, *J. Chem. Phys.*, 1966, **45**, 678.
79. J. A. Allkins and P. J. Hendra, *Spectrochim. Acta*, 1967, **23A**, 1671.
80. J. Shiro *et al.*, *Bull. Chem. Soc. Japan*, 1970, **43**, 612.
81. G. Bergson, *Arkiv Kemi*, 1959, **13**, 11.
82. B. A. Gingras *et al.*, *Can. J. Chem.*, 1965, **43**, 1650.
83. R. von Paetzols *et al.*, *Z. Anorg. Allg. Chem.*, 1967, **352**, 295.
84. M. G. King and G. P. McQuillan, *J. Chem. Soc. A*, 1967, 898.
85. K. A. Jensen and L. Hendriksen, *Acta Chem. Scand.*, 1970, **24**, 3213.
86. M. Därger and G. Gatlow, *Chem. Ber.*, 1971, **104**, 1429.
87. K. A. Jensen and P. H. Nielsen, *Acta Chem. Scand.*, 1963, **17**, 549.
88. U. Anthoni *et al.*, *Acta Chem. Scand.*, 1970, **29**, 959.
89. P. Bassignana *et al.*, *Spectrochim. Acta*, 1965, **21**, 605.
90. R. V. Kendall and R. A. Olofsun, *J. Org. Chem.*, 1970, **35**, 806.
91. R. von Paetzold and H. Amoulong, *Z. Chem.*, 1966, **6**, 29.
92. F. A. Billing, *Intra-Sci. Chem. Rep.*, 1967, **1**, 225.
93. F. A. Billing and N. Kharasch, *Quart. Rep. Sulphur Chem.*, 1966, **1**, 1189.
94. B. Nagel and A. B. Remizov, *Zh. Obbshsch. Khim.*, 1978, **45**, 1189.
95. G. Paulson *et al.*, *Biol. Med. Mass Spectrom.*, 1978, **5**, 128.
96. L. Cazaux *et al.*, *Spectrochim. Acta*, 1979, **35A**, 15.
97. Y. Tanaka *et al.*, *Spectrochim. Acta*, 1983, **39A**, 159

17 Organic Phosphorus Compounds

P–H and P–C Vibrations

The stretching vibration of the P–H group[6–8,10,11] gives rise to a sharp band of medium intensity in the region 2500–2225 cm^{-1} (4.00–4.49 μm). For aliphatic and aryl phosphines, this band occurs in a much narrower region: 2285–2265 (4.38–4.42 μm).

The stretching vibration of the P–C bond gives a medium-to-strong band in the region 795–650 cm^{-1} (12.58–15.38 μm).

P–OH and P–O Vibrations

Compounds with the P–OH group,[10] for which, of course, hydrogen bonding normally occurs, have two broad bands of weak-to-medium intensity at 2700–2560 cm^{-1} (3.70–3.90 μm) and 2300–2100 cm^{-1} (4.35–4.76 μm) which are due to the O–H stretching vibrations and a medium-to-strong, broad band at 1040–910 cm^{-1} (9.62–10.99 μm) due to the P–O stretching vibration. However, since most phosphorus compounds absorb in this latter region, this band is of little value. Those compounds which also contain the P=O group[12,18,21,22] have a broad band near 1680 cm^{-1} (5.92 μm), e.g. dialkyl phosphoric acids, phosphorous acids. For phosphoric acids, the band near 2600 cm^{-1} (3.85 μm) is stronger than those near 2200 cm^{-1} (4.55 μm) and 1680 cm^{-1} (5.95 μm) whereas for phosphinic acids, the band near 1680 cm^{-1} is the strongest of the three and for phosphonic acids all three bands have about the same relative intensity.

Acid salts containing the P–OH group have broad bands in the regions 2725–2525 cm^{-1} (3.76–3.94 μm) and 2500–1600 cm^{-1} (4.00–6.25 μm).

P–O–C Vibrations

For aliphatic compounds, the asymmetric stretching vibration of the P–O–C group[12,15,18,25] gives a very strong broad band, normally found in the region 1050–970 cm^{-1} (9.52–10.31 μm). In the case of pentavalent and trivalent methoxy compounds, this band is sharp and strong, occurring at 1090–1010 cm^{-1} (9.17–9.90 μm) and 1035–1015 cm^{-1} (9.67–9.85 μm) respectively, the characteristic symmetric methyl deformation band near 1380 cm^{-1} (7.25 μm) being absent in some cases.

In general, the band due to the asymmetric stretching vibration of the P–O–C group of pentavalent phosphorus occurs at lower frequencies than that for the trivalent compound. Pentavalent ethoxy compounds have an additional strong band at 985–940 cm^{-1} (10.15–10.64 μm), which may be weak for higher alkoxy compounds. Methoxy and ethoxy compounds have a strong band at 830–740 cm^{-1} (12.05–13.51 μm) which is probably due to the symmetric stretching of the P–O–C group. However, this band is usually absent in other alkoxy compounds. Methoxy compounds have a weak, sharp band near 1190 cm^{-1} (8.40 μm). Other alkoxy phosphorus compounds have a medium-intensity band near 1165 cm^{-1} (8.59 μm). For compounds which have only ethoxy groups (i.e. no other alkyl groups), two characteristic doublets are observed in the region 1500–1350 cm^{-1} (6.67–7.41 μm) due to the C–H deformation vibrations. For aromatic compounds, P–O–phenyl, the band due to the P–O–C asymmetric stretching vibration occurs at 995–855 cm^{-1} (10.05–11.70 μm).

P=O Vibrations

The band due to the stretching vibration of the P=O group[7,13–24,37] is strong and in the region 1350–1150 cm^{-1} (7.41–8.70 μm). Due to the size of the phosphorus atom, the frequency of the P=O stretching vibration is almost independent of the type of compound in which the group occurs and of the size of the substituents. However, it is influenced by the number of electronegative substituents directly bonded to it, as well as being very sensitive to association effects.[23,24] For instance, a phase change results in a shift in band position of about 60 cm^{-1}. The P=O band may sometimes appear as a doublet,[14]

Chart 17.1 The positions and intensities of bands observed in the infrared spectra of phosphorus compounds

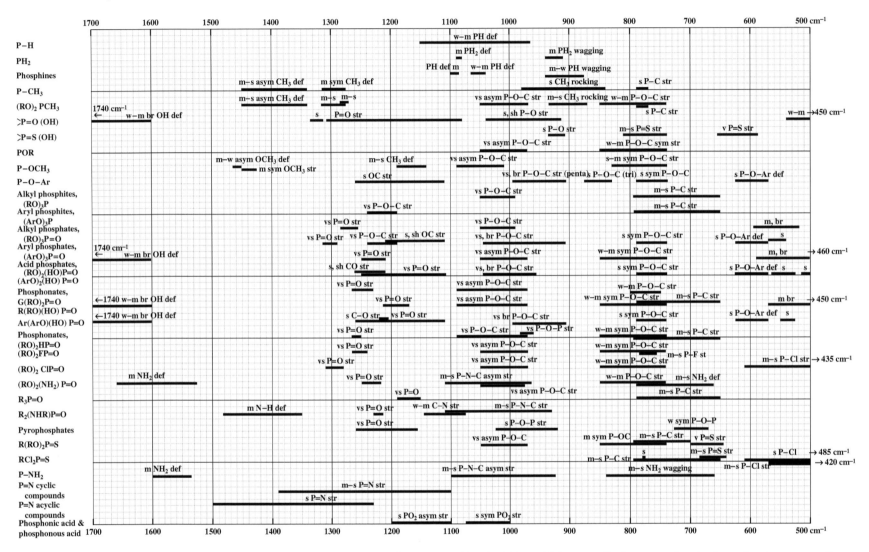

the separation either being small, as for some triaryl phosphates, or as large as 50 cm^{-1}. This splitting is believed, in some cases, to be partly due to Fermi resonance and, in others, such as some substituted triaryl phosphates, to rotational isomerism.

Pyrophosphates,[17] O=P–O–P=O, have only one P=O band, unless the pyrophosphate is a non-symmetrical compound. Therefore, unlike carboxylic acid anhydrides which generally have two bands that arise due to coupling between the C=O groups, no coupling appears to exist between the two P=O groups in pyrophosphates.

Phosphoric acids have extremely strong intermolecular hydrogen bonds which are present even in very dilute solution in inert solvents and result in the P=O band usually being about 50 cm^{-1} lower than for the corresponding ester.

Chart 17.1 (*continued*)

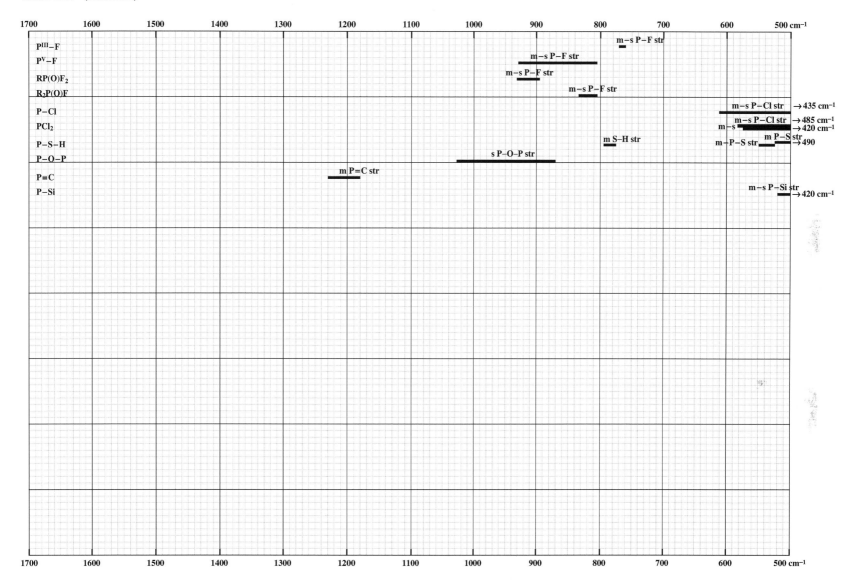

The position of the band due to the P=O stretching vibration is dependent on the sum of the electronegativities of the attached groups. Electronegative groups tend to withdraw electrons from the phosphorus atom thus competing with the oxygen which would otherwise have a tendency to form P^+-O^-, therefore resulting in a stronger bond and hence in a higher vibration frequency. Similarly, hydrogen bonding tends to lower the frequency of the P=O stretching vibration and broaden the band. The frequency[7,13] of this band may be calculated for different compounds with reasonable accuracy

Table 17.1 Organic phosphorus compounds. (The data given are, except where stated, for condensed phase spectra, i.e. liquids or solids, measured in nujol or as discs)

Functional Groups	Region cm^{-1}	Region μm	Intensity IR	Intensity Raman	Comments
P–H *vibrations*:					
P–H	2500–2225	4.00–4.49	m	m–w	P–H str
	1150–965	8.70–10.36	w–m	m–w	P–H def vib
PH$_2$	2440–2275	4.10–4.40	m	m–w	PH str
	1090–1080	9.17–9.26	m	m–w	PH def vib
	940–910	10.64–10.99	m	w	PH wagging vib
Alkyl phosphines, P–H	2320–2265	4.31–4-42	m	m–w	P–H str
	1100–1085	9.09–9.21	m	m–w	P–H$_2$ scissoring vib
	1065–1040	9.39–9.62	w–m	m–w	P–H def vib
	940–910	10.64–10.99	m–w	w	PH$_2$ wagging vib
Aryl phosphines, P–H	2285–2270	4.38–4.41	m	m–w	P–H str, see ref. 35
	1100–1085	9.09–9.21	m	m–w	P–H def vib
	~885	~11.30	m–w	w	PH wagging vib
Phosphonates, (GO)$_2$HP=O	2455–2400	4.07–4.17	m	m–w	P–H str
	980–960	10.20–10.41	m–s	w	PH wagging vib
Phosphine oxides, G$_2$HP=O	2380–2280	4.20–4.39	m	m–w	P–H str
	990–965	10.10–10.36	m–w	w	P–H wagging vib
G$_2$HP=S	2340–2280	4.27–4.39	m	m–w	P–H str
	950–910	10.53–10.99	m–s	w	P–H wagging vib
Phosphonates, (RO)$_2$HP=O	2450–2380	4.08–4.20	m	m–w	PH str
	980–960	10.20–10.42	vs	w	Probably due to interaction between P–O–P stretching and P–H wagging vib
P–D	1795–1650	5.57–6.06	m	m–w	P–D str
	745–615	13.42–16.26	w–m	w	P–D bending vib
P–C *and* PC–H *vibrations*:					
P–C	795–650	12.58–15.38	m–s		P–C str
P–CH$_3$	1450–1390	6.90–7.19	m–s	w	asym CH$_3$ def vib
	1345–1275	7.49–7.85	m–s	w	sym CH$_3$ def vib
	980–840	10.20–11.90	s	w	CH$_3$ rocking def vib, often doublet (for P^{V} compounds 935–870 cm^{-1}, for P^{III} at 905–860 cm^{-1}, for PHCH$_3$ ~845 cm^{-1})
	790–770	12.66–12.99	s		⟩P—C str
⟩PHCH$_3$	850–840	11.76–11.90	m–s		
P(CH$_3$)$_2$	960–835	10.42–10.70	m–s		Two or three bands
(RO)$_2$PCH$_3$	1285–1270	7.77–7.87	m–s		} P–CH$_3$ bands
	870–865	11.49–11.56	s		
CH$_3$(RO)HP=O	1300–1295	7.69–7.72	m–s		} P–CH$_3$ bands
	850–840	11.76–11.90	s		
CH$_3$(RO)$_2$P=O	1320–1305	7.58–7.66	m–s		} P–CH$_3$ bands
	930–885	10.75–11.30	s		

Table 17.1 (*continued*)

Functional Groups	Region cm^{-1}	Region μm	Intensity IR	Intensity Raman	Comments
CH$_3$(RO)P—O$^-$ ‖ O	1310–1280	7.63–7.81	m–s		⎫ P–CH$_3$ bands
	900–875	11.11–11.43	s		⎭
CH$_3$(RO)ClP=O	1315–1300	7.60–7.69	m–s		⎫ P–CH$_3$ bands P–C$_2$H$_5$
	925–885	10.81–11.30	s		⎭
	1285–1225	7.78–8.17	w		Doublet (P^{III} compounds also have medium intensity band at 1235–1205 cm^{-1})
P–CH$_2$–P	845–780	11.83–12.82	m–s		asym P–C–P vib
	770–720	12.99–13.89	m–s		sym P–C–P vib
P–CH$_2$–	1440–1405	6.94–7.12	m	m	CH$_2$ def vib
	780–760	12.82–13.16	s		P–C str
P–CH$_2$–Ar	795–740	12.58–13.51	s		P–C str
P–Ar	~3050	~3.33	m–w	m–s	C–H str
	~1600	~6.25	m–w	m–s	Aromatic ring in-plane str
	~1500	~6.67	m–w	m	Aromatic ring in-plane str
	1455–1425	6.90–7.02	m–s	m	Aromatic ring in-plane str
	1010–990	9.09–10.10	m–s		Interaction between aromatic ring vib and P–C str
	560–480	17.86–20.83	m–s		
P–Ph	1130–1090	8.85–9.71	s–m	w	P–C str
	750–680	13.33–14.71	s	w	Out-of-plane CH def vib
P–N–Ph	1425–1380	7.02–7.25	w–m		
P–O–H *vibrations*: ﹥P⟨$^O_{OH}$	2725–2525	3.76–3.96	w–m	w	br, OH str, hydrogen bonded
	2350–2080	4.26–4.81	w–m	w	br, may be doublet for aromatic phosphorus acids
	1740–1600	5.75–6.25	w–m	w	br, OH def vib
	1335–1080	7.55–9.26	s	m–w	P=O str
	1040–910	9.62–10.99	s	m–w	sh, P–O str, dependent on inductive effect of substituent
	540–450	18.52–22.22	w–m		Often a doublet
R(OH)$_2$P=O	1030–970	9.71–10.31	s	m–w	
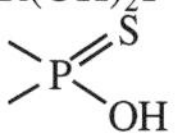 ﹥P⟨$^S_{OH}$	3100–3000	3.23–3.33	w	w	br, OH str
	2360–2200	4.24–4.55	w	w	br, OH str
	935–910	10.70–10.99	s	m–w	P–O str
	810–750	12.35–13.33	m–s	m–w	P=S str
	655–585	15.27–17.12	v	m–w	P=S str
P–O–C *vibrations*: P–O–R	1050–970	9.52–10.31	vs	m–w	asym P–O–C str (see ref. 15), (for phosphonium compounds, range extends to 1090 cm^{-1})

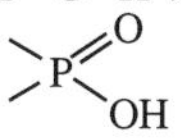

(continued overleaf)

Table 17.1 (*continued*)

Functional Groups	Region		Intensity		Comments
	cm^{-1}	μm	IR	Raman	
P–O–CH$_3$	850–740	11.76–13.51	w–m		Sometimes very weak
	1465–1450	6.83–6.90	m	m	asym CH$_3$ def vib
	1450–1435	6.90–6.97	m	w	sym CH$_3$ def vib
	1190–1140	8.40–8.77	m–s	w	CH$_3$ def vib
	1090–1010	9.17–9.90	vs	m–w	asym P–O–C def vib (trivalent P 1035–1015 cm^{-1})
	830–740	12.05–13.51	s–m		sym P–O–C str (asym str ~1050 cm^{-1})
P–O–C$_2$H$_5$	1485–1470	6.73–6.80	m–w	m	OCH$_2$ def vib
	1450–1445	6.90–6.92	m–w	m–w	CH$_3$ def vib
	1400–1390	7.14–7.19	m	m	OCH$_2$
	1375–1370	7.27–7.30	m–w	m	CH$_3$ def vib
	1165–1155	8.59–8.68	w–m	w	CH$_3$ rocking vib
	1105–1095	9.05–9.13	m	w	CH$_3$ rocking vib
	1045–1005	9.57–9.95	s		
	988–920	10.15–10.87	s		
	830–740	12.05–13.51	m–s		sym P–O–C str
P–O–CH$_2$R	1170–1100	8.55–9.09	w–m	m–w	Number of bands
	1045–985	9.57–10.15	s		
Isopropyl–O–P	1190–1170	8.40–8.55	w	m–w	
	1150–1135	8.70–8.81	w	m–w	
	1115–1100	8.97–9.09	w		
P–O–Ar	1460–1445	6.85–6.92	w–m		
	1260–1110	7.94–9.01	s	w	sh, mainly O–C str
	995–905	10.05–11.05	vs	m–w	br, P–O–C str (pentavalent)
	875–830	11.43–12.05	s	m–w	P–O–C str (trivalent)
	790–740	12.66–13.51	s		sym P–O–C str
	625–570	16.00–17.54	s		P–O–Ar def vib
Alkyl phosphites (RO)$_3$P	1050–990	9.52–10.10	vs	m–w	P–O–C str
Aryl phosphites (ArO)$_3$P	1240–1190	8.07–8.40	vs	m–w	P–O–C str
Phosphites (GO)$_3$P	580–510	17.24–19.61	m		
	580–400	17.24–25.00	s		
	400–295	25.00–33.90	s		
Hydrogen phosphites	560–545	17.86–18.35	s		
	540–500	18.52–20.00	w–m		
P=O *vibrations*:					
P=O (unassociated)	1350–1175	7.41–8.51	vs	m–w	P=O str
P=O (associated)	1250–1150	8.00–8.70	vs	m–w	P=O str, see refs 13, 15
Alkyl phosphates, (RO)$_3$P=O	1285–1255	7.78–7.97	vs	m–w	
	1050–990	9.52–10.10	vs	m–w	P–O–C str
	595–520	16.81–19.23	m		br
	495–465	20.20–21.51	m		br
	430–415	23.26–24.10	w		
	395–360	25.32–27.78			
Aryl phosphates, (ArO)$_3$P=O	1315–1290	7.61–7.75	vs	m–w	P=O str
	1240–1190	8.07–8.40	vs	m–w	P–O–C str
	625–575	16.00–17.39	s		

Table 17.1 *(continued)*

Functional Groups	Region		Intensity		Comments
	cm^{-1}	μm	IR	Raman	
	570–540	17.54–18.52	s		
	510–490	19.61–20.41			
	460–430	21.74–23.26	m–w		
Acid phosphates (RO)$_2$(HO)P=O	1250–1210	8.00–8.26	vs	m–w	P=O str, see ref. 32
	590–460	16.95–21.74	m		br
	400–380	25.00–26.32	w		Not observed for phosphonates
(ArO)$_2$(HO)P=O	565–535	17.70–18.69	s		
	515–500	19.42–20.00	s		
	490–470	20.41–21.28	w		
	400–380	25.00–26.32	w		
(RO)(HO)$_2$P=O	~1250	~8.00	vs	m–w	P=O str (aryl compounds ~1200 cm^{-1})
Phosphonates, G(RO)$_2$P=O	1265–1230	7.91–8.13	vs	m–w	P=O str. see ref. 32
	800–750	12.50–13.33	w–m		P–O–C str
Alkyl phosphonates, R(RO)$_2$P=O	570–500	15.54–20.00	m		br
	490–410	20.41–24.39	m		br
	440–400	22.73–25.00	w		
Aryl phosphonates Ar(ArO)$_2$P=O	620–600	16.13–16.67	m		see ref. 32
	535–515	18.69–19.42	s		
	500–480	20.00–20.83	vw		
	425–415	23.53–24.10	vw		
Dialkyl aryl phosphonates (RO)$_2$ArP=O	585–565	17.09–17.70	s		See ref. 32
	530–520	18.87–19.23	s		
	435–420	22.99–23.81	w		
	320–310	31.25–32.26	w		
Hydrogen phosphonates R'(RO)(HO)P=O	1215–1170	8.23–8.55	vs	m–w	P=O str, see ref. 32
	570–540	17.54–18.52	m		br
	500–450	20.00–22.22	m		br
	320–300	31.25–33.33	w		
Ar'(ArO)(OH)P=O	1220–1205	8.20–8.30	vs	m–w	P=O str, see ref. 32
	605–570	16.53–17.54	s		
	550–535	18.18–18.69	s		
	495–485	20.20–20.62	m		
	~460	~21.74	m		
	430–420	21.26–21.81	m		
	370–350	27.03–28.57	w		
	315–290	31.75–34.48	w		
(RO)(HO)HP=O	1215–1200	8.23–8.33	vs	m–w	P=O str
(RO)$_2$HP=O	1265–1250	7.97–8.00	vs	m–w	P=O str
(RO)$_2$FP=O	1315–1290	7.61–7.75	vs	m–w	P=O str
(ArO)$_2$FP=O	1330–1325	7.52–7.55	vs	m–w	P=O str

(continued overleaf)

Table 17.1 (*continued*)

Functional Groups	Region		Intensity		Comments
	cm^{-1}	μm	IR	Raman	
(RO)$_2$ClP=O	1310–1280	7.63–7.81	vs	m–w	P=O str (CN-substituted compounds at ~1290 cm^{-1})
(RO)$_2$(RS)P=O	1270–1245	7.87–8.06	vs	m–w	P=O str
(RO)$_2$(NH$_2$)P=O	1250–1220	8.00–8.20	s	m–w	P=O str, see ref. 34
(ArO)$_2$(NH$_2$)P=O	~1250	~8.00	vs	m–w	P=O str
(RO)$_2$(NHR)P=O	1260–1195	7.94–8.36	vs	m–w	P=O str
(RO)$_2$(NR$_2$)P=O	1275–1250	7.84–8.00	vs	m–w	P=O str
R$_2$(R'O)P=O	1220–1180	8.20–8.48	vs	m–w	P=O str
R$_2$(HO)P=O	1190–1140	8.40–8.77	vs	m–w	P=O str
Ar$_2$(HO)P=O	1205–1085	8.30–9.21	vs	m–w	P=O str
R(HO)HP=O	1175–1135	8.51–8.81	vs	m–w	P=O str
R$_3$P=O	1185–1150	8.44–8.70	vs	m–w	P=O str
Ar$_3$P=O	1190–1175	8.40–8.51	vs	m–w	P=O str
R$_2$HP=O	~1155	~8.66	vs	m–w	P=O str
Ar$_2$HP=O	1185–1170	8.44–8.55	vs	m–w	P=O str
R$_2$ClP=O	~1215	~8.23	vs	m–w	P=O str (for dichloro-, see ref. 36)
Ar$_2$ClP=O	~1235	~8.10	vs	m–w	P=O str
G$_2$BrP=O	~1250	~8.00	vs	m–w	P=O str
(RS)$_3$P=O	~1200	~8.00	vs	m–w	P=O str
(ArS)$_3$P=O	~1210	~8.26	vs	m–w	P=O str
R$_2$(RS)P=O	~1200	~8.33	vs	m–w	P=O str
(RHN)$_3$P=O	1230–1215	8.18–8.23	vs	m–w	P=O str
(R$_2$N)$_3$P=O	1245–1190	8.03–8.40	vs	m–w	P=O str
R$_2$(NHR)P=O	1180–1150	8.48–8.66	vs	m–w	P=O str
R(NHR)$_2$P=O	1220–1160	8.20–8.62	vs	m–w	P=O str
Pyrophosphates, diagram P–O–P	1310–1205	7.63–8.30	vs	m–w	P=O str (see ref. 17); usually one band, two bands for unsymmetrical pyrophosphates
Alkyl pyrophosphates, (RO)$_2$P—O—P(RO)$_2$ ‖ ‖ O O	1240–1205	8.07–7.63	vs	m–w	P=O str, usually one band
	1310–1280	7.63–7.81	vs	m–w	P=O str
Phosphonic anhydrides, R(RO)P—O—P(RO)R ‖ ‖ O O	1270–1250	7.87–8.00	vs	m–w	P=O str
	930–915	10.75–10.93	s		br, asym P–O–P str
P–O–P	1025–870	9.85–11.49	s		Usually broad, asym str often found in region 945–925 cm^{-1} (a weak band also near 700 cm^{-1})
P—S—C ‖ O	615–555	16.26–18.18	m		
	575–510	17.39–19.61	m		

Table 17.1 (*continued*)

Functional Groups	Region		Intensity		Comments
	cm^{-1}	μm	IR	Raman	
P=S *vibrations*:					
P=S	865–655	11.56–15.27	m–s	s, p	See refs 27–29. May be found in region 895–300 cm^{-1}
	750–530	13.33–18.87	v	v	P=S str
(OH)P=S	810–750	12.35–13.33	m	s	OH bands 3100–3000 cm^{-1}, 2360–200 cm^{-1} and P–O str 935–905 cm^{-1}
P–OP (S=)	655–585	15.27–17.09	v	v	
	865–770	11.56–12.99	m	s	
P=S (X=F or Cl)	610–585	15.27–17.09	v	v	
	835–750	11.98–13.33	m–s		
P–N (S=)	860–725	11.63–13.79	m–s	s	
—P(Cl)—N (S=)	715–550	13.99–18.18	v		
	810–765	12.35–13.07	m–s	s	
	675–600	14.81–16.67	v		
P=Se	590–515	16.95–19.42	m	s	P=Se str, see refs 27–29
	535–420	18.69–23.81	m	v	
P=Te	470–400	21.28–25.00			P–Te str
R(RO)$_2$P=S	805–770	12.42–12.99	m	s	
	650–595	15.38–16.81	v	v	
R$_2$(RO)P=S	795–770	12.58–12.99	m	s	
	610–565	16.39–17.70	v	v	
(RO)$_2$(RS)P=S	835–790	11.98–12.66	m	s	
	665–645	15.04–15.50	m–s	v	
(RO)$_2$(RS)P=Se	~590	~16.95	s	s	P=Se str
(RO)$_2$(SH)P=S	865–835	11.56–11.98	m	s	S–H bending vib
	780–730	12.82–13.70	m	v	
	660–650	15.15–15.38	m–s		
R$_2$ClP=S	775–750	12.90–12.33	m	s	
	650–590	15.38–16.95	m–s	v	
RCl$_2$P=S	780–775	12.82–12.90	s	s	((RO)Cl$_2$P=S ~830 cm^{-1})
	685–640	14.60–15.63	m–s	v	
(R$_2$N)$_3$P=S	840–790	11.90–12.66	m	s	
	715–690	13.99–14.49	m–s	v	
Metal phosphorodithioates, (MS)(RO)$_2$P=S M=Zn, Cd, Ni	660–635	15.15–15.75	s		Probably due to P=S, see ref. 25

(*continued overleaf*)

Table 17.1   (*continued*)

Functional Groups	Region		Intensity		Comments
	cm^{-1}	μm	IR	Raman	
	555–535	18.02–18.69	s		Possibly due to P–S–M group
PN *vibrations*:					
P–N	1110–930	9.01–10.75	m–s		Probably asym P–N–C str, see ref. 20
	750–680	13.33–14.71	m–s		sym P–N–C str
$P^{III}N$	1010–790	9.90–12.66	m–s		See ref. 33
$P–N–CH_3$	1320–1260	7.58–7.94	m		
	1205–1155	8.30–8.66	w–m		
	1080–1050	9.26–9.52	w–m		
	1010–935	9.90–10.70	s		
$P–N(C_2H_5)_2$	1225–1190	8.16–8.40	m–s		
	1190–1155	8.40–8.66	m		for P^V, w for P^{III}
	1110–1085	9.01–9.22	w–m		for P^V, w for P^{III}
	1075–1055	9.30–9.48	w–m		
	1050–1015	9.52–9.85	m–s		
	975–930	10.26–10.75	m–s		
	930–915	10.75–10.93	w		
$P–NH_2$	3330–3100	3.00–3.23	m	m–w	NH_2 str
	1600–1535	6.25–6.51	m	w	NH_2 def vib
	1110–920	9.01–10.87	m–s		P–N–C asym str
	840–660	11.91–15.15	m–s	w	NH_2 wagging vib
P–NH	3200–2900	3.13–3.45	m	m	NH str
	1145–1075	8.73–9.30	w–m	m	C–N str
	1110–930	9.01–10.75	m–s		P–N–C asym str
P=N (cyclic compounds)	1440–1100	6.94–9.09	m–s		P=N str, see ref. 15. Trimer 1300–1155 cm^{-1}, tetramer 1420–1180 cm^{-1}.
P=N (acylic compounds)	1500–1230	6.67–8.13	s		P=N str
$(RO)_3P$=N–Ar and $R(RO)_2P$=N–Ar	1385–1325	7.22–7.55	s		P=N str
$R–NH–P(O)Cl_2$	~560	~17.86			
$–O–NH–P(O)Cl_2$	545–520	18.35–19.23			
$\backslash N–P(S)Cl_2 \diagup$	525–490	19.05–20.41			
$–O–O–PCl_2$	510–495	19.61–20.20			
$–O–O–P(O)Cl_2$	~590	~16.95			
$–O–O–P(S)Cl_2$	560–535	17.86–18.69			
$(RO)Cl_2P$=O	~570	~17.54			
$(RO)Cl_2P$=S	~535	~18.69	s		
Phosphinic acid, $R_2PO_2^-$ and phosphonous acid, $RHPO_2^-$	1200–1100	8.33–9.09	s	s	asym PO_2^- str
	1075–1000	9.30–10.00	s	s	sym PO_2^- str
$(RO)_2PO_2^-$ (salt)	1285–1120	9.78–8.93	s	s	asym PO_2^- str
	1120–1050	8.93–9.52	s	s	sym PO_2^- str
$R(RO)PO_2^-$	1245–1150	8.03–8.70	s	s	asym PO_2^- str
	1110–1050	9.01–9.52	s	s	sym PO_2^- str
RPO_3^{2-}	1125–970	8.89–10.31	s	s	asym PO_3^{2-} str
	1000–960	10.00–10.42	m	s	sym PO_3^{2-} str

Table 17.1   (*continued*)

Functional Groups	Region		Intensity		Comments
	cm^{-1}	µm	IR	Raman	
ROPO$_3^{2-}$	1140–1055	8.77–9.48	s	s	asym PO$_3^{2-}$ str
	995–945	10.05–10.58	m	s	sym PO$_3^{2-}$ str
R$_2$POS$^-$	1140–1050	8.77–9.52	s	m–w	P–O str
	570–545	17.54–18.35	m		P–O str
(RO)$_2$POS$^-$ and R(RO)POS$^-$	1215–1110	8.24–9.01	s	m–w	P–O str
	660–575	15.15–17.39	m		P–S str
Inorganic salts; PO$_2^-$	1300–1150	7.69–8.70	s	vs	
Inorganic salts, PO$_3^{2-}$	1090–970	9.17–10.31	s	s	asym str
Inorganic salts, PO$_4^{3-}$	1100–1000	9.09–10.00	s	s	see ref 37
PS$_2^-$	585–545	17.09–18.35			
P^{III}–F	770–760	12.99–13.10	m–s	w	Usually strong
P^{V}F	930–805	10.75–12.42	m–s	w	Usually strong
R$_2$P(O)F	835–805	11.98–12.42	m–s	w	P–F str, see ref. 30
RP(O)F$_2$	930–895	10.75–11.17	m–s	w	P–F str, see ref. 30
P–Cl	610–435	16.39–22.99	m–s		see refs 15, 31
P–Br	485–400	20.62–25.00	m–s		P–Br str
PCl$_2$	610–485	16.39–20.62	s	s–m	P–Cl str
	570–420	17.54–23.81	m–s	s	P–Cl str
P–Cl where P is bonded to O, C, or F atom	565–440	17.70–22.73	m–s		
P–Cl where P is bonded to N or S atom	540–435	18.52–22.99	m–s		
P–S–H	550–525	18.18–19.05	m		P–S str
	525–490	19.05–20.41	m		P–S str
P–S–C	1050–970	9.52–10.31			Observed for aliphatic compounds
	565–550	17.70–18.18	m		
	490–440	20.41–22.73	m		
>P–S—C ‖ S	560–495	17.86–20.20	m–s		P–S–C str
	545–470	18.35–21.28	m		
P–S–P	550–400	18.18–25.00	m		
	495–460	20.20–21.74	m		
P–Se	570–520	17.54–19.23	m		
	535–470	18.69–21.28	m		
P–O–S	930–815	10.75–12.27			asym P–O–S str
	765–700	13.07–14.29			
P–O–Si	1070–855	9.35–11.70			
P–C=C	1645–1595	6.08–6.27	m	s	C=C str
P–C–C=C	1660–1630	6.02–6.14	m	s	C=C str
P–C≡N	2220–2180	4.51–4.59	m	m–s	C≡N str
P–Sn	360–280	27.78–35.71			
P–Ga	370–310	22.03–32.26			
P–Si	520–420	19.23–23.81			
P–Ge	400–300	25.00–31.33			

from the relationship

$$\nu = 930 + 40 \sum \pi$$

where π is the phosphorus inductive constant of a given substituent group. It should be noted that the frequency of the P=O stretching vibration has also been correlated with Taft σ values.[7,13]

Spectral changes for $\diagdown$P=O compounds (a) where the P=O group acts as a good proton acceptor, (b) where coordination occurs, are given elsewhere.[26]

Other Bands

The band due to the P=S stretching vibration[25,27–29] occurs at 865–655 cm^{-1} (11.56–15.27 μm) and is of medium-to-strong intensity. Also, a band of variable intensity occurs at 730–550 cm^{-1} (13.70–18.18 μm), possibly due to the P–S bond stretching vibration. Like the phosphonyl group (P=O), the position of the P=S band is affected by the electronegativity of adjacent groups although this effect is not so marked as for the phosphonyl group since the P=S group has less ionic character. The P=S band may consist of a doublet due to the presence of rotational isomerism. Normally, the band is difficult to identify since there are many other groups which have bands in the same region.

Compounds containing a phenyl–P bond have a band due to an aromatic ring vibration occurring at 1455–1425 cm^{-1} (6.90–7.02 μm) which is of medium-to-strong intensity. This band is useful since it occurs in a region normally free from absorptions by phosphorus compounds.

Compounds containing the P–Cl bond[20] have a medium-to-strong absorption at 605–435 cm^{-1} (16.53–22.99 μm) due to the P–Cl stretching vibration. The position of the band due to the P–X (X=F or Cl) stretching vibration[30,31] is affected by the oxidation state of the phosphorus atom. In the presence of more than one halogen atom directly attached to the phosphorus atom, two peaks are observed due to the asymmetric and symmetric stretching vibrations respectively. Difluorides of the type –P(O)F$_2$ absorb at 930–895 cm^{-1} (10.75–11.17 μm) and 890–870 cm^{-1} (11.24–11.49 μm).

Reviews have been published dealing with the infrared spectra of organic phosphorus compounds,[1–18] as has a correlation chart[9] for inorganic phosphorus compounds.

References

1. E. Steger, *Z. Chem.*, 1972, **12**, 52.
2. R. A. Nyquist and W. J. Potts, in *Analytical Chemistry of Phosphorus Compounds*, M. Halmann (ed.), Interscience, New York, 1972, pp. 189–293.
3. J. R. Ferraro, *Prog. Infrared Spectrosc.*, 1964, **2**, 127.
4. D. E. C. Corbridge, *Topics Phosphorus Chem.*, 1970, **6**, 235.
5. L. C. Thomas, *Interpretation of the Infrared Spectra of Organo-phosphorus Compounds*. Heyden, London, 1974.
6. D. E. C. Corbridge, *J. Appl. Chem*, 1956, **6**, 456.
7. L. C. Thomas and R. A. Chittenden, *Chem. Ind.*, **1961**, 1913.
8. R. A. Chittenden and L. C. Thomas, *Spectrochim. Acta*, 1965, **21**, 861.
9. D. E. C. Corbridge and E. J. Lowe, *J. Chem. Soc.*, **1954**, 493 and 4555.
10. L. C. Thomas and K. P. Clark, *Nature*, 1963, **198**, 855.
11. R. Wolf *et al.*, *Bull. Soc. Chim. France*, **1963**, 825.
12. U. Dietze, *J. Prakt. Chem.*, 1974, **316**, 293.
13. L. C. Thomas and R. A. Chittenden, *Spectrochim. Acta*, 1964, **20**, 467.
14. R. A. Nyquist and W. W. Muelder, *Spectrochim. Acta*, 1966, **22**, 1563.
15. E. M. Popov *et al.*, *Adv. Chem. Moscow*, 1961, **30**, 362.
16. R. A. Jones and A. R. Katritzky, *J. Chem. Soc.*, **1960**, 4376.
17. A. N. Lazarev and V. S. Akselrod, *Opt. Spectrosc.*, 1960, **9**, 170.
18. N. A. Slovochotova *et al.*, *Bull. Acad. Sci. URSS, Ser. Chim.*, **1961**, 62.
19. N. P. Greckin and R. R. Sagidullin, *Bull. Acad. Sci. URSS*, **1960**, 2135.
20. R. A. McIvor and C. E. Hubley, *Can. J. Chem.*, 1959, **37**, 869.
21. D. F. Peppard *et al.*, *J. Inorg. Nucl. Chem.*, 1961, **16**, 246.
22. J. R. Ferraro and C. M. Andrejasich, *J. Inorg. Nucl. Chem.*, 1964, **26**, 377.
23. T. Gramstad, *Spectrochim. Acta*, 1964, **20**, 729.
24. U. Blindheim and T. Gramstad, *Spectrochim. Acta*, 1965, **21**, 1073.
25. J. Rockett, *Appl. Spectrosc.*, 1962, **16**, 39.
26. D. M. L. Goodgame, *J. Chem. Soc.*, **1961**, 2298 and 3735.
27. R. A. Chittenden and L. C. Thomas, *Spectrochim. Acta*, 1964, **20**, 1679.
28. S. Husebye, *Acta Chem. Scand.*, 1965, **19**, 774.
29. R. A. Zingaro and R. M. Hedges, *J. Phys. Chem.*, 1961, **65**, 1132.
30. R. J. Schmutzler, *J. Inorg. Nucl. Chem.*, 1963, **25**, 335.
31. R. R. Holmes *et al.*, *Spectrochim. Acta*, 1973, **29A**, 665.
32. J. R. Ferraro *et al.*, *Spectrochim. Acta*, 1963, **19**, 811.
33. P. R. Mathis *et al.*, *Spectrochim. Acta*, 1974, **30A**, 357.
34. L. A. Strait and M. K. Hrenoff, *Spectrosc. Lett.*, 1975, **8**, 165.
35. M. I. Kabachnik, *Austral. J. Chem.*, 1975, **28**, 755.
36. O. A. Raevskii *et al.*, *Izv. Acad. Nauk. SSSR Ser. Khim.*, 1978, **3**, 614.
37. A. Rulmont *et al.*, *Eur. J. Solid State Inorg. Chem.*, 1991, **28**, 207.

18 Organic Silicon Compounds

Due to the mass and size of the silicon atom, the infrared spectra of organo-silicon compounds,[1,2] to a first approximation, consist of essentially independent group vibrations. In general, similar absorption bands to those of the corresponding carbon compounds are observed except that they are at lower frequencies and are usually more intense than their carbon analogues (due to the difference in electronegativity between carbon and silicon).

Si–H Vibrations

For organic silanes, a strong absorption band due to the Si–H stretching vibration[3-11] is found at $2250-2100 \, cm^{-1}$ ($4.44-4.76 \, \mu m$). In general, the frequency of this band tends to increase with increase in the electronegativity of the substituents on the silicon atom. It has been observed that as the number of hydrogen atoms directly bonded to the silicon atom decreases so does the frequency of the Si–H stretching vibration. Alkyl substituents on the silicon atom also tend to lower this frequency whereas aryl substituents tend to raise it.

The band due to the deformation vibration of the Si–H group occurs in the range $985-800 \, cm^{-1}$ ($10.15-12.50 \, \mu m$). The $-SiH_3$ group has two bands due to deformation vibrations in the region $945-910 \, cm^{-1}$ ($10.58-10.99 \, \mu m$), whereas deformation vibrations of the $>SiH_2$ group give rise to only one strong band in the region $950-930 \, cm^{-1}$ ($10.53-10.75 \, \mu m$) and $\equiv SiH$ has no strong band in this region. However, due to strong interactions between vibrational modes, it may be difficult to identify Si–H_n ($n=1-3$) deformation bands. For some molecules normal coordinate calculations show a high degree of mixing between modes.

Methyl–Silicon Compounds, Si–CH$_3$

Methyl groups attached to silicon atoms[7] have a characteristic, very sharp band at $1290-1240 \, cm^{-1}$ ($7.75-8.06 \, \mu m$) due to the symmetric deformation vibration of the CH_3 group. Electropositive groups or atoms (e.g. metals) directly bonded to the silicon atom make the band due to the symmetric CH_3 deformation vibration tend to the higher end of this range whereas for silanes and siloxanes the band occurs near the lower end. When there are three methyl groups attached to a silicon atom, the band due to the symmetric deformation often splits into two components of unequal intensity. The asymmetric deformation vibration of the CH_3 group results in a weak band near $1410 \, cm^{-1}$ ($7.09 \, \mu m$).

The frequencies of the stretching vibrations of the methyl group are not affected much by being bonded to a silicon atom rather than to a carbon atom. The bands due to the methyl rocking vibration and the Si–C stretching vibration occur in the region $890-740 \, cm^{-1}$ ($11.26-13.51 \, \mu m$).

Ethyl–Silicon Compounds

Ethyl-substituted silicon compounds[20] have a characteristic band of medium intensity at $1250-1220 \, cm^{-1}$ ($8.00-8.20 \, \mu m$) and two other useful bands at $1020-1000 \, cm^{-1}$ ($9.80-10.00 \, \mu m$) and $970-945 \, cm^{-1}$ ($10.31-10.58 \, \mu m$).

Alkyl–Silicon Compounds

The band due to the $SiCH_2R$ deformation vibration, which occurs at $1250-1175 \, cm^{-1}$ ($8.00-8.51 \, \mu m$), tends to decrease in intensity as the length of the aliphatic chain increases, the frequency of the vibration decreasing also. Obviously, as the chain length increases, the smaller becomes the influence of the silicon atom on the terminal C–H vibrations. Hence, the bands near $2950 \, cm^{-1}$ ($3.39 \, \mu m$), $1470 \, cm^{-1}$ ($6.80 \, \mu m$), and $1390 \, cm^{-1}$ ($7.19 \, \mu m$) increase in intensity with increase in the paraffin chain length.

Aryl–Silicon Compounds

The bands due to the aromatic ring vibration, which are normally found in the region $1600-1450 \, cm^{-1}$ ($6.25-6.90 \, \mu m$), are displaced to lower wavenumbers

Table 18.1   Organic silicon compounds

Functional Groups	Region cm^{-1}	μm	Intensity IR	Raman	Comments
Silanes					
Si–H	2250–2100	4.44–4.76	m–s	m–s, p	Si–H str, general range
	985–800	10.15–12.50	m–s	w	Si–H def vib, general range
RSiH$_3$	2155–2140	4.64–4.67	s	m–s, p	Si–H str
	945–930	10.52–10.75	m–s	w	Si–H asym def vib
	930–910	10.75–10.99	m–s	w	Si–H sym def vib
	680–540	14.71–18.52	s	w	rocking vib
R$_2$SiH$_2$	2140–2115	4.67–4.73	s	m–s	Si–H str
	950–930	10.53–10.75	m–s	w	Si–H def vib
	900–885	11.11–11.30	m–s	w	Si–H wagging vib
	745–560	13.42–17.56	m–s	w	twisting vib
	600–460	16.67–21.74	m–s	w	sh rocking vib
R$_3$SiH	2110–2090	4.74–4.78	s	m–s, p	Si–H str
	845–800	11.83–12.60	s	m	Si–H wagging vib
ArSiH$_3$	2160–2150	4.63–4.65	s	m–s	Si–H str
	945–930	10.52–10.75	m–s	w	Si–H asym def vib
	930–910	10.75–10.99	m–s	w	Si–H sym def vib
	680–540	14.71–18.52	s	w	Rocking vib
Ar$_2$SiH$_2$	2150–2130	4.65–4.69	s	m–s	Si–H str
	950–925	10.53–10.81	m–s	w	Si–H def vib
	870–840	11.49–11.91	m–s	w	Si–H wagging vib
	740–625	13.51–16.00	w	w	Twisting vib
	600–460	16.67–21.74	m	w	sh rocking vib
Ar$_3$SiH	2135–2110	4.68–4.74	s	m	Si–H str
	845–800	11.83–12.50	s	w	Si–H wagging vib
SiH$_3$C≡C–	2190–2170	4.57–4.61	s	m–s	Si–H str
Si–D	1690–1570	5.92–6.37	s	m–s	Si–D str
	710–665	14.08–15.04	s	w	Si–D def vib
R$_4$Si	1280–1240	7.81–8.07	s		sym Si–C bending vib
	850–800	11.76–12.50	s	w	Si–C rocking vib
	760–750	13.10–13.33	s	w	Si–C rocking vib
Si(CH$_3$)$_n$, n = 1, 3, or 4	1290–1240	7.75–8.06	s–m	m–w	sh, sym CH$_3$ def vib
	870–760	11.49–13.61	s–m	w–m	Si.CH$_3$ rocking vib
	~765	~13.07	s–m	s	Si–C str see ref. 5
\\—SiCH$_3$/					
\\Si(CH$_3$)$_2$/	860–845	11.63–11.83	v	s	Si–C str
	815–800	12.27–12.50	v	s	Si–C str
-Si(CH$_3$)$_3$	860–840	11.63–11.90	v	s	Si–C str
	770–750	12.99–13.33	v	s	Si–C str
	660–485	15.15–20.62	w	s	Si–C str
	330–240	30.30–41.67	w		Si(CH$_3$)$_3$ rocking vib
Si–C$_2$H$_5$	1250–1220	8.00–8.20	m	w–m	CH$_2$ wagging vib
	1020–1000	9.80–10.00	m		
	970–945	10.31–10.58	m		

Table 18.1   (*continued*)

Functional Groups	Region cm⁻¹	Region μm	Intensity IR	Intensity Raman	Comments
Si–CH₂R	1250–1175	8.00–8.51	w–m	w	Long-chain alphatics absorb at low frequency end of range
	760–670	13.10–14.93	m	w	CH₂ rocking vib
Si–CH=CH₂	~1925	~5.20	w	–	Overtone
	1615–1590	6.19–6.29	m	s	C=C str, see ref. 16
	1410–1390	7.09–7.19	s–m	w	CH₂ in-plane def vib
	1020–1000	9.80–10.00	m	w	*Trans* CH wagging vib
	980–940	10.20–10.64	s–m	w	CH₂ wagging vib
	580–515	17.24–19.42	w	w	Hydrogen out-of-plane def vib
Si–Ph	3080–3030	3.25–3.30	m	s	C–H str
	~1600	~6.25	m	s–m	Ring vib, usually stronger than band near ~1430 cm⁻¹
	1480–1425	6.99–7.02	m–s	m	sh, ring vib
	1125–1090	8.98–9.17	vs		X-sensitive band, Si–Ph str
	~730	~13.70	s–m	w	Ring vib out-of-plane CH
	700–690	14.29–14.49	s–m	w	Out-of-plane C–H vib
R₃SiPh	670–625	14.93–16.00	w		Ring in-plane bending vib
	490–445	20.41–22.47	s		Si–C–C out-of-plane bending vib
	405–345	24.69–28.99	w		Si–C str and ring in-plane def vib
	~290	~34.48	v		Si–Ph in-plane def vib
R₂SiPh₂	635–605	15.75–16.53	w		Ring in-plane bending
	495–470	20.20–21.28	s		S–C–C out-of-plane bending vib
	445–500	22.47–25.00	w		asym Si–C str
	380–305	26.32–32.79	w		sym Si–C str
RSiPh₃	625–605	16.00–16.53	w		Ring in-plane bending vib
	515–485	19.42–20.62	s		Si–C–C out-of-plane bending vib
	445–420	22.47–23.81	w		asym Si–Ph str
	~330	~30.30	v		sym Si–Ph str
Si–CH₂–Si	1180–1040	9.26–9.62	s		
Cyclopentamethylene dialkylsilanes	495–480	20.20–20.83	m–s		Probably due to heterocyclic ring, but cyclopentamethylene silane and diphenyl derivatives do not exhibit this band
Silanols					
Si–OH	3700–3200	2.70–3.13	m	w	May be br, O–H str
	1040–1020	9.62–9.80	m–w		Si–OH def vib
	955–830	10.47–12.05	s		Si–O str, for condensed-phase samples a br, m–w band occurs near 1030 cm⁻¹ due to SiOH def vib
Silyl esters and ethers					
RCOSiR₃	~1620	~6.17	s	w–m	C=O str
Si–O–R	1110–1000	9.01–10.00	vs	w	asym Si–O–C str, at least one band, Si–O–Si also absorbs in this region
	990–945	10.10–10.58	s	w	SiOC str
Si–O–CH₃	~2860	~3.50	m	m–s	sym CH₃ str
	~1190	~8.40	s	w	CH₃ rocking vib
	~1100	~9.00	vs	w	asym Si–O–C str
	810–800	11.76–12.50	s–m	w	sym Si–O–C str
⟩Si(OCH₃)₂	390–360	25.64–27.78	s	s–m	asym Si–O–C def vib

(continued overleaf)

Table 18.1 *(continued)*

Functional Groups	Region		Intensity		Comments
	cm^{-1}	μm	IR	Raman	
$-Si(OCH_3)_3$	480–440	20.83–22.73	s		asym Si–O–C def vib
	470–330	21.28–30.30	w		
Si–O–CH$_2$–	1190–1140	8.40–8.77	s		
	1100–1070	9.09–9.35	vs	w	asym Si–O–C str, usually a doublet
	990–945	10.10–10.64	s–m	s–m	sym Si–O–C str
$\diagdown Si(OC_2H_5)_2$ $\diagup$	475–405	21.05–24.69	w		asym Si–O–C def vib
$-Si(OC_2H_5)_3$	500–440	20.00–22.73	s		
Si–O–Ar	1135–1090	8.81–9.17	vs	w	Several sh bands, probably Si–O–C str
	970–920	10.31–10.87	s	s–m	Si–O str
Si–O–Si	1090–1010	9.17–9.90	vs	w	Si–O str, two bands of almost equal intensity, siloxane chains absorb near 1085 cm^{-1} and 1020 cm^{-1} increasing in intensity with increased chain length, cyclic siloxanes have only one strong band although a second band is sometimes observed for tetramers and larger rings
Cyclic trimers, (SiO)$_3$	1020–1010	9.80–9.90	vs	w	Si–O str
Siloxanes $-(SiO)_n-$	1100–1000	9.09–10.00	s	w	
	~800	~12.50	m	s	Si–C str
Disiloxanes					
Si–O–Si	625–480	16.00–20.83	w	vs	br, sym Si–O–Si str band occurs at lower frequencies for substituted disiloxanes and linear polymeric siloxanes
Siloxanes					
$-OSiCH_3$ (end group)	850–840	11.76–11.90	s	s	Si–C str
$-OSiC_2H_5$ (linear polymer)	810–800	12.35–12.50	s	s	Si–C str
$-OSiCH_3$ (cyclic compounds)	820–780	12.20–12.82	s	s	
Silyl amines					
Si–NH$_2$	3570–3475	2.80–2.88	m	m–w	NH$_2$ str
	3410–3390	2.93–2.95	m	m	NH$_2$ str
	1550–1530	6.45–6.54	m	w	NH$_2$ def vib
Si–NH–Si	~3400	~2.94	m	m–w	NH str
	~1175	~8.51	m–s		
	950–910	10.53–10.99	m–s		
Aminosilanes, $H_2N\!-\!\underset{\vert}{\overset{\vert}{Si}}\!-\!NH_2$	880–835	11.36–11.98	s		asym N–Si–N str
	800–785	12.50–12.74	m		sym N–Si–N str
Silicon halides					
$\diagdown \overline{}SiF$ $\diagup$	920–820	10.87–12.22	m–s	m–w	Si–F str (general ranges: Si–F str, 1000–800 cm^{-1}; Si–F def vib, 425–265 cm^{-1})
$\diagdown SiF_2$ $\diagup$	945–915	10.58–10.93	m–s	m–w	asym str
	910–870	10.99–11.49	m	m–w	sym str

Table 18.1 (*continued*)

Functional Groups	Region		Intensity		Comments
	cm^{-1}	μm	IR	Raman	
$-SiF_3$	980–945	10.20–10.58	s	m–w	asym str
	910–860	10.99–11.63	m	m–w	sym str
$>SiCl$	550–470	18.18–21.28	s	s	Si–Cl str (Si–Cl def vib, 250–150 cm^{-1} – general range)
$>SiCl_2$	600–535	16.67–18.69	s	s	asym Si–Cl str
	540–460	18.52–21.74	m	s	sym Si–Cl str
$-SiCl_3$	625–570	16.00–17.54	s	s	asym Si–Cl str
	535–450	18.69–22.22	m	s	sym Si–Cl str
$>SiBr$	430–360	23.26–27.78	w	s	Si–Br str
$>Si-Br_2$	460–425	21.74–23.53	w	s	asym Si–Br str
	395–330	25.32–30.30	w	s	sym Si–Br str
$-SiBr_3$	480–450	20.83–22.22	w	s	asym Si–Br str
	360–300	27.78–33.33	w	s	sym Si–Br str
$>SiI$	365–280	27.40–35.71	w	s	sym Si–I str
$>SiI_2$	390–330	25.64–30.30	w	s	Si–I str
	325–275	30.77–26.36	w		
$-SiI_3$	410–365	24.39–27.40	w	s	Si–I str
	280–220	35.71–45.45	w		
$>Si-Si<$	~425	~23.53		s	Si–Si str
Other groups					
Si–Ph	1125–1090	8.89–9.17	s		see ref. 15
Ge–Ph	~1080	~9.26	s		see ref. 15
Sn–Ph	1080–1050	9.26–9.52	s		usually 1065 cm^{-1}, see ref. 15
Pb–Ph	~1050	~9.52	s		see ref. 15
Ge–H	2160–1990	4.63–5.03	m		Ge–H str
Sn–H	1910–1790	5.24–5.59	m		Sn–H str
Al–H	1910–1675	5.24–5.97	m		Al–H str
Organogermanium, Ge–O–Ge	900–700	11.11–14.29	s		asym Ge–O–Ge str; cyclic trimers ~850 cm^{-1}, cyclic tetramers ~860 cm^{-1}, linear polymers 870 cm^{-1}
Organotin, Sn–O	780–580	12.82–17.74	s		br
Organolead, Pb–O	~625	~16.00			

by about 20 cm^{-1} for phenyl–silicon compounds.[12] One of these bands, which is sharp and of medium intensity, is almost always found at 1430 cm^{-1} (6.99 μm). The band is broadened or altogether absent when the ring is substituted by an additional group.

Phenyl–silicon compounds have a strong, characteristic band at about 1100 cm^{-1} (9.09 μm) which often splits into two when two phenyl groups are attached to the one silicon atom, but appears as a single band in the case of three phenyl groups. In addition, phenyl–silicon compounds have two weak bands, one near 1030 cm^{-1} (9.71 μm) and the other near 1000 cm^{-1} (10.00 μm).

The band pattern normally observed in the overtone region 2000–1660 cm^{-1} (5.00–6.02 μm) cannot be relied upon for the determination of the substitution pattern, although it is satisfactory for a large number of aryl silanes.

Si–O Vibrations[13,17–19]

The band due to the asymmetric Si–O–Si stretching vibration is normally in the region 1100–1000 cm^{-1} (9.09–10.00 μm) and, due to the greater ionic character of the Si–O group, this band is much more intense than the corresponding C–O band for ether. The band pattern may be used to distinguish between cyclic and linear polysiloxanes.[13] Long chain siloxanes have two broad bands in the region 1100–1000 cm^{-1} (9.09–10.00 μm).

Due to the influence of ring strain, cyclic siloxane trimers absorb at 1020–1010 cm^{-1} (9.80–9.90 μm), which is about 60 cm^{-1} less than other cyclic siloxanes, whereas tetramers (which have less ring strain) absorb at 1090–1070 cm^{-1} (9.17–9.35 μm) along with higher cyclic siloxanes. It is difficult to distinguish between other cyclic siloxanes and the region of absorption overlaps, in fact, that of linear polysiloxanes.

Linear small-chain siloxanes tend to absorb at about 1050 cm^{-1} (9.52 μm) and with increase in molecular weight this band gradually broadens to occupy the region 1100–1000 cm^{-1} (9.09–10.00 μm). For long-chain polymers, a broad, strong band with maxima at about 1085 cm^{-1} (9.21 μm) and 1025 cm^{-1} (9.76 μm) is observed.[13]

Silicon–Nitrogen Compounds

The band due to the asymmetric Si–N–Si stretching vibration occurs at about 900 cm^{-1} (11.11 μm) and is of strong intensity, whereas, due to the influence of ring strain, cyclic disilazanes absorb at about 870 cm^{-1} (11.49 μm), cyclic trimers absorb at about 920 cm^{-1} (10.87 μm), and cyclic tetramers at about 940 cm^{-1} (10.64 μm). This behaviour is similar to that observed for siloxanes. A band which has been assigned to the N–H deformation vibration occurs at about 1150 cm^{-1} (8.70 μm) for cyclic trisilazanes and at about 1180 cm^{-1} (8.48 μm) for cyclic tetrasilazanes.

As might be expected, primary silyl amines[14] have two weak bands in the region 3580–3380 cm^{-1} (2.79–2.96 μm) due to the asymmetric and symmetric N–H stretching vibrations. Secondary silyl amine compounds have only one weak band, at about 3400 cm^{-1} (2.94 μm). Primary silyl amines also have a medium-to-strong intensity band at about 1530 cm^{-1} (6.54 μm) and linear secondary silyl amines have a medium-to-strong band at about 1175 cm^{-1} (8.51 μm) due to the N–H deformation vibration. This band is found about 30–40 cm^{-1} lower for cyclic silyl amines where the nitrogen atom forms part of the ring than for linear secondary silyl amines.

Silicon–Halide Compounds

Chloro-, bromo-, and iodosilanes, in the presence of moisture, hydrolyse to form siloxanes and hydrogen halide, so that care must be exercised in handling samples.

Characteristic silicon–chloride stretching vibration bands are observed in the far infrared region below 600 cm^{-1} (above 16.67 μm).

Silicon compounds with more than one chlorine atom exhibit two bands due to the asymmetric and symmetric vibrations. The asymmetric band, which of course occurs at higher frequencies than the symmetric case, is generally the more intense of the two.

Hydroxyl–Silicon Compounds

The band due to the O–H stretching vibration occurs in the same region as that for alcohols, phenols, etc. However, the band due to the O–H deformation vibration occurs at 870–820 cm^{-1} (11.49–12.19 μm) when the hydroxyl group is bonded to a silicon atom, whereas it is near 1050 cm^{-1} (9.52 μm) when bonded to a carbon atom.

References

1. K. Licht and P. Reich, *Literature Data for Infrared, Raman and N.M.R. Spectra of Silicon, Germanium, Tin and Lead Organic Compounds*, Dent. Verlag. Wiss., Berlin, 1971.
2. A. L. Smith, *Spectrochim. Acta*, 1960, **16**, 87.
3. G. J. Janz and Y. Mikawa, *Bull. Chem. Soc. Jpn*, 1961, **34**, 1495.
4. A. L. Smith and J. A. McHard, *Anal. Chem.*, 1959, **31**, 1174.
5. H. G. Kuivita and P. L. Maxfield, *J. Organonmet. Chem.*, 1967, **10**, 41.
6. G. Kessler and H. Kriegsmann, *Z. Anorg. Allgem. Chem.*, 1966, **342**, 53.
7. I. F. Kovalev, *Opt. Spectrosc.*, 1960, **8**, 166.
8. S. D. Gokhale and W. L. Jolly, *Inorg. Chem.*, 1964, **3**, 946.
9. E. A. Groschwitz *et al.*, *J. Organomet. Chem.*, 1967, **9**, 421.
10. A. L. Smith, *Spectrochim. Acta*, 1963, **19**, 849.
11. R. N. Kinseley *et al.*, *Spectrochim. Acta*, 1959, **15**, 651.
12. M. C. Harvey and W. H. Nebergall, *Appl. Spectrosc.*, 1962, **16**, 12.
13. W. Noll, *Angew. Chem. Int. Ed. Engl.*, 1963, **2**, 73.
14. W. Fink, *Helv. Chim. Acta*, 1962, **45**, 1081.
15. F. J. Bajer and H. W. Post, *J. Org. Chem.*, 1962, **27**, 1422.
16. V. F. Mironov and N. A. Chumaevskii, *Dok. Acad. Nauk SSSR*, 1962, **146**, 1117.
17. P. Tarte *et al.*, *Spectrochim. Acta*, 1973, **29A**, 1017.
18. J. Chiosnet *et al.*, *Spectrochim. Acta*, 1975, **31A**, 1023.
19. E. D. Lipp and A. L. Smith, in *The Analytical Chemistry of Silicones* (A. L. Smith, ed.), Wiley, New York, 1991.
20. G. A. Giurgis *et al.*, *J. Mol. Struct.*, 1999, **510(1–3)**, 13.

19 Boron Compounds

Boron compounds generally have intense bands, as, for example, those due to the B–H,[1-2] B–halogen,[3-8] B–O, and B–N[9,10] groups. The position and intensity of certain bands give information not only on the boron-containing group itself but frequently also on its environment. The bands due to certain boron-containing groups often appear as doublets, this being due to the presence of two naturally-abundant isotopes of boron.

Bands due to the B–H stretching vibrations[1,2] occur at $2640-2350\,cm^{-1}$ ($3.79-4.26\,\mu m$) for the groups BH and BH_2 in which the hydrogen atom is free. By free, it is meant that the hydrogen atom is a terminal, or exo, atom.

No isotope band-splitting is observed for compounds containing a single, free (exo) B–H group whereas it does occur for free (exo) BH_2 groups (in gas-phase spectra).

The B–H stretching vibration of samples enriched in ^{10}B is at slightly higher frequencies than for samples with the naturally-occurring ratio of boron isotopes.

In some cases, the band due to the B–H stretching vibration of borane–amine complexes exhibits isotope-splitting. For alkyl boranes, the band tends to lower frequencies with increasing substitution.

In many boron compounds, two boron atoms are bridged by a hydrogen atom. In compounds bridged by two hydrogens, four B–H stretching vibration modes are possible:

| Symmetric | Symmetric | Asymmetric | Asymmetric |
| in-phase | out-of-phase | in-phase | out-of-phase |

Compounds with the $B\cdots H\cdots B$ bridge have a series of weak-to-medium intensity bands in the region $2140-1710\,cm^{-1}$ ($4.67-5.85\,\mu m$) and a strong

Table 19.1 Boron compounds

| Functional Groups | Region | | Intensity | | Comments |
	cm^{-1}	μm	IR	Raman	
B–H (free hydrogens)	2565–2480	3.90–4.03	m–s	m–w	B–H str
	1180–1110	8.48–9.01	s		B–H in-plane def vib
	920–900	10.87–11.11	m–w		Out-of-plane bending vib
Alkyl diboranes (free hydrogens)	2640–2570	3.79–3.89	m–s	m–w	sym BH_2 str
	2535–2485	3.95–4.02	m–s	m–w	asym BH_2 str
	1205–1140	8.30–8.77	m–s		sometimes br, BH_2 def vib
	975–920	10.26–10.87	m		BH_2 wagging vib
Alkyl diboranes	2140–2080	4.67–4.81	w–m		sym in-phase motion of H atom
$B\cdots H\cdots B$ (bridged hydrogen)	1990–1850	5.03–5.41	w		sym out-of-plane motion of H atom, several bands
	1800–1710	5.56–5.85	w–m		asym out-of-phase motion of H atom
	1610–1525	6.21–6.56	vs		asym in-phase motion of H atom
Borazines, borazoles	3500–3400	2.86–2.94	m	m	N–H str, see refs 11–14

(continued overleaf)

Table 19.1 (*continued*)

Functional Groups	Region cm⁻¹	Region μm	Intensity IR	Intensity Raman	Comments
	2580–2450	3.88–4.08	m	m–w	B–H str
	1465–1330	6.83–7.52	s	m	B–N str
	700–680	14.29–14.71	m		B–N out-of-plane def vib, doublet
Boron hydride salts and amine–borane complexes (with boron octet complete)	2400–2200	4.17–4.55	m	m–w	B–H str
Borane BH_3 (in complexes)	2380–2315	4.20–4.32	s	m–w	asym B–H str
	2285–2265	4.38–4.42	s	m–w	sym B–H str
	~1165	~8.58	s		BH_3 def vib
BH_4^- ion	2400–2195	4.17–4.56	s	m–w	B–H str, two bands (one due to Fermi resonance)
	1150–1000	11.70–10.00	s		BH_2 def vib
(structure: M–B bridged with H atoms)	2600–2400	3.85–4.17	s		Doublet split 80–40 cm⁻¹
	2150–1950	4.65–5.13	s		May have shoulder
	1500–1300	6.67–7.69	s		br
(structure: M–H–B bridged with H atoms)	2600–2450	3.85–4.08	s		
	2200–2100	4.55–4.76	s		Doublet split 80–50 cm⁻¹
	1250–1150	8.00–8.70			
B–OH, boric acid, boronic acids (solid phase)	3300–3200	3.03–3.13	m	w	br, O-H str
B–OH, aryl boronic acids	~1000	~10.00	m		br ⎱Not present in anhydrides
	800–700	12.50–14.29	m		br ⎰
1,1-Dialkyl diboranes and trialkyl	1185–1100	8.44–9.09	s		asym C–B–C str (isotope splitting large, ~10–30 cm⁻¹)
	845–770	11.83–12.98	m		sym C–B–C str
$B–CH_3$	1460–1405	6.85–7.12	m	m	asym CH_3 def
	1330–1280	7.52–7.81	m	m–w	sym CH_3 def vib
B–R	1270–620	7.87–16.13	v		B–C str, isotopic splitting sometimes observed. For BR_3 compounds, one strong band due to asym C–B–C str and one weak band (sometimes absent) due to sym C–B–C str
Monomethyl boranes	1010–835	9.90–11.98	m		B–C str (isotopic splitting observed)
Di- and trimethyl boranes	1240–1140	8.07–8.77	vs		asym C–B–C str
	720–675	12.20–14.18	m–w		sym C–B–C str, infrared-inactive for symmetrical compounds
Alkyl boranes (other than methyl)	1135–1110	8.81–9.01	m		asym C–B–C str (isotopic splitting of 20 cm⁻¹)
	675–620	14.82–16.22	w		sym C–B–C str often absent
B–Ar	1440–1430	6.94–6.99	m–s		sh, ring vib
	1280–1250	7.81–8.00	s		X-sensitive band

Table 19.1 (*continued*)

Functional Groups	Region		Intensity		Comments
	cm^{-1}	μm	IR	Raman	
	~760	~13.16	s		Ring C–H out-of-plane def vib, for phenyl compounds only, doublet if more than one phenyl group on boron atom (~$20\,cm^{-1}$ separation)
bis-(Alkyl amino) phenyl boron compounds 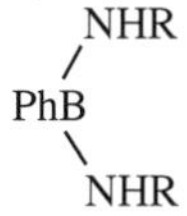	1450–1440	6.90–6.94	s		B–C str
NHR					
/					
PhB					
\					
NHR					
bis-Phenyl boron compounds, Ph$_2$B–	1260–1250	7.94–8.00	s		B–C str
Aryl boron dihalides, ArBX$_2$ (X═halide)	1270–1215	7.87–8.23	s		B–C str, isotopic splitting present
B–O, borates, boronates, boronites, boronic anhydrides, boronic acids, borinic acids	1380–1310	7.25–7.63	s		B–O str, weak band when boron octet complete e.g. compounds with a nitrogen coordinate to the boron
Trialkyl borates, B(RO)$_3$	1350–1310	7.41–7.63	vs		br, also have strong band at 1070–$1040\,cm^{-1}$ probably due to C–O str
Dialkyl phenyl boronates, (RO)$_2$BPh	1435–1425	6.97–7.02	m–s		B–C str
	1330–1310	7.52–7.63	s		asym C–O–B–O–C str
	1180–1120	8.48–8.93	s		sym C–O–B–O–C str
	675–600	14.81–16.67	m–s		B–O def vib, isotopic splitting present
Boronites, R$_2$BOR	1350–1310	7.41–7.63	s		B–O str
B-alkoxyl borazoles, 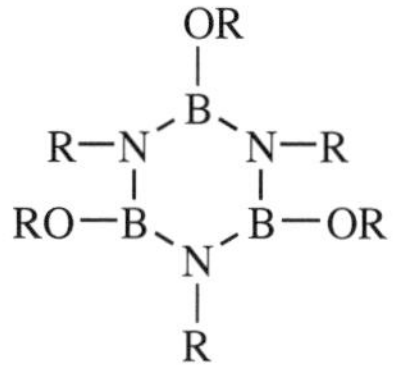	1500–1435	6.67–6.97	s		B–N str

OR
|
B
R—N   N—R
| |
RO—B   B—OR
N
|
R

| | 1330–1310 | 7.52–7.63 | m–s | | B–O str |
| Alkyl and aryl metaborates, 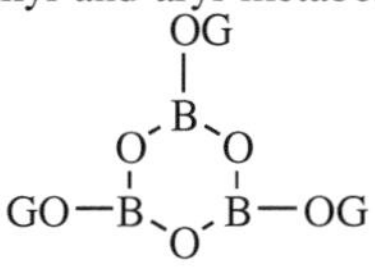 | 1380–1335 | 7.25–7.49 | s | | B–O str |

OG
|
B
O   O
| |
GO—B   B—OG
O

| | 1225–1080 | 8.16–9.26 | s | w | C–O str, higher frequencies for aryl compounds, lower for *n*-alkyl compounds |
| 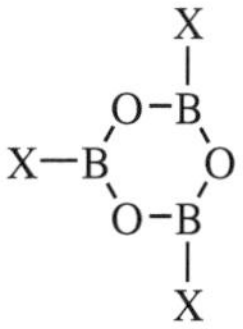 Haloboroxines, | 1470–1180 | 6.80–8.48 | s | | B–O str, isotopic splitting present |

X
|
O—B
X—B   O
O—B
|
X   (X = halogen)

| Fluoroboroxine | ~970 | ~10.31 | m | | asym B–F str (sym B–F str infrared-inactive) |

(*continued overleaf*)

Table 19.1   (*continued*)

Functional Groups	Region cm⁻¹	Region μm	Intensity IR	Intensity Raman	Comments
Chloroboroxine	~760	~13.16	s		asym B–Cl str, isotopic splitting present, see ref. 16
Aryl boronic acid esters,	1360–1330	7.35–7.52	s		asym B–O str, isotopic splitting present
	1240–1235	8.07–8.10	s	w	asym C–O str
	1075–1065	9.30–9.39	s		sym B–O str, isotopic splitting present
	1030–1020	9.71–9.80	s	s–m	sym C–O str
Boronic acid anhydrides,	1390–1355	7.19–7.38	s		B–O str
	1255–1145	7.97–8.74	m		B–C str, isotopic splitting present
Metallic orthoborates, $M_x(BO_3)_7$	1280–1200	7.82–8.34	s		br, asym B–O str, isotopic splitting present
	~900	~11.11	w		sym B–O str, often absent, see ref. 15
	580–550	17.24–18.18	m–w		
$BX_3 (X = F)$, (complexes of acids, esters, and ethers)	725–610	13.79–16.39	s		br, B⋯O str, isotopic splitting present
Covalent boron-nitrogen compounds	1550–1330	6.45–7.52	s		B–N str (general range), isotopic splitting present
Amine–borane complexes	780–680	12.82–14.71	m–s		B⋯N str (general range), isotopic splitting present see ref. 18
N-Alkyl B-halo borazoles, (X = halogen)	1510–1400	6.62–7.14	s		B–N str
	720–635	13.89–15.75			B–N def vib
N-Alkyl B-chloro borazoles	1090–960	9.17–10.42	s		B–Cl str
N-Alkyl B-bromo borazoles	1075–950	9.30–10.53	s		B–Br str
N-Alkyl amino borazoles,	1520–1490	6.58–6.71	s		B–N str, see refs 12, 14

Table 19.1 (*continued*)

Functional Groups	Region		Intensity		Comments
	cm^{-1}	μm	IR	Raman	
N-methyl B-aryl borazoles, Ar CH$_3$ 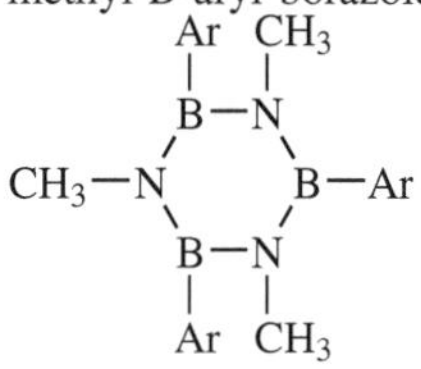	1470–1440	6.80–6.94	s		B–N str, isotopic splitting – ^{10}B shoulder present see ref 13
	750–720	13.33–13.89			B–N def vib
N-Alkyl B-aryl borazoles	1430–1410	6.99–7.09	s		B–N str, isotopic splitting – ^{10}B shoulder present see ref. 13
	750–720	13.33–13.89			B–N def vib
N-Aryl B-methyl borazoles	1400–1375	7.14–7.27	s		B–N str, isotopic splitting – ^{10}B shoulder present see ref. 13
Alkyl borazenes, (CH$_3$)$_2$B–NR$_1$R$_2$	1550–1330	6.45–7.52	s		B–N str
bis-Dimethylamino boranes, –B[N(CH$_3$)$_2$]$_2$	1550–1500	6.45–6.67	s		asym B–N str, isotopic splitting present
	1415–1375	7.07–7.32	s		sym B–N str, isotopic splitting present
Boron–fluorine compounds	1500–840	6.67–11.90	v		B–F str (general range), usually strong, isotopic splitting present
Boron difluorides, XBF$_2$ (in boron trihalides)	1500–1410	6.67–7.09	s		asym B–F str
	1300–1200	7.69–8.33	s		sym B–F str (for BF$_3$ this vib is infrared-inactive (Raman ~885 cm^{-1}))
Boron monofluorides, X$_2$BF (in boron trihalides)	1360–1300	7.35–7.69	s		B–F str, see ref. 3
BF$_3$ complexes	1260–1125	7.94–8.89	s		asym B–F str ⎱ Isotopic splitting present, band may
	1030–800	9.71–12.50	s		sym B-F str ⎰ be split further, see refs 5, 6, 8, 9
Tetrafluoroborate ion, BF$_4^-$	1160–760	8.62–13.56	vs		asym B–F str, shoulder ~1060 cm^{-1} (sym B–F str infrared-inactive), see ref. 5
Chlorotrifluoroborate ion, ClBF$_3^-$	1080–1025	9.26–9.76	s		asym B–F str, doublet
	890–840	11.24–11.90	w		sym B–F str, doublet
Boron–chlorine compounds	1090–290	9.17–34.48	v		B–Cl str (general range), isotopic splitting present, higher frequency end of range for trichloroborazoles, lower end for BCl$_3$ complexes
Boron dihalides (in boron trihalides)	1030–950	9.71–10.53	s		asym B–Cl str, isotopic splitting present (for BCl$_3$, band at ~955 cm^{-1})
	920–470	10.87–21.28	s		sym B–Cl str, isotopic splitting present (vib infrared-inactive for BCl$_3$)
Boron monochlorides (in boron trihalides)	955–690	10.47–14.49	s		B–Cl str
Alkyl aryl chloroboronites	1220–1195	8.20–8.36	s		Probably asym C–B–C str
	910–890	10.99–11.24	s		B–Cl str
BCl$_3$ in complexes	785–660	12.74–15.15	s		asym B–Cl str, isotopic splitting present, see ref. 8

(*continued overleaf*)

Table 19.1 (*continued*)

Functional Groups	Region		Intensity		Comments
	cm^{-1}	μm	IR	Raman	
	540–290	18.52–34.48	s		sym B–Cl str
	290–200	34.48–50.00	m–s		B–Cl def vib, several bands
Tetrachloroborate ion, BCl$_4^-$	760–645	13.16–15.50	s		br, asym B–Cl str, several peaks, (sym B–Cl str vib infrared-inactive)
Aryl boron dichlorides	970–915	10.31–10.93	s		asym B–Cl$_2$ str, isotopic splitting present
	645–630	15.50–15.87	s		sym BCl$_2$ str
	585–550	17.09–18.18	w		BCl$_2$ out-of-plane def vib
	~340	~29.41	s		BCl$_2$ rocking vib
	~230	~43.48	s		BCl$_2$ scissoring vib
	~130	~76.92	s		BCl$_2$ torsional vib
Boron–bromine compounds	1080–240	9.26–41.67	v		B–Br str (general range), isotopic splitting often present, higher frequency end of range for bromoboronazoles, lower end for BBr$_3$ comlexes
Boron dibromides (in boron trihalides)	910–820	10.99–12.20	s		asym B–Br str, (BBr$_3$ band at ~820 cm^{-1} with shoulder at 855 cm^{-1} due to isotopic splitting)
	420–275	23.81–36.36	s		sym BBr str, (infrared- inactive for BBr$_3$ (Raman ~280 cm^{-1}))
BBr$_3$ in amine complexes	~700	~14.29	s		asym B–Br str, isotopic splitting sometimes present, see ref. 8
	~250	~40.00	s		sym B–Br str, isotopic splitting sometimes present
	~200	~50.00	s		asym B–Br def vib
	~175	~57.14	s		
	~125	~80.00	vs		B–Br rocking vib
Tetrabromoborate ion, BBr$_4^-$	~600	~16.67	s		asym B–Br str, isotopic splitting present
	~240	~41.67	w		sym B–Br str
	~165	~60.61	m		B–Br def vib
Aryl boron dibromides	890–865	11.24–11.56	s		asym BBr$_2$ isotopic splitting present
	~620	~16.13	s		sym BBr$_2$ str
	~525	~19.05			out-of-plane BBr$_2$ vib
	~270	~37.04	s		BBr$_2$ rocking vib
	~160	~62.50	m		BBr$_2$ scissoring vib
Thio-orthoborate esters (symmetrical), —S—B—S— $\mid$ S	955–905	10.47–11.05	s		asym B-S str, several peaks due to isotopic splitting (sym B-S infrared-inactive)

band at 1610–1525 cm^{-1} (6.21–6.56 μm). The band due to the B–H stretching vibration of compounds for which the boron atom has a complete octet of electrons occurs in the range 2400–2200 cm^{-1} (4.17–4.55 μm).

The asymmetric and symmetric methyl deformation vibrations of B–CH$_3$ occur at 1460–1405 cm^{-1} (6.85–7.12 μm) and 1330–1280 cm^{-1} (7.52–7.81 μm) respectively.

Compounds with the B–aryl group have a strong, sharp band, due to the ring vibration, at 1440–1430 cm^{-1} (6.94–6.99 μm). Compounds with a B–phenyl group have a strong band at about 760 cm^{-1} (13.16 μm) due to the ring CH wagging motion,

A review of the infrared spectra of inorganic boron compounds has been published.[17]

Boron trifluoride absorbs in the following regions 1500–1445 cm^{-1} (6.67–6.92 μm) (two bands being observed due to the two isotopes ^{10}B and ^{11}B), ~890 cm^{-1} (~11.24 μm), 720–690 cm^{-1} (13.89–14.49 μm) (two bands due to the two isotopes) and at about 480 cm^{-1} (20.83 μm).

References

1. W. Gerrard, *Organic Boron Compounds*, Academic Press, New York, 1961, p. 223.
2. W. J. Lenhmann and I. Shapiro, *Spectrochim. Acta*, 1961, **17**, 396.
3. L. P. Lindemann and M. R. Wilson, *J. Chem. Phys.*, 1956, **24**, 242.
4. R. L. Amster and R. C. Taylor, *Spectro. chim. Acta*, 1964, **20**, 1487.
5. T. C. Waddington and F. Klanberg, *J. Chem. Soc.*, **1960**, 2339.
6. M. Taillander *et al.*, *J. Mol. Struct.*, 1968, **2**, 437.
7. W. Kynaston *et al.*, *J. Chem. Soc.*, **1960**, 1772.
8. P. G. Davies *et al.*, *Inorg. Nucl. Chem. Lett.*, 1967, **3**, 249.
9. A. Meller, *Organometall. Chem. Rev.*, 1967, **2**, 1.
10. W. Sawodny and J. Goubeau, *Z. Phys. Chem. N.F.*, 1965, **44**, 227.
11. J. M. Butcher *et al.*, *Spectrochim. Acta*, 1962, **18**, 1487.
12. D. W. Aubrey *et al.*, *J. Chem. Soc.*, **1961**, 1931.
13. J. E. Burch *et al.*, *Spectrochim. Acta*, 1963, **19**, 889.
14. W. Gerrard *et al.*, *Spectrochim. Acta*, 1962, **18**, 149.
15. A. Mitchell, *Trans. Faraday Soc.*, 1968, **62**, 530.
16. B. Latimer and J. R. Devlin, *Spectrochim. Acta*, 1965, **21**, 1437.
17. H. J. Becher and F. Thevenot, *Z. Anorg. Allg. Chem.*, 1974, **410**, 274.
18. M. T. Forel *et al.*, *Colloq. Int. Cent. Rech. Sci.*, 1970, **191**, 167.

20 The Near Infrared Region

The near infrared region, $14\,000-4000\,cm^{-1}$ $(0.7-2.5\,\mu m)$, is more akin to the ultraviolet and visible regions than the normal infrared region and hence longer path-length cells are employed. This means that the cells are easy to clean and more robust. Also, being made of glass with quartz windows, or of silica, they are not attacked by water.

The most useful solvents are those not containing hydrogen. For example, carbon tetrachloride has no strong absorptions in this region. Carbon disulphide can also be used, as it, too, is transparent in the near infrared region (see Chart 1.3).

In general, bands in the near infrared region are due to the overtones or combinations of fundamental bands occurring in the region $3500-1600\,cm^{-1}$ $(2.8-6.2\,\mu m)$. Therefore, qualitatively, this spectral region is not as characteristic as the 'fingerprint' region.[1-7] Although a fair amount of investigation still needs to be done in this region, it is obvious that the straightforward compilation of spectra will not, in the main, yield the type of qualitative information to which we are accustomed in the normal infrared range. Hence, the near infrared region is used primarily for quantitative measurements, such as those normally required for product quality assurance.

Since intensity measurements are reliable and relatively easy to make, both band position and accurate values of intensity are usually quoted in the literature. Often, in the near infrared region, relatively broad, overlapping bands are observed for samples. Since absorptions in this region originate from combinations or overtones of fundamental bands in the mid-infrared range or from electronic transitions in heavy atoms, the pathlength of the sample must be increased in order to examine successfully the higher frequency part of the range.

The near infrared region has been found extremely useful in the assignment of particular groups containing hydrogen.[4,13-15]

Bands due to CH, NH and OH are responsible for the majority of the absorption features observed in the near infrared. Much of the basic work in this field has been directed towards quantitative measurements involving water, alcohols, amines and any substance containing the CH, NH and OH groups.

Often, little sample preparation is required for simple solids, liquids or gases. Simultaneous multi-component analysis is usually possible, although many special, specific devices have been developed for the examination of foods and many of these can operate completely automatically. Quantitative analysis is usually fast, although it may involve the use of statistical techniques such as multiple linear regression analysis, principal component analysis, discriminant analysis, partial least squares and principal regression analysis. Multivariate regression analysis is often applied to the derivatives (first, second,...) of spectra (in other words, derivatisation is often applied as a pretreatment to the regression analysis). An analysis for a single component may involve a number of absorption positions so that corrections can be applied. The automatic correction background and interference can be performed by modern instruments using computer algorithms. Obviously, quantitative analysis cannot be carried out where corrections for very strong solvent absorptions have to be made. The detection limits are dependent on the particular band or bands used for the analysis, the nature of the sample and its environment, etc., but, in general, detection limits can be made low. For biological and medical applications, near infrared techniques can be non-invasive and non-destructive. For such applications, the use of microscope techniques and fibre optics is increasing.

Various publications[8,18,21] review near infrared spectroscopy in many fields. An atlas of near infrared spectra has been published by Sadtler.[3]

Useful reviews of the application of near infrared[8,18,21,26] to the study of various classes of compound are as follows: organic compounds,[4] polymers,[9] silicon compounds,[10] pharmaceuticals,[11] food,[12,32] petrochemicals,[15,30] agricultural products,[16,17] surface hydrolysis of cellulose,[31] biological and medical.[27]

Carbon–Hydrogen Groups

Strong combination bands associated with C–H groups occur in the region $5000-4000\,cm^{-1}$ $(2.00-2.50\,\mu m)$ and first and second overtones of the C–H stretching vibration are observed at $6250-5550\,cm^{-1}$ $(1.60-1.80\,\mu m)$ and

9090–8200 cm^{-1} (1.10–1.22 μm) respectively.[15] Methyl groups absorb in the region 8375–8360 cm^{-1} (~1.195 μm), methylene groups at 8255–8220 cm^{-1} (1.21–1.22 μm). Compounds containing aromatic C–H bonds absorb near 6000 cm^{-1} (1.67 μm) due to the overtone of the C–H stretching vibration and at 8740–8670 cm^{-1} (1.14–1.15 μm). Aldehydes have a characteristic band in the region 4760–4520 cm^{-1} (2.10–2.21 μm) which probably arises from a combination of the C=O and C–H stretching vibrations. Aromatic aldehydes have characteristic bands near 4525 cm^{-1} (2.21 μm), 4445 cm^{-1} (2.25 μm), and 8000 cm^{-1} (1.25 μm).

Terminal epoxide groups have absorption bands near 6060 cm^{-1} (1.65 μm) and 4550 cm^{-1} (2.20 μm), these positions being similar to those of terminal methylene groups discussed below. However, the epoxide bands are not so complicated and are much more intense. Cyclopropanes also have similar absorptions at 6160–6060 cm^{-1} (1.62–1.65 μm) and 4500–4400 cm^{-1} (2.22–2.27 μm).

Terminal methylene groups, $>$C=CH$_2$, absorb near 6200 cm^{-1} (1.613 μm) and near 4750 cm^{-1} (2.11 μm). The terminal methylene groups of vinyl ethers, –O–CH=CH$_2$, absorb near 6190 cm^{-1} (1.616 μm) and those of α,β-unsaturated ketones, –CO–CH=CH$_2$, near 6175 cm^{-1} (1.619 μm) whereas for unsaturated hydrocarbons this absorption occurs near 6135 cm^{-1} (1.630 μm).

Cis-alkenes, –CH=CH–, have at least three bands in the near infrared region, one of which is near 4650 cm^{-1} (2.15 μm), whereas the *trans-* isomers have no strong absorptions in the near infrared region.

For terminal methyne groups, –C≡CH, the band due to the C–H stretching vibration, as discussed previously, occurs near 3330 cm^{-1} (3.00 μm), the overtone of this band being found near 6535 cm^{-1} (1.53 μm). Both of these bands are sharp and may easily be distinguished from the absorptions of amino groups, which also occur at these positions, since the C–H overtone band has almost twice the molar absorptivity of the N–H absorption.

Compounds with the CH$_2$CHC≡N–group have a combination band (due to CH and CN stretching vibrations) near 5230 cm^{-1} (1.91 μm).

Oxygen–Hydrogen Groups

In dilute carbon tetrachloride solution, primary alcohols absorb near 3635 cm^{-1} (2.751 μm), secondary alcohols near 3625 cm^{-1} (2.759 μm), and tertiary alcohols near 3615 cm^{-1} (2.766 μm). Aryl and unsaturated alcohols, in which the hydroxyl group may interact with the π-electrons of the system, normally have their maximum intensity absorptions near 3615 cm^{-1} (2.766 μm), with a shoulder near 3635 cm^{-1} (2.751 μm), in dilute solution spectra. The greater the interaction, the smaller the intensity of the shoulder.

Carboxylic acids, depending on their degree of association, have several bands in the region 3700–3330 cm^{-1} (2.70–3.00 μm). Even in dilute solutions, carboxylic acids exist in a high proportion as dimers. However, resulting from the fundamental stretching vibration of the OH groups of monomers, combination and overtone bands are observed near 3570 cm^{-1} (2.80 μm), 4750 cm^{-1} (2.11 μm), and 6900 cm^{-1} (1.45 μm).

Hydroperoxides absorb near 4800 cm^{-1} (2.08 μm) and 6850 cm^{-1} (1.46 μm).

Water has combination and overtone bands, due to the stretching and deformation vibrations of the OH group, near 7140 cm^{-1} (1.40 μm) and 5150 cm^{-1} (1.94 μm). The latter band may be used to determine the water content of a substance.

Carbonyl Groups

Carbonyl groups have an overtone band, due to the C=O stretching vibration, in the region 3600–3330 cm^{-1} (2.78–3.00 μm). This band may easily be distinguished from those due to N–H and O–H groups, which may also occur in this region, due to its comparatively low intensity. The position of this carbonyl overtone band follows the pattern observed for the position of bands due to the C=O stretching vibration – that is, in general, esters absorb at higher frequencies than aliphatic ketones which in turn absorb at higher frequencies than aromatic ketones.

Nitrogen–Hydrogen Groups

Primary, secondary, and tertiary amines may be distinguished on examination of the spectra of their dilute solutions in this region. The fundamental N–H vibrations have been discussed previously in the section dealing with amines. Primary amines have two bands in the region 3500–3300 cm^{-1} (2.86–3.03 μm) due to their fundamental N–H stretching vibrations. In the first overtone region, 7000–6500 cm^{-1} (1.43–1.54 μm), they have two bands and there is, in addition, a single band near 10 000 cm^{-1} (1.00 μm). Secondary amines have single bands in each of these regions and since tertiary amines have no NH group they do not, of course, absorb at all in these regions.

Primary amines also have a band resulting from the combination of the N–H bending and stretching modes which appears near 5000 cm^{-1} (2.00 μm) whereas secondary amines do not.

Alkyl and aryl amines may also be easily distinguished as the latter have, in general, the more intense absorptions in the near infrared region.

Chart 20.1   Near infrared region. The absorption ranges and corresponding intensities in terms of average molar absorptivity, cm^2 mol^{-1}

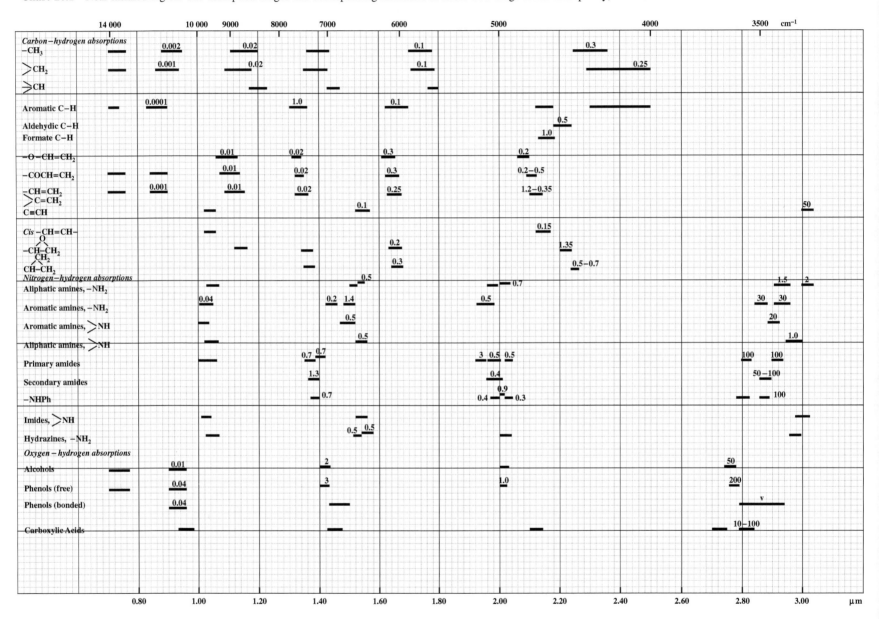

Chart 20.1 *(continued)*

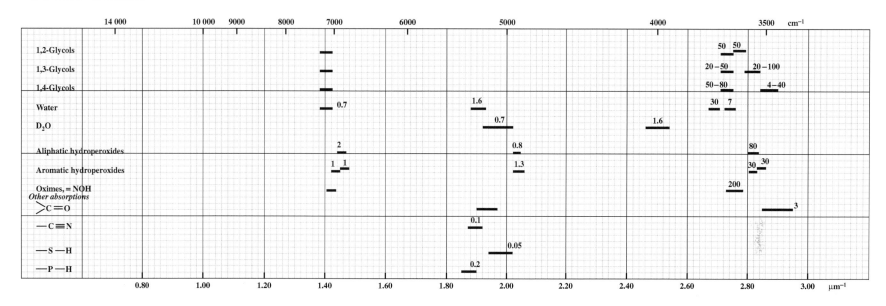

Primary amine hydrohalides have broad bands at 4600–4500 cm⁻¹ (2.17–2.22 µm) which may be the result of the combination of NH_3 bending and CH_2 stretching vibrations.

Polymers

Most polymers have absorptions in the near infrared region. The majority of these absorptions are due to the combination of vibrations.

For example, methylmethacrylate monomer absorbs near 5890 cm⁻¹ (~1.70 µm). Both the vinyl and vinylidene groups absorb strongly near 4775 cm⁻¹ (~2.09 µm) and have a much weaker band near 6180 cm⁻¹ (~1.62 µm). *Cis*-vinylene has a weak band at 4660 cm⁻¹ (~2.15 µm). Compounds containing the epoxy group absorb strongly near 4530 cm⁻¹ (~2.21 µm) and have a weaker band at 6060 cm⁻¹ (~1.65 µm). The curing of epoxy resins may be monitored by following the relative intensity of the band near 4530 cm⁻¹ (~2.21 µm).

In the near infrared region, polystyrene has absorptions near 5950 cm⁻¹ (~1.68 µm) and 4610 cm⁻¹ (~2.17 µm). Phenolic polymers absorb in the region 3640–3600 cm⁻¹ (2.75–2.78 µm).

It is possible to make use of the overtone and combination bands which are observed in the near infrared to determine the compositions of copolymers. The intensities of bands associated with particular monomer components of the copolymer can be measured relative to one another and a calibration graph constructed or the absorptivity of the bands can be used or determined. In the case of vinylchloride–vinylacetate copolymer, a band near 4650 cm⁻¹ (~2.15 µm), which may be associated with the carbonyl group, can be used for determining the proportion of vinylacetate. The absorptivity should be determined first or, alternatively, a calibration graph composed. This band may be used for the determination of the compositions of other copolymers involving esters.

The composition of polystyrene–butadiene copolymer may be determined by making use of bands near 4250 cm⁻¹ (~2.35 µm) and 4580 cm⁻¹ (~2.18 µm) which may be associated with the aliphatic CH and aromatic CH respectively. The structural isomersm of polyisoprene may be studied by making use of a sharp band that is observed for *cis*-1,4-polyisoprene near 4060 cm⁻¹ (~2.46 µm), the *trans*-1,4-polyisoprene has only a very weak absorption at this position.

Biological, Medical, and Food Applications

Instrumental advances have meant that for biological[21,22,26], pharmaceutical[30] and medical applications, near infrared techniques can be non-invasive and non-destructive. Hence, it is possible to examine live tissues and even living animals including man.[28] For such applications, the use of microscope techniques and fibre optics is increasing. For *in vivo* near infrared spectroscopy, high water absorbance of tissues, light scattering, the overlap of absorptions, temperature dependent absorptions and light-sensitive absorptions may all be encountered and result in problems for the analyst. Fibre optic catheters and probes have been used for various applications, both medical and biological, and for food analyses.[21,26] Instrumental advances have meant that near infrared video cameras and tunable light sources have made it possible to identify lesions in living arteries of patients. For example, near infrared can be used at different wavelengths to obtain images of blood vessels, the images being obtained by making use of the different absorption levels at the different wavelengths.

Near infrared has been used to determine the amount of water in various foods, for example, fruits, vegetables and dairy products. The band near $5160\,cm^{-1}$ ($\sim 1.94\,\mu m$) has often been used for this purpose.[12,21]

In cases where the spectra are broad or show only subtle differences, these differences can be highlighted by making use of algorithms which perform simple calculus and convert the normal absorption spectrum into one which is the first, second or even higher derivative. In this way, small changes in the slope of the absorption curve can be highlighted. For example, the second derivative of the near infrared absorption curve of a sample may be used to monitor the water content's state of hydrogen bonding. The proportion of free water molecules to those involved in hydrogen bonding with one or two hydroxyl groups can be estimated from the peaks near $7000\,cm^{-1}$ ($\sim 1.43\,\mu m$), $7005\,cm^{-1}$ ($\sim 1.46\,\mu m$) and $6620\,cm^{-1}$ ($\sim 1.51\,\mu m$) respectively.

A strong band near $4760\,cm^{-1}$ ($\sim 2.10\,\mu m$) is observed in foods containing starch,[23] such as rice and maize. The hydrolysis of starch may be followed by making use of absorptions near $4975\,cm^{-1}$ ($\sim 2.01\,\mu m$) and $4650\,cm^{-1}$ ($\sim 2.15\,\mu m$). These bands have been assigned as combination bands of OH/CO and CH/CO deformations respectively. Unsaturated vegetable oils have two bands at $4340\,cm^{-1}$ ($\sim 2.30\,\mu m$) and $4265\,cm^{-1}$ ($\sim 2.34\,\mu m$). Proteins have an absorption near $4590\,cm^{-1}$ ($\sim 2.18\,\mu m$), the intensity of which not only depends on the concentration of the protein in the sample but also on its structure and conformation. It has been observed that the intensity of this band decreases as the disulphide linkage of a protein is reduced. The denaturation of proteins has also been studied.[29]

The iodine number of oils and fats can be determined reasonably quickly and accurately by making use of absorption bands in the near infrared. With change in iodine number, significant changes are observed for the bands near $5815\,cm^{-1}$ ($\sim 1.72\,\mu m$) and $4675\,cm^{-1}$ ($\sim 2.14\,\mu m$). These bands have been attributed to the CH_2 and $-CH=CH-$ groups respectively. The sugar content[24,25] of fruit can also be determined using near infrared.

Lipoproteins have characteristic bands[28] near $5715\,cm^{-1}$ ($\sim 1.75\,\mu m$) and $4330\,cm^{-1}$ ($\sim 2.31\,\mu m$).

References

1. R. F. Goddu and D. A. Delker, *Anal. Chem.*, 1960, **32**, 140.
2. K. B. Whetsel, *Appl. Spectrosc. Rev.*, 1968, **2(1)**, 1.
3. *The Atlas of Near Infrared Spectra*, Sadtler Res. Labs., Philadelphia, PA, 1981.
4. L. G. Weyer, *Appl. Spectrosc. Rev.*, 1985, **21(1)**, 1.
5. E. Stark, K. Luchter and M. Margoshes, *Appl. Spectrosc. Rev.*, 1986, **22(44)**, 335.
6. E. G. Kramer and R. A. Lodder, *Crit. Rev. Anal. Chem.*, 1991, **22(6)**, 443.
7. M. Iwamoto and S. Kawano (Eds), *The Proceedings of the Second International Near I.R. Conference*, Tsukuba, Japan, Korin Pub., Tokyo, 1990.
8. D. A. Burnes and E. W. Ciurczak, *Handbook of Near Infrared Analysis*, in *Pract. Spectrosc.*, 1992, **13**.
9. C. E. Miller, *Appl. Spectrosc. Rev.*, 1991, **26(4)**, 227.
10. E. D. Lipp, *Appl. Spectrosc. Rev.*, 1992, **27(4)**, 385.
11. E. W. Ciurcsak, *Appl. Spectrosc. Rev.*, 1987, **23(2)**, 147.
12. B. G. Osborne and T. Fearn, *Food Analysis*, Longman Sci. and Technology. 1986.
13. H. Mark, *Pract. Spectrosc.*, 1992, **13** (*Handbook of Near Infrared Analysis*), 329.
14. J. J. Workman Jr. *Pract. Spectrosc.*, 1992, **13** (*Handbook of Near Infrared Analysis*), 247.
15. B. Buchanan, *Pract. Spectrosc.*, 1992, **13** (*Handbook of Near Infrared Analysis*), 643.
16. J. S. Shenk *et al.*, Pract. Spectrosc. 1992, **13** (*Handbook of Near Infrared Analysis*).
17. K. I. Hildrum, T. Isakasson, T. Naes and A. Tandberg, *Bridging the Gap Between Data Analysis and NIR Applications*, Ellis Horwood Chichester: 1992.
18. I. Murray and I. A. Cowe, *Making Light Work: Advances in Near Infrared Spectroscopy, Proceedings 4th International Conference on Near Infrared*, VCH, Weinheim, 1992.
19. K. A. B. Lee, *Appl. Spectrosc. Rev.*, 1993, **28(3)**, 231.
20. H. J. Gold, *Food Technol.*, 1964, **18**, 586.
21. P. Williams and K. Norris (eds), *Near Infrared Technology in the Agricultural and Food Industries*, American Assoc. of Cereal Chemists, 1987.
22. H. K. Yamashita *et al.*, *Nippon Shokuhin Kogyo Gakkaishi*, 1994, **41**, 61.
23. K. Nishinari *et al.*, *Starch*, 1989, **41**, 110.
24. S. Kawano *et al.*, *J. Jpn. Hort. Sci.*, 1992, **61**, 445.
25. S. Kawano *et al.*, *J. Jpn Hort. Sci.*, 1993, **62**, 465.
26. S. Kawano, in *Characterisation of Food*, (A. G. Gaonkar, ed.), Elsevier, Amsterdam 1995, 185.
27. R. J. Dempsey *et al.*, *Appl. Spectrosc.*, 1996, **50(2)**, 18A.
28. R. A. Lodder *et al.*, *Talanta*, 1989, **36**, 193.
29. K. Murayama *et al.*, *Nippon Kagaku Kaishi*, 1999, **10**, 637.
30. M. Ulmschneider and E. Penigautt, *Analysis*, 2000, **28(2)**, 136.
31. V. Svedas, *Appl. Spectrosc.*, 2000, **54(3)**, 420.
32. Y. Wu *et al.*, *J. Phys. Chem. B*, 2000, **104(24)**, 5840.

21   Polymers – Macromolecules

Introduction

The purpose of this chapter is help those interested in the characterisation/identification of polymers. It is not the intention of this chapter to deal with the theoretical aspects of the vibrational spectroscopy of polymers (infrared[1-8,38] or Raman) nor to deal with the sampling methods for the two techniques.[1,5-8,39,50] There are many good books dealing with these aspects. However, as will be appreciated, it is not possible to deal with the characterisation of polymers without some mention of these aspects but this will be kept to a minimum. For example, in dealing with sampling techniques, the aim is merely to give an idea of what commonly-available techniques may be applied. It is also not the intention of this chapter to list recent developments in the field.

The vast majority of functional groups present in polymers give rise to bands in the infrared region.[1-8,38,44] Hence, vibrational spectra can be used to identify polymers through the use of group frequencies or simply by attempting to compare the spectrum of an unknown with that of reference spectra. This latter approach can run into difficulties when dealing with copolymers or polymers that have been modified in one way or another – for example, by the addition of fillers or by being chemically modified, or where there are crystallinity differences between samples. In addition to providing data to enable the identification of polymers, vibrational spectroscopy can also yield valuable information on the microstructure of a polymer. This includes configurational and conformational information on the structure, how successive monomer units were added to the chain in both homo- and copolymers and the identification of end-groups and of defects.

It must be emphasised that infrared and Raman spectroscopy should not be used to the exclusion of other techniques such as ^{1}H and ^{13}C nuclear magnetic resonance, which are particularly useful characterisation techniques. Other useful techniques are mass spectroscopy, ultraviolet–visible spectroscopy, chromatography, thermo-analytical techniques (such as differential scanning calorimetry (DSC), thermal gravimetry (TG) etc.), or combined techniques such as GC–MS (gas chromatography combined with mass spectrometry) or chromatography (liquid)combined with mass spectroscopy etc. Such techniques may either yield additional information or provide the confirmation of a group or some other aspect that is required. As an example, nuclear magnetic resonance (NMR) and DSC may be used to distinguish between a blend or copolymer of two amorphous polymers, whereas this cannot easily be done using either infrared or Raman spectroscopy. Simple techniques should also not be ignored as they can save a great deal of time, for example, density, copper-wire flame test, etc.

A question often asked is, 'For polymer analysis, which is the better technique, infrared or Raman?' There is no simple answer since it depends on the task in hand. Even though great improvements have been made in laser Raman spectroscopy, the technique is still considered to be inferior to infrared spectroscopy for the characterisation and analysis of polymers. Some of the reasons for this are as follows:

1. Raman spectrometers tend to be more expensive than those of infrared and so are less common and therefore not readily available to the analyst. On the other hand, infrared is generally available in most laboratories for routine analysis and is a very versatile technique.
2. If good Raman spectra are to be obtained, more skill is required by the instrument operator and analyst than is the case for infrared, both in the experimental aspects and in the interpretation of the spectra obtained, although in many cases, sample preparation for Raman spectra is simpler than for infrared.
3. Infrared spectrometers and techniques and accessories are more established than those of Raman.
4. The acquisition of Raman spectral data has, in the past, been relatively slow, although in recent years great improvements have been made in this area.
5. One major advantage of the use of infrared spectra is that there is a vast base of reference spectra which can easily be referred to. In the case of Raman spectra, the reference libraries, although much better nowadays, still do not compare with those available for infrared.
6. Fluorescence has been a major source of difficulty for those using Raman spectra. Historically, this has led to the acquisition of poor spectra or in

some cases no spectra at all. Of course, techniques are now available to minimise the effects of this problem. Techniques to burn out the fluorescence can be used in the case of some samples. In many cases, the use of near-infrared, Fourier-transform, Raman spectrometers has proved invaluable in overcoming the difficulties involved in obtaining the Raman spectra of many polymers. Removing or cleaning the surface layer of a polymer can reduce/remove the fluorescence observed.

7. If the sample absorbs the radiation used for excitation, this may result in poor Raman spectra being obtained. Localised heating may occur and this may result in numerous problems – phase changes, decomposition, etc. – if care is not taken. This heating effect may be a problem when using Raman techniques to examine coloured samples.

8. Quantitative measurements are a little more involved in Raman spectroscopy. In infrared spectroscopy, the concentration of a functional group is linearly dependent on the absorbance of its related band, absorbance being the logarithmic ratio of the intensities of the incident and transmitted radiation. This means that both short and long term fluctuations in the intensity of the radiation source are irrelevant. However, in the case of Raman spectroscopy, the intensity of a band is linearly dependent on the concentration of its related functional group. This means that direct measurements are required and this is not always possible, hence ratio techniques are commonly used and, in the case of solutions, an internal standard may be added to the solution.

On the other hand, it should be noted that:

1. Polymers usually contain a large number of additives, fillers, pigments, etc. Many of these substances may result in interference in the infrared spectrum or present other problems such as requiring prior removal or special sample preparative techniques, or other special techniques. Many of the pigments used in the polymer industry, with the exception of carbon black, are poor Raman scatterers, although some may exhibit fluorescence. Glass fibres are also poor Raman scatterers and hence samples containing these can often be examined without prior treatment. In general, sampling techniques are often not as involved as those for infrared since there may be no need to remove the additives, fillers, etc., before examination. However, despite this advantage, it may still not be possible to obtain a suitable Raman spectrum.

2. Over the years, both techniques have become more automated. However, since Raman frequently requires little or no sample preparation, pre-alignment sampling techniques have meant that often little operator skill is needed – the operator simply places the sample in its compartment and starts the scan. However, if the sample exhibits fluorescence or absorbs the radiation used or has certain other problems, no spectrum will be obtained.

It is fair to say that most organic substances exhibit fluorescence to some degree and it may be impurities in the polymer sample that are responsible for the observation of fluorescence

3. These days it is possible, for a relatively small additional expenditure, to purchase dual purpose instruments: infrared/Raman spectrometers. However, the operation of such instruments is slightly more involved than for a straight infrared or Raman spectrometer. In addition, dual-purpose instruments do not have available the same high specifications as those using a single technique.

4. Raman has an advantage in the study of some samples in that glass cell and aqueous solutions may be used but then most soluble polymers require an organic solvent and infrared can easily be used.

5. Another point to bear in mind is that the infrared and Raman spectra of a given sample may differ considerably and hence each can be used to gain a different insight into the structure and properties of the sample. Often in the Raman spectra of polymers, the skeletal vibrations give characteristic bands which are usually very weak and of not much use for characterisation in the infrared. For example, the intensities and positions of the bands due to the skeletal vibrations are very characteristic of the different types of aliphatic nylon available and may be used for identification purposes.

6. Certain bands which are weak or inactive in the infrared, for example, those due to the stretching vibrations of $C=C$, $C\equiv C$, $C\equiv N$, $C-S$, $S-S$, $N=N$ and $O-O$ functional groups, exhibit strong bands when examined by Raman spectroscopy. Of course, the opposite is also true – certain vibrational motions of some groups which have weak bands in Raman may have strong bands in the infrared. However, the $C=C$, $C\equiv C$, $C\equiv N$, $C-S$, $S-S$, $N=N$ and $O-O$ functional groups, which, as mentioned, result in strong bands in Raman spectra, are to be found in many polymers and so, in this respect, Raman may sometimes have an advantage over infrared. Bands due to the following groups: OH, $C=O$, $C-O$, $S=O$, SO_2, $P=O$, PO_2, NO_2, etc. are strong in infrared. It should be pointed out that, for aromatics, the type of substitution present can normally be easily determined by infrared, the strong bands due to the CH out-of-plane vibrations and the overtone–combination bands being used. These vibrations result in weak (or no) bands in Raman spectra. However, other bands may be used in Raman spectroscopy to assist in the identification of the nature of the ring substitution. Although not always true, as a general rule, bands that are strong in infrared spectra are often weak in Raman spectra and bands that are strong in Raman spectra are often found to be weak in infrared spectra.

7. In some cases, in the infrared spectrum, bands occur in regions where they are overlapped by bands due to other groups making characterisation difficult/impossible (the same is true of Raman). By making use of Raman

spectroscopy, it is possible to examine bands which occur in relatively interference-free regions. For example, the alkene C=C stretching vibration band occurs near $1640\,cm^{-1}$ ($\sim6.10\,\mu m$) where few other functional groups absorb. The C=C stretching vibration band is strong in Raman spectra. On the other hand, the alkene CH out-of-plane vibration bands are often overlapped in the infrared spectra of polymers, making assignments using these bands difficult/impossible.

8. In some cases, the infrared sample preparation techniques that may be required for the examination of a particular sample may destroy or modify the characteristics of interest. For Raman, very little, if any, sample preparation may be required.

9. Raman spectrometers are capable of covering lower wavenumbers (down to $100\,cm^{-1}$ or lower) than those of infrared (400 or $200\,cm^{-1}$) and so can reveal information relating to polymer structure (see below) not easily available using other techniques.

Certainly for the routine analysis of polymers, infrared is by far the more popular but both techniques have their advantages and disadvantages.

Pretreatment of Samples

Polymers are difficult to characterise, not because their infrared or Raman spectra are complicated, difficult to interpret or consist of broad overlapping bands, as, in general, they do not, but because so many different substances are added to polymers for one reason or another. For example, fillers may be added to modify the physical or chemical properties of the basic polymer or its appearance. Fillers may be added to alter the mechanical, thermal, electrical or magnetic properties of the final product. Of course, fillers may also be used simply to make the product cheaper by the addition of inexpensive substances such as chalk, glass, wood shavings, silica, or air (gas bubbles). In addition, other substances such as lubricants (to assist in the processing of the polymer), heat stabilisers (to prevent thermal degradation during processing), pigments (for colouring), plasticisers, antioxidants, UV stabilisers, fire retardants, etc., are added to polymers. The list is endless, basically, just about anything may be added to polymers. In some cases, it may be difficult to believe that two samples are based on the same polymer, for example, polyvinylchloride (PVC) plasticised and unplasticised.

It must also be remembered that, different preparative techniques may be used by different manufacturers with very different conditions. These may lead to the same polymer having similar but quite different characteristics. Different catalysts may be involved and, in some cases, the catalyst may still be bound/remain in the polymer. Different catalysts may be used to form stereo-regular polymers. Also, the basic polymer unit may be modified chemically eg, polyethylene and chlorinated polyethylene.

It must also be remembered that, unlike organic or inorganic compounds, the 'molecules' are not all identical, they do not all have the same relative mass (molecular weight) (ignoring isotopic variation). Copolymers are also commonly encountered in many everyday products. The proportions of the monomers and/or the sequencing employed may vary from manufacturer to manufacturer or be varied in order to obtain the properties required. The stereochemical nature of the polymer, its microstructure, crystallinity, may have a bearing on the spectrum obtained.

Polymers are very versatile materials and are used in many different products. For example, they are encountered in many different guises: fibres, paints, coatings, rubbers, adhesives, packaging, and are even used in food products, etc. Some products may appear to be a single polymer but are actually laminates or composed of mixtures of polymers. Composites may, for example, contain fibres or materials in another form and these substances may be organic or inorganic in nature.

Some polymers exist in equilibrium with water. In such polymers, there may be as much as 2% water so that additional bands due to water will normally be observed in their spectra. For example, for polyetherketones and polyethersulphones, bands near $3650\,cm^{-1}$ ($2.74\,\mu m$) and $3550\,cm^{-1}$ ($2.82\,\mu m$) may be observed. In order to avoid complications due to the presence of water, particularly if quantitative measurements are to be made, some polymers may need to be thoroughly dried before their spectrum is recorded.

It is seen that the spectrum of a sample may consist of the basic polymer spectrum on which are superimposed the spectra of the additives, fillers, lubricants, fire retardants, catalyst residues, contaminants, etc. in proportions relating to their concentrations in the sample. Hence, there are many reasons why difficulties may be encountered when examining polymeric samples. Obviously, there are advantages if the problem can be made simpler, perhaps by separation, eg. use of chromatography or solvent extraction etc. With certain polymeric samples, it may be possible to use selective solvent extraction but, in some of these samples, this may result in fine particles of carbon black remaining in suspension and being difficult to remove.

A simple technique used to determine the inorganic filler incorporated in a polymer is simply to burn the sample (place the sample in a furnace) and spectroscopically examine the residue. It may also be possible to determine the polymer by pyrolysing the sample and examining the pyrolysate.

In Raman spectroscopy, polymer samples often exhibit fluorescence due to contaminants on their surfaces. Wiping the sample with a solvent, e.g. acetone or alcohol, can reduce this fluorescence. Alternatively, taking a thin slice off the surface of the sample can also be helpful.

Sample Preparation

Of course, as always, care must be taken not to contaminate samples or cells and not to use preparative techniques which affect the characteristics of the sample which are of interest.

Some polymeric samples can be examined directly without prior treatment. For example, thin polymeric films may be used for infrared transmission spectra and samples with glass-fibre fillers may be examined by Raman directly.

The sampling technique chosen is dependent not only on the availability of a spectrometer (infrared or Raman) and the facilities available on it, but also on the nature of the sample and the type of information required. In some instances, all that is required is confirmation that the sample is the same as that previously examined or is of a specific class of polymer, in which case a simple fingerprint may be sufficient to achieve this.

In many cases, the properties which make polymers attractive may actually make sampling difficult. For example, thermoplastics cannot easily be ground to form a powder for use in infrared, dispersive sampling techniques and many polymers exhibit fluorescence themselves (or the substances introduced to them are fluorescent) which can result in problems when attempting to obtain a Raman spectrum. Raman spectroscopy has two great advantages in that samples often need little, if any, preparation and samples of varying shapes and sizes can be examined.

Basic Techniques – Liquid, Solution, Dispersion

The techniques used in the study of low molecular weight organic and inorganic samples can, in many cases, be simply applied to polymers. For example, for infrared spectra, liquids may be examined in thin cells (small pathlength) having transparent windows over the frequency region of interest, or liquids (non-volatile) may be held by surface tension between transparent plates. Solids can be dissolved in suitable solvents and examined in the same way, perhaps also making use of compensating techniques. It should be noted that for polymeric fibres and powders, diffuse reflectance techniques, DRIFT, can be applied. In diffuse reflectance techniques, the radiation penetrates the sample and interacts with it, being partly scattered, reflected and absorbed. Hence, the emergent sample beam has the characteristic absorptions of the sample. Difference techniques can also be applied to many aspects of polymer spectroscopic studies. Reflection techniques, such multiple internal reflection, can be applied.

Dispersive Techniques

Polymeric samples, already in powder form, may, bearing in mind any particle size restrictions that might apply, also be examined by dispersive techniques, for example by preparing mulls or discs. Mulls may be prepared using liquid paraffin or polyfluorinated paraffin or some other suitable liquid. Discs may be prepared using potassium bromide or some other pure substance that is transparent over the spectral region of interest. If the polymeric specimen being examined is not already a powder then the solid may be ground to give a powder of the correct particle size. This may not be a problem if the material is brittle but if it is a thermoplastic or a rubber (elastomer) then it has to be cooled below its glass transition temperature before grinding can be successfully employed – low temperature grinding.

Films, Solvent Cast, Hot Press, Microtome

If the polymeric sample being examined is actually a sufficiently thin film, then it may be introduced directly into the sample compartment with no further preparation and examined by infrared transmission techniques, for example. It must be borne in mind that, with films examined by straight transmission, an interference pattern is often observed superimposed on the actual spectrum of the sample. Just as with low molecular weight organic substances, the variation in the band intensities observed for different functional groups means that, in order to obtain the optimum spectrum, the pathlength may differ significantly from sample to sample. The thickness of the film may need to be adjusted in order to obtain the best spectrum.

If the polymeric specimen is a thermoplastic then it is possible to use a hot press (a temperature-controlled hydraulic press) to prepare a thin film which may then be directly examined. If the polymer is soluble in a suitable, relatively-volatile solvent, then solvent casting may be used to prepare a film, this being similar to the technique of casting solid deposits on a transparent plate. Of course, with the latter technique, care must be taken to ensure that the solvent is completely removed or it will appear in the spectrum recorded. Special techniques may be required to remove the film from the base on which the film was cast. In some cases, it may be advantageous to cast the film on an infrared transparent plate. Thin films may also be prepared by using a microtome. Some thermoplastics may need a cryogenic microtome or, at least, to be cooled below their glass transition temperature so that they are sufficiently brittle to be sectioned. With some polymeric samples(eg. rubbers), solvent swelling prior to the use of the microtome may be beneficial.

Attenuated Total Reflection, Multiple Internal Reflection and Other Reflection Techniques

If the sample has a smooth, planar surface so that good physical contact with an infrared-transparent, higher refractive index prism/plate may be achieved, it can be examined by an infrared reflection technique, for example, one using attenuated total reflection, multiple internal reflection. It should be borne in mind, when examining polymeric films, that the sample may be a laminate and hence examination may give very different spectra from reflection at the two surfaces and also from transmission techniques. A coating or paint may be directly applied to a transparent plate for transmission or reflection techniques.

Pyrolysis, Microscope, etc.

Different pyrolytic techniques may be used, the pyrosylates being examined by infrared, Raman spectroscopy or other techniques. Microscope techniques[35] have improved over the last few years and this now means that very small regions of a sample may be successfully examined. The use of infrared microscopes has proved to be invaluable for the examination of laminates. It is often possible to microtome a thin cross-section from a laminate (for example, methylmethacrylate resin is sometimes used for this purpose) and then to examine individual layers in transmission using an infrared or Raman microscope system.

Other Techniques

It must not be forgotten that microscope techniques,[34,35,43] infrared and Raman can be used to examine small samples or single fibres. For example, single fibres of the aromatic polyamide, Kevlar[TM], have been studied.[26,27] In general, fibres have 'cylindrical' cross-sections. Synthetic polymer fibres[33] are manufactured by extrusion from spinnerets. In some cases, the surface of the fibre may be treated and hence have an outer skin. The transmission spectra of fibres have three components: (1) stray light which has passed by the fibre without coming into contact with it, (2) radiation reflected from the surface of the fibre and (3) radiation transmitted through the fibre which has different pathlengths. Hence, strictly speaking the Beer–Lambert law does not hold in most cases, although the opposite is often assumed for certain bands in the spectra of fibres.

Glass fibres are often surface coated with an agent to help adhesion to the polymer matrix, for example, a silane coupling agent.

As a result of the manufacturing process, the molecules of synthetic polymer fibres are more or less oriented along the axis of the fibre. The degree of orientation affects the physical properties of the fibres. Just as in the case of drawn films etc., dichroism may be observed for fibres when they are examined using linearly polarised infrared radiation, an example where dichroic behaviour is observed being that of Kevlar[TM]. Dichroism may also be measured by making use of Raman spectroscopy. The examination of fibres can often be made much simpler and more informative using Raman spectroscopy rather than infrared. Also, since often there is little required by way of sample preparation, the Raman technique can prove invaluable.

Theoretical Aspects – Simplified Explanations

General Introduction

In general, the infrared and Raman spectra of very large molecules are broad and so it is often difficult to identify the origins of particular bands. This is particularly true of large naturally-occurring substances such as proteins, carbohydrates, cell tissues. Even with these large molecules (biological), many advances have been made in both techniques and band identification.

Although it is helpful to have some basic understanding of the theoretical aspects of polymer vibrational spectroscopy, it is not absolutely essential in order to be able to identify or characterise polymers at a basic level. However, an understanding is useful in at least appreciating the origins of bands. The aspects covered below give the minimum required for a reasonable understanding, as applied to group frequency characterisation. The approaches given are relatively simple, so for a thorough understanding of the theory, the reader should turn to one of the many excellent texts available.

Consider a sample of a commercial polymer, such as polyethylene. Each of its chains will consist of a large number of atoms, on average 12 000 – at least. Hence, applying $3N - 6$ for a non-linear molecule, it can be seen that approximately 36 000 fundamental vibrations would be expected, which is a very large number. Therefore, it might be expected that, as with many large, natural molecules, the spectra of synthetic polymers would consist of broad absorptions with few discernable features. However, in general, fortunately, the fundamental vibrations occur in relatively narrow ranges. Thus, unlike the spectra of many large, naturally-occurring molecules, the spectra of most synthetic polymers usually consist of sharp bands to which the normal group frequency approach may be applied. For some polymers, the spectra obtained are often very closely related to those of the monomers involved (with the addition of end-groups). Where this is the case, the bands in the polymer spectrum may be sharper than those in the spectra of the monomers. It is also true to say that, in other cases, the polymer spectrum observed bears little resemblance to the spectra of the starting materials.

In simple terms, it may be considered that the polymer chains are so long that the vast majority of functional groups experience very similar environments and interactions and therefore their vibrational motions are very similar so that they occur over narrow ranges.

Looked at another way, considering, say, polyethylene, the vast majority of CH_2 groups experience an averaged-out environment, so that the CH_2 groups in the middle of the chain will not experience environments very different from each other. In order to make such a statement, various assumptions/approximations are made, for example that chain folding does not have an influence on the vibrations of the group, that interactions between chains do not occur, etc. To simplify matters, consider that each polymer molecule is isolated from its neighbours or, alternatively, that all polymer molecules (and hence the repeat unit or functional groups) experience an averaged-out environment or interaction. The motions of each polymer functional group may be considered to be independent of its neighbours. Therefore, a change in the electric dipole or polarisability induced in one part of the polymer molecule may be cancelled by the opposite effect elsewhere in the chain. Hence, it is only when the vibrations of the functional groups are in phase that a net change in dipole or polarisability would occur and a band in the infrared or Raman spectra would be observed, that is, the vibration would be infrared or Raman active. As a result of this, the infrared and Raman spectra of polymers generally consist of sharp bands.

However, this would not be the situation when dealing with low molecular weight polymers or if a polymer is partly crystalline in nature but has numerous defects. For these situations, many functional groups within the molecule would differ from each other and therefore their vibrations would be infrared or Raman active. In fact, a similar effect is observed for long chain paraffins, where there is a general broadening of bands occurring as the chain length increases. For polymers, this broadening effect is not necessarily symmetrical about the central band position.

Crystalline Polymers

In general, it is true to say that the spectrum of a crystalline substance contains sharp discrete bands whereas that of non-crystalline materials contains broad, diffuse bands. In general, the vibrational spectra of crystalline polymers also exhibit a high degree of definition,[7] since as mentioned previously, it is only the in-phase vibrational motions that result in active (infrared or Raman) spectral bands.

For theoretical purposes, just as with any other substances, the vibrational modes of crystalline polymers may be considered in terms of their unit cell and the symmetry associated with this cell. The number of atoms in the unit cell determines the maximum number of fundamental vibrations that may occur, rather than the number of atoms in the polymer repeat unit. Hence, since more than one polymer chain is often involved as part of the unit cell, the number of fundamental vibrations that may occur is almost always greater than that determined by considering the number of atoms in the isolated repeat unit. For example, two chains are involved in the unit cell of polyethylene and, for isotactic polypropylene, three chains are involved for each rotation of its helix.

The vibrational motions of a crystalline polymer may be considered as having two origins, internal and lattice. Lattice modes of vibration are those due to polymer chains moving relative to each other and occur at low wavenumbers, generally below $150\,cm^{-1}$ (above $66.67\,\mu m$). Internal vibrational modes are those due to the motions of the atoms of a chain relative to each other and, in general, these occur in the region $4000–150\,cm^{-1}$ ($5.00–66.67\,\mu m$).

It is a simple matter to distinguish between the two modes of vibration. If the temperature of the sample is lowered then lattice vibrational frequencies increase since the distance between chains decreases, the force between the chains increases and this is directly related to the vibration frequency. On the other hand, internal vibrational modes are very little affected by temperature.

As mentioned earlier, in a crystalline polymer, more internal modes of vibration can occur than if a polymer molecule were considered as an isolated entity. The number is dependent on the structure of the unit cell, that is, it is dependent on the number chains involved in the unit cell. The internal modes of vibration of the chains in the unit cell may be in phase or out of phase with each other. Due to the intermolecular interactions of the chains, the in-phase and out-of-phase vibrations occur at different frequencies and so their associated internal vibrations occur at definite and fixed values. For example, for an isolated polyethylene chain, the CH_2 wagging vibration would be expected to occur at about $725\,cm^{-1}$ ($13.79\,\mu m$) but, in the crystalline phase, a doublet is observed in the infrared for this vibration, the bands occurring near $720\,cm^{-1}$ ($13.89\,\mu m$) and $730\,cm^{-1}$ ($13.70\,\mu m$). The components of this doublet are not necessarily of equal intensity since the absorptivities (infrared) or scattering cross-sections (Raman) may be different for the two vibrations. Another example is that of crystalline isotactic polypropylene where, due to the high degree of the symmetry of the unit cell, a number of additional bands, above those expected for an isolated chain, are observed.

Another approach to explain the observation that there are often more bands observed for crystalline polymers than expected by considering an isolated unit alone, is simply to consider that crystallinity results in a perturbation of the vibrational modes. Hence, using this approach, it may easily be appreciated that the intensity ratios of bands is related to the degree of crystallinity of the polymer.

As seen, the vibrational spectra of crystalline polymers have a high degree of definition. If the crystallinity of a particular sample is decreased, then various spectral changes are observed, the bands become broader and often new bands appear. These new bands are due to the vibrational motions of different conformations and/or rotational configurations of the parts of the polymer chains present in the disordered phases.

Heating a polymer sample will, in general, result in the broadening of bands as the crystalline arrangement is destroyed. The opposite is also true: as a polymer is cooled, and hence crystallises, its bands become narrower. It is very important to bear in mind, when examining the spectrum of a polymer, that a band should not be assigned as originating from the crystalline arrangement unless (a) it disappears on melting, (b) it is predicted by group theory and can be shown to depend on the presence of the crystal lattice. However, it is not always possible to ascertain that these conditions have been met. Hence, vibrational spectroscopy cannot always be thought of as a good method for determining the crystallinity of a polymer. It should also be borne in mind that conformational regularity may also be associated with amorphous regions and, for example, with orientated, but not necessarily crystalline, arrangements. When in doubt, either use another technique, such as X-ray diffraction or use such a technique to justify the infrared (or Raman) approach to be adopted. Even when it can be shown that bands are a result of the crystallinity of the polymer, their intensities cannot be relied on to be a good measure of the degree of crystallinity and a calibration plot must be made. In addition, different crystalline arrangements of a polymer may, in fact, have common absorption bands. Hence, if a polymer is polymorphic, care must be taken in making assignments and determinations.

If a polymer sample absorbs the exciting radiation significantly, it will become hotter and hence the bands will become broader. If no account is taken of the fact that radiation is strongly absorbed by the sample, it is possible to determine incorrectly the phase transition (crystalline–amorphous) temperature.

The degree of crystallinity and the amorphous content of polyethylene can be determined.[1] The degree of crystallinity may be determined from the integrated intensity of the band near $1415 \, \text{cm}^{-1}$ ($7.07 \, \mu\text{m}$) and the amorphous content from the intensities of the bands near $1300 \, \text{cm}^{-1}$ ($7.69 \, \mu\text{m}$) and $1080 \, \text{cm}^{-1}$ ($9.26 \, \mu\text{m}$) . However, these days, the degree of crystallinity of a polymer is often determined by correlating whole spectra or spectral regions with X-ray or differential scanning calorimetry (DSC) measurements, using such techniques as partial least squares for calibration purposes.[51–55] For example, the density of polyethylene may be determined by using a partial least-squares calibration employing micro-Raman spectroscopy.[53] For the determination of the amorphous content of polytetrafluoroethylene, PTFE, a univariate method based on peak heights in the infrared region can be used.[51,52] Fourier transform Raman spectroscopy may be used to measure the crystallinity of polyetheretherketone in isotropic and uniaxial samples using univariate and partial least-squares calibrations.[54,55] Fourier transform Raman spectroscopy has been used to examine the crystallinity of polyethyleneteraphthalate.[55]

All crystalline polymers experience low-frequency vibrations along their chain – in effect, the chain is compressed and extended. The forces restraining this motion act along the axis of the chain and are very much smaller than those of the internal vibrations of groups. The frequencies of these vibrations are dependent on the Young's modulus of the crystal along the axis of the chain. Since these motions occur at very low frequencies, they are referred to as longitudinal acoustic vibrations.[45] It should be pointed out that these motions can also occur along the transverse axes as well. In general, the frequencies of these motions are well below $200 \, \text{cm}^{-1}$ and the bands are not really of any use in the identification or characterisation of polymers. However, these vibrations can yield information relating to the morphology of polymers. Normally Raman techniques are employed to observe these acoustic vibrations.

Non-crystalline Polymers

An individual, non-crystalline polymer chain may adopt a large number of rapidly interchanging rotational conformations relative to itself and its neighbours and hence the theoretical analysis of the polymer for spectroscopic purposes, using symmetry and fundamental vibrational modes, is impossible. The only approach which may be adopted for the analysis of such spectra is that based on the examination of the repeat unit, plus the end-groups, and treatment of the polymer as a liquid. As mentioned earlier, the spectra of non-crystalline polymers tend to involve bands that are broader than might be expected if the polymer had been crystalline.

If a vibration involves hydrogen bonding, or is affected by the conformational changes that occur, then the band may be very broad. On the other hand, if the band is relatively insensitive to external influences then the band may be quite sharp. For example, the spectrum of high density polyethylene (HDPE) has relatively sharp bands when compared with that of low density polyethylene (LDPE). The spectrum of non-crystalline, atactic polystyrene has bands due to the aromatic ring which are relatively sharp whereas other bands tend to be a little broader than for the crystalline, isotactic form.

It should be noted that, due to their lack of a uniform consistent structure, non-crystalline polymers do not exhibit lattice or acoustic bands.

Band Intensities

The intensities of bands are related to the concentrations of the functional groups producing them, allowing quantitative analysis if required. Provided the normal precautions are taken and calibration is feasible, good results may be obtained. It must be borne in mind that, in general, for various reasons, using Raman for quantitative analysis is a little more difficult than using infrared which is quite straightforward. These days, it is fair to say that Raman excitation sources, lasers, are very much more stable and there is very little problem in making quantitative measurements provided the usual precautions are taken.

An oriented sample, such as a drawn polymer film, exhibits different vibrational spectra when the orientation of the sample relative to the direction of linear polarised electromagnetic radiation is altered. In other words, it should be borne in mind that, in the presence of polarised radiation, the relative intensities of bands may be affected. The interaction between the polarised electric field of the radiation and the dipole moment associated with the vibration becomes a maximum or minimum depending on the angle between these two vectors, 0° or 90°. Hence, in polarised light, the spectra of stretch-oriented polymers exhibit dichroism.[4] Dichroism may also be observed in the stressed areas of a polymeric sample. The dichroic behaviour of a sample can provide information on (a) the direction of the vibrational modes, (b) the orientation of the functional group in the crystalline lattice and (c) the fraction of the perfect orientation in the oriented sample. The monitoring of the dichroism can be used to monitor the production of oriented polymeric films. This is commercially important as the physical properties of drawn samples are related to the degree of orientation.

As an example of dichroic behaviour, consider the infrared spectra of polyethylene. Polyethylene may be considered as a long chain of CH_2 units with its end groups, branching and any double bonds being ignored. Polyethylene molecules align themselves along the drawn axis and, as a result, the intensities of the bands due to the asymmetric and symmetric stretching vibrations reach a maximum when the electric field of the polarised radiation is perpendicular to the drawn axis, whereas the band associated with the wagging vibration reaches a maximum when these two are parallel. Other bands in the infrared spectra of polyethylene exhibit similar behaviour with regard to orientation and polarised light. It should be noted that polyacetylenes exhibit anomalous dichroic behaviour.[9]

Applications – Some Examples

Introduction

There are so many polymers and copolymers and so many possible variations that the account given below can do no more than give selected examples in order to indicate the characterisation possibilities. When examining commercial products and artefacts, it must be borne in mind that, as already mentioned, the base polymers may contain numerous other products – stabilisers, fillers, etc. It should also be borne in mind that the relative intensities of bands in the spectra of copolymers are dependent on the proportions and sometimes the sequencing of the components present in the copolymer unless, of course, the bands are common to both units.

The difference in intensities observed for various compositions of a particular type of copolymer may be used to determine the composition of the copolymer, that is, the relative amounts of each monomer unit present. In the simplest case, where a particular band is due solely to one component of the copolymer, then either the absorptivity may be determined or a calibration graph constructed for this purpose. For systems where a band position free from the absorptions of other components of the copolymer cannot be found, a slightly lengthier approach is required. The absorptivities at various suitable locations in the spectrum must be determined for each component and then, by taking measurements for a variety of concentrations of the components in the copolymer at these different locations, equations can be constructed to determine the composition of an unknown copolymer. An example of this approach is the determination of the individual isomers of butadiene copolymers, *cis*-1,4-, *trans*-1,4- and 1,2-polybutadiene . From the solution spectra (using a suitable solvent such as carbon disulphide) of the individual components, which may be obtained separately, the absorptivities of each isomer may be determined at suitable points in the spectrum of the copolymer and hence used directly in the three equations required for the quantitative determination. In a similar manner, the isomer compositions of isoprene[26,47] and chloroprene may be determined.

By making use of infrared, the pathlength of a liquid cell may be determined by measurement of the interference pattern observed. In the same way, the interference pattern often observed in the infrared spectra of thin polymeric films may be used to determine the thickness of the film. A knowledge of the refractive index of the polymer is required for this determination. If the incidence of the radiation is not normal then the angle of incidence is also required, see the equation below.

$$d = \frac{N}{2(\nu_2 - \nu_1)(n^2 - \sin^2 \theta)^{1/2}},$$

where d is the thickness measured in cm,
n is the refractive index of the film,
N is the number of peaks between the wavenumbers ν_1 and ν_2 measured in cm^{-1},
and θ is the incident angle of the radiation.

The interference pattern in the infrared spectra of thin polymeric films may give rise to difficulties when attempting to observe weak bands. This problem can be overcome in several ways, the simplest being to place the film on an infrared transparent (in the region of interest) plate or simply, when casting a film, to leave it adhered to the transparent plate and to examine it directly allowing for compensation if necessary. It should be borne in mind that the infrared spectra of laminates may also exhibit an interference as a result of the interaction of reflections at boundaries and radiation transmitted directly.

Depending on the nature of the sample and the information required, it may be advantageous to use infrared, or Raman, or both techniques in a study. Remember that groups that have weak bands in Raman may have strong bands in the infrared, for example, OH, $C=O$, $C-O$, $S=O$, SO_2, $P=O$, PO_2, NO_2, etc are all strong in infrared. The $C=C$, $C=C$, $C\equiv N$, $C-S$, $S-S$, $N=N$ and $O-O$ functional groups result in strong bands in Raman spectra and usually weak bands in infrared. All these groups are commonly found in many polymers.

Stereoregularity, Configurations and Conformations

These days catalysts exist for the preparation of many stereoregular polymers. The mechanical properties and spectra of the different stereoregular isomers of a particular polymer and also its atactic form may differ significantly. In general, it is obvious that stereoregular polymers may easily form crystals when they solidify. From the commercial point of view, since the different configurations of a polymer have different properties, it is essential to know the isomeric composition of a sample bearing in mind its intended application. When compared with the vibrational spectrum of the atactic form, the spectra of stereoregular isomers appear to have more bands and many of the bands are sharper. For example, the spectrum of isotactic polypropylene[32,41] has numerous additional sharp bands in the region $1350–800 cm^{-1}$ $(7.41–12.50 \mu m)$. The spectrum of isotactic polystyrene differs significantly from that of the atactic form, which has broader bands. The same is true of the syndiotactic forms. Hence, in general, the spectra of stereoregular isomers have additional sharp bands when compared with spectra of the atactic form.

For some polymers, different conformers may be possible, for example, polyethylene terephthalate[40] has two conformational isomers *gauche* and *trans*. For the *gauche* isomer, the $-O-CH_2-O-$ group has its oxygen atoms slightly displaced from each other, whereas in the case of the *trans* form, the oxygen atoms are opposite each other. The spectra of both forms are quite different, additional bands being observed for each form. Two characteristic bands for the *gauche* form occur near $1140 cm^{-1}$ $(\sim 8.77 \mu m)$ and $890 cm^{-1}$

$(\sim 11.24 \mu m)$. Two characteristic bands for the *trans* form occur near $970 cm^{-1}$ $(\sim 10.31 \mu m)$ and $840 cm^{-1}$ $(\sim 11.90 \mu m)$.

Some polymers, in addition to having different conformers, may also have configurational isomers associated with some of the conformers. These different arrangements may all be observed in their vibrational spectra. Polyvinylchloride is an example where bands originating from different isomers are commonly observed. In its infrared spectrum, a broad absorption is observed in the region $750–550 cm^{-1}$ $(13.33–18.18 \mu m)$, this being due to a number of overlapping bands some of which can be quite distinct.

Morphology – Lamellae and Spherulites

Some polymers, when they solidify from a melt, form crystals which have the appearance of being composed of thin, flat platelets, lamellae which are about 0.1 nm thick and many micrometres wide. In some cases, the polymer crystallites may be arranged in groups with their axes arranged radially. These groups form features known as spherulites. Spherulites are often many times larger than crystallites and can sometimes be seen by the naked eye. The morphology of a polymer has a great bearing on its mechanical strength and stability and hence is of great interest. Infrared and Raman spectroscopy may be used to study the morphology of polymers.

The frequency bands due to longitudinal acoustic vibrations which, as mentioned previously, are not observed in infrared spectra but occur in Raman spectra, are inversely proportional to lamella thickness. These bands are usually difficult to observe. For example, a low frequency band due to a longitudinal acoustic vibration has been found in the Raman spectra of polyethylene and polypropylene which is related to the chain length and the lamellar thickness.[2,3] The longitudinal acoustic vibration is dependent on the force constant (dependent on the chain's longitudinal Young's modulus), the interlamellar forces, structure of the chain folding sequence, the proportions of the amorphous and crystalline components and the density of the polymer.

$C=C$ Stretching Band

An advantage of using Raman spectroscopy is that the $C=C$ group, which is commonly found in many polymers, has a stretching vibration resulting in a strong band (in infrared this band is, generally, weak or in inactive) and hence can often be used to determine polymer conformations in which it occurs, determine the extent of curing or cross-linking, or follow the chemical kinetics of polymerisation.[20] For example, the different isomers of butadiene may be distinguished by using Raman and examining the band due to the $C=C$ stretching vibration.[18,19,22] Bearing in mind that in certain instances fluorescence may present problems, polymers containing aromatic groups may also be easily examined by the use of Raman techniques. Of course, the same is true of polymers which are similar, such as isoprene[26,48] and chloroprene.

Thermal and Photochemical Degradation

As a result of thermal degradation, both polyethylene and polypropylene form hydroperoxide groups. These groups are not easy to detect by infrared, especially as the O–O stretching vibrations result in a very weak band and, in addition, the concentration of the hydroperoxide is low, although it should be borne in mind that peroxides give a strong Raman band. The OH stretching band may also be difficult to observe as it only results in a medium intensity band. Fortunately for infrared analysts, hydroperoxide groups react to form a variety of carbonyl-containing compounds. It is usually possible to detect bands due to ketones which absorb near $1720\,cm^{-1}$ ($\sim5.81\,\mu m$), aldehydes, near $1735\,cm^{-1}$ ($\sim5.76\,\mu m$) and carboxylic acids, near $1710\,cm^{-1}$ ($\sim5.85\,\mu m$). In a given sample, these bands are often observed to overlap one another. The carboxylic acid band near $1710\,cm^{-1}$ ($\sim5.85\,\mu m$) may be removed by converting the acid into a salt by treating the sample with a relatively strong alkali. The band due to the salt CO_2^- group, occurs near $1610\,cm^{-1}$ ($\sim6.21\,\mu m$). In the case of photochemically decomposed samples, in addition to bands due to the various carbonyl groups, bands due to the vinyl group are also observed, occurring near $910\,cm^{-1}$ ($\sim10.99\,\mu m$) and $990\,cm^{-1}$ ($\sim10.10\,\mu m$). By making use of an infrared microscope, it is possible, for example, to monitor the effects of weathering on a polymeric sheet by examining a cross-section of the polymer sample at positions relating to various depths.

Due to the sensitivity of Raman spectroscopy to the $-C{=}C-$vibrational motion, a strong signal being observed, the degradation of polyvinylchloride, PVC, may be studied by making use of resonance Raman techniques. The degradation of PVC results in the loss of hydrogen chloride gas and the production of carbon–carbon double bonds, leading eventually to sequences of conjugated polyenes.

Polyethylene and Polypropylene

Polyethylene has strong bands in its infrared spectrum near $2950\,cm^{-1}$ ($\sim3.39\,\mu m$) and $1460\,cm^{-1}$ ($\sim6.85\,\mu m$) and a band of medium intensity, which is often a doublet, near $725\,cm^{-1}$ ($\sim13.79\,\mu m$). These bands are due to the CH stretching, deformation and rocking vibrations. If the polymer has significant branching then additional weak bands near $1380\,cm^{-1}$ ($\sim7.25\,\mu m$)

and $1365\,cm^{-1}$ ($\sim7.33\,\mu m$) are observed.[42] These bands are usually observed in the spectra of samples of low density polyethylene, LDPE.

The methyl groups of chain-branched polyethylene have, as mentioned above, a weak band near $1380\,cm^{-1}$ ($\sim7.25\,\mu m$). On the other hand, methylene groups have a stronger band at $1365\,cm^{-1}$ ($\sim7.33\,\mu m$) which overlaps this methyl band. By making use of the spectra of linear polyethylene and deconvolution techniques, it is possible to determine the degree of branching.

Linear low density polyethylenes (LLDPE), are low-concentration α-olefin modified polyethylenes. The olefines usually used are propylene, butene, hexene, octene, 4-methyl pentene-1. Due to the relatively high concentration of methyl groups in linear low density polyethylenes, greater intensities are normally observed for the bands associated with the CH_3 group than are observed for high density polyethylenes (HDPE) so care must be exercised when making assignments.

As a result of its commercial preparation, the chemical structure of polyethylene may also contain double bonds. The percentage of double bonds may be estimated by making use, in the infrared, of the band due to the vinyl CH out-of-plane deformation vibration which occurs near $910\,cm^{-1}$ ($\sim10.99\,\mu m$) and, in the Raman, of the $C{=}C$ stretching band near $1640\,cm^{-1}$ ($\sim6.10\,\mu m$) and determining the ratio of the intensities of these bands compared with other bands in the polyethylene spectrum. Some commercial low-density polyethylene, prepared using high pressures, contains vinylidene groups which, in the infrared, absorb near $890\,cm^{-1}$ ($\sim11.24\,\mu m$) due to the CH_2 out-of-plane deformation vibration. Polyethylene prepared using Ziegler catalysts often contains defects resulting in the presence of vinyl, vinylidene and *trans*-vinylene groups. The positions of the CH out-of-plane deformation vibration bands for vinyl and vinylidene have been mentioned above, that for *trans*-vinylene is near $965\,cm^{-1}$ ($\sim10.36\,\mu m$).

It is often true that the infrared and Raman spectra of samples have few similarities. In the Raman spectrum of polyethylene, the C–H stretching vibration bands are very strong and those due to rocking vibrations, near $725\,cm^{-1}$ ($\sim13.79\,\mu m$), are very weak or absent. In addition, in Raman spectra, the skeletal vibrations give characteristic bands near $1300\,cm^{-1}$ ($\sim7.69\,\mu m$), $1130\,cm^{-1}$ ($\sim8.85\,\mu m$) and $1070\,cm^{-1}$ ($\sim9.35\,\mu m$).

The infrared spectrum of polypropylene[31] has strong bands near $2950\,cm^{-1}$ ($\sim3.39\,\mu m$), $1460\,cm^{-1}$ ($\sim6.85\,\mu m$) and $1380\,cm^{-1}$ ($\sim7.25\,\mu m$). In addition, bands of medium intensity are observed near $1155\,cm^{-1}$ ($\sim8.66\,\mu m$) and $970\,cm^{-1}$ ($\sim10.31\,\mu m$). For isotactic polypropylene,[32,41] a number of sharp bands of medium intensity are observed in the region 1250–$835\,cm^{-1}$ (8.00–$11.98\,\mu m$). In the spectra of the molten, or atactic form, of polypropylene, most of these sharp bands disappear, except for the bands near $1155\,cm^{-1}$ ($\sim8.66\,\mu m$) and $970\,cm^{-1}$ ($\sim10.31\,\mu m$). For some samples of polypropylene, a band near $885\,cm^{-1}$ ($\sim11.30\,\mu m$) is observed which may be due to the CH out-of-plane motions of an end group, $-(CH_3)C{=}CH_2$.

Some ionomers are based on polyethylene with carboxyl groups located along the carbon chain. These carboxyl groups allow for the cross-linking of chains to occur by means of ionic bonds. Metal ions, such as sodium, potassium, magnesium and zinc, form the cationic link. The infrared spectra of ionomers are composed of the bands due to polyethylene mentioned above. In addition, bands are observed for the carboxylate portion near $1640\,cm^{-1}$ ($\sim6.10\,\mu m$), $1560\,cm^{-1}$ ($\sim6.41\,\mu m$) and $1400\,cm^{-1}$ ($\sim7.14\,\mu m$). Bands are also observed in the region $1350{-}1100\,cm^{-1}$ ($7.41{-}9.09\,\mu m$) which have their origin in the CH_2-acid salt structure.

Polystyrenes

In its vibrational spectra, polystyrene has a strong band due to the $={=}C{-}H$ stretching vibration between 3100 and $3000\,cm^{-1}$ ($3.23{-}3.33\,\mu m$). In general, this band may be observed for aromatic or olefinic components (or both). Its presence in the polystyrene spectrum, together with that of a medium intensity band at $1600\,cm^{-1}$ ($\sim6.25\,\mu m$), indicates aromatic, rather than olefinic, components. This band is due to one of the aromatic ring-stretching vibrations which occur in the region $1600{-}1430\,cm^{-1}$ ($6.25{-}6.99\,\mu m$).

The very strong bands observed in the infrared spectrum near $760\,cm^{-1}$ ($\sim13.16\,\mu m$) and $690\,cm^{-1}$ ($\sim14.49\,\mu m$) confirm the presence of a monosubstituted aromatic group. These bands are due to the CH out-of-plane vibration and a ring out-of-plane deformation respectively. The overtone and combination bands which occur in the region $2000{-}1660\,cm^{-1}$ ($5.00{-}6.02\,\mu m$) also indicate the presence of a monosubstituted aromatic. The positions of these bands are approximately $1940\,cm^{-1}$ ($\sim5.15\,\mu m$), $1870\,cm^{-1}$ ($\sim5.35\,\mu m$), $1800\,cm^{-1}$ ($\sim5.56\,\mu m$), $1740\,cm^{-1}$ ($\sim5.75\,\mu m$) and $1670\,cm^{-1}$ ($\sim5.99\,\mu m$). The band due to the C–H stretching vibration of the aliphatic group occurs between $3000{-}2800\,cm^{-1}$ ($3.33{-}3.57\,\mu m$). The bands due to the aliphatic CH deformation vibrations are in their typical positions.

In addition to the bands due to polystyrene, the spectrum of styrene-butadiene copolymer contains bands which may be associated with the butadiene component. A band near $1640\,cm^{-1}$ ($\sim6.10\,\mu m$), due to the C=C stretching vibration, and strong bands near $965\,cm^{-1}$ ($\sim10.36\,\mu m$) and $910\,cm^{-1}$ ($\sim10.99\,\mu m$), due to the CH out-of-plane vibrations, are observed in the infrared spectra of this copolymer. Bands due to the different isomers of butadiene may also be observed. The *cis*-1,4-butadiene isomer which absorbs weakly near $730\,cm^{-1}$ ($\sim13.70\,\mu m$) is often overlooked due to the presence of the strong bands of styrene. The 1,2-isomer and the *trans*-1,4-isomer absorb strongly near $965\,cm^{-1}$ ($\sim10.36\,\mu m$) and $910\,cm^{-1}$ ($\sim10.99\,\mu m$). Hence, the

relative proportions of the 1,2- and the *trans*-1,4-isomers present in the sample affect the spectral region $1000{-}900\,cm^{-1}$ ($10.00{-}11.11\,\mu m$).

In addition to the bands mentioned in the previous paragraph, the infrared spectra of acrylonitrile-butadiene-styrene copolymers will contain bands due to the acrylonitrile component. The additional presence of the characteristic band due to the nitrile group, which occurs near $2240\,cm^{-1}$ ($\sim4.46\,\mu m$), in a relatively band-free region of the infrared range, is a good indicator for this copolymer. It should be noted that the nitrile group gives a strong band in Raman spectra. Conformers of polyacrylonitrile[57] have been studied.

Polyvinylchloride, Polyvinylidenechloride, Polyvinylfluoride, and Polytetrafluoroethylene

The infrared spectrum of polyvinylchloride contains the bands typical of aliphatic CH groups, except that the band due to the CH_2 deformation vibration is shifted by about $30\,cm^{-1}$ to lower wavenumbers, to near $1430\,cm^{-1}$ ($\sim6.99\,\mu m$). In addition to the aliphatic CH bands, the spectra of PVC contain contributions due to the C–Cl vibrations.[31] For example, a broad, strong band is observed in the region $710{-}590\,cm^{-1}$ ($14.08{-}16.95\,\mu m$) due to the C–Cl stretching vibration. Since there are a very large number of additives possible, great care needs to be taken in the analysis of PVC samples. A band near $1720\,cm^{-1}$ ($\sim5.81\,\mu m$) is often observed in the infrared spectra of commercial samples of PVC. This band may be assigned to a carbonyl group present in the plasticiser employed and hence is assigned to the C=O stretching vibration.

Polyvinylidenechloride has a strong doublet in its infrared spectrum near $1060\,cm^{-1}$ ($\sim9.43\,\mu m$) and strong bands due the $={=}CCl_2$ stretching vibrations near $660\,cm^{-1}$ ($\sim15.15\,\mu m$) and $600\,cm^{-1}$ ($\sim16.67\,\mu m$). A band of medium intensity is also observed near $1420\,cm^{-1}$ ($\sim7.04\,\mu m$) due to CH deformation vibrations.

Polyvinylfluoride has a strong band near $1085\,cm^{-1}$ ($\sim9.22\,\mu m$) due to the C–F stretching vibration. The bands near $2940\,cm^{-1}$ ($\sim3.40\,\mu m$) and $1430\,cm^{-1}$ ($\sim6.99\,\mu m$) are due to CH stretching and deformation vibrations respectively. Polytetrafluoroethylene[31] has a strong absorption in the region $1250{-}1100\,cm^{-1}$ ($8.00{-}9.09\,\mu m$) apart from which the region above $650\,cm^{-1}$ (below $15.38\,\mu m$) is relatively free of absorptions. The weak band near $2330\,cm^{-1}$ ($\sim4.29\,\mu m$) is due to the overtone of the CF_2 stretching vibration. Polyvinylidenefluoride has a relatively weak band near $2940\,cm^{-1}$ ($\sim3.40\,\mu m$) due to the CH stretching vibration. However, the CH_2 deformation vibration is stronger than might be expected. The spectrum of polyvinylidenefluoride is greatly affected by the sample preparation techniques used.

Polyesters, Polyvinylacetate

In general, the infrared spectra of all polyesters contain bands due to the ester group, that is, bands which may be associated with the carbonyl, C=O, and C–O functional groups. The positions of these bands are characteristic of the basic nature of the particular polyester. Hence, strong bands in the region 1800–1700 cm^{-1} (5.56–5.88 μm) due to the carbonyl group and also in the region 1300–1000 cm^{-1} (7.69–10.00 μm) are expected.

The spectrum of polyvinylacetate contains bands typical of ester groups and in particular of the acetate group. Strong bands are observed near 1740 cm^{-1} (~5.75 μm) due to the C=O stretching vibration and 1250 cm^{-1} (~8.00 μm) due to the asymmetric stretching of the acetate C–O–C group. Of course, bands due to the aliphatic portion of the polymer are also present in the spectrum at their typical positions. It should be noted that carbonates also have bands at the two positions mentioned above but do not have the strong band found near 1020 cm^{-1} (~9.80 μm).

The spectra of copolymers of vinylacetate with other monomers will, of course, contain the bands of the spectra of both components superimposed in the proportions in which they are present in the copolymer.

Acrylates have two characteristic strong bands due to the C–O stretching vibration, one near 1260 cm^{-1} (~7.94 μm) and the other near 1170 cm^{-1} (~8.55 μm), this latter band being the stronger of the two. Polymethylmethacrylate also has an additional band near 1200 cm^{-1} (~8.33 μm) which, together with a band near 835 cm^{-1} (~11.98 μm), can be used to identify it. The ratio of the intensities of the CH deformation bands, which appear in the region 1470–1370 cm^{-1} (6.80–7.30 μm), may be used to distinguish polymethylmethacrylate from polyvinylacetate (the CH$_3$O group may be distinguished from the CH$_3$C group in that the former absorbs in the region 1475–1440 cm^{-1} (6.78–6.94 μm)) . Other features which may be of assistance in this task are that polymethylmethacrylate usually has a doublet of medium intensity in the region 1500–1425 cm^{-1} (6.67–7.02 μm), a medium-to-strong band near 1150 cm^{-1} (~8.70 μm) and a medium-intensity band at 750–725 cm^{-1} (~13.33–13.79 μm), which are not normally evident in spectra of polyvinylacetate. Polymethylmethacrylate has a sharp Raman band at 800 cm^{-1} (~12.50 μm) whereas polyvinylacetate absorbs near 650 cm^{-1} (~15.38 μm). The spectrum of polyvinylacetate is very similar to that of cellulose acetate but these two polymers may be distinguished by examination of the region below 1000 cm^{-1} (above 10.00 μm). Polyethylmethacrylate has strong bands near 1025 cm^{-1} (~9.76 μm) and 850 cm^{-1} (~11.76 μm).

The infrared spectrum of polyethylene terephthalate contains a band due to the carbonyl group near 1740 cm^{-1} (~5.75 μm) and two strong bands, typical of aromatic esters, near 1260 cm^{-1} (~7.94 μm) and 1130 cm^{-1} (~8.85 μm), due to the asymmetric and symmetric stretching vibrations of the C–O–C

functional group. In addition, bands due to the aliphatic and aromatic portions of the polymer are present in the spectrum, most of which are in their normal positions. The strong band due to the aromatic ring out-of-plane deformation is not in its normal position for *para*-substituted aromatics, instead it is found at slightly higher wavenumbers, 730 cm^{-1} (~13.70 μm). This shift is attributed to an interaction of the ester group with the aromatic ring. The vibrational spectra of polyethylene terephthalate are greatly influenced by both the crystallinity and molecular orientation of the polymer. If not thoroughly dried , bands due to water may be observed and these may make any quantitative measurements of the end-groups, hydroxyl and carboxyl groups (–OH and –COOH) difficult. The OH and COOH groups absorb near 3450 cm^{-1} (~2.90 μm) and 3260 cm^{-1} (~3.07 μm) respectively. In Raman spectra, polyethylene terephthalate has a very strong band due to the aromatic component near 1000 cm^{-1} (~10.00 μm).

Polyamides and Polyimides

The infrared spectra of polyamides[31,37] have a number of bands due to the amide group. The amide I, II and III bands occur near 1640 cm^{-1} (~6.10 μm), 1540 cm^{-1} (~6.49 μm) and 1280 cm^{-1} (~7.81 μm) respectively. In addition, bands of medium-to-weak intensity, due to the secondary NH group, may be observed near 3310 cm^{-1} (~3.02 μm) and 3070 cm^{-1} (~3.26 μm). These bands are generally broad due to the presence of hydrogen bonding. Individual aliphatic polyamides, nylons, may be identified by the careful examination of the relatively weak bands in the region 1500–900 cm^{-1} (6.67–11.11 μm), although care is needed as the crystallinity of the polymer affects its spectrum.

Hence, for a series of aliphatic polyamides where the number of methylene groups is increased, differences in their vibrational spectra may be observed in the region 1500–900 cm^{-1} (6.67–11.11 μm).[21,23] These differences are mainly due to bands resulting from CH$_2$ bending, twisting and wagging vibrations and the skeletal motions of the C–C backbone. It is possible to identify particular polyamides, as with any polymer, by simply comparing a spectrum with the spectra of known examples. It also possible to identify particular polyamides by measuring the relative intensities of bands. The bands normally used are those due to the CH$_2$ bending vibration at approximately 1440 cm^{-1} (~6.94 μm) and that of the amide I band at 1640 cm^{-1} (~6.10 μm). These two bands are used since there is a linear dependence of the ratio on the number of methylene groups present in the polyamide.

As seen, there are several types of polyamide which have different chemical structures. In addition, crystalline isomers, α-form, β-form and γ-forms, may result in slightly different spectra for a given polyamide. Hence, the identification of polyamides requires great care. It has been suggested that to identify certain simple polyamides[44] the following band positions may be used:

polyamide-6: $1465 \, cm^{-1}$ ($\sim 6.83 \, \mu m$), $1265 \, cm^{-1}$ ($\sim 7.91 \, \mu m$), $960 \, cm^{-1}$ ($\sim 10.42 \, \mu m$), $925 \, cm^{-1}$ ($\sim 10.81 \, \mu m$)

polyamide-66: $1480 \, cm^{-1}$ ($\sim 6.76 \, \mu m$), $1280 \, cm^{-1}$ ($\sim 7.81 \, \mu m$), $935 \, cm^{-1}$ ($\sim 10.70 \, \mu m$)

polyamide-610: $1480 \, cm^{-1}$ ($\sim 6.76 \, \mu m$), $1245 \, cm^{-1}$ ($\sim 8.03 \, \mu m$), $940 \, cm^{-1}$ ($\sim 10.64 \, \mu m$)

polyamide-11: $1475 \, cm^{-1}$ ($\sim 6.78 \, \mu m$), $940 \, cm^{-1}$ ($\sim 10.64 \, \mu m$), $720 \, cm^{-1}$ ($\sim 13.89 \, \mu m$)

In addition to the bands mentioned above, the spectra of aromatic polyamides, such as Kevlar[TM] [49] and Nomex[TM], contain bands due to the aromatic components.

Polyimides have a characteristic doublet near $1780 \, cm^{-1}$ ($\sim 5.61 \, \mu m$) and $1720 \, cm^{-1}$ ($\sim 5.81 \, \mu m$) which is due to the carbonyl group of the imide ring, the latter band being broader and stronger than the former band which tends to be relatively sharp. Due to their aromatic ring nature, polyimides have a number of sharp absorptions which may be associated with CH and CC vibrations. The Raman spectra of polyimides have been studied.[29]

Polyvinyl Alcohol

The infrared spectra of polyvinyl alcohols contain characteristic bands due to the OH stretching vibrations near $3400 \, cm^{-1}$ ($\sim 2.94 \, \mu m$) which is of strong intensity and due to the C–O stretching vibration near $1100 \, cm^{-1}$ ($\sim 9.09 \, \mu m$) which is of medium-to-strong intensity. Bands are also observed due to the CH stretching and deformation vibrations near $2940 \, cm^{-1}$ ($\sim 3.40 \, \mu m$) and $1420 \, cm^{-1}$ ($\sim 7.04 \, \mu m$) respectively.

Polycarbonates

The infrared spectra of polycarbonates have strong characteristic bands near $1785 \, cm^{-1}$ ($\sim 5.60 \, \mu m$) and $1250 \, cm^{-1}$ ($\sim 8.00 \, \mu m$) due to the C=O and C–O–C stretching vibrations respectively. Aromatic polycarbonates also contain a band of medium-to-strong intensity due to CH out-of-plane vibrations indicating the presence of a *p*-substituted aromatic, near $860 \, cm^{-1}$ ($\sim 11.63 \, \mu m$). Often, different aromatic polycarbonates may be identified by careful examination of the region $1100–900 \, cm^{-1}$ ($9.09–11.11 \, \mu m$).

Polyethers

The spectra of aromatic and aliphatic polyethers contain bands that may be associated with the ether linkage, C–O–C. There is a strong absorption in the infrared spectra of aliphatic polyethers in the region $1150–1060 \, cm^{-1}$ ($8.70–9.43 \, \mu m$) and for aromatic polyethers, $1270–1230 \, cm^{-1}$ ($7.87–8.13 \, \mu m$) due to the C–O–C asymmetric stretching vibration. In Raman spectra, the aliphatic polyethers absorb strongly at $1140–820 \, cm^{-1}$ ($8.77–12.20 \, \mu m$) and aromatic polyethers at $1120–1020 \, cm^{-1}$ ($8.93–9.80 \, \mu m$). Of course, bands associated with the other components of the polymer, aliphatic and aromatic, are also present.

Polyetherketone and Polyetheretherketone

The chemical structure of polyetherketones is such that the functional groups, the ether and ketone, are separated in the chain by an aryl group and should properly be named poly (aryl ether ketones) and poly (aryl ether ether ketones).

Polyetherketones may exist in equilibrium with water – as much as 2% water may be present in a sample – so bands due to water will normally also be observed in the spectra of polyetherketones and polyetheretherketones. For example, bands near $3650 \, cm^{-1}$ ($2.74 \, \mu m$) and $3550 \, cm^{-1}$ ($2.82 \, \mu m$) may be observed.

The crystallinity of polyetherketone may be determined by measurement of the relative intensities of the doublet observed due to the C=C stretching vibration[24] in the Raman spectrum of the polymer. The relative intensity of the band near $1595 \, cm^{-1}$ ($\sim 6.27 \, \mu m$) increases compared with that of the band near $1605 \, cm^{-1}$ ($\sim 6.23 \, \mu m$) as the crystallinity of the polymer increases. The relationship of the ratio of the intensities to crystallinity appears to be linear. However, it has been found that for uniaxially-oriented polyetheretherketone, this intensity ratio is also dependent on the alignment of the sample to the laser beam.[25] Other bands in the spectra of these types of polymer may be used for the determination of crystallinity.

Polyethersulphone and Polyetherethersulphone

The chemical structure of polyethersulphones is such that the functional groups, the ether and ketone, are separated in the chain by an aryl group and should properly be named poly (aryl ether sulphones) and poly (aryl ether ether sulphones).[36]

Polyethersulphones may exist in equilibrium with water – as much as 2% water may be present in a sample – so bands due to water will normally also be observed in the spectra of polyethersulphones and polyetherethersulphones. For example, bands near $3650 \, cm^{-1}$ ($2.74 \, \mu m$) and $3550 \, cm^{-1}$ ($2.82 \, \mu m$) may be observed.

The relative compositions of copolymers of polyethersulphone and polyetherethersulphone can be determined.[28] Polysulphones are amorphous and hence there are no changes due to crystallinity to result in problems determining relative compositions. In Raman spectra, the intensity ratio of the components of the doublet near $1600 \, cm^{-1}$ ($\sim 6.25 \, \mu m$) is related to the composition of the copolymer. The ratio of the intensities of bands near $1200 \, cm^{-1}$ ($\sim 8.33 \, \mu m$) and $1070 \, cm^{-1}$ ($\sim 9.35 \, \mu m$) may be used in a similar way.

Polyconjugated Molecules

Polyconjugated systems have been studied in undoped, doped (with electron donors or acceptors) and photoexcited states.[16] The spectra obtained for a particular polymer in these three different states are quite different. Some of the spectral features observed are common to all polyconjugated polymers in these different states. Some of the features are related to the existence of a network of delocalised π electrons which can be considered to be along a one-dimensional lattice. An important parameter which affects the spectral features observed is the conjugation length.[10] As might be expected for polyconjugated polymers,[9] as with most polymers, stretch-oriented samples in the different states exhibit dichroism.[4] Various polyconjugated systems have been studied – polyacetylene, polythiophene, phenylpolyene, etc.[16]

The infrared spectra of undoped polymers do not show unusual features. Their spectra may be interpreted in the normal way by using group frequency correlations or vibrational analysis. In general, the vibrational frequencies of groups are independent of the number of mers in an oligomer. As the number of mers is increased, the spectra become simpler, since the relative intensity of the bands due to the end-groups decreases compared with those due to the majority of units in the chain 'core'. The chain length may be determined by measuring the ratio of the intensities of bands due to vibrations from these two types of group – end and core. Doped and photoexcited systems exhibit dichroism.[12] The Raman spectra[8] of undoped polyconjugated polymers, such as polyacetylenes, are simple – in general, only a few bands are observed. These bands are due to the C=C stretching vibration, which occurs in the region $1500-1400 \, cm^{-1}$ ($6.67-7.14 \, \mu m$), and have a relatively strong intensity, and bands due to C–C and CH wagging vibrations, which are observed in the approximate region $1200-900 \, cm^{-1}$ ($8.33-11.11 \, \mu m$). In general, the strongest bands in the Raman spectra exhibit some degree of dispersion,[11] the frequency decreasing with increase in the number of conjugated units.[5] For polyconjugated systems, the Raman scattering cross-sections of some groups have been observed to increase rapidly with increase in the number of conjugated units.[10]

The infrared and Raman spectra of doped or photoexcited polyconjugated systems are very different from those of the equivalent undoped material.[6,7] New bands which are extremely strong and complex are observed. These new bands result in a broad, poorly-defined pattern in the region $1600-700 \, cm^{-1}$ ($6.25-14.29 \, \mu m$). The position (frequency) of the bands in the infrared spectra of doped samples appears to be independent of the doping species,[14] although band intensities may decrease with increased doping level. The frequencies of vibrations do not appear to alter with concentration of the doping species.[13] In a few oligomers, the frequency of some bands observed in the spectra of doped species and photoexcited species decrease with increase in the number of conjugated units.

In general, it has been found that the infrared spectra of photoexcited and doped materials are very similar.[6] It has been found that for some samples, for example trans-polyacetylene, bands in the spectrum of the photoexcited species experience a red-shift with respect to the doped material. In general, however, the spectra of the two are almost identical. Very slightly doped materials usually have very similar Raman spectra to those of the equivalent undoped system. The Raman spectra of doped and photoexcited species usually have broad, weak bands and are often not observed unless resonance enhancement conditions can be achieved. For example, for polyacetylene,[15] at large doping levels, new bands appear near 1600 ($\sim 6.25 \, \mu m$) and $1270 \, cm^{-1}$ ($\sim 7.87 \, \mu m$). At very high doping levels, some characteristic bands of undoped polyacetylene become weak and Raman scattering becomes weak.

The vibrational spectra of polyconjugated systems can only be interpreted by taking both the molecular structure and the electronic structure (π electrons) into account.

Polyacetylene in the cis-form undergoes a solid state thermal isomerisation to give the trans-isomer. The infrared spectra of the cis- and trans-isomers are quite different.[6] The out-of-plane CH vibration of the cis-isomer results in a strong band which occurs near $735 \, cm^{-1}$ ($\sim 13.61 \, \mu m$) at room temperature. When time-dependent spectroscopy is used to examine the cis-isomer at an elevated temperature, this band is observed to decrease in intensity with time and move to slightly higher wavenumbers, approximately $+7 \, cm^{-1}$. Similarly, the trans-isomer, which has its out-of-plane CH vibration near $1015 \, cm^{-1}$ ($\sim 9.85 \, \mu m$), when examined at elevated temperature, increases in intensity and moves to lower wavenumbers, approximately $-5 \, cm^{-1}$. Other bands in the spectra of these cis- and trans-isomers are also observed to change position slightly.

Resins

The infrared spectra of phenol formaldehyde resins have a broad, strong band at about $3350 \, cm^{-1}$ ($\sim 2.99 \, \mu m$) due to the OH stretching vibration of the phenolic group. Another strong band is observed near $1230 \, cm^{-1}$ ($\sim 8.13 \, \mu m$) due to the C–O stretching vibration. A doublet is usually observed at $1600 \, cm^{-1}$ ($\sim 6.25 \, \mu m$) due to a stretching vibration of the aromatic ring. Strong bands are also observed in the infrared spectra of phenol formaldehyde resins at about $760\,cm^{-1}$ ($\sim$13.16 μm) and $820\,cm^{-1}$ ($\sim$12.20 μm) due to the aromatic CH out-of-plane vibrations. These last two bands indicate that the aromatic ring has formed both *ortho* and *para* bonds. Novolak resins have only one strong band in their infrared spectra in the region 900–$730\,cm^{-1}$ (11.11–13.70 μm), this occurring at about $760\,cm^{-1}$ ($\sim$13.16 μm).

Uncured resole resins have a strong band near $1010\,cm^{-1}$ ($\sim$9.90 μm) due to the CO stretching vibration of the methylol group. As curing takes place, the intensity of this band decreases. Care needs to be exercised, as hexamine, which is often used as the cross-linking agent, has a weak absorption in the same region.

The infrared spectra of melamine–formaldehyde resins contain a band due to the OH stretching vibration near $3350\,cm^{-1}$ ($\sim$2.99 μm), a strong, broad band near $1560\,cm^{-1}$ ($\sim$6.41 μm), principally due to the stretching motions of the triazine ring, a broad, medium-to-strong band near $1040\,cm^{-1}$ ($\sim$9.62 μm), due to the C–O stretching vibration, and a band near $820\,cm^{-1}$ ($\sim$6.41 μm), again due to the triazine ring. The infrared spectra of urea–formaldehyde have a broad, strong band at about $3350\,cm^{-1}$ ($\sim$2.99 μm) due to the OH stretching vibration, a broad, strong band near $1570\,cm^{-1}$ ($\sim$6.37 μm), a strong band near $1040\,cm^{-1}$ ($\sim$9.62 μm), due to the C–O stretching vibration and a broad band near $625\,cm^{-1}$ ($\sim$16.00 μm).

Coatings and Alkyd Resins

For environmental and other reasons, these days, many coatings/paints are water based. This means that coatings must either be first dried or cells with water-stable windows must be used if infrared spectra[4] are required. In addition, large regions in which water absorbs strongly may obscure sample bands that need to be observed. Heavy water can be used to expose the regions in which the water bands cause concern but this, in general, is not helpful if commercial samples need to be examined. On the other hand, water is not a problem if Raman spectra can suffice. Polymerisation and curing reactions may be followed spectroscopically.[48]

With dried coatings, reflection techniques can be employed to obtain infrared spectra. An emulsion or latex can be examined by similar techniques to those described above. Solvent-based coatings can be examined either directly in liquid cells or as dried films. Usually the evaporation of the solvent can be monitored spectroscopically. Alkyd resins which are solvent based still form a substantial part of the commercially available coatings. The band due to the C=C stretching vibration may be used to follow the curing process.

Elastomers

Elastomers[30,36,56] present numerous problems with regard to the acquisition of vibrational spectra. For infrared spectra, special techniques must be used

for sample preparation.[1] In the past, using Raman, fluorescence has restricted the number of elastomers and the manner in which they could be studied. It was only for pure, unvulcanized elastomers that spectra could be obtained. Certainly, near infrared Fourier transform Raman spectrometers have helped resolve these difficulties. Of course, samples containing high proportions of carbon black can still present problems, especially with regard to Raman spectra. In the case of Raman spectroscopy, the absorption of the laser excitation source by the carbon can lead to rapid heating of the sample and degradation of the elastomer or, when a spectrum is obtained, a high baseline is observed, resulting in a poor spectrum where weak bands are lost.

It should be borne in mind that the cooling or stretching of certain elastomers can lead to crystallisation. Hence, the morphology of such samples can be studied using vibrational spectroscopy.

When spectra can be obtained by Raman, the bands due to the groups C=C and S–S, which occur in many elastomers, are strong and easy to observe. The band due to the C=C stretching vibration occurs near $1600\,cm^{-1}$ ($\sim$6.25 μm) and that due to the S–S stretching vibration occurs near $480\,cm^{-1}$ ($\sim$20.83 μm). By making use of Raman spectroscopy, it is possible to study vulcanisation and identify different types of sulphur linkage, for example disulphide, polysulphide, thioalkane and thioalkene.

Polyisobutylene[58] has a strong, sharp band near $1220\,cm^{-1}$ ($\sim$9.20 μm) and a characteristic absorption due to the two CH_3 groups, near $1385\,cm^{-1}$ ($\sim$9.20 μm) and $1365\,cm^{-1}$ ($\sim$7.33 μm).

Silicone rubbers have a very strong, broad absorption in their infrared spectra at 1100–$1000\,cm^{-1}$ (9.09–10.00 μm) due to the Si–O–Si stretching vibration. The band may be split into two broad peaks. The symmetric CH_3Si deformation vibration occurs at $1265\,cm^{-1}$ ($\sim$7.91 μm) and, because it is strong and sharp, it is easily identified even in the presence of other functional groups/substances. Another useful band occurs near $810\,cm^{-1}$ ($\sim$12.35 μm) and is due to the Si–C stretching vibration and CH_3 deformation vibration. The OH group, which is found in some silicone rubbers, absorbs near $3340\,cm^{-1}$ ($\sim$2.99 μm) due to the O–H stretching vibration. The Si–O stretching vibration results in a broad band in the region 900–$835\,cm^{-1}$ (11.11–11.98 μm). However, this band may be obscured by bands associated with the $SiCH_3$ group.

Plasticisers

Strongest Band(s) in the Infrared Spectrum

In this section, the characteristics bands are given for various common polymer plasticisers.

Strongest bands near 2940 cm^{-1} *(∼3.40 μm) and* 1475 cm^{-1} *(∼6.78 μm)* Substances whose strongest bands appear near 2940 cm^{-1} (∼3.40 μm) and 1475 cm^{-1} (∼6.78 μm) are primarily aliphatic hydrocarbons. Hydrocarbon oils have additional weak bands near 835 cm^{-1} (∼11.98 μm) and 715 cm^{-1} (∼13.99 μm). In addition to the two strong bands mentioned above, pure paraffin wax has a doublet near 725 cm^{-1} (∼13.79 μm). Similar spectra are also observed for aliphatic chlorinated hydrocarbon mixtures but with an additional broad band near 1260 cm^{-1} (∼7.94 μm).

Strongest band near 1000 cm^{-1} *(∼10.00 μm)* Phosphates, which are commonly found in many plasticisers, have their strongest bands near 1000 cm^{-1} (∼10.00 μm) in their infrared spectra. For aliphatic phosphates, this band is usually at slightly higher wavenumbers whereas, for aromatic phosphates, this band is usually at lower wavenumbers, near 970 cm^{-1} (∼10.31 μm). It should be borne in mind that some phosphate plasticisers are mixtures and hence their vibrational spectra may differ from sample to sample. For example, some aromatic phosphate plasticisers are mixtures of isomers and the relative intensities of their bands in the region 835–715 cm^{-1} (11.98–13.99 μm) will vary according to the composition of the plasticisers.

Strongest band near 1100 cm^{-1} *(∼9.09 μm)* Ethers and alcohol–ethers usually have the strongest band in their infrared spectra near 1100 cm^{-1} (∼9.09 μm). In the case of alcohol–ethers, a band due to the hydroxyl group is also normally observed near 3340 cm^{-1} (∼2.99 μm). Numerous alcohols are used, some being long chain aliphatics, such as lauryl and stearyl alcohols, others being polyhydric in nature, such as sorbitol and sucrose.

If the long chain alcohols can be examined spectroscopically in the crystalline phase, then the weak bands in the region 1000–910 cm^{-1} (10.00–10.99 μm) may possibly be used to identify them. With regard to polyhydric alcohols, the region 1100–1000 cm^{-1} (9.09–10.00 μm) may sometimes be used for identification purposes.

Strongest band in the region 835–715 cm^{-1} *(11.98–13.99 μm)* The infrared spectra of aromatic hydrocarbons and chlorinated aromatic hydrocarbons normally have their strongest band in the region 835–715 cm^{-1} (11.98–13.99 μm). Aromatic hydrocarbons are often encountered as resins and may be of such low molecular weight as to be viscous liquids. Examples of aromatic hydrocarbons are methyl styrene and styrene/butadiene copolymer and polycumarone/indene, the latter having an intense band near 750 cm^{-1} (∼13.33 μm). It should be borne in mind that, for copolymers, the intensities observed throughout the spectrum will be dependent on the composition of the copolymer. These days, the hazardous nature of certain chlorinated aromatic hydrocarbons is known.

Characteristic Absorption Patterns of Functional Groups Present in Plasticisers The characteristic absorption patterns of some common functional groups that appear in plasticisers are discussed below.

Carbonyl groups Carbonyl groups in one chemical form or another are commonly found in many plasticisers. Plasticisers containing carbonyl groups are discussed below.

Carboxylic acids Ether extracts often contain carboxylic acids either as free acids or as salts. The carboxyl group absorbs strongly near 1700 cm^{-1} (∼5.88 μm) and has a broad band due to the OH stretching vibration. In addition, as most plasticisers are long chain aliphatic acids, a band is observed near 2940 cm^{-1} (∼3.40 μm), due to CH stretching. In the case of high molecular weight aliphatic acids which are obtained in a reasonably pure state, their spectra are quite distinctive. The principal differences occur in the region 1280–1180 cm^{-1} (7.81–8.47 μm), where they have weak absorptions, and these differences may be used for identification purposes. For example, the infrared spectrum of lauric acid has three weak, clear, sharp bands in this region whereas that of stearic acid has five weak bands in this region.

Carboxylic acid salts In their infrared spectra, carboxylic acid salts have two relatively strong bands near 1590 cm^{-1} (∼6.29 μm) and 1410 cm^{-1} (∼7.09 μm). The position and shape of the band near 1590 cm^{-1} (∼6.29 μm) is dependent on the anion but there is little to recommend infrared as a means of identifying the metal anion by this method as there are simpler and more positive methods, such as atomic emission spectroscopy.

Care should be exercised with *o*-hydroxyl benzophenone, as it has a strong, broad absorption in its infrared spectrum near 1590 cm^{-1} (∼6.29 μm) but no strong absorption near 1410 cm^{-1} (∼7.09 μm), with a hydroxyl band near 3340 cm^{-1} (∼2.99 μm) also being observed, in addition to the aromatic bands.

Ortho-Phthalates Phthalates absorb at the positions give in the table below, Table 21.1. It should be borne in mind that some solvent extracts from polymeric samples may be due to the presence of *o*-phthalate resins rather than plasticisers.

Since *o*-phthalates have a very distinctive infrared spectrum, they are easily recognised. If, in the infrared spectrum of a sample, there are no additional bands having a significant intensity in the region 1500–600 cm^{-1} (6.67–16.67 μm) then the substance is a simple alkyl phthalate. In a relatively

Table 21.1 Phthalates

Region cm^{-1}	µm	Intensity	Comment
3090–3075	3.24–3.25	m	CH str
3045–3035	3.28–3.31	w	CH str
2000–1660	5.00–6.00	w	Ortho overtone pattern
1740–1705	5.75–5.87	vs	C=O str. Overtone ~3550 cm^{-1}
1610–1600	6.21–6.25	w–m	Ring str
1590–1580	6.29–6.33	w–m	Ring str
1500–1485	6.67–6.73	m	Ring str
1310–1250	7.63–8.00	vs	asym COC str
1170–1110	8.55–9.09	s	sym COC str
1070–1040	9.35–9.62	m	
770–735	12.99–13.61	s	Out-of-plane CH.def, usually 745 cm^{-1}
410–400	24.39–25.00		

pure state, the lower members of the alkyl series can be distinguished by careful examination of the weak bands in the region 1000–835 cm^{-1} (10.00–11.98 µm). Examples of these simple alkyl phthalates are dimethyl, diethyl, di-*n*-butyl, ... di-*sec*-octyl etc. The higher alkyl esters are difficult to identify unambiguously unless they possess some significant structural feature. For example, phthalates containing a gem-dimethyl group have a doublet due to the CH$_3$ deformation vibrations at 1385–1365 cm^{-1} (7.22–7.25 µm). Hence, in general, the esters of higher members of the alkyl series are more difficult to identify by infrared alone. They usually have a single broad band in the region 950 cm^{-1} (~10.53 µm).

In addition to the usual bands in the region 1000–900 cm^{-1} (10.00–11.11 µm), di-allyl phthalate has a sharp band near 1650 cm^{-1} (~6.06 µm) due to the C=C stretching vibration.

The infrared spectra of samples exhibiting the normal distinctive *o*-phthalate pattern but, in addition, possessing sharp weak bands in the region 1110–835 cm^{-1} (9.01–11.98 µm) may be associated with the alicyclic esters. For example, the spectra of cyclohexyl esters possess a number of sharp bands of medium intensity in the region 1430–715 cm^{-1} (6.99–13.99 µm).

Phthalates derived from aromatic alcohols or phenols, such as diphenyl phthalate and dibenzyl phthalate, are relatively easy to distinguish. In addition to the *o*-phthalate pattern, their infrared spectra also contain bands due to the aromatic substitution pattern in the region 835–670 cm^{-1} (11.98–14.93 µm) and the band due to CH stretching vibration near 2940 cm^{-1} (~3.40 µm) is very weak, especially when compared with the relative band intensity at this position for the alkyl phthalates. However, the possibility of mixed alkyl–aryl esters should be borne in mind when considering the band near 2940 cm^{-1} (~3.40 µm) – for example, butyl benzyl phthalate has a band of medium intensity due to the CH stretching vibration. In the case of this phthalate, the region,

1000–900 cm^{-1} (10.00–11.11 µm) may be examined for bands characteristic of the butyl group.

The infrared spectra of complex phthalates possess the characteristic *o*-phthalate bands but, in addition, usually have strong bands in the region 1430–1000 cm^{-1} (6.99–10.00 µm). For example, the spectra alkyl phthalyl alkyl glycollates, such as methyl phthalyl methyl glycollate, have a strong near 1200 cm^{-1} (~8.33 µm) due to the C–O stretching vibration. It should be noted that this band appears near 1160 cm^{-1} (~8.62 µm) for simple mixtures of phthalates with aliphatic esters. In the case of some glycol phthalates, one hydroxyl group is esterified with phthalic acid and the other hydroxyl group is condensed with an aliphatic alcohol to form an ether, for example di-methoxy ethyl phthalate. The C–O–C stretching vibration results in an additional band of medium-to-strong intensity near 1125 cm^{-1} (~8.62 µm). This band is, generally, narrower and at higher wavenumbers than those of simple alkyl ethers.

Aliphatic esters   Although the infrared spectra of aliphatic esters allows them to be distinguished from other carbonyl-containing compounds, it is often difficult to differentiate between esters which are similar in chemical structure. It is not sufficient to compare the spectrum of an unknown with that of a reference and thus to conclude, because the spectra are similar, even after careful attention to the positions and relative intensities of bands, that an identification has been positively made. It is often necessary to carry out hydrolysis of the sample and to examine the alcohol and acid fragments separately, or, alternatively, another spectroscopic technique may be used to identify the ester directly.

All long chain aliphatic esters have a band near 725 cm^{-1} (~13.79 µm) which may, in some samples, appear as doublet. Typical compounds are ethyl palmitate and glyceryl di-stearate. Both of these substances exhibit a band near 3340 cm^{-1} (~2.99 µm) due to the presence of hydroxyl groups. The observation of this band is useful in that esters derived from poly-functional alcohols generally exhibit this absorption.

Plasticisers containing the acetyl group, such as glyceryl triacetate, have a strong band in their infrared spectra near 1230 cm^{-1} (~8.13 µm).

Esters containing the epoxy group have a band, often a doublet, of weak-to-medium intensity near 835 cm^{-1} (~11.98 µm).

The spectra of different esters based on the same dibasic, such as alkyl adipates or sebacates, are very similar. Replacing the alcohol, in the case of simple low molecular weight alcohols, has a greater effect on the infrared spectrum than changing the acid.

As mentioned above, *o*-phthalate esters also containing an ether linkage normally exhibit a medium-to-strong band near 1125 cm^{-1} (~8.89 µm). Other plasticisers containing the aliphatic ester group and the ether group, for example those based on di- and tri-ethylene glycol and monocarboxylic acids, have a broad absorption due to the ether C–O–C stretching vibration near 1110 cm^{-1} ($\sim$9.01 µm). In this latter case, the absorption is similar to that observed for polyethylene oxide derivatives.

Some plasticisers based on dicarboxylic acids may not only be esters but also salts. The infrared spectra of these plasticisers obviously contain the bands associated with both esters and carboxylic acid salts. This means that it is difficult to distinguish between a compound and a mixture of an ester and salt. In a similar fashion, it is difficult to distinguish between plasticisers based on dicarboxylic acids and diols and polyesters based on similar compounds. There are no easily recognisable spectral features to distinguish between monomeric esters and an equivalent polyester.

Aromatic esters   The most common plasticisers based on aromatic esters are benzoates of one type or another. The infrared spectra of benzoates all have a strong band near 715 cm^{-1} ($\sim$13.99 µm). Another reasonably common type of aromatic ester is that based on salicylic acid. The spectra of salicylates, in common with other aromatic compounds which have a hydroxyl group adjacent to a carbonyl group (i.e. at the *ortho* position), do not have the band near 3340 cm^{-1} ($\sim$2.99 µm).

Sulphonamides, sulphates and sulphonates   Sulphonamides are easily recognised by the strong bands in their infrared spectra near 1315 cm^{-1} ($\sim$7.60 µm) and 1165 cm^{-1} ($\sim$8.58 µm). Different sulphonamides may be distinguished by the number bands, positions and intensities, in the region 850–650 cm^{-1} (11.76–15.38 µm). In addition, *N*-substituted sulphonamides have bands associated with NH rather than NH$_2$, the band-structure of the former being simpler (see the chapter dealing with amines).

Sulphonic acid esters   Alkyl aryl sulphonic acid esters, such as ethyl *p*-toluene sulphonate, have two strong characteristic bands near 1350 cm^{-1} ($\sim$7.41 µm) and 1180 cm^{-1} ($\sim$8.47 µm) in their infrared spectra. Aryl esters of alkyl sulphonic acids normally have a strong absorption near 865–650 cm^{-1} (11.56–15.38 µm) due to aromatic CH out-of-plane deformation vibrations and in this respect their spectra are similar to those of aryl sulphonamides but, of course, sulphonamides have additional bands due to their NH stretching vibrations.

Characteristic Bands of Other Commonly Found Substances

The solvent extracts of resins may contain antioxidants, some of which may be based on aromatic amines, examples of antioxidants being derivatives of diphenyl amine and *p*-phenylene diamine. These substances have spectra which are similar to those of sulphonamides. They have one or two bands due to NH or NH$_2$ stretching vibrations respectively near 3340 cm^{-1} ($\sim$2.99 µm) and a strong band near 1300 cm^{-1} ($\sim$7.69 µm). However, since these compounds do not have the second strong band near 1165 cm^{-1} ($\sim$8.58 µm), this is a reliable way of distinguishing between them and sulphonamides.

Sulphates and sulphonates have strong absorptions in their infrared spectra between 1250 and 1110 cm^{-1} (8.00–9.01 µm).

A strong band near 3340 cm^{-1} ($\sim$2.99 µm), with one or more strong bands near 1250 cm^{-1} ($\sim$8.00 µm), indicates a possible phenolic constituent. It should be borne in mind that epoxy compounds, which are often extracted from resins, have spectra similar to those of phenols. Since they are usually of low molecular weight and have only terminal hydroxyl groups, the O–H stretching vibration band near 3340 cm^{-1} ($\sim$2.99 µm) is usually of moderate intensity. In addition, it should be noted that *p*-aromatic epoxy compounds have a prominent band near 835 cm^{-1} ($\sim$11.98 µm).

Common Inorganic Additives and Fillers

The Inorganic chapter of this book, and some earlier chapters, should also be studied for relevant information concerning many of the inorganic additives and fillers commonly found in polymers. The chapters dealing with silicon, boron and phosphorus may also contain information relevant to the inorganic compounds found in a particular polymer of interest. As mentioned earlier and in the Inorganic chapter, Raman spectroscopy is particularly useful in the characterisation of inorganic compounds that are commonly found in commercial polymer samples.

Carbonates

The infrared spectra of inorganic carbonates consist of a strong broad band at 1530–1320 cm^{-1} (6.54–7.58 µm) (which in Raman is of weak-to-medium intensity and is often found near 1450 cm^{-1} ($\sim$6.90 µm)), a band of medium intensity near 1160 cm^{-1} ($\sim$8.62 µm), a weak band at 1100–1040 cm^{-1} (9.09–9.62 µm) (which is of strong-to-medium intensity in Raman), a band of medium intensity at 890–800 cm^{-1} (11.24–12.50 µm) and a band of variable intensity at 745–670 cm^{-1} (13.42–14.93 µm) (which is of weak intensity in Raman). For calcium carbonate, a strong Raman band is observed near 1085 cm^{-1} ($\sim$9.26 µm).

Table 21.2   Calcium carbonate

Functional Groups	Region cm^{-1}	Region μm	Intensity IR	Intensity Raman	Comments
CaCO$_3$	2530–2500	3.95–4.00			
	1815–1770	5.51–5.65	w		
	1495–1410	6.69–7.09	vs	m–s	
	~1160	~8.62	m		
	1090–1080	9.17–9.26	w	s	May be absent
	885–870	11.30–11.49	m–s		
	860–845	11.63–11.83	m		Sharp
	~715	~13.99	m–w	w	
	705–695	14.18–14.39	m–w	w	
	~330	~30.30	vs		Broad
	~230	~43.48	s		Sharp

Sulphates

In the infrared, sulphates have medium-intensity bands at 1200–1140 cm^{-1} (8.33–8.77 μm) and 680–580 cm^{-1} (14.71–17.42 μm) (these bands are of medium-to-strong intensity in Raman), a strong band at 1130–1080 cm^{-1} (8.85–9.26 μm) (which is of medium-to-strong intensity in Raman) and weak bands at 1065–955 cm^{-1} (9.39–10.47 μm) and 530–405 cm^{-1} (18.87–24.69 μm) (in Raman these bands are both of strong intensity). The Raman spectrum of barium sulphate has a strong band near 985 cm^{-1}

Table 21.3   Barium sulphate

Functional Groups	Region cm^{-1}	Region μm	Intensity IR	Intensity Raman	Comments
BaSO$_4$	~3430	~2.92	m	w	Broad
	~2350	~4.26	w		
	~1650	~6.06	w		
	~1470	~6.80	w		Broad
	~1330	~7.52	w		
	1200–1180	8.33–8.47	s	m–s	Sharp
	1130–1110	8.85–9.01	vs	m–s	Doublet
	1090–1070	9.17–9.35	m–w	s	
	1000–980	10.00–10.20	m–w	s	
	680–580	14.70–17.24	m	m–s	
	~725	~13.79	w	w	
	640–630	15.62–15.87	m–w	m–s	Sharp } Broad
	615–605	16.26–16.53	m–w	m–s	Doublet
	~460	~21.74	w	w	
	~415	~24.10	w	w	

(~10.15 μm) and weak bands near 1160 cm^{-1} (~8.62 μm), 1135 cm^{-1} (~8.81 μm), 645 cm^{-1} (~15.50 μm), 615 cm^{-1} (~16.26 μm), 460 cm^{-1} (~21.74 μm) and 450 cm^{-1} (~22.22 μm).

Talc

Talc has strong bands at 1030–1005 cm^{-1} (9.71–9.95 μm), 675–665 cm^{-1} (14.81–15.04 μm), 540–530 cm^{-1} (18.52–18.87 μm) and 455–445 cm^{-1} (21.98–22.47 μm).

Clays

Most clays usually have strong bands in the region 3670–3600 cm^{-1} (2.72–2.78 μm), 3450–3400 cm^{-1} (2.90–2.94 μm) and 500–450 cm^{-1} (20.00–22.22 μm), a very strong band at 1075–1050 cm^{-1} (9.30–9.52 μm), a band of medium-to-weak intensity near 1640 cm^{-1} (~6.10 μm), a band at 945–905 cm^{-1} (10.58–11.05 μm), a weak band at 885–800 cm^{-1} (11.30–12.50 μm) and a band of variable intensity at 440–420 cm^{-1} (22.73–23.81 μm).

Kaolin normally contains water of crystallisation and as a result has a distinctive absorption pattern in the infrared near 3600 cm^{-1} (~2.78 μm).

Table 21.4   Talc

Functional Groups	Region cm^{-1}	Region μm	Intensity IR	Comments
Talc	~3685	~2.71		
	~3675	~2.72		
	~3660	~2.73		
	1640–1620	6.10–6.17	w	
	1050–1040	9.52–9.61		Sharp
	1030–1005	9.71–9.95	vs	Broad
	785–770	12.74–12.99	w	
	~740	~13.51		
	700–690	14.29–14.49		Sharp
	675–665	14.81–15.04	s	
	540–530	18.52–18.87	s	
	~500	~20.00		
	475–455	21.05–21.98	m	Sharp
	455–445	21.98–22.47	vs	
	~440	~22.73		
	~425	~23.53	m	Sharp

Table 21.5   Kaolin

Functional Groups	Region cm⁻¹	Region μm	Intensity IR	Comments
Kaolin	3710–3695	2.69–2.71	s	
	3670–3650	2.72–2.74	m–s	
	3655–3645	2.74–2.74	v	Usually s
	3630–3620	2.75–2.76	m–s	
	1650–1640	6.06–6.10	w	
	1120–1090	8.93–9.17	s	Sharp
	1050–1000	9.52–10.00	s	
	1020–995	9.80–10.05	vs	
	960–935	10.42–10.70	m	Sharp
	920–905	10.87–11.05	s	Sharp
	800–780	12.50–12.82	w	
	760–745	13.16–13.42	w	
	700–685	14.29–14.60	m	
	~605	~16.53	w	
	550–515	18.18–19.42	s	
	475–460	21.05–21.74	s	
	435–415	22.99–24.10	s	
	~345	~28.99	w	
	~275	~36.36	w	
	~200	~50.00	w	
	~190	52.63	w	

Titanium Dioxide

Titanium dioxide has strong absorptions at 700–660 cm⁻¹ (14.29–15.15 µm) and 525–460 (19.05–21.74 µm), a medium-to-strong intensity band at 360–320 cm⁻¹ (27.78–31.25 µm) and weak bands at 185–170 cm⁻¹ (54.05–58.82 µm) and 100–80 cm⁻¹ (100.00–125.00 µm). The region below 200 cm⁻¹ (above 50 µm), which is easily accessible in Raman, can be used to distinguish between rutile and anatase.

Silica

Table 21.6   Silica

Functional Groups	Region cm⁻¹	Region µm	Intensity IR	Comments
Silica	1225–1200	8.16–8.33	m–w	
	1175–1150	8.51–8.70	m–w	Sharp
	1100–1075	9.09–9.30	vs	
	805–785	12.42–12.74	m	

Table 21.6   (continued)

Functional Groups	Region cm⁻¹	Region µm	Intensity IR	Comments
	795–775	12.58–12.90	m	
	725–700	13.95–14.29	m	
	670–595	17.93–16.81	w	
	525–500	19.05–20.00	m–s	
	490–475	20.41–21.05	m–s	
	465–440	21.50–22.73		
	~435	~22.99	s	
	395–370	25.32–27.07	m	
	~260	~38.46	w	

Table 21.7   Antimony trioxide

Functional Groups	Region cm⁻¹	Region µm	Intensity IR	Comments
Sb₂O₃	770–740	12.99–13.51	s	
	~685	~14.60	s	
	~590	~16.95	v	
	415–395	24.10–25.32	m–w	
	385–355	25.97–28.17	s	
	~385	~25.97	s	
	~345	~28.99	w	
	~320	~31.25	w	
	~265	~37.74	s	Broad
	~180	~55.56	w	

Antimony Trioxide

Antimony trioxide has strong absorption bands at 770–740 cm⁻¹ (12.99–13.51 µm) and 385–355 cm⁻¹ (25.97–28.17 µm), a medium-to-weak intensity band at 415–395 cm⁻¹ (24.10–25.32 µm) and a weak band at 200–180 cm⁻¹ (50.00–54.05 µm).

Infrared Flowcharts

The flowcharts given below have been based on strong bands, bands which occur in relatively interference free regions, or bands that are easy to identify. However, in using the flowcharts, it should be borne in mind that the spectra of polymers may differ from those on which the flowcharts have been based. This is especially true where copolymers are concerned. With copolymers, the spectra observed are dependent on the percentages of the individual components present. For example, some styrene–butadiene copolymers contain certain small amounts of acrylonitrile, this resulting in a band near $2220\,\mathrm{cm}^{-1}$ ($4.50\,\mu\mathrm{m}$). The presence of the band near $2220\,\mathrm{cm}^{-1}$ ($4.50\,\mu\mathrm{m}$) could be misleading. It should also be borne in mind that polymers prepared by different methods, or using different catalysts, may have slightly different spectra.

If polymers are examined spectroscopically without removing additives such as fillers, plasticisers, stabilisers, lubricants, etc. then their infrared spectra may be affected drastically by the presence of these substances. Also, if care has not been taken during the preparation of a sample, bands due to contaminants such as water, silicate, phthalates, polypropylene (from laboratory ware), etc, may appear in the spectra and so result in some confusion. Hence, the flowcharts given below should be used with some degree of caution. In order to confirm an assignment made by use of the flow chart, it is important finally to make use of known infrared reference spectra. However, it should be borne in mind that stereoregular polymers may have spectra which differ from their atactic form and that sample preparative techniques may also affect the spectrum obtained for a particular polymeric sample.

Chart 21.1 Infrared – polymer flowchart I

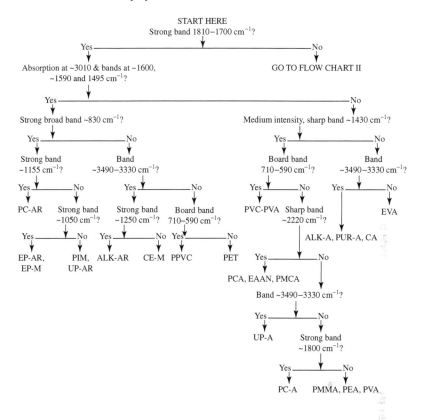

Table 21.8 List of polymers used in flowcharts

Name	Abbreviation
Acrylonitrile–butadiene–styrene	ABS
Alkyd resin – aliphatic	ALK-A
Alkyd resin – aromatic	ALK-AR
Aramide	AR
Butadiene acrylonitrile (Nitrile rubber)	NBR
Butyl rubber	BUTYL
Cellolose film	CF
Cellulose acetate	CA
Cellulose ether modified Ar	CE-M
Cellulose ether	CE
Cellulose nitrate	CN
Epoxy	EP
Epoxy – Aliphatic	EP-A
Epoxy – Aromatic	EP-AR
Epoxy – Aromatic (Modified)	EP-M
Ethyl cellulose	EC
Ethylacrylate acrylonitrile	EAAN
Ethylene vinylacetate	EVA
Ethylene polysulphone	EPS
Ionomer	ION
Melamine–formaldehyde	MF
Methyl cellulose	MC

Table 21.8 (*continued*)

Name	Abbreviation
Neoprene	NP
Nitrated polystyrene	NPS
Nylon-11	N11
Nylon-6, 10	N610
O,O-Novolac	OONOV
Phenol–formaldehyde	PF
Phenolic resin	PHR
Plasticised polyvinylchloride/vinylidenechloride	PVC-PVDC
Plasticised polyvinylchloride	PPVC
Poly(4-methyl penten-1)	TPX

(*continued overleaf*)

Chart 21.2   Infrared – polymer flowchart II

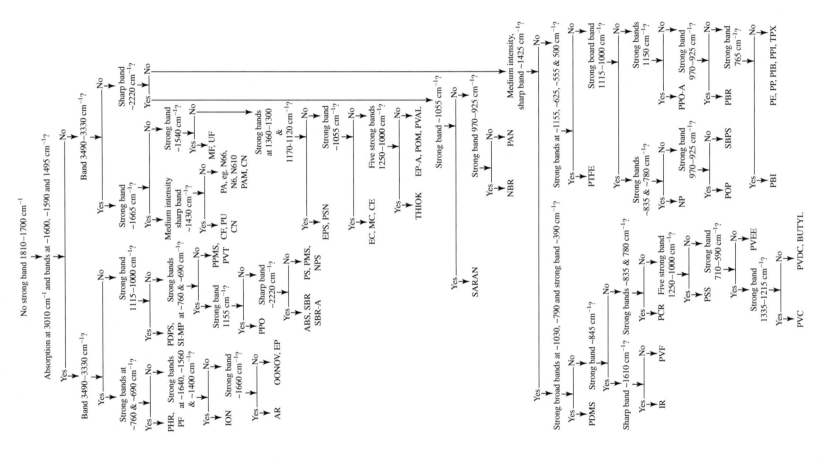

Table 21.8 (*continued*)

Name	Abbreviation
Poly-*p*-isopropylstyrene	PPIPS
Poly-*p*-methylstyrene	PPMS
Polyacrylamide	PAM
Polyacrylonitrile	PAN
Polyamide – aromatic	PA-AR
Polyamide – aliphatic	PA-A
Polyanhydride	PAH
Polybutadiene	PBR
Polybutene-1	PB1
Polycarbonate – aliphatic	PC-A
Polycarbonate – aromatic	PC-AR
Polycarpolactam–Nylon-6	N6
Polychloroprene	PCR
Polycyanoacrylate	PCA
Polydimethylsiloxane	PDMS
Polydiphenylsiloxane	PDPS
Polyester – aromatic	UP-AR
Polyester–aliphatic amine	UP-NH
Polyester – aliphatic	UP-A
Polyether – aliphatic	POE
Polyethylacrylate	PEA
Polyethylene terephthalate	PET
Polyethylene	PE
Polyhexamethylene adipamide–Nylon 66	N66
Polyimide	PIM
Polyisobutylene	PIB
Polyisoprene	IR
Polyisoprene 1,4 *cis*	IRC
Polymethylcyanoacrylate	PMCA
Polymethylmethacrylate	PMMA
Polymethylphenylsiloxane	SI-MP
Polymethylstyrene	PMS
Polyoxymethylene–Polyacetal	POM
Polyoxypropylene	POP
Polypentene-1	PP1
Polyphenylene oxide	PPO
Polyphospine oxide – aliphatic aromatic	PPO-AAR
Polyphospine oxide – aliphatic	PPO-A
Polyphospine oxide – aromatic	PPO-AR
Polypropylene	PP
Polystyrene	PS
Polysulphide	PSS
Polysulphide–formal	THIOK
Polysulphone	PSN
Polytetrafluoroethylene	PTFE
Polyurea	PU
Polyurethane	PUR-AR

Table 21.8 (*continued*)

Name	Abbreviation
Polyurethane – aliphatic	PUR-A
Polyvinyformal	PVFL
Polyvinyl fluoride	PVF
Polyvinylacetate	PVA
Polyvinylalcohol	PVAL
Polyvinylbutyral	PVB
Polyvinylchloride	PVC
Polyvinylchloride vinylacetate copolymer	PVC-PVA
Polyvinylethylether	PVEE
Polyvinylidene chloride	PVDC
Polyvinylidene chloride,acrylonitrile,vinylacetate	SARAN
Polyvinylidene fluoride	PVDF
Polyvinylpyrrolidone	PVP
Polyvinyltoluene-butadiene	PVTB
Polyvinyltoluene,*p*65%,*o*33%	PVT
Sec-butylpolysilicate	SBPS
Styrene–acrylonitrile	SAN
Styrene–butadiene	SBR
Tri(chloroethyl)phosphate	TCEP
Urea formaldehyde	UF

References

1. J. Haslam *et al.*, *Identification and Analysis of Plastics*, Iliffe, London, 1972.
2. D. O. Hummel, *Infrared Analysis of Polymers, Resins and Additives: An Atlas*, Wiley, New York, 1972.
3. D. O. Hummel (ed.), *Polymers Spectroscopy*, VCH, Weinheim, 1974.
4. R. Zibinden, *Infrared Spectroscopy of High Polymers*, Academic Press, New York, 1964.
5. H. W. Siesler and K. Holland-Moritz, *Infrared and Raman Spectroscopy of Polymers*, Marcel Dekker, New York, 1980.
6. C. J. Henniker, *Infrared Spectroscopy of High Polymers*, Academic Press, New York, 1967.
7. D. I. Bower and W. F. Maddams, *The Vibrational Spectroscopy of Polymers*, Cambridge University Press, Cambridge, 1992.
8. P. C. Painter *et al.*, *The Theory of Vibrational Spectroscopy and its Application to Polymeric Materials*, Wiley, New York, 1982.
9. G. R. Strobl and W. Hagedorn, *J. Polym. Sci. Polym. Phys. Educ.*, 1978, **16**, 1181.
10. S. L. Hsu and S. Krimm, *J. Polym. Sci. Polym. Phys. Educ.*, 1978, **17**, 2105.
11. A. Peterlin, *J. Mater. Sci.*, 1979, **14**, 2994.
12. Chicago Soc. For Paint Tech., *An Infrared Spectroscopy Atlas for Coatings Industry*, Fed. of Soc. Paint Tech., 1315 Walnut St., Philadelphia, PA, 1980
13. V. Hernandez *et al.*, *Phys Rev. B*, 1984, **50**, 9815.
14. M. Gussoni *et al.*, in *Spectroscopy. of Advanced Materials.*, R.T.H. Clark and R.E. Hester (eds), Wiley, New York, 1991, p.251.

15. G. Zerbi, in *Conjugated Polymers*, J.L. Bredas and R. Silbey (eds), Kluwer, Amsterdam, 1991, p.435.
16. C. Rumi *et al.*, *Chem. Phys.* 1997, **106**, 24.
17. C. Castilioni *et al.*, *Solid State Commun.*, 1985, **56**, 863.
18. B. Tiam *et al.*, *Chem. Phys.*, 1991, **95**, 3191.
19. V. Hernandez *et al.*, *Phys. Rev. B*, 1994, **50**, 9815.
20. P. Piaggio *et al.*, *Solid State Commun.*, 1984, **50**, 947.
21. D. B. Tanner *et al.*, *Synth. Met.*, 1989, **28**, 141.
22. C. Rumi *et al.*, *Chem. Phys. Lett.*, 1994, **231**, 70.
23. H. Eckhardt *et al.*, *Mol. Cryst. Liq. Cryst.*, 1985, **117**, 401.
24. T. A. Skotheim *et al.*, *Handbook of Conducting Polymers*, 1998, Marcel Dekker.
25. E. Villa *et al.*, *J. Chem. Phys.*, 1996, **105**, 9461.
26. S. W. Cornell and J. L. Koenig, *Macromol.* 1969, **2**, 540.
27. J. L. Koenig, *Chem. Technol.*, 1972, **2**, 411.
28. C. Pretty and R. Bennet, *Spectrochim. Acta*, 1990, **46A**, 331.
29. W. F. Maddams and L. A. M. Royaud, *Spectrochim. Acta*, 1990, **46A**, 309.
30. K. D. O. Jackson *et al.*, *Spectrochim. Acta*, 1990, **46A**, 217.
31. P. J. Hendra *et al.*, *J. Chem. Soc., Chem Commun.*, **1970**, 1048.
32. J. Agbenyega *et al.*, *Spectrochim. Acta*, 1990, **46A**, 197.
33. N. J. Everall *et al.*, *Spectrochim. Acta*, 1991, **47A**, 1305.
34. R. G. Messerschmidt and D. B. Chase, *Appl. Spectrosc.*, 1989, **43**, 11.
35. F. J. Bergin, *Spectrochim. Acta*, 1990, **46A**, 153.
36. G. Ellis *et al.*, *J. Molec. Struct.*, 1991, **247**, 385.
37. C. Johnson and S. L. Wunder, *SAMPE J.*, 1990, **26**, 19.
38. A. Elliot, *Infrared Spectra and Structure of Long-Chain Polymers*, Edward Arnold, 1969.
39. R. G. J. Miller and B. C. Stacey, *Laboratory Methods in Infrared Spectroscopy*, Heyden 1972.
40. G. Ellis *et al.*, *Spectrochim. Acta*, 1995, **51A**, 2139.
41. M. Arruebaraena de Baez, *et al.*, *Spectrochim. Acta*, 1995, **51A**, 2117.
42. S. F. Parker, *Spectrochim. Acta*, 1997, **53A**, 119.
43. R. G Messerschmidt and M. A. Harthcock (ed.), *Infrared Microspectroscopy*, Marcell Dekker, New York, 1988.
44. T. Ogawa, *Handbook for Polymer Analysis*, Japan Soc. For Anal. Chem., Asakura Shoten, Tokyo, 1985.
45. D. L. Drapcho *et al.*, *Mikrochim. Acta Suppl.*, 1997, **14**, 585.
46. M. Arruebaraena de Baez *et al.*, *Spectrochim. Acta*, 1995, **51A**, 2117.
47. J. Wang *et al.*, *Polymer*, 1989, **30**, 524
48. G. Ellis *et al.*, *Spectrochim. Acta*, 1990, **46A**, 227
49. L. Penn and F. Milanovich, *Polymer*, 1979, **20**, 31
50. H. A. Willis, J. H. van der Mass and R. G. J. Miller, *Laboratory Methods in Vibrational Spectroscopy*, 3rd edn, Wiley, Chichester, 1985.
51. H. W. Starkweather, *et al.*, *Macromolecules*, 1985, **18**, 1684.
52. R. J. Lehnert *et al.*, *Polymer*, 1997, **38(7)**, 1521.
53. K. P. J. Williams and N. J. Everall, *J. Raman Spectrosc.*, 1995, **26**, 427.
54. N. J. Everall *et al.*, *J. Raman Spectrosc.*, 1994, **25**, 43.
55. J. M. Chalmers and N. J. Everall, *Trends Anal. Chem.*, 1996, **15(1)**, 18.
56. R. N. Data *et al.*, *Rubber Chem. Technol.*, 1999, **72(5)**, 829.
57. M. Minagawa *et al.*, *Macromolecules*, 2000, **33(12)**, 4526.
58. J. Yang and Y-S. Huang, *Appl. Spectrosc.*, 2000, **54(2)**, 202.

22 Inorganic Compounds and Coordination Complexes

Since many inorganic compounds and complexes contain groups or atoms dealt with previously, the earlier chapters of this book should also be studied for any relevant information. For example, the chapters on silicon, boron, phosphorus and polymers contain a great deal of information relevant to inorganic compounds. Also, if interest is in, say, metal–olefin compounds, then sections dealing with alkenes should be examined, not only because the band positions of the free ligand should be known but also because some bands for these complexes may also be included in these sections.

The infrared study of inorganic compounds presents some difficulties in that the use of conventional organic solvents is not always possible and the use of aqueous solutions is generally precluded by energy considerations. Therefore, the use of solids (as powders) in dispersive sampling techniques is extensive. Polyethylene cells may be used in the far infrared region. For many inorganic substances, the use of Fourier transform Raman spectroscopy has a number of advantages over the use of infrared spectroscopy. Some of these advantages are that (a) for water soluble substances, aqueous samples may easily be prepared and studied (effects due to hydrogen bonding need to be borne in mind), (b) often, little or no sample preparation is required, (c) bands are often sharper and hence a better–defined spectrum is obtained, (d) glass sample cells may be used and (e) low wavenumbers, below $200\,cm^{-1}$ (above $50\,\mu m$), are accessible. For many inorganic substances, those with relatively heavy atoms, this low wavenumber accessibility can be very useful in characterisation.

Unfortunately, the infrared spectra of inorganic substances may not always be reproducible for a given sample since the extensive grinding necessary for some sampling techniques may result in (a) decomposition of the sample, (b) the crystal lattices being strained, (c) polymorphic changes, (d) varying degrees of hydration (or solvation) or (e) differences in particle size, all of which may result in spectral changes.

In general, for inorganic substances, band intensities have been less extensively studied than frequencies so it is important to realise that the absence of information in a column of a table does not necessarily indicate the absence of a band but rather suggests the absence of definitive data in the literature.

As mentioned, the infrared spectra of inorganic substance are mostly obtained in the crystalline state by using a dispersive technique e.g. a mull or a KBr disc. The structure of a substance, e.g. a metal complex, in the crystalline state may be quite different from that in solution or vapour phase. In the crystalline state, the configuration around a metal atom of a complex may become distorted or changed by coordination to neighbouring molecules. In extreme cases, dimerisation or polymerisation may occur. Even where this is not the situation, molecules or ions in crystals are in crystal fields which may cause bands to shift from the positions where they are to be found in solution or gaseous phase spectra. In general, it is difficult to make general predictions about crystal field effects on band positions. Hydrogen bonding too may cause significant band shifts. Bands due to water occur very frequently in the infrared spectra of inorganic compounds and this also needs to be borne in mind.

Compared with the spectra of organic compounds, those of inorganic compounds often consist of a relatively small number of broad bands, the exceptions being the spectra of organometallic compounds. It must be appreciated that the situation relating to infrared frequency correlations for inorganic compounds is very different from that for organic molecules where the range of atomic masses and force constants is severely restricted and there are not such numerous structural possibilities. In addition, vibrational interactions may further complicate the situation. Inorganic infrared spectroscopy should really be considered in terms of molecular modes in which many of the bonds and bond angles may change, rather than vibrations being localised in one bond or group of atoms. Hence, within the ranges given below, some limited correlations hold but they must be used with caution if applied further afield.

A number of books[1-7] and useful reviews[7-14] of a general or specific nature may be found in the literature. References are given which deal with ionic crystals,[1-5] complexes,[1,3,4,7,10-14] carbonyl compounds,[1,10,11,13,33] transition element compounds,[1,3,4,10,11] and minerals.[2,5]

Ions[1,2,4,5,11,13]

The spectra of ionic solids composed of monatomic ions, such as sodium chloride and potassium bromide, consist of broad bands. The only vibrations which can occur are the vibrations of individual ions and these are dependent on their nearest neighbouring ions. Vibrations of this type are referred to as lattice vibrations. With increase in the atomic weights of the ions involved, these vibrations occur at lower frequencies. Lattice vibrations may result in bands in both infrared and Raman spectra. In general, these bands occur below $200\,cm^{-1}$ and their intensity is variable.

Chart 22.1 Infrared – band positions of ions

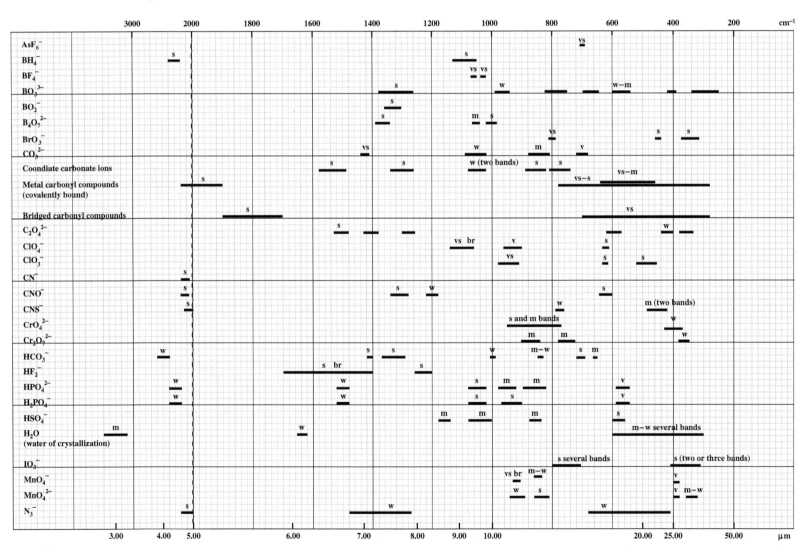

Chart 22.1 (*continued*)

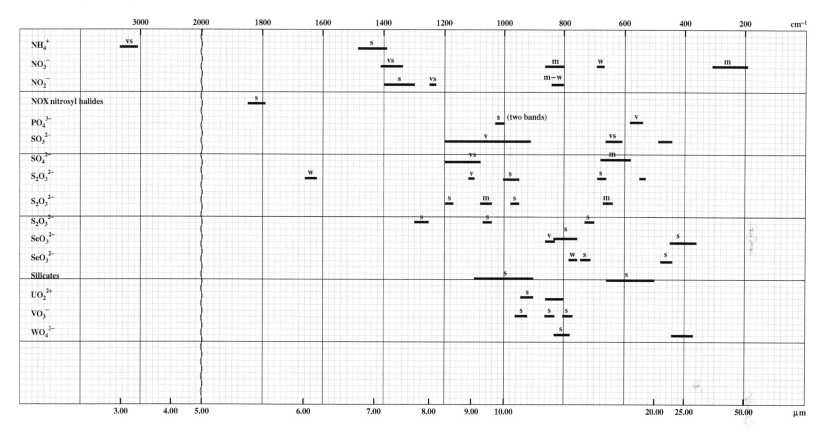

Absorption by polyatomic ions may be due to (a) internal vibrations of the ion, (b) torsional oscillations of water or other solvation molecules and (c) lattice vibrations.

The internal vibrations of the ions are independent of the sample phase and of the associated ion(s) and dependent only on the atomic structure of the polyatomic ion. These vibrations are similar to those occurring in organic substances and are characteristic of the particular ion. For example, carbonate or sulphate ions have characteristic vibration frequencies which are very nearly independent of the cation. The sulphate ion SO_4^{2-} has two characteristic bands – a very strong broad band at $1130-1080\,cm^{-1}$ ($8.85-9.26\,\mu m$) and a less intense band at $680-580\,cm^{-1}$ ($14.71-17.24\,\mu m$). The nitrate ion has a very strong absorption at $1410-1340\,cm^{-1}$ ($7.09-7.46\,\mu m$), a sharp, less intense band at $860-800\,cm^{-1}$ ($11.63-12.50\,\mu m$) and a weak band in the region $740-725\,cm^{-1}$ ($13.51-13.79\,\mu m$).

Torsional oscillations result from the water or other solvent molecules being restricted in their rotational motion, hence resulting in bands due to these torsional vibrations which may complicate the infrared spectrum.

Lattice vibrations for inorganic compounds are due to the translational and rotational motion of molecules or ions within the crystalline lattice and normally result in absorptions below $200\,cm^{-1}$ (above $50.00\,\mu m$).

Small shifts in band position may be observed for different cations. The various radii and charges of different cations alter the electrical environment of polyatomic anions and hence affect their vibrational frequencies. Obviously, different crystalline arrangements may result when the cation is altered. Normally, with increase in mass of the cation there is a shift to lower frequency. The characteristic bands of particular polyatomic ions are given in Table 22.1

Table 22.1   Free inorganic ions and coordinated ions

Functional Groups	Region cm^{-1}	Region μm	Intensity IR	Intensity Raman	Comments
AlO$_2^-$	920–800	10.87–12.50	m		br
	670–620	14.93–16.13	w		
	560–515	17.86–19.42	w		
	480–450	20.83–22.22	w		
	380–370	26.31–27.02	w		
AsO$_4^{3-}$, orthoarsenate	910–890	10.99–11.24		m–s	
	890–800	11.24–12.50	s	s–m	
	390–325	25.64–30.77		w	
AsF$_4^-$, hexafluoroarsenate	705–680	14.18–14.71	vs	s	
	~375	~26.67		m	
BH$_4^-$	2400–2195	4.17–4.56	s		Two bands. For BD$_4^-$: 1710–1570 cm^{-1}
	1150–1000	8.70–10.00	s		
BN$^-$	~1390	~7.19	s		
	~810	~12.35	w		
BBr$_4^-$, tetrabromoborate	600–240	16.67–41.67	vs		
BF$_4^-$, tetrafluoroborate	~1125	~8.89			
	~1060	~9.43	vs		
	~1030	~9.71	vs		
	780–760	12.82–13.16	w	s	
	560–510	17.86–19.60	w	m	sh
	~350	~28.57		m	
BO$_2^-$	~1175	~8.51	m		
	1350–1300	7.41–7.69	s		
	~925	~10.81			Broad
BO$_3^{3-}$, borates	1490–1260	6.71–7.94	s–m		
	1030–1010	9.71–9.09	s		
	~950	~10.53	w		
	830–750	12.05–13.33		s	Two Raman bands: one strong, one medium intensity.
	700–650	14.29–15.39			
	590–540	16.95–18.52	w–m		
	420–400	23.81–25.00		w–m	
	350–250	28.57–40.00			
B$_4$O$_7^{2-}$, tetraborate	1380–1330	7.25–7.52	s		
	1150–1100	8.70–9.09	w		
	1080–1040	9.26–9.62	m	w	
	1020–980	9.80–10.20	s	m–w	
	990–900	10.10–11.11			
	~830	~12.05			
	600–565	16.67–17.70		s	
	545–520	18.35–19.23	w		
	~500	~20.00	w		
	470–450	21.28–22.22	w		
	415–350	24.10–28.57			Several bands
BrO$_3^-$, bromate	850–760	11.77–13.16	vs	s	
	450–430	22.22–23.26	s	m	sh
	370–355	27.03–28.17	m–w	m	sh

Table 22.1 (*continued*)

Functional Groups	Region cm⁻¹	Region μm	Intensity IR	Intensity Raman	Comments
$CO_3{}^{2-}$, carbonate	1530–1320	6.54–7.58	vs	m	br
	~1160	~8.62	m		
	1100–1020	9.09–9.80	w	s–m	
	890–800	11.24–12.50	m		May be split in basic carbonates
	745–670	13.42–14.93	v	w	Not always present (anhydrous rare earth carbonates absorb at 405–305 cm⁻¹)
$CS_3{}^{2-}$, thiocarbonate	~1050	~9.52		w	
	~910	~10.99		w	
	~505	~19.80		s	
	~325	~30.77		m–w	
$ClO_4{}^{-}$, perchlorate	1170–1040	8.55–9.62	s	w	br, one or two bands observed
	955–930	10.47–10.75	v	s	
	630–620	15.87–16.13	s	m–w	sh
	490–420	20.41–23.81		m–s	
$ClO_3{}^{-}$, chlorate	1100–900	9..09–11.11	m–s	m	2–3 bands
	980–910	10.20–10.99	vs	s	
	630–615	18.87–16.26	s	w	sh
	510–480	19.61–20.83	s	m–s	sh, 1 or 2 bands
$C_2O_4{}^{2-}$, oxalate ion	1730–1680	5.78–5.95	s	m–s	C=O str
	1490–1400	6.71–7.14		s	C–O and C–C band
	1300–1260	7.69–7.94			C–O and O–C=O band
	900–800	11.11–12.50		m–s	Two Raman bands
	600–520	16.67–19.23		m–s	
	500–415	20.00–24.10	w		Ring and O–C=O def vib
	~365	~27.40			O–C=O def vib
CN^-	2240–2070	4.46–4.83	m–s	s	sh, usually 2080–2070 cm⁻¹
Cyanide, cyanate and thiocyanate: CN^-, CNO^- and SCN^-	2250–2000	4.44–5.00	s	s	CN^- has no bands below 700 cm⁻¹ SCN^- has a sharp doublet at 520–425 cm⁻¹ separation ~30 cm⁻¹
CNO^-, cyanate	~2175	~45.98	s	m	
	~1300	~7.69		s	
	~635	~15.74		w	
	~625	~16.00		w	
$CrO_4{}^{2-}$, chromate	960–770	10.42–12.99	s	s	Several bands, not all strong (except for complexes – all strong)
	420–300	23.81–33.33	w	s–m	Two bands
$Cr_2O_7{}^{2-}$, dichromate	1000–900	10.00–11.11		s	Several bands in Raman strong-to-medium intensity
	900–840	11.11–11.90	m	s	
	800–730	12.50–13.70	m	w	
	600–515	16.67–19.42		w	
	380–350	26.32–28.57	w	m	
	290–200	34.48–50.00		m	
$Fe(CN)_6{}^{4-}$	2130–2010	4.69–4.98	m	s	
	610–580	16.39–17.24	w		
	500–410	20.00–24.39	w		

(*continued overleaf*)

Table 22.1 (*continued*)

Functional Groups	Region cm^{-1}	Region μm	Intensity IR	Intensity Raman	Comments
Fe(CN)$_6$$^{3-}$	~2100	~4.76	m	s	
Fe$_2$O$_4$$^{2-}$	610–550	16.39–18.18	s		Broad
	450–400	22.22–25.00	m		Broad
HCO$_3$$^-$, bicarbonate	3300–2000	3.03–5.00	br		Number of broad bands
	1700–1600	5.88–6.25	s		Number of bands
	1420–1400	7.04–7.14	s		
	1370–1290	7.30–7.75	s	m–w	br
	1000–990	10.00–10.10	w–m	s	br
	840–830	11.90–11.05	w		sh
	710–590	14.08–14.49	m–s	m	br
	665–655	15.04–15.27	m	w–m	
HF$_2$$^-$	2125–2050	4.71–4.88	m		br
	1700–1400	5.88–7.14	s–m		Very br, max ~1450 cm^{-1} (6.90 m)
	1260–1200	7.94–8.33	s		
HPO$_4$$^{2-}$, dibasic phosphate	2900–2750	3.45–3.64	w	m–w	br
	2500–2150	4.00–4.65	w	s	br
	1900–1600	5.26–6.25	w		br
	1410–1200	7.09–8.33	w		br
	1220–1100	8.20–9.09		m–s	
	1150–1000	8.69–10.00	s	s	Broad, may be a doublet
	1110–925	9.01–10.81	m–w	m	Broad
	920–825	10.86–12.12	m–w	w	Broad
	580–450	17.24–22.22	v		Not always present
	430–390	23.26–25.64	w		Broad
H$_2$PO$_4$$^-$	~2700	~3.70	w	m–w	
	2400–2200	4.17–4.77	w	s	br
	~1700	~5.88	w		br
	~1250	~8.00	s–m		br
	1200–950	8.33–10.53	s	m–s	
	950–850	10.53–11.76	s	s	
	580–540	17.24–18.52	v	m	Not always present
	450–350	22.22–28.57		m–s	
HSO$_4$$^-$, bisulphate	3400–1900	2.94–5.26			br, number of maxima
	1190–1160	8.40–8.62	s–m	s	
	1080–1000	9.26–10.00	s		
	880–840	11.36–11.90	m	m	
	~600	~16.67		m	
GeO$_4$$^{2-}$	800–700	12.50–13.33			
	350–300	28.57–33.33			
IO$_3$$^-$, iodate	1650–1625	6.06–6.15	w		
	830–690	12.05–14.49	s–m	s	Several bands
	420–310	23.81–32.26	s	m–w	Two or three sh bands
Iodates (covalently bonded or coordinated)	810–755	12.35–13.25	m–s		
	795–715	12.58–13.99	s		
	690–630	14.49–15.87	s		

Table 22.1 (*continued*)

Functional Groups	Region cm^{-1}	Region μm	Intensity IR	Intensity Raman	Comments
	480–420	20.83–23.81	m		
MnO$_4^-$	950–870	11.76–11.49	vs		br
	850–750	11.76–13.33	m–s		
	400–380	25.00–26.32	v		Often weak doublet
MnO$_4^{2-}$	900–800	11.11–12.50	s		More than one band
MoO$_4^-$	935–890	10.70–11.24	w–m	s	Several bands
	850–810	11.76–12.35	s	w–m	
	400–380	25.00–26.32	v	w–m	
	350–310	28.57–32.26	m	m–w	
N$_3^-$, azide,	2170–2030	4.61–4.93	s	w	
	1375–1175	7.27–8.51	w	s	Strong Raman band ~1360 cm^{-1}, not infrared active. A weak Raman band occurs ~1270 cm^{-1}.
	680–410	14.71–24.39	w–m	w	May be a doublet
NH$_4^+$, ammonium	3335–3030	3.00–3.30	vs	m	Broad
	1490–1325	6.71–7.55	s	w	(No bands below 700 cm^{-1})
NO$_3^-$, nitrate	1800–1700	5.56–5.88	w		Number of bands
	1520–1280	6.58–7.81	vs	m–w	br
	1070–1015	9.35–9.85	w	s	
	860–800	11.63–12.50	m–w	m–s	sh (no bands below 700 cm^{-1})
	770–700	12.99–14.29	m–w	m–w	
	315–190	31.75–52.63	m		
NO$_2^-$, nitrite	1400–1300	7.14–7.69	s	s	Two bands for nitrite complexes. Raman band ~1320 cm^{-1}
	1285–1185	7.78–8.44	vs	w–m	
	860–800	11.62–12.50	m–w	m–s	(No bands below 700 cm^{-1})
	~750	~13.33			
Nitric oxide (monomer)	~1885	~5.31			(*cis* dimer, ~1860 and ~1765 cm^{-1}; *trans* dimer, ~1740 cm^{-1})
NO$^+$	2370–2230	4.22–4.48	s		(Nitric acid, ~2220 cm^{-1})
NO$^+$ (coordinated M–NO)	1945–1500	5.14–6.67	s		
Nitrosyl halides, NOX	1850–1790	5.41–5.59	s		
NCO$^-$, cyanates	2225–2100	4.49–4.76	s	s–m	Out-of-plane CNO str
	1335–1290	7.49–7.75	s	s	In-plane CNO str
	1295–1180	7.72–8.47	w		Overtone. Raman strong-to-medium band ~1210 cm^{-1}.
	650–590	15.39–16.95	s–m		Bending vib
NCS$^-$, thiocyanates	2190–2030	4.57–4.93	s	s	asym str
	~950	~10.53	w		
	760–740	13.16–13.51	w	w–m	
	470–420	~21.28–23.81	w–m	w	
OsO$_4^{2-}$	350–300	28.57–33.33	v		
PF$_6^-$, hexafluorophosphate	~915	~10.93	m		
	850–840	11.76–11.90	vs		
	750–745	13.33–13.42		s	
	580–555	17.24–18.02	m	w	

(*continued overleaf*)

Table 22.1 (*continued*)

Functional Groups	Region cm^{-1}	Region µm	Intensity IR	Intensity Raman	Comments
	~475	~21.05		m	
PO$_4^{3-}$, phosphate	1180–1000	8.48–10.00	s	s	Often complex structure, broad
	1000–900	10.00–9.09		s	
	580–540	17.24–18.52	v	w–m	Not always present
	415–380	24.01–26.32		w–m	
PO$_3^-$, metaphosphate	1305–1105	7.66–9.05		m–w	
	1250–1150	8.00–8.70		s	
	700–670	14.29–14.93		m–s	
P$_2$O$_7^{4-}$, pyrophosphate,	1220–1060	8.20–9.43	s	m–s	
	1060–960	9.43–10.42	w–m	s	
	980–850	10.20–11.76	m	w	
	770–700	12.99–14.29	m–w	w–m	
	600–500	16.67–20.00	m–s	w–m	
	530–400	17.24–25.00		w–m	
	355–315	28.17–31.75		v	
ReO$_4^-$	950–890	10.53–11.24			
RuO$_4^{2-}$	350–300	28.57–33.33	v		
S$_2$O$_3^{2-}$, thiosulphate	1660–1620	6.02–6.17	w		
	1200–1100	8.33–9.09	v	m–s	Usually strong
	~1010	~9.90		s	
	1000–950	10.00–10.53	s		
	695–660	14.39–15.15	s	m–s	
	550–530	18.18–18.87		w	
	~450	~22.22		s	
S$_2$O$_5^{2-}$, pyrosulphite	1250–1200	8.00–8.33		w–m	
	1190–1170	8.40–8.55	s	w–m	
	1100–1040	9.09–9.92	m	s	
	990–950	10.10–10.53	m–s	m–w	
	670–640	14.93–15.63	m	m–s	
	570–560	17.54–17.86	m		
	540–510	18.52–19.61	m		
	450–440	22.22–22.73	m		
	280–250	35.71–40.00		s	
S$_2$O$_8^{2-}$, peroxysulphate	1310–1250	7.63–8.00	s		
	1070–1040	9.35–9.62	s	s	sh
	~815	~12.27		m–s	
	740–690	13.51–14.49	s–m		
	600–580	16.67–17.24	w–m		sh
	~560	~17.86	m		
	470–440	21.28–22.73	w		
S$_2$O$_6^{2-}$, dithionate	1230–1200	8.13–8.33		w	
	1100–1000	9.09–10.00		s	
	760–690	13.16–14.49		m–s	
	570–500	17.54–20.00		m–w	
	350–300	28.57–33.33		m–w	Two bands
SO$_4^{2-}$, sulphate	1200–1140	8.33–8.77	m	m–s	

Table 22.1 *(continued)*

Functional Groups	Region cm⁻¹	Region μm	Intensity IR	Intensity Raman	Comments
	1130–1080	8.85–9.26	vs	m–s	Broad band with shoulders
	1065–955	9.39–10.47	w	s	sh, not always present
	680–580	14.71–17.24	m	m–s	Several bands
	530–405	18.87–24.69		m–s	Doublet
SO_3^{2-}, sulphite	~1215	~8.23	w		
	~1135	~8.81	w		
	1010–900	9.90–11.11	v	s	Often strong, broad. Usually two bands.
	660–615	15.15–16.26	m	m	
	495–450	20.20–22.22		m–s	
SbF_6^-	~695	~16.43	m–s		
Selenate, SeO_4^{2-}	935–830	10.70–12.05	v	m–s	Often strong
	830–750	12.05–13.33	s	s	
	450–350	22.22–28.57	s	m	
	370–300	27.02–33.33		m	
Selenites SeO_3^{2-}	780–750	12.82–13.33	w		
	740–710	13.51–14.08	s		
	480–450	20.83–22.22	s		
SiF_6^-, hexafluorosilicate	~725	~13.79	s		
	665–645	15.04–15.50	w	s	
	490–360	20.41–27.78		w–m	Not observed in infrared
Silicate	1100–900	9.09–11.11	s	s	
SiO_3^{2-}	~1165	~8.58	m		
	1030–960	9.71–10.42	s		
	790–750	12.66–13.33	w		
	500–450	20.00–22.22	s		
SiO_4^{4-}, orthosilicate	1180–860	8.47–11.63	s		Number of bands
	540–470	18.52–21.28	s		
Metasilicate	750–730	13.33–13.70	w		
	470–460	21.28–21.74	s		
TeO_4^-	940–860	10.64–11.63			
	320–300	31.25–33.33			
TeO_4^{2-}	650–600	15.39–16.67			
	340–280	29.41–35.71			
TiO_3^{2-}, titanate	700–500	14.28–20.00	s	s	br
	450–360	22.22–27.78	s	s–m	br
	400–200	25.00–50.00	br	s	Strong Raman band
	280–250	35.71–40.00	w		
	100–60	100.00–166.67	w		
UO_2^+	940–900	10.64–11.11	s		
$U_2O_7^{2-}$	900–880	11.11–11.36	s		
	480–470	20.83–21.28	m–s		
	280–270	35.71–37.04	w		
VO_3^-, metavanadate	1010–920	9.90–10.87	s	s	
	890–830	11.24–12.05	s	s–m	
	800–770	12.50–12.99	s		
	~650	~15.38		m	

(continued overleaf)

Table 22.1   (*continued*)

Functional Groups	Region cm^{-1}	Region μm	Intensity IR	Intensity Raman	Comments
	540–490	18.52–20.41		w	
	~250	~40.00		m	
VO$_4$$^{3-}$, orthovanadate	~1000	~10.00		s	
	875–825	11.43–12.12	s		
	350–300	28.57–33.33		w	
WO$_4$$^{2-}$, tungstate	960–780	10.42–12.82	s	s	
	900–770	11.11–12.99	s	m–s	1 to 3 bands
	~490	~20.41	w	w	
	~350	~28.57	w	m	
ZrO$_3$$^{2-}$	770–700	12.99–14.29	w		
	600–450	16.67–22.22	s	s	
	500–300	20.00–33.33	s	s–m	
	240–230	41.67–43.48	w	s	

Coordination Complexes[1,3,10–13]

When a ligand coordinates to a metal atom M, new modes of vibration, not present in the free ligand, may become infrared or Raman active. For example, for a coordinated water molecule, rocking, twisting and wagging modes become possible and, of course, this is in addition to the metal–ligand stretching vibration (e.g. M–O). The ammonia molecule exhibits bands in its spectrum which result from rocking and M–N stretching vibrations not observed in the free molecule. This general behaviour is true of other ligands, e.g. NO$_2$, PH$_3$.

In general, the frequencies of these new bands not only depend on the ligand involved but are also sensitive to the nature of the metal atom – its size, charge, etc. On coordination to a metal atom, the infrared bands of a ligand alter position and intensity when compared with the free ligand. Unfortunately, changes in band intensities have been very much less extensively studied than those of frequency. The direction of a particular band shift is dependent on the structure of the complex. In a series of metal complexes having the same structure, the shift in position of the ligand bands increases with coordination bond strength. Also, the nature of a normal vibration and the effect of coordination on it affects the direction of the corresponding shift. For example, the band due to N–H stretching vibration of the glycino ion, [NH$_2$·CH$_2$·CO$_2$]$^-$, shifts to lower frequencies on coordination to a metal atom whereas in the free ligand the bands due to the CO$_2$$^-$ stretching vibrations (asymmetric and symmetric) which occur near 1620 cm^{-1} (6.17 μm) and 1400 cm^{-1} (7.14 μm) respectively are replaced in the complex by a single band due to the C=O stretching vibration which appears near 1740 cm^{-1} (5.75 μm). The M–O and C–O vibrations are dependent on the nature of the metal atom, its charge, oxidation state and other ligands bonded to it.

In general, other factors being equal, for a given metal atom, the atom–ligand vibration frequencies increase with the oxidation state of the atom and the atom–ligand frequencies decrease with increasing coordination number. This last effect is independent of the charge of the complex.

Other parameters being equal, the factors affecting metal–ligand stretching frequencies are given in Table 22.2.

When different substituents are attached to a common moiety, the atom–ligand frequencies show small but systematic variations, the nature of which are partly dependent on the nature of the bonding of the moiety to the atom. However, the magnitude of these variations is much smaller than those due to the oxidation state or coordination number of the atom.

Ligands resulting in a strong *trans*-effect significantly weaken the metal–ligand bond *trans* to them, hence the corresponding stretching vibration of this bond occurs at lower frequencies than might otherwise be expected. This effect is particularly easy to observe in Pt(II) complexes with ligands exhibiting a strong *trans*-effect such as H, CH$_3$, CO, C$_2$H$_4$.

Chart 22.2 Infrared – band positions of hydrides

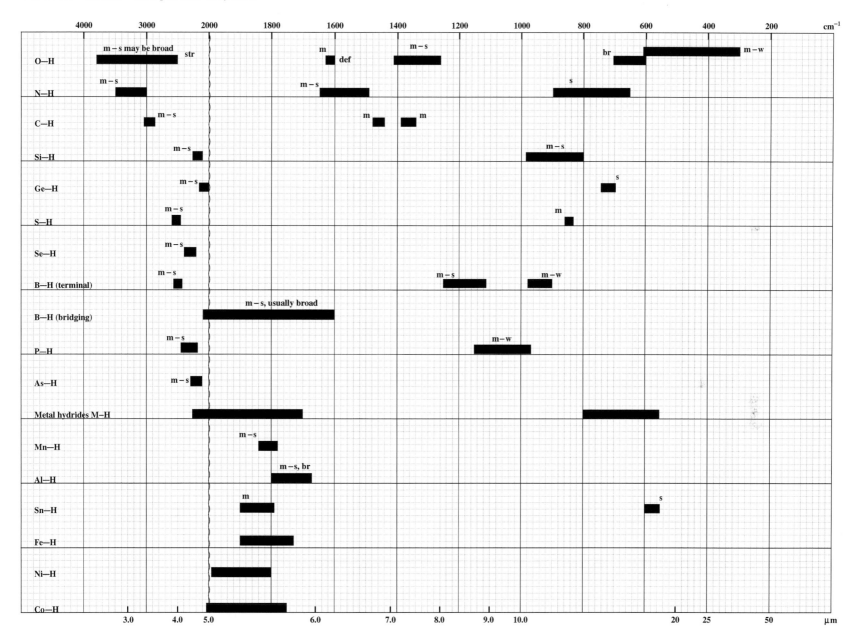

Chart 22.2 (*continued*)

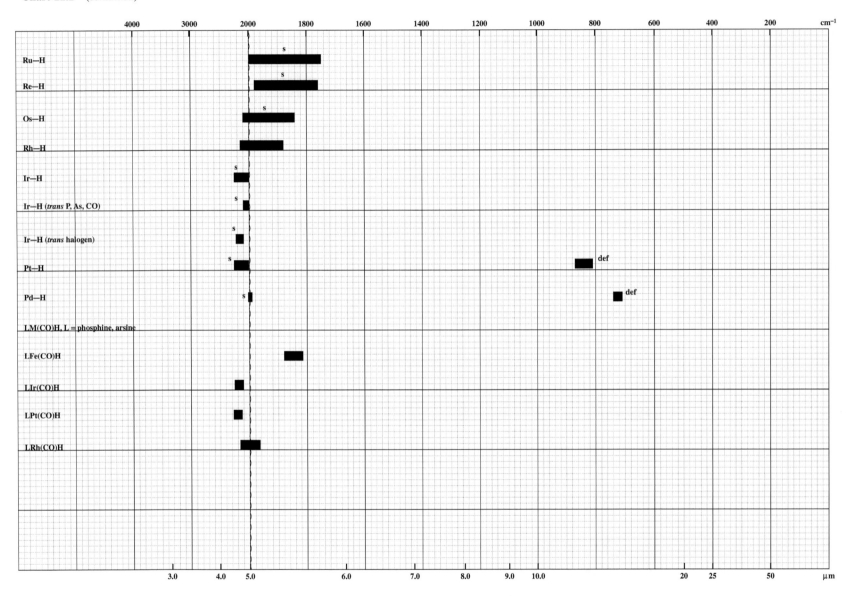

Chart 22.3 Infrared – band positions of complexes, ligands and other groups

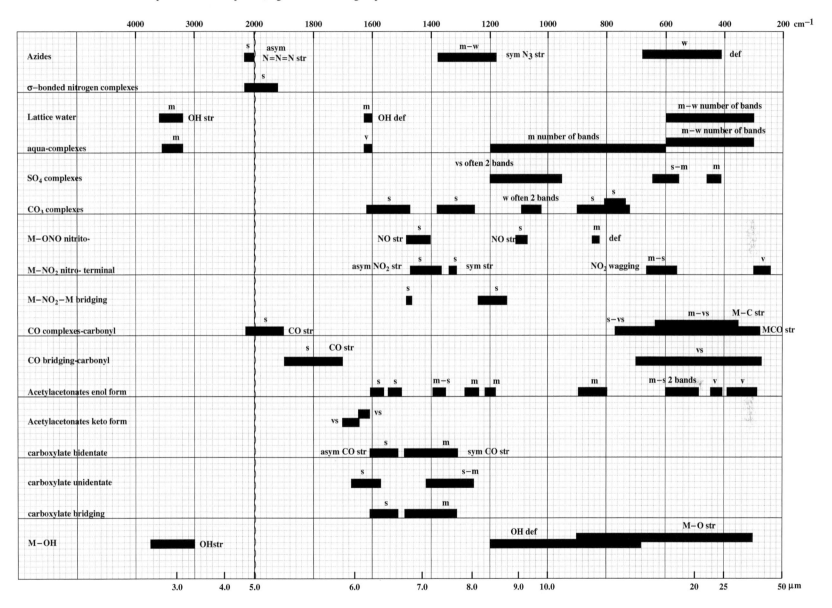

Chart 22.3 (*continued*)

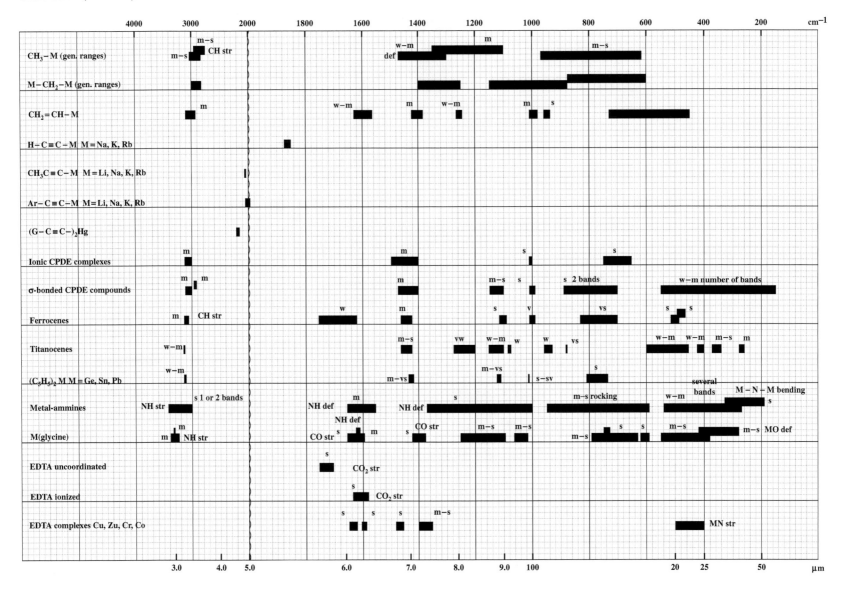

Chart 22.3 (*continued*)

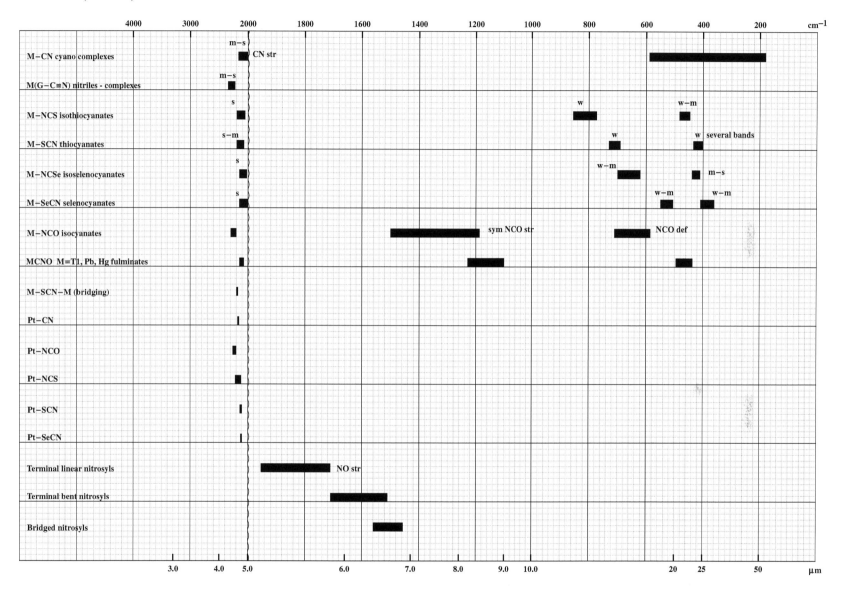

Chart 22.3 (*continued*)

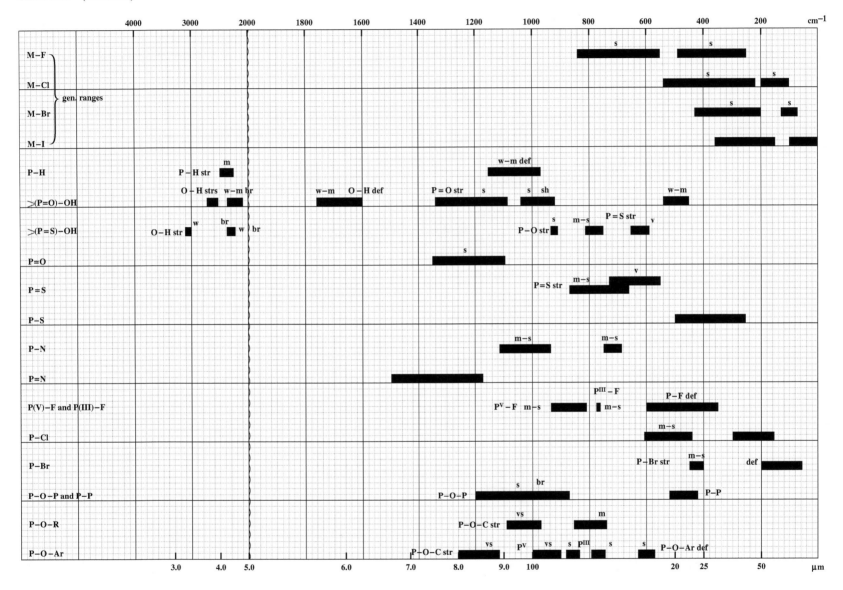

In general, coordination complexes of CO_2, SO_2, NO and other oxygen-containing ligands have strong bands. Polyene and polyenyl complexes, in general, do not give characteristic absorptions.

As with a polyatomic ligand, when a free ion becomes covalently bound, there is a decrease in its structural symmetry. This results in the removal of the degeneracy of some vibrations, causing new bands to be observed in the infrared (and Raman) region. Hence, infrared spectroscopy may be used to distinguish between ionic and covalent bonds in coordination complexes.

In general, the positions of the absorption bands for a particular ligand depend on the metal atom(s) to which coordination occurs. However, the position may also be dependent on the crystalline environment.

Isotopic Substitution

A problem often encountered in the assignment of infrared bands for a complex is that of ambiguity. For example, a band near $2000\,cm^{-1}$ ($5.00\,\mu m$) may be due to M–H or C=O stretching vibrations. To resolve the problem, isotopic substitution may be used. For example, the deuterated equivalent complex will have the band resulting from its M–H vibration shifted to lower frequency. This frequency shift may easily be estimated using reduced masses and hence the presence of the group confirmed or otherwise.

If the mass of a metal atom is much greater than one, as is generally the case, then

$$\frac{\mu(M-D)}{\mu(M-H)} \approx 2$$

where D indicates deuterium and μ is the reduced mass.

Table 22.2 Metal–ligand factors

Factors affecting metal–ligand stretching vibrations	M–ligand stretching vibrations
Oxidation state of metal atom	An increase in the oxidation number increases the frequency
Coordination number	The greater the coordination number the lower is the frequency
Coordination bond strength	The stronger the bond the higher is the frequency
Mass of metal atom	The larger the mass the lower is the frequency
Base strength of ligand (i.e. proton affinity)	The greater the base strength of the ligand the higher is the frequency
Bridging ligand	Bridging ligands have lower frequency metal–ligand stretching vibrations than equivalent terminal metal–ligands

This means that the metal–deuteron stretching vibration, ν_{M-D} occurs at $\nu_{M-H}/\sqrt{2}$. On deuteration, other bands in the spectra may shift slightly to lower frequency but not by this amount. Note that there are reasons why the shift may not be precisely that given above (see page 314).

Isotopic considerations may help reduce ambiguity or simplify spectra. For example, instead of observing the boron spectra of compounds containing both naturally abundant isotopes ^{10}B and ^{11}B, only one isotope form may be used in the chemical preparation of a compound. There may be occasions when it is necessary to bear in mind that chlorine has two naturally occurring isotopes, ^{35}Cl and ^{37}Cl.

Coordination of Free Ions having Tetrahedral Symmetry

Typical of this class of ion are sulphates, perchlorates and phosphates. The sulphate ion, for instance, may coordinate to a metal atom as a unidentate ligand $M-OSO_3$, as a chelating bidentate ligand

or as a bridging bidentate ligand $M-O-SO_2-O-M$.

Free ions with tetrahedral symmetry, T_d, have four fundamental vibrations, only two of which are infrared active (one stretching mode and one bending mode). For unidentate coordination, the symmetry is reduced to C_{3v}, each of the bands for the free ion being split into two bands with, in addition, the two previously only Raman active vibrations now becoming infrared active. Therefore, three bands due to stretching vibrations and three due to bending vibrations are expected. For bidentate coordination, the symmetry is reduced to C_{2v} and each of the bands due to the two modes of vibration of the free ion is now split into three, so that, taking into account the bands which were inactive for the free ion, four bands due to stretching vibrations and four due to bending vibrations are observed. The above applies equally to all other ions with tetrahedral symmetry.

Coordination of Free Ions having Trigonal–Planar Symmetry

Typical of this class of ion are carbonates and nitrates. These ions may form complexes as a unidentate or bidentate ligand. The free ion, which has D_{3h} symmetry, has one stretching and one in-plane deformation vibration which are infrared active, the bands each being split into two in the case of both unidentate and bidentate coordination. In addition, a band due to the symmetric stretching vibration, which previously appeared only in the Raman spectrum, appears in the infrared spectrum, this vibration now being infrared active. Unidentate

Table 22.3 Sulphate and carbonate ion complexes

Functional Groups	Region		Intensity		Comments
	cm^{-1}	μm	IR	Raman	
Sulphate ion complexes SO$_4$ complexes (including bridging)	1200–950	8.33–10.53	vs–s	m–s	Sometimes two bands SO$_4$ str
	650–550	15.39–18.18	s–m	m–s	SO$_4$ def vib
	460–410	21.74–24.39	m		
Carbonate ion complexes CO$_3$ complexes (including bridging)	1620–1450	6.21–6.90	s	m	
	1380–1250	7.25–8.00	s		
	1090–1020	9.17–9.80	w	m–s	Usually two bands
	900–720	11.11–13.89	s		
	810–735	12.35–13.61	s		

coordination may be distinguished from bidentate coordination since the separation of two of the bands due to stretching vibrations is larger for the latter.

For carbonato complexes, bands due to the Pt–O and Co–O stretching vibrations occur near 390 cm^{-1} (25.64 μm) and 430–360 cm^{-1} (23.26–27.78 μm) respectively.

Anhydrous covalent nitrates which have C_{2v} symmetry exhibit strong bands due to metal–oxygen stretching vibrations in the region 350–250 cm^{-1} (28.57–50.00 μm), whereas ionic nitrates (D_{3h} symmetry) do not have bands in this region. Anhydrous rare earth nitrates absorb in the region 270–180 cm^{-1} (37.04–55.56 μm).

Coordination of Free Ions having Pyramidal Structure

The sulphite ion, SO$_3{}^{2-}$, has a pyramidal structure and may coordinate with a metal atom as a uni- or bi-dentate ligand. It may also act as a bridging ligand.

The free sulphite ion has C_{3v} symmetry, exhibiting two bands due to stretching vibrations near 1010 cm^{-1} (9.90 μm) and 960 cm^{-1} (10.42 μm) and two bands due to bending vibrations at about 635 cm^{-1} (15.75 μm) and

495 cm^{-1} (20.20 μm). Coordination through the sulphur atom does not alter the symmetry. However, coordination through an oxygen atom reduces the symmetry and, as a result, both the bands near 960 cm^{-1} (10.42 μm) and 495 cm^{-1} (20.20 μm) split into two components. Coordination through the sulphur atom results in the bands due to SO stretching vibrations being shifted towards higher frequencies as compared with the free ion whereas coordination through the oxygen atom results in a shift to lower frequencies as compared with the free ion.

Coordinate Bond Vibration Modes

Bands due to the stretching or deforming of the coordinate bond are generally found at the low-frequency end of the infrared range, both the heavy metal atom and the nature of the coordinate bond being responsible for this.

Coordination complexes frequently contain metal–oxygen or metal–nitrogen bonds, but the absorption bands associated with these bonds are normally difficult to assign empirically since their position is dependent not only on the metal but also on the ligand and, in addition, coupling with other vibration modes often occurs.

By comparing the spectrum of the free ligand with that of the complex, metal–ligand vibrations may often be identified although, since some ligand vibrations may become infrared or Raman active on forming the complex (as explained previously), it is not uncommon for no clear assignments to be made by this comparison.

Metal–ligand vibrations may sometimes be identified by changing the central metal atom or its valency state. This technique is useful when a series of complexes with the same structure is being studied.

Isotopic substitution of the ligand in order to observe isotopic shifts in spectral bands may also be used for the study of metal–ligand vibrations,

Table 22.4 Aquo complexes etc

Functional Groups	Region		Intensity		Comments
	cm^{-1}	μm	IR	Raman	
Lattice water	3600–3200	2.82–3.13	m	w	H–O–H str
	1630–1600	6.14–6.25	m	w	H–O–H def vib
	600–300	16.67–33.33	m–w	w	Number of bands
Aquo complexes	3550–3200	2.82–3.13	m	w	H–O–H str. Due to hydrogen bonding these bands may be observed at even lower frequencies, may be broad
	1630–1600	6.14–6.25	m–w	w	H–O–H def vib
	1200–600	8.50–16.67	m	w	Number of bands
	600–300	16.67–33.33	m–w	w	Number of bands. M–O bands also observed for true aquo complexes
Hydroxo complexes M–OH	3760–3000	2.66–3.33	m	w	O–H str
	1200–700	8.33–13.33	m		MOH def vib
	900–300	11.11–33.33			M–O str
$M^{2+}OH_2$, M=transition metal	580–530	17.21–18.87			M–OH$_2$ wagging vib
	450–35	22.22–285.71			M–O str
$M^{2+}OH_2$, M=rare earth	480–430	20.83–23.26			M–OH$_2$ wagging vib
	450–400	22.22–20.00			M–O str

although caution must be exercised as shifts in other ligand bands are also observed. Isotopic substitution of the metal atom is preferable, if possible, since only bands due to vibrations involving the metal atom will be shifted. The magnitude of the isotopic shift is usually small (2–10 cm^{-1}), depending, of course, on the relative mass difference.

Similar difficulties to those mentioned above arise in assignments of bands due to metal–carbon vibrations. Cyano complexes fall into this category and are extremely important and also common. The back donation of electrons by the metal atom to the ligand may complicate matters further by altering the character of the M–C bond. Carbonyl complexes also fall into this category.

Structural Isomerism

Cis–trans isomerism

Infrared spectroscopy may be used to distinguish between *cis* and *trans* isomers of compounds. The structural symmetry of the molecule is used to determine the point group, the vibrational selection rules then being applied to determine which vibration bands are observed.

Lattice Water and Aquo Complexes[1,10–13]

In true aquo complexes, the water molecule is firmly bound to the metal atom by means of a partial covalent bond and is known as coordinated water. However, in some other cases, the metal–oxygen bond may be almost ionic

in nature and in these hydrates the water molecule may be considered as crystal or lattice water. Since this type of water molecule is trapped, certain rotational and vibrational motions become partially hindered by environmental interactions and may in fact become infrared (and Raman) active. The resulting absorption bands are observed in the region 600–300 cm^{-1} (16.67–33.33 μm). It should be borne in mind that, although bands may be identified as due to a particular mode of vibration, coupling of vibrational modes may occur as is true in this case. The positions of the bands are sensitive to the anions present since hydrogen bonding also occurs.

In a similar manner, the vibrational modes of coordinated water molecules, such as wagging, twisting and rocking (which cannot occur in lattice water molecules) become infrared active, the resulting bands occurring in the region 880–650 cm^{-1} (11.36–15.38 μm). (For large water clusters see reference 36.)

In addition to the above vibrations which may become infrared active, bands due to asymmetric and symmetric H–O–H stretching vibrations are observed in the region 3550–3200 cm^{-1} (2.82–3.13 μm) and bands due to the H–O–H bending vibrations occur in the region 1630–1600 cm^{-1} (6.13–6.25 μm). As might be expected, the vibration frequencies of coordinated water are affected by the metal ion's charge and mass. It should be noted that the band near 1600 cm^{-1} is not exhibited by hydroxo complexes, M–OH, instead a band due to the M–O–H bending vibration is observed below 1200 cm^{-1} (8.33 μm). In hydroxo complexes,[35] where the OH group forms a bridge, this bending vibration occurs at about 950 cm^{-1}.

Table 22.5 Metal alkyl compounds

Functional Groups	Region		Intensity		Comments
	cm^{-1}	μm	IR	Raman	
CH$_3$–M (general ranges)	3050–2810	3.28–3.56	m–s	m	asym CH$_3$ str, intensity dependent on metal atom
	2950–2750	3.39–3.64	m–s	m	sym CH$_3$ str
	1475–1300	6.78–7.69	m–w	m	asym CH$_3$ def vib
	1350–1100	7.41–9.09	m–s	m–w	sym CH$_3$ def, sharp
	975–620	10.26–16.13	m–s	w	CH$_3$ rocking vib
Bridging CH$_3$ groups					
Li–CH$_3$–Li	~2840	~3.52	m–s	m	CH$_3$ str
	~2780	~3.60	m–s	m	CH$_3$ str
	~1480	~6.76	m	m–w	asym CH$_3$ def vib
	~1425	~7.02	m	m–w	asym CH$_3$ def vib
	~1095	~9.13	m	w	sym CH$_3$ def vib
	~1060	~9.43	w	w	sym CH$_3$ def vib
Be–CH$_3$–Be	~2910	~3.44	m–s	m	CH$_3$ str. No bands 1500–1400 cm^{-1}
	~2855	~3.50	m–s	m	CH$_3$ str
	~1255	~7.97	m	m–w	sym CH$_3$ def vib
	~1245	~8.03	m	m–w	sym CH$_3$ def vib
Mg–CH$_3$–Mg	~2850	~3.51	m	m	CH$_3$ str. No bands 1500–1400 cm^{-1}
	~2780	~3.60	m	m	CH$_3$ str
	~1200	~8.33	m	m	sym CH$_3$ def vib
	~1185	~8.44	m	w	sym CH$_3$ def vib
	~710	~14.09	w	w	CH$_3$ rocking vib
Bridging methylene groups					
M–CH$_2$–M (general ranges)	3000–2800	3.33–3.57	m	m	CH$_2$ str
	1400–1250	7.14–8.00	m	m–w	CH$_2$ def vib
	1150–875	8.70–11.43	w	w	CH$_2$ rocking vib
	850–600	11.77–16.67	w	w	CH$_2$ rocking vib
Co^{III}–C	535–355	18.69–28.17			
Cr^{III}–C	460–330	21.74–30.30			
Cu–C	365–285	27.40–35.09			
Sn–C	610–500	16.39–20.00			Sn–C asym str
	530–450	18.87–22.22			Sn–C sym str
Hg–C	580–330	17.24–30.30			Hg–C str
Mo–C	405–365	24.69–27.40			
Pb–C	500–420	20.00–23.81			Pb–C str
Al–C	775–505	12.90–19.80			Al–C str
Pt–C	580–505	17.24–19.80			Pt–C str, square planar complexes
Pd–C	535–435	18.69–22.99			Pd–C str, square planar complexes
Au–C	545–475	18.35–21.05			Au–C str, square planar complexes

Metal–Alkyl Compounds[10–13]

Compounds containing methyl–metal atom bonds give rise to the normal CH$_3$ vibrations but, in addition, bands due to M–C and C–M–C stretching and deformation vibrations may be observed. More than one band may be observed in the regions given in Table 22.5 due to the combination of vibrational modes which occur if more than one methyl group is bonded to the metal atom or if, in the solid phase, molecular distortion occurs.

Table 22.6 Approximate stretching vibration frequencies for tetrahedral halogen compounds (AX$_4$)

Substance	Region cm^{-1}	Region µm	Substance	Region cm^{-1}	Region µm	Substance	Region cm^{-1}	Region µm
CF$_4$	~1280	~7.81	GeF$_4$	~800	~12.50	TiF$_4$	~795	~12.58
	~910	~10.99		~740	~13.51		~710	~14.09
CCl$_4$	~775	~12.99	GeCl$_4$	~460	~22.22	TiCl$_4$	~490	~20.41
	~460	~21.74		~400	~25.00		~390	~25.64
CBr$_4$	~675	~14.81	GeBr$_4$	~330	~30.30	TiBr$_4$	~385	~25.31
	~270	~37.04		~235	~42.55		~230	~43.48
CI$_4$	~555	~18.02	GeI$_4$	~265	~37.74	TiI$_4$	~320	~31.25
	~180	~55.56		~160	~62.50		~160	~62.50
SiF$_4$	~1025	~9.80	SnCl$_4$	~405	~24.69	VCl$_4$	~480	~20.83
	~800	~12.50		~370	~27.03		~385	~25.97
SiCl$_4$	~610	~16.39	SnBr$_4$	~280	~35.71	ZrF$_4$	~670	~14.93
	~425	~23.53		~220	~45.45	ZrCl$_4$	~385	~25.97
SiBr$_4$	~485	~20.62	SnI$_4$	~215	~46.51		~120	~83.33
	~250	~40.00		~150	~66.67	ThF$_4$	~520	~19.23
SiI$_4$	~405	~24.69	PbCl$_4$	~350	~28.57	ThCl$_4$	~335	~29.85
	~165	~60.60		~330	~30.30	HfF$_4$	~645	~15.50
						HfCl$_4$	~395	~25.32

In general, the positions of bands due to stretching and asymmetric deformation vibrations are little affected by changing the metal atom, whereas the symmetric deformation and rocking vibrations are much more sensitive to such changes. In a given series, both the latter vibrations move to lower frequencies with increase in mass of the metal atom. With increase in mass of the metal atom, both the M–C stretching and C–M–C deformation frequencies decrease, as does the separation of the corresponding bands.

For transition metal–methyl groups, the symmetric deformation vibration is at substantially lower frequencies than for organic compounds, occurring in the region 1245–1170 cm^{-1} (8.03–8.55 µm), and the associated absorption band is characteristically intense and sharp. In addition to the bands mentioned for the methyl group, ethyl–metal compounds have bands associated with C–C stretching, C–C–M bending and methyl torsional vibrations. For cycloalkyl groups bonded directly to a metal atom or an other atom (e.g. P, S or Si), the C–H stretching vibration frequency increases as the ring size decreases. In the case of heterocyclic compounds, in general, considerable mixing of vibrational modes occurs.

Metal–carbon stretching frequencies occur in the region 775–420 cm^{-1} (12.90–23.81 µm) for metal–alkyl and metal–alkenyl bonds, aluminium absorbing at the high end of this range.

Metal Halides[1,3,10–13,15]

The bands due to metal–halogen vibrations occur in the following regions:

	F	Cl	Br	I
M–X str (cm^{-1})	945–550	610–220	430–200	360–150
M–X def (cm^{-1})	490–250	200–100	130–50	100–30

As might be expected from previous discussions, the position of a metal–halogen absorption band is dependent on (a) the strength of the bond, (b) the mass of the metal atom, and (c) the valence state of the metal atom. Other factors being equal, the frequency decreases with increase in mass of the metal atom. In general, the bands due to M–X stretching vibrations are of strong intensity, the intensity increasing as the electronegativity of the halogen increases. The spectra of Group II metal dihalides[15] and of Group III–V metal fluorides[17,18] have been reviewed.

As mentioned earlier, other factors being equal, the higher the oxidation state of an atom, the greater the atom–ligand stretching frequency. The metal–halogen stretching frequencies are dependent on oxidation number and the stereochemistry of the compound. As a rough guide, for the transition metals, the other ligands being the same, $\nu_{M-Br}/\nu_{M-Cl} = 0.77$.

Metal halides of the MX_6 and MX_4 type (the former being octahedral and the latter either square planar or tetrahedral) should have only one band due to stretching vibrations, although more than one may be observed if the symmetry is lowered. This may be due either to molecular distortion because of stereochemical considerations or to interactions with neighbouring molecules in a particular crystalline environment. The effect of molecular distortion is much larger than that due to the crystalline phase, the latter often being too small to be observed. The broadening of the band(s) may also be observed for chlorine and bromine due to the presence of the different naturally occurring isotopes. As might be expected, in different crystalline environments, frequency shifts are observed for ionic metal halides, e.g. $MCl_6{}^{2-}$, the shift being dependent, for a given halide, on the nature of the cation and its size. In general, as expected from simple infrared theory, the M–X stretching frequency decreases with increase in cation size.

For tetrahedral AX_4 compounds, it has been found that, other factors being equal, for any atom, the ratio of the symmetric or asymmetric stretching frequencies for any two given halogens is approximately constant. For example, for asymmetric stretching, ν_{C-Cl}/ν_{C-F} and ν_{Si-Cl}/ν_{Si-F} are both approximately 0.51.

For transition metal complexes, the metal–halogen stretching frequency has been found to be dependent on the *trans* influence of the ligand in the *trans* position (see page 292).

Increasing the coordination number of a metal atom generally results in a decrease in the metal–halogen stretching frequency. For example, the Ti–Cl stretch in $TiCl_4$ occurs at about $490\,cm^{-1}$ ($20.41\,\mu m$) but when the titanium has coordination numbers of 6 and 8, the Ti–Cl frequency decreases to about $375\,cm^{-1}$ ($26.67\,\mu m$) and $315\,cm^{-1}$ ($31.75\,\mu m$) respectively. Parallel changes are observed for vanadium VCl_4 which has the band due to its V–Cl stretching vibration at about $480\,cm^{-1}$ ($20.83\,\mu m$) whereas for coordination numbers of 6 and 8, the bands are similar in position to those given for titanium above.

Halogen atoms may act as bridging ligands between two metal atoms, so bands due to both bridging and terminal halogens may be observed for binuclear complexes. The metal–halogen–metal bond angle and the degree of interaction between bond stretching vibration modes affect the position of the observed bands for bridging halogens. In general, terminal and bridging halogens may be distinguished since the bridging halogen has a lower stretching frequency than in the corresponding terminal position. Planar binuclear complexes of the type

$$
\begin{array}{ccccc}
X & & X & & X \\
 & \diagdown \diagup & & \diagdown \diagup & \\
 & M & & M & \\
 & \diagup \diagdown & & \diagup \diagdown & \\
X & & X & & X \\
\end{array}
$$

Table 22.7 Band positions of metal halide ions

Group	Position (cm^{-1})	Group	Position (cm^{-1})
$AlCl_4{}^-$	580–345	$NiF_6{}^{2-}$	655–560
$AuBr_4{}^-$	255–195	$OsCl_6{}^{2-}$	350–240
$AuCl_4{}^-$	360–320	$PaF_6{}^-$	310–305
$AuI_4{}^-$	195–110	$PdBr_4{}^{2-}$	260–165
$BBr_4{}^-$	600–240	$PF_6{}^-$	920–740
$CdBr_4{}^{2-}$	185–165	$PtBr_4{}^{2-}$	235–190
$CoBr_4{}^{2-}$	~230	$PtBr_6{}^{2-}$	245–205
$CuBr_2{}^-$	~190	$PtCl_4{}^{2-}$	335–305
$CuBr_4{}^{2-}$	220–170	$PtCl_6{}^{2-}$	345–330
$FeBr_4{}^-$	~290	$PtI_4{}^{2-}$	180–125
$FeBr_4{}^{2-}$	~220	$ReBr_6{}^{2-}$	220–210
$FeCl^{2-}$	~285	$ReCl_6{}^{2-}$	350–300
$FeCl_4{}^-$	385–330	$SiF_6{}^{2-}$	~725
$GaBr_4{}^-$	290–205	$SnBr_3{}^-$	215–180
$GaCl_4{}^-$	455–345	$SnBr_6{}^{2-}$	185–135
$GeCl_3{}^-$	320–250	$SnCl_6{}^{2-}$	320–295
$HgBr_4{}^{2-}$	170–165	$TiCl_6{}^{2-}$	465–330
$InBr_4{}^-$	240–194	$TlBr_4{}^-$	210–190
$InCl_6{}^{2-}$	275–245	$WCl_6{}^-$	330–305
$InI_4{}^-$	185–135	$WCl_6{}^{2-}$	325–305
$MnBr_4{}^{2-}$	~220	$ZnBr_2{}^{2-}$	210–170
$MnI_4{}^{2-}$	~185	$ZnBr_6{}^{2-}$	380–195
$NiBr_4{}^-$	235–220	$ZnI_4{}^{2-}$	170–120

exhibit two bands due to bridging halogen–metal stretching vibrations and one due to the end–halogen stretching vibration. For example, in the vapour phase, Al_2Cl_6 has three strong bands near $625\,cm^{-1}$ ($16.00\,\mu m$), $485\,cm^{-1}$ ($20.62\,\mu m$) and $420\,cm^{-1}$ ($23.81\,\mu m$). (See Chart 22.4 and Tables 22.6–22.12.)

Metal–π-Bond and Metal–σ-Bond Complexes – Alkenes, Alkynes, etc.[10–13]

Alkenes

The infrared spectra of organometallic coordination compounds containing olefins and similar ligands show great similarity to the spectra of the free ligand. In the infrared, the band due to the C=C stretching vibration is weak or absent and that due to the CH_2 in-plane deformation is also of weak intensity. However, the band due to the C=C stretching vibration is relatively strong in Raman spectra.

Table 22.8 Positions of metal halide stretching vibrations

Bond	Position (cm^{-1})	Bond	Position (cm^{-1})
Al–Br	480–405	Pb^{IV}–Cl	350–325
Al–Cl	625–350	Pd^{II}–Cl	370–285
As–Br	285–275	Pt^{II}–Cl	365–295
As–Cl	415–305	Pu–F	630–520
As–F	740–640	Re–F	755–595
As–I	230–200	Rh–F	725–630
Be–Br	~1010	Ru–F	735–675
Be–Cl	~1115	S–Cl	545–360
Be–F	1555–1520	Sb–Br	255–205
Bi–Br	200–165	Sb–F	720–260
Bi–Cl	290–240	Se–Br	295–200
Bi–I	145–115	Se–Cl	590–285
Cd^{II}–Br	315–120	Se–F	780–660
Co^{II}–Br	400–230	Si–Br	490–245
Cr–F	780–725	Si–Cl	650–370
Cu^{II}–Cl	500–290	Si–F	1035–800
Fe^{II}–Cl	500–265	Si–I	405–165
Ge–Br	330–230	Sn–F	535–490
Ge–Cl	455–355	Sn^{IV}–Br	315–220
Ge–F	800–685	Sn^{IV}–Cl	560–365
Ge–I	265–155	Sn^{IV}–I	220–145
Hf–Cl	395–345	Ta–F	550–510
Hf–F	~645	Tc–F	705–550
Hg^{II}–Br	295–205	Te–Br	250–220
Hg^{II}–Cl	415–290	Te–Cl	380–340
Hg^{II}–I	240–110	Te–F	755–670
In–I	185–135	Te–I	175–140
Kr–F	590–445	Th–F	~520
Mg^{II}–Cl	~595	Ti–Br	415–225
Mn^{II}–Cl	475–225	Ti–Cl	510–370
Mo–F	745–490	Ti–F	600–520
N–Cl	805–535	U–F	670–530
Nb–Cl	500–315	V–Cl	505–365
Nb–F	685–540	V–F	805–555
Ni^{II}–Cl	525–220	W–Cl	410–300
Ni^{IV}–Cl	410–405	W–F	775–480
Np–F	650–525	Xe–Cl	~315
O–Cl	1115–685	Xe–F	590–500
Os–Cl	325–285	Zn^{II}–Br	450–205
Os–F	735–630	Zn^{II}–Cl	520–280
P–Br	495–300	Zn^{II}–F	760–750
P–Cl	610–390	Zn^{II}–I	350–335
P–F	1025–695	Zr^{IV}–Cl	425–385
Pb^{II}–Cl	355–300	Zr^{IV}–F	725–600

Table 22.9 Approximate positions of metal hexafluoro compounds MF$_6$ M–F stretching vibration bands

Substance	Region	
	cm^{-1}	μm
PtF$_6$	~705	~14.18
IrF$_6$	~715	~13.99
OsF$_6$	~720	~13.89
ReF$_6$	~715	~13.99
RhF$_6$	~725	~13.79
RuF$_6$	~735	~13.60
CrF$_6$	~790	~12.66
MoF$_6$	~740	~13.51
IrF$_6$	~720	~13.89
UF$_6$	~625	~16.00
PuF$_6$	~615	~16.26
WF$_6$	~710	~14.09
NpF$_6$	~625	~16.00
K$_2$MF$_6$ M=Ti, V, Cr, Mn, Ni, Ru Rh, Pt, Re, Os, Ir	655–540	15.27–18.52

In the case of ethylene coordinated to a metal atom, most of the bands are at the positions observed for free ethylene except that due to the C=C stretching vibration, which is found at 1580–1500 cm^{-1} (6.33–6.67 μm) or near 1220 cm^{-1} (8.20 μm) instead of near 1625 cm^{-1} (6.15 μm). Also, the band due to the CH$_2$ in-plane deformation which occurs near 1340 cm^{-1} (7.46 μm) in the spectra of free ethylene may occur in the region 1530–1500 cm^{-1} (6.54–6.67 μm) or near 1320 cm^{-1} (7.58 μm) and that due to the CH$_2$ twisting vibration occurs near 730 cm^{-1} (13.70 μm) instead of near 1005 cm^{-1} (9.95 μm) as in free ethylene.

The position of the bands due to the C=C stretching vibration and the CH$_2$ in-plane deformation vibration is dependent on the strength of the metal–π-bond. This is due to the effect of coupling between the C=C stretching and the CH$_2$ deformation vibrations (of course, coupling occurs to some extent in free ethylene). Hence, it must be borne in mind that, in the complex, the band attributed to the C=C stretching vibration is not pure. The C=C stretching vibration is influenced not only by the strength of the metal–ethylene bond but also by the coupling with the CH$_2$ in-plane deformation vibration. Coordination through π-bonding to a metal atom reduces the C=C stretching frequency and brings it closer to the CH$_2$ deformation frequency which allows increased coupling to occur.

A strong metal–olefin interaction results in a situation where there is cross–over in the nature of the bands assigned to C=C stretching and the CH$_2$ in-plane deformation. Typical examples of complexes where a strong metal–olefin interaction occurs are KC$_2$H$_4$PtCl$_3$ and C$_2$H$_4$Fe(CO)$_4$ – in these complexes, the

Table 22.10 Approximate positions of M–X and M–X–M stretching vibration bands for M_2X_6 and $(RMX_2)_2$

Functional Groups	Region cm^{-1}	μm	Comments
Al_2Cl_6	~625	~16.00	Terminal M–X str (AlF$_3$ ~945 cm^{-1})
	~485	~20.62	Terminal M–X str
	~420	~23.81	Bridging M–X–M str
	~285	~35.09	Bridging M–X–M str
Al_2Br_6	~500	~20.00	Terminal M–X str
	~375	~26.67	Terminal M–X str
	~340	~29.41	Bridging M–X–M str
	~200	~50.00	Bridging M–X–M str
Al_2I_6	~415	~24.10	Terminal M–X str
	~320	~31.25	Terminal M–X str
	~290	~34.48	Bridging M–X–M str
	~140	~71.43	Bridging M–X–M str
Ga_2Cl_6	~475	~21.05	Terminal
	~400	~25.00	Terminal
	~310	~32.26	Bridging
	~280	~35.71	Bridging
Ga_2Br_6	~345	~28.99	Terminal
	~270	~37.04	Terminal
	~230	~43.48	Bridging
	~200	~50.00	Bridging
Ga_2I_6	~275	~36.36	Terminal
	~215	~46.51	Terminal
	~200	~50.00	Bridging
	~135	~70.07	Bridging
$(CH_3AlCl_2)_2$	~495	~20.20	M–C str ~705 and ~700 cm^{-1}
	~485	~20.62	
	~380	~26.32	
	~345	~28.99	
$(CH_3AlBr_2)_2$	~515	~19.42	M–C str ~690 and ~675 cm^{-1}
	~400	~25.00	
	~360	~27.78	
	~350	~28.57	
$(CH_3GaCl_2)_2$	~400	~25.00	M–C str ~610 and ~605 cm^{-1}
	~380	~26.32	
	~310	~32.26	
	~290	~34.48	

band near 1500 cm^{-1} (6.67 μm) is predominantly CH$_2$ in character and that near 1220 cm^{-1} (8.20 μm) is predominantly C=C in character. The silver–olefin bond is relatively weak and the decrease in C=C stretching frequency on coordination is not large enough to produce significant coupling and hence cross–over in the character of C=C and CH$_2$ modes. This means that the band due to the C=C vibration occurs near 1580 cm^{-1} (6.33 μm) and that due to the CH$_2$ deformation vibration near 1320 cm^{-1} (7.58 μm).

Table 22.11 Transition metal halides

Substance	Region cm^{-1}	μm	Comments
Pt(II)–Cl†	365–340	27.40–29.41	
Pt(II)–Br†	265–225	37.74–44.44	
Pt(II)–I†	200–170	50.00–58.82	
Pd(II)–Cl†	370–345	27.03–28.99	
Pd(II)–Br†	285–265	35.09–37.74	
Ir–Cl	350–245	28.57–40.82	
Ir–Cl *trans* to phosphorus or arsenic	280–260	35.71–38.46	
Ir–Cl *trans* to hydrogen	250–245	40.00–40.82	
Ir–Cl *trans* to chlorine	320–300	31.25–33.33	
Ru–Cl *trans* to chlorine	350–300	28.57–33.33	
Ru–Cl *trans* to CO	315–265	31.75–37.74	
Ru–Cl *trans* to phosphorus	265–225	37.74–44.44	
Ru–Cl *trans* to arsenic	~270	~37.04	
CoCl$_2$·2Py	~320	~31.25	Note CoCl$_2$ ~ 430 cm^{-1}
CoCl$_2$·4Py	~230	~43.48	
TiCl$_4$ coordination No. 6	~375	~26.67	Strong band
TiCl$_4$ coordination No. 8	~315	~31.75	Strong band
TiBr$_4$L$_2$	330–290	30.30–34.48	Strong band, Ti–Br
Cis octahedral complexes			
MCl$_4$L$_2$ M=Ti or V	390–365	25.64–27.40	Strong band
	340–280	29.41–35.71	Weak band. (TiBr str 330–290 cm^{-1})
Trans MX$_4$L$_2$			
Ti–Cl	395–370	25.32–27.03	Strong band
	330–275	30.30–36.36	Weak band
Ti–Br	330–290	30.30–34.48	Strong band
ZrCl$_4$L$_2$	355–295	28.17–33.90	Strong band
ZrBr$_4$L$_2$	270–260	37.04–38.46	Strong band
Trans NiCl$_2$L$_2$	~495	~24.69	asym str
NiBr$_2$L$_2^†$	~330	~30.30	*trans* form
	270–230	37.04–43.48	Tetrahedral form
SnX$_2$R$_2$			
X=Cl	~355	~28.17	asym str. Position solvent sensitive
	~350	~28.57	sym str (def ~120 cm^{-1})
X=Br	~250	~39.22	asym str
	~240	~41.67	sym str
X=I	~200	~50.00	asym str
	~180	~55.56	sym str
Trans RhCl$_2$L$_2$	~270	~37.04	
Cis RhCl$_2$L$_2$	~290	~34.48	
	~260	~38.46	

$†$ = Phosphine complexes.

Table 22.12 Bridging halides

Functional Groups	Region	
	cm^{-1}	μm
	M–X–M stretching vibration	
Pt(II)–Cl†	335–310	29.85–32.26
	295–250	33.90–40.00
Pt(II)–Br†	230–205	43.48–48.78
	190–175	52.63–57.14
Pt(II)–I†	190–150	52.63–66.62
	150–135	66.67–74.07
Pd(II)–Cl†	310–300	32.26–33.33
	280–250	35.71–40.00
Pd(II)–Br†	220–185	45.45–54.05
	200–165	50.00–60.61
Rh(II)–Cl†	290–260	34.48–38.46
Rh(II)–Br†	200–170	50.00–58.82

† = Phosphine complexes.

In deuterated olefins, the separation of the C=C stretching frequency and the CH$_2$ in-plane deformation frequency is much greater than in the normal olefin and hence, on coordination of the deuterated form, cross–over does not occur (since there is less coupling), the band due to the C=C stretching vibration occurring near 1500 cm^{-1} (6.67 μm) and that due to the CD$_2$ deformation vibration near 980 cm^{-1} (10.20 μm). Also, in complexes where the olefin is totally substituted (i.e. the hydrogens are replaced) and hence coupling cannot occur, the band due to the C=C vibration occurs in its normal position i.e. near 1500 cm^{-1} (6.67 μm).

The metal–olefin stretching vibration for platinum and palladium complexes results in a band at about 500 cm^{-1} (20.00 μm) and 400 cm^{-1} (25.00 μm) respectively. For iron–olefin complexes, a band near 360 cm^{-1} (27.78 μm) has been assigned to iron–olefin stretching vibrations and bands at 400 cm^{-1} (25.00 μm) and 300 cm^{-1} (33.33 μm) to tilting vibrations. Rhodium complexes have two bands near 400 cm^{-1} (25.00 μm).

Olefinic complexes which simultaneously exhibit both σ- and π-bonding are known, e.g. R$_2$SnCH=CH$_2$·2CuCl. The bands due to the C=C stretching vibration for the vinyl–tin bond occurs in the region 1595–1575 cm^{-1} (6.27–6.35 μm), when coordination of the copper atom to the C=C bond occurs this results in this band appearing in the region 1510–1490 cm^{-1} (6.62–6.71 μm).

Cyclopropenone complexes of the type [(C$_6$H$_5$)$_2$C$_3$O]$_2$M$_i$X$_j$ (where M=Zn, Cu, Co, Ru, Pt, Pd and X=Cl, Br, I, ClO$_4$) have a band near 1590 cm^{-1} (6.29 μm) instead of near 1630 cm^{-1} (6.14 μm) as for the free cyclopropenone. A band near 1850 cm^{-1} (5.41 μm) is also observed, this being at the same position as for the free cyclopropenone. These results indicate that, for these complexes, the metal atom coordinates via the oxygen atom and not through the C=C bond.

The band due to the C=C stretching vibration of free cyclopentene occurs near 1620 cm^{-1} (6.17 μm). On forming complexes of the type (C$_5$H$_8$MX$_2$)$_2$, where M=Pt, Pd and X=Cl, Br, the frequency of this C=C vibration decreases by approximately 200 cm^{-1}.

Alkynes

Alkynes can form both σ- and π-bonds to metals. In general, for the σ-type, the C≡C stretching vibration occurs at lower frequencies, and the band is also usually stronger, than for the free alkyne. The C≡C frequency decreases with increase in mass of the metal atom. The band due to the C≡C stretching vibration may appear up to about 120 cm^{-1} lower than that for the corresponding free alkyne if there is no metal–alkyne π interaction (many metal derivatives are found to exhibit both σ- and π-bonds simultaneously). If π interaction also occurs, such as

$$-C \equiv C - M \cdots \overset{\displaystyle |}{\underset{\displaystyle |}{\overset{\displaystyle C}{\underset{\displaystyle C}{\|}}}} \\ M$$

which is observed for some derivatives (e.g. Cu), the frequency may be even lower, down to about 220 cm^{-1}.

More than one band due to the C≡C stretching vibration may be observed for some complexes, e.g. alkyne–metal phosphines, this being due to the different C≡C groups which are found in the crystalline structure.

Transition elements are found with both pure σ-type and pure π-type structures. For some σ-types, the band due to the C≡C stretching vibration is of strong-to-moderate intensity whereas, for the π-type, this band may be weak or absent from the infrared spectrum. For the π-type compounds, the shift in the C≡C stretching frequency relative to that for the free alkyne is much greater than for the σ-type. In the case of pure π-type, the C≡C frequency is lower by approximately 230–130 cm^{-1}. However, if the alkyne structure is distorted by the coordination (the C≡C bond order being reduced), the vibration frequency may be lowered by approximately 500 cm^{-1}. This type of complex is stabilised by electron–withdrawing substituents on the alkyne. Many π-bonded alkyne complexes have structures which lie between these two extremes.

Chart 22.4   Transition metal halides stretching vibrations

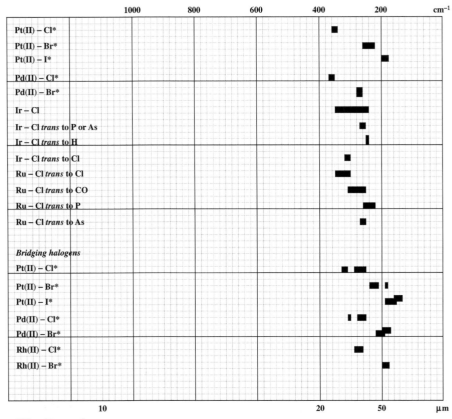

* Phosphine complexes

For platinum–acetylene complexes, the C≡C stretching vibration band is found at about 2000 cm^{-1} (5.00 μm) instead of near 2230 cm^{-1} (4.48 μm) as for free acetylene. In σ-bonded complexes, the band due to the C≡C stretching vibration occurs in the range 2055–2000 cm^{-1} (4.87–5.00 μm), whereas π-bonding reduces the stretching frequency which may occur as low as 1400 cm^{-1} (7.14 μm).

Cyclopentadienes

The cyclopentadiene ligand may be bound to a metal (a) ionically, e.g. KC$_5$H$_5$, (b) by means of σ-bonds, e.g. Mg(C$_5$H$_5$)$_2$ and Pb(C$_5$H$_5$)$_2$, and (c) by means of π-bonds, e.g. Fe(C$_5$H$_5$)$_2$ and Co(C$_5$H$_5$)$_2$ (sandwich–type complexes).

In case (a), four bands are observed in the infrared spectra, these being due to C–H stretching, ring deformation, C–H deformation, and C–H out-of-plane vibrations. In case (b), seven infrared bands are expected, these being due to two C–H stretching, two ring deformation, one C–H deformation, and two out-of-plane C–H bending vibrations. In case (c), in addition to the first seven infrared bands of (b), other bands due to ring–metal vibrations are observed: asymmetric ring tilting, asymmetric metal–ring stretching, and metal–ring deformation vibrations. Complexes with two rings parallel to each other exhibit one band due to metal–ring stretching vibrations and one due to tilting vibrations. These bands are observed below 550 cm^{-1} (above 18.18 μm). Usually strong intensity bands are observed for metal–ring stretching vibrations and medium-to-weak bands for ring tilting vibrations see Table 22.13.

Ferrocenes have several characteristic absorptions: a typical band due to the C–H stretching vibration near $3075 \, cm^{-1}$ $(3.25 \, \mu m)$, a band of medium intensity near $1440 \, cm^{-1}$ $(6.94 \, \mu m)$ due to the C–C stretching vibration and strong bands near $1110 \, cm^{-1}$ $(9.01 \, \mu m)$ and $1005 \, cm^{-1}$ $(9.95 \, \mu m)$ due to asymmetric ring-in-plane and C–H out-of-plane vibrations respectively. The last two bands are absent in the case of disubstituted ferrocenes. Cyclopentadienyl complexes also have from three to six weak bands, possibly overtones, in the region $1750–1615 \, cm^{-1}$ $(5.71–6.19 \, \mu m)$. Almost all ferrocenes have two strong bands in the region $515–465 \, cm^{-1}$ $(19.42–21.51 \, \mu m)$ which for solid samples are normally sharp and for liquids, broad.

In general, compounds with a heavy metal atom coordinated between two parallel cyclopentadiene or benzene rings (or other suitable ligands such as cyclo–octadiene and norbornadiene) have strong bands at $530–375 \, cm^{-1}$ $(18.87–26.67 \, \mu m)$, due to the asymmetric ring tilting motion, at $460–305 \, cm^{-1}$ $(21.74–32.79 \, \mu m)$, due to the heavy metal atom moving perpendicular to the two rings in an asymmetric stretching vibration, and at $185–125 \, cm^{-1}$ $(54.05–80.00 \, \mu m)$, due to the metal atom moving parallel to the two rings. This last band is not always observed and, in benzene sandwich compounds, the former two bands are very sensitive to the nature of the metal atom.

Phthalocyanine has bands at $620 \, cm^{-1}$ $(16.13 \, \mu m)$, $616 \, cm^{-1}$ $(16.23 \, \mu m)$, and $557 \, cm^{-1}$ $(17.95 \, \mu m)$. Metal phthalocyanine compounds absorb near $645 \, cm^{-1}$ $(15.50 \, \mu m)$, this band often appearing as a double peak, and near $555 \, cm^{-1}$ $(18.02 \, \mu m)$. A band of medium-to-weak intensity is also observed at about $435 \, cm^{-1}$ $(22.99 \, \mu m)$.

σ-Bonded metal N phenyl compounds have a band in the region $1120–1050 \, cm^{-1}$ $(8.93–9.52 \, \mu m)$, the position of which is dependent on the metal atom.

Metal–Cyano and Nitrile Complexes[1,3,4,8,10–13]

In metal–cyano complexes, the C≡N group may act as a terminal or bridging group. Terminal C≡N groups exhibit a sharp band in the region $2250–2000 \, cm^{-1}$ $(4.44–5.00 \, \mu m)$ whereas bridging C≡N groups absorb near $2130 \, cm^{-1}$ $(4.69 \, \mu m)$. Absorption bands are also observed in the ranges $570–180 \, cm^{-1}$ $(17.54–55.55 \, \mu m)$ and $450–295 \, cm^{-1}$ $(22.22–33.90 \, \mu m)$. Cyano complexes exhibit bands due to M–C stretching in the region $600–350 \, cm^{-1}$ $(16.67–28.57 \, \mu m)$, due to M–CN deformation in the region $500–350 \, cm^{-1}$ $(20.00–28.57 \, \mu m)$ and due to NC–M–CN deformation in the region $130–60 \, cm^{-1}$ $(76.92–166.67 \, \mu m)$ see Table 22.14.

For linear dicyanides the M–CN stretching vibration occurs at $455–360 \, cm^{-1}$ $(21.98–27.78 \, \mu m)$ and the MCN deformation vibration at $360–250 \, cm^{-1}$ $(27.78–40.00 \, \mu m)$. For octahedral cyanides of the type $M(CN)_6$ and $ML(CN)_5$ the MCN deformation band occurs at higher frequencies than the M–CN stretching vibration, $585–465 \, cm^{-1}$ $(17.09–21.51 \, \mu m)$ and $430–365 \, cm^{-1}$ $(23.26–27.40 \, \mu m)$ respectively.

In aqueous solution, the free CN^- ion absorbs near $2080 \, cm^{-1}$ $(4.81 \, \mu m)$ (general range, $2250–2000 \, cm^{-1}$, covalently bonded cyanide compounds absorb in the region $2250–2170 \, cm^{-1}$). The CN^- ion may coordinate to a metal atom by σ-donation, which increases the frequency of the CN stretching vibration, or by π-donation from the metal, which reduces the CN stretching frequency. Since CN^- is a good σ-donor and a poor π-acceptor, the CN stretching frequency generally increases on coordination.

As expected from the general introduction on all ligands given at the beginning of this chapter, the CN stretching frequency is governed by the oxidation state, coordination number and the electronegativity of the metal atom as well as by the other ligands attached to the atom. Lower electronegativity means poorer σ-donation and hence a lower CN frequency than might otherwise be expected. The higher the oxidation state of the metal, the stronger the σ-bonding and hence the higher the CN stretching frequency. An increase in the coordination number of a metal means a smaller positive charge and hence a weaker σ-bond which in turn means a lower CN stretching frequency than might be expected. For bridged cyano complexes, M–C≡N–M′, the CN stretching frequency increases (whereas that of the M–CN decreases), the opposite trend is observed for carbonyl complexes.

On coordination to a metal through the nitrogen atom, nitrile compounds, such as G–C≡N, G≡R or Ar, exhibit an increase in the C≡N stretching vibration unless there is strong π back–donation from the metal. In general, there is also an increase in intensity of the CN band for simple nitriles. The band due to the C≡N vibration for simple nitrile complexes is at $2360–2225 \, cm^{-1}$ $(4.23–4.47 \, \mu m)$. When coordination occurs through the triple bond, the CN stretching frequency is lower than that for the free nitrile. For acetonitrile complexes, the band due to the M–N stretching vibration occurs in the region $450–160 \, cm^{-1}$ $(22.22–62.50 \, \mu m)$.

Ammine, Amido, Urea and Related Complexes[1,3,4,10–13]

Ammine complexes have bands due to NH_3 stretching vibrations (asymmetric and symmetric) in the region $3400–3000 \, cm^{-1}$ $(2.94–3.33 \, \mu m)$. Due to the coordination and hence subsequent weakening of the N–H bond, these frequencies are lower than those for free NH_3. Two bands due to deformation vibrations are observed at $1650–1550 \, cm^{-1}$ $(6.06–6.45 \, \mu m)$ and $1370–1000 \, cm^{-1}$ $(7.30–10.00 \, \mu m)$ and a band due to the NH_3 rocking vibration occurs at $950–590 \, cm^{-1}$ $(10.53–16.95 \, \mu m)$ see Table 22.15.

Metal–ammine complexes also have a number of weak to medium intensity bands in the region $535–275 \, cm^{-1}$ $(18.69–36.36 \, \mu m)$ and a strong band in the region $330–190 \, cm^{-1}$ $(30.30–52.63 \, \mu m)$. The first of these is due to a

Table 22.13 Cyclopentadienyl, alkene and alkyne complexes

Functional Groups	Region		Intensity		Comments
	cm^{-1}	μm	IR	Raman	
Ionic cyclopentadienyl complexes	3100–3000	3.23–3.33	m	m	
	~2905	~3.44	w	m	C–H str
	1500–1400	6.67–7.14	m	m–s	C–C str
	1010–1000	9.90–10.00	s	m	C–H def vib
	750–650	13.33–15.85	s	w	Usually vs, often broad
σ-Bonded cyclopentadienyl complexes	3100–3000	3.23–3.33	m	m	C–H str
	2950–2900	3.39–3.45	m	m	C–C str
	1450–1400	6.90–7.14	m	m–s	C–C str
	1150–1100	8.70–9.09	m–s	m–s	C–C str
	1010–990	9.90–10.10	s	m–s	C–C str
	890–700	11.24–14.29	s	m–w	Doublet
	620–610	16.13–16.39	m		
	550–150	18.18–66.67	w–m		Several bands
$(C_5H_5)_2Hg$ and C_5H_5HgX X=Cl, Br, I	3100–3000	3.23–3.33	m–w	m	=C–H str
	3000–2800	3.33–3.57	m–w	m	C–H str
	1600–1500	6.25–6.67	w	s	C=C str
	1460–1300	6.85–7.69	m	m–s	
	940–900	10.64–11.11	s–vs	m–w	
	760–690	13.16–14.49	s–vs	m–s	Ring vib
	360–330	27.78–30.30	s		Hg–C str
	300–185	33.33–54.05	s		Hg–X str
Ferrocenes	3110–3025	3.22–3.32	m	m	C–H str
	1750–1615	5.71–6.19	w	m	3 to 6 bands
	~1440	~6.94	m	s	C=C str
	1115–1090	8.97–8.17	s	m–s	sh. C=C str
	1010–990	9.90–10.10	v	m–s	sh. Asym ring in-plane vib.
	830–700	12.05–14.29	vs	m	sh. Out-of-plane CH def vib
	515–485	19.42–20.62	s	m–s	asym ring tilt vib (not always present)
	495–465	20.20–21.51	s		Ring–Fe vib
$(C_5H_5)_2M$ M=Sn, Ge, Pb	3100–3060	3.23–3.27	m–w	m	CH str
	1430–1415	6.99–7.07	m–vs	m–s	CC str
	1120–1110	8.93–9.01	m–vs	m	Ring vib
	1010–1005	9.90–9.95	s–vs	m	CH def vib
	810–735	12.35–13.61	s	w	CH def vib (2 bands)
Titanocene derivatives $(C_5H_5)_n$ Ti(III) and (IV) n=1 or 2	3120–2095	3.21–4.77	w–m	m	C–H str
	1460–1420	6.85–7.04	m–s	m–s	C–C str
	1275–1195	7.84–8.37	vw	m	C–H bending vib
	1150–1105	8.70–9.04	w–m	m–s	Ring str
	1085–1075	9.21–9.30	w	w–m	C–H bending vib
	960–930	10.42–10.75	w	w–m	C–C bending vib, combination band
	880–795	11.36–12.58	vs–s	m	Out-of-plane C–H bending vib, may have shoulders
	605–455	16.53–21.98	w–m		Number of bands

Table 22.13   (*continued*)

Functional Groups	Region		Intensity		Comments
	cm^{-1}	μm	IR	Raman	
	~415	~24.10	w−m		
	~360	~27.40	m−s		M−(C$_5$H$_5$) asym str
	280−260	35.09−38.46	m		
σ-Bonded olefinic metal compounds CH$_2$=CH−M	3100−2900	3.22−3.45	m	m	asym. and sym. CH$_2$ and CH str
	1630−1565	6.14−7.19	w−m	s	C=C str but often intense
	1425−1385	7.02−7.22	m	m	CH$_2$ def vib
	1265−1245	7.91−8.03	w−m	w−m	CH rocking vib
	1010−985	9.90−10.15	m	w−m	CH wagging vib
	960−940	10.42−10.64	s	w−m	CH$_2$ wagging vib
	730−450	13.70−22.22	m−s	w	CH def vib
Trans-(CH$_3$)$_3$M−CH=CH−CH$_3$ M=Si, Ge, Sn	1620−1605	6.17−6.23	w−m	s	C=C str
	990−980	10.10−10.20	s	m−w	
Cis-(CH$_3$)$_3$M−CH=CH−CH$_3$	1610−1605	6.21−6.23	w−m	s	C=C str, no bands in region 1050−925 cm^{-1}
Trans-ClCH=CH−M−Cl M=Hg, Tl, Sn	1160−1140	8.62−8.77		m	
	950−935	10.52−10.70			
Cis-ClCH=CH−M−Cl	1275−1260	7.84−7.94		m	
	920−915	10.87−10.93			
π-bonded allyl complexes	1510−1375	6.62−7.27	m−s		Three bands, metal−olefin bands at 570−320 cm^{-1}
Phenyl−metal compounds C$_6$H$_5$−M	1500−1460	6.67−6.85	m−s	m−s	Ring vib
	1440−1415	6.94−7.07	vs	m−s	Ring vib
	1120−1050	8.93−9.52	m		Position of band metal sensitive
	750−720	13.33−13.89	vs	w	CH def vib
	~300	~33.33	w		CM str
Triphenyl compounds Ph$_3$MA (M=Sn or Ge; A=H, Cl, Br, I)	460−440	21.74−22.73	m−s		X-Sensitive band
	375−235	26.67−42.55	m		M−A str
Tetraphenyl compounds Ph$_4$M (M=Sn, Ge, Pb)	480−440	20.83−22.73	w		C−H str
3,4-Dihydroxy-3-cyclobutene-1,2-dione [(C$_4$O$_4$)Ti(C$_5$H$_5$)2]$_n$	~3130	~3.19	w	m	C−H str
	~1795	~5.57	m−s	m−w	sym C=O str
	~1695	~5.90	s	m−w	asym C=O str
	~1555	~6.43	vs	m−s	sym C=O, C=C str occurs in other similar complexes
	~1405	~7.12	vs	m	asym C=O, C=C str
	~1065	~9.39	m	m−s	C−C str
	~745	~13.42	m	m−s	Ring vib
	~680	~14.71	m	m−w	asym C=O def vib
	~435	~22.99	s		sym Ti−O str
	~405	~24.69	w		asym Ti−O str

(*continued overleaf*)

Table 22.13 (*continued*)

Functional Groups	Region		Intensity		Comments
	cm^{-1}	μm	IR	Raman	
M–C≡C–H	3305–3280	3.02–3.05	m–s	m	C–H str
	2055–2000	4.87–5.00	w	m–s	(M–C≡CCH$_3$ 2200–2170 cm^{-1} s)
	710–675	14.08–14.81	s	w	C–H bending vib
	665–575	15.04–17.39	s	m–w	C–H bending vib
H–C≡C–M, M=Na, K, Rb, Cs	1870–1840	5.35–5.42	s–m	s	C≡C str
CH$_3$–C≡C–M, M=Li, Na, K, Rb, Cs, P, As, Sn, Ge	2055–2010	4.87–4.98	s–m	s	C≡C str
Ar–C≡C–M, M=Li, Na, K, Rb	2040–1990	4.90–5.03	s–m	s	C≡C str
(G–C≡C)$_2$Hg	2190–2140	4.57–4.67	s–m	s	C≡C str
X–C≡CCH$_3$, X=P, As, Sn, Ge	2200–2170	4.55–4.61	s–m	s	C≡C str

Table 22.14 Cyano and nitrile complexes

Functional Groups	Region		Intensity		Comments
	cm^{-1}	μm	IR	Raman	
M–CN	2250–2000	4.44–5.00	m–s		
	570–180	17.54–55.56		m–s	MC str, MCN def (2 or more bands)
	130–60	76.92–166.67			CMC def vib
Bridging M–CN–M	~2130	~4.69	m–s	m–s	
Ni–C≡N–BF$_3$	~2250	~4.44	m–s	m–s	
Fe(II)–CN–Cr(III)	~2090	~4.78	m–s	m–s	
Cr(III)–CN–Fe(II)	~2170	~4.61	m–s	m–s	
	~2115	~4.73	m–s	m–s	
Cyano complexes					
Mn(CN)$_6$$^{3-}$	~2120	~4.72	m–s	m–s	
Mn(CN)$_6$$^{4-}$	~2060	~4.85	m–s	m–s	
Mn(CN)$_6$$^{5-}$	~2050	~4.88	m–s	m–s	
Cr(CN)$_6$$^{3-}$	~2130	~4.69	m–s	m–s	
Fe(CN)$_6$$^{3-}$	~2120	~4.72	m	m–s	
Co(CN)$_6$$^{3-}$	~2130	~4.69	m–s	m–s	
Nitrile complexes M(G–C≡N)	2360–2235	4.23–4.47	m–s	m–s	

M–N stretching vibration and the second due to a deformation vibration. In metal–ammine complexes, the NH$_3$ vibrations are affected by the nature of the anion. This has been attributed to N–H···X hydrogen bonding. Other factors being equal, the stronger the hydrogen bonding, the lower the frequency of the NH$_3$ stretching vibrations (and hence the greater the M–N stretching frequency) and the higher those due to NH$_3$ rocking. The intensity of the band due to the M–N stretching vibration increases as the M–N bond becomes more ionic in nature, also the lower the M–N frequency the stronger the band observed. For NH$_2$ complexes, the bands due to M–N stretching vibrations occur below 700 cm^{-1} (above 14.29 μm).

Coordination of the urea molecule to a metal atom may occur through either the oxygen or the nitrogen atoms. The electronic structure of urea may be considered as a hybrid of the three resonance structures:

(a) NH$_2$–C=O, NH$_2$ (b) N$^+$H$_2$=C–O$^-$, NH$_2$ (c) NH$_2$–C–O$^-$, N$^+$H$_2$

Table 22.15 Ammine complexes

Functional Groups	Region		Intensity		Comments
	cm^{-1}	μm	IR	Raman	
Metal–Ammine	3400–3000	2.94–3.33	s	w	NH$_3$ str (1 or 2 bands). The stronger the M–N bond, the lower the NH str frequency.
	1650–1550	6.06–6.45	m	w	NH$_3$ def vib
	1370–1000	7.30–10.00	s	m–w	NH$_3$ def vib
	950–590	10.53–16.95	m–s	w	NH$_3$ rocking vib
	540–275	18.52–36.36	w–m		M–N str triplet
	330–190	30.30–52.63	s		N–M–N in-plane bending vib
	M–N stretching bands				
Co(NH$_3$)$_6$$^{2+}$	~325	~30.77			
Co(NH$_3$)$_6$$^{3+}$	~475	~21.05			
Hg(NH$_3$)$_2$$^{2+}$	~470	~21.28			
Hg(NH$_3$)$_4$$^{2+}$	~410	~24.39			
Zn(NH$_3$)$_4$$^{2+}$	~435	~22.99			
Zn(NH$_3$)$_6$$^{2+}$	~300	~33.33			
Cd(NH$_3$)$_4$$^{2+}$	~380	~26.32			
Cd(NH$_3$)$_6$$^{2+}$	~300	~33.33			
Ni(NH$_3$)$_6$$^{2+}$	~335	~29.85			
Fe(NH$_3$)$_6$$^{2+}$	~320	~31.25			
Rh(NH$_3$)$_6$$^{2+}$	~470	~21.28			
Ir(NH$_3$)$_6$$^{3+}$	~475	~20.05			
Pt(NH$_3$)$_6$$^{4+}$	~535	~18.69			
M(glycine)$_2$	3350–3200	2.99–3.13	m	w	NH$_2$ str
	3300–3080	3.03–3.25	m	w	NH$_2$ str
	1650–1590	6.06–6.29	s	m–w	C=O str
	1620–1605	6.17–6.23	m	w	NH$_2$ def vib
	1420–1370	7.04–7.30	s	m	C–O str
	1250–1095	8.00–9.13	m–s	m	NH$_2$ rocking vib
	1060–1020	9.43–9.80	m–s	m	NH$_2$ rocking vib
	795–630	12.58–15.87	m–s	m	NH$_2$ rocking vib
	750–725	13.33–13.79	s	m–w	C=O def vib
	620–590	16.13–16.95	s	m–w	C=O def vib
	550–380	18.18–26.32	m–s		M–N str
	420–290	23.81–34.48	m–s		M–O str

If coordination through the oxygen atom occurs, the contribution by structure (a) will be small. Therefore, coordination through the oxygen atom tends to decrease the frequency of the CO stretching vibration and increase that of the C–N stretching vibration relative to that observed for 'free' urea, for which the bands due to the CO and CN stretching vibrations are observed near 1685 cm^{-1} (5.93 μm) and 1470 cm^{-1} (6.80 μm) respectively. Coordination through the nitrogen atom tends to have the reverse effect. Similar observations apply to thiourea. Hence, the donor atom may be determined, and linkage isomers distinguished, by infrared or Raman spectroscopy.

In the case of ligands containing the NH$_2$–(CO) group, the N–bonded complexes have bands due to their N–H bending vibrations near 1265 cm^{-1} (7.90 μm) whereas for free N–H (i.e. non–coordinated), these bands occur at about 1685 cm^{-1} (5.93 μm) and 1120 cm^{-1} (8.93 μm). The N–H stretching frequency decreases with increase in the metal–nitrogen bond strength.

Glycine and other amino acids exist as Zwitter ions, so that, for the free amino acid, bands due to CO_2^- and NH_3^+ may be observed in the solid state. Glycine may coordinate to a metal atom as a uni- or bidentate ligand, i.e.

$$M-NH_2 \cdot CH_2 \cdot COOH \quad \text{or} \quad M \underset{O-C}{\overset{NH_2}{\underset{\displaystyle \|}{\overset{\displaystyle}{\bigwedge}}}} \begin{matrix} \\ CH_2 \\ \end{matrix}$$

The unidentate ligand may be ionised, in which case a strong band due to the CO_2^- asymmetric stretching vibration is observed near $1610\,cm^{-1}$ ($6.21\,\mu m$). If un-ionised, there is a strong absorption near $1710\,cm^{-1}$ ($5.85\,\mu m$) due to the C=O stretching vibration of the COOH group. The coordinated bidentate glycine ligand absorbs in the range $1650-1590\,cm^{-1}$ ($6.06-6.29\,\mu m$). Note that the NH_2 deformation vibration may result in a band near $1610\,cm^{-1}$ ($6.21\,\mu m$). Obviously, when the glycine bonds through the oxygen atom, a band due to the M–O stretching vibration is expected which would otherwise be absent.

Square planar bis (glycino) complexes may have *cis* or *trans* configurations. In general, the *cis* isomer exhibits two bands for each of the M–N and M–O stretching vibrations.

For alkyl–metal pyridine complexes, no significant differences between the infrared spectra of the free ligand and that of the coordinated pyridine above $650\,cm^{-1}$ (below $15.39\,\mu m$) are observed (in Raman spectra the bands near $990\,cm^{-1}$ ($10.10\,\mu m$) and $1030\,cm^{-1}$ ($9.71\,\mu m$) are replaced by an intense band near $1020\,cm^{-1}$ ($9.80\,\mu m$) and a weaker band near $1050\,cm^{-1}$ ($9.52\,\mu m$)). The pyridine bands near $605\,cm^{-1}$ ($16.53\,\mu m$) and $405\,cm^{-1}$ ($24.69\,\mu m$) due to the in-plane ring and out-of-plane ring deformation vibrations respectively shift to higher wavenumbers on complex formation. In some, but not all, complexes, the M–C stretching frequency decreases from that observed for the free metal–alkyl compound.

Metal Carbonyl Compounds[1,3,4,10–13,16]

The carbonyl groups of metal (and some non-metal) carbonyl compounds absorb strongly at $2180-1700\,cm^{-1}$ ($4.59-5.88\,\mu m$) due to the CO stretching vibration (most carbonyl complexes have a strong, sharp band in the region $2100-1800\,cm^{-1}$).

Free carbon monoxide, CO, absorbs at about $2150\,cm^{-1}$ ($4.65\,\mu m$), whereas in metal carbonyl complexes this stretching vibration decreases by a hundred or more wavenumbers. This indicates that the bond is via π-orbitals, thus weakening the bond, rather than through σ-orbitals as in $CO \cdot BH_3$, which absorbs at about $2180\,cm^{-1}$ ($4.59\,\mu m$). Strong back-donation accentuates this

effect. A positive charge tends to increase the stretching frequency whereas a net negative charge tends to decrease it.

Metal complexes with a single CO group have a strong band in the region $2100-1700\,cm^{-1}$ ($4.76-5.88\,\mu m$). In complexes with more than one carbonyl, the CO vibrations generally couple. This can be seen in octahedral dicarbonyl complexes which may have *cis* or *trans* configurations. When the carbonyls are in the *trans* position, the symmetric stretching vibration, which occurs at a higher frequency than the asymmetric, results in a weak band.

| Symmetric vib | Asymmetric vib | Symmetric vib | Asymmetric vib |
| *Trans* position | | *Cis* position | |

It might be expected that this vibration would be infrared inactive but there is generally sufficient mixing with other bond vibrations to make it observable. In the *cis* position, the intensities of the symmetric (higher frequency) band and the asymmetric band are similar. These observations may be transferred to other carbonyl complexes with more than two CO groups. Using the above ideas, it can be seen that tricarbonyl octahedral complexes where two carbonyls are in the *trans* position to each other have three bands due to CO stretching vibrations, a symmetric and two asymmetric, whereas when the carbonyls are all in *cis* positions relative to each other, only two bands are observed, the band due to the symmetric stretching vibration being of lower intensity than that due to the asymmetric stretching vibration.

The position of the bands due to the CO stretching vibrations in the ranges given is dependent on the metal atom involved, the nature of the other ligands and the net ionic charge of the complex. For example, strong donor ligands, a net negative charge or a π-basic metal all result in back donation thus weakening the CO bond which means that the bands are observed at lower frequencies.

In transition metal hydridocarbonyl complexes, Fermi resonance interaction occurs between the carbonyl and metal–hydride stretching vibration. Hence, a significant shift of the CO stretching frequency may be observed on deuteration of a complex, e.g. $30\,cm^{-1}$, and anomalous ν_{M-H}/ν_{M-D} ratios are observed: instead of being $\sqrt{2}$, the ratio is less.

The stretching frequencies of terminal carbonyls in metal carbonyl complexes are usually found in the range $2170-1900\,cm^{-1}$ ($4.61-5.26\,\mu m$). However, as mentioned above, increasing the negative charge on a metal

Table 22.16 Carbonyl complexes

Functional Groups	Region cm^{-1}	Region µm	Intensity	Comments
CO complexes M–CO	2170–1790	4.60–5.59	s	Terminal CO, CO covalently bonded to metal atom. (Usually above 1900 cm^{-1}, negative charge lowers frequency)
	790–275	12.66–36.36	s–vs	MCO bending vib
	640–340	15.63–29.41	m–vs	M–C str
CO bridging complexes M–CO–M	1900–1700	5.26–5.88	s	Bridging CO. (Band at higher frequencies due to terminal CO see above)
	700–275	14.29–36.36	vs	
Transition metal thiocarbonyls M–CS	1400–1150	7.14–8.70		Terminal CS
	1150–1100	8.70–9.09		Bridging CS

Metal carbonyl compounds	Approximate CO stretching vibration frequency cm^{-1}	Approximate CO stretching vibration frequency µm	Approximate M–C stretching & MCO bending vibration frequencies (cm^{-1})
Mn(CO)$_6$$^+$	∼2090	∼4.79	
Ni(CO)$_4$	∼2050	∼4.88	420, 390
Co(CO)$_4$$^-$	∼1890	∼5.29	555(s), 530(w), 440
Fe(CO)$_4$$^{2-}$	∼1790	∼5.59	550, 465
Ni(CO)$_4$	∼2075	∼4.82	
Fe(CO)$_5$	∼2045	∼4.90	395, 110, 80
Cr(CO)$_6$	∼2030	∼4.93	400
Re(CO)$_6$$^+$	∼2200	∼4.54	450
W(CO)$_6$	∼1980	∼5.05	
Mo(CO)$_6$	∼1990	∼5.03	

complex lowers the CO frequency, which may be as low as 1790 cm^{-1} (5.59 µm) (Table 22.16).

Bridging carbonyl compounds, in which a carbonyl group is associated with two metal atoms, absorb at 1900–1700 cm^{-1} (5.26–5.88 µm), that is, at lower frequencies than those for terminal CO groups. As mentioned, most other carbonyl groups absorb above 1900 cm^{-1} (below 5.26 µm) except in the case of complexes with strong electron–donor ligands or with a large negative charge.

The in-plane bending vibration of the M–CO group of metal carbonyl compounds gives rise to a band of very strong or strong intensity at 790–275 cm^{-1} (12.66–36.36 µm) and the stretching vibration of the M–C group of these compounds gives a band of very-strong-to-medium intensity at 640–340 cm^{-1} (15.63–29.41 µm). In general, for any given molecule or

ion, the M–CO bending vibrations are at higher frequencies than the M–C stretching frequencies, the exception being in the case of tetrahedral species for which the reverse is true. In the case of neutral carbonyls, the ranges are shorter than those given above, the band due to the M–C stretching vibration occurring within 480–355 cm^{-1} (20.83–28.17 µm), and that due to the MCO bending vibration at 755–460 cm^{-1} (13.25–21.74 µm), again except for tetrahedral species where the band due to the bending vibration may be as low as 275 cm^{-1} (36.36 µm). Octahedral metal carbonyls absorb in the regions 790–465 cm^{-1} (12.66–21.51 µm) and 430–365 cm^{-1} (23.26–27.40 µm) for the MCO bending and M–C stretching vibrations respectively.

CO adsorbed on nickel gives rise to two bands, one at 2080–2045 cm^{-1} (4.80–4.89 µm) due to the mono dentate Ni–C≡O group and the other at about 1935 cm^{-1} (5.17 µm) due to the bridging structure Ni–CO–Ni.

Table 22.17 Acetylacetonates

Functional Groups	Region		Intensity		Comments
	cm^{-1}	μm	IR	Raman	
Acetylacetonates, CH$_3$·CO / HC〈 〉H / CH$_3$·CO	1605–1560	6.23–6.41	s		C⋯O and C⋯C str
	1550–1500	6.45–6.67	s	m–s	C⋯O and C⋯C str
	1390–1350	7.19–7.41	m–s	m	CH$_3$ str
	1290–1240	7.75–8.07	m	m–s	C⋯C and C–CH$_3$ str
	~1195	~8.37	m	w	CH bending vib
	~850	~11.77	m	w	CH bending vib
	600–490	16.67–20.41	m–s		Two bands
	~430	~23.26	v		
	390–290	25.64–34.48	v		
M–OCH$_2$COR (keto form) (general range)	1700–1650	5.88–6.06	vs	w–m	C=O str
	1650–1610	6.06–6.21	vs	w–m	
M–OCH$_2$COR M=Si, Ge	1700–1670	5.88–5.99	s	w–m	C=O str
M–OCR=CH$_2$ M=Si, Ge	1655–1620	6.04–6.17	m	s	C=C str

Table 22.18 Carboxylates

Functional Groups	Region		Intensity		Comments
	cm^{-1}	μm	IR	Raman	
Bidentate carboxylate	1610–1515	6.21–6.60	s	m–s	asym CO str
	1495–1315	6.69–7.60	m	w	sym CO str
Unidentate carboxylate	1675–1575	5.97–6.35	s	m–w	C=O str, see text
	1420–1260	7.04–7.35	s–m	w	C–O str
Bridging carboxylate	1610–1515	6.21–6.60	s	m–s	asym CO str
	1495–1315	6.69–7.60	m	w	sym CO str
Trimethyl cyclobutene-1,2-dione [(CH$_3$)$_3$M]$_2$C$_4$O$_4$ M=Al, Ga, In	~1560	~6.41	s	m	C⋯O str
	1155–1090	8.66–9.17	m–s	m	Number of bands, C⋯C str
	760–745	13.16–13.42	m–s	m–s	Ring breathing vib
	665–550	15.04–18.18	vs–s	m	C⋯O def vib
	490–270	20.41–37.04	s–m		May be br., number of bands
EDTA uncoordinated	1750–1700	5.71–5.88	s	m–w	CO$_2$ str
EDTA ionised	1630–1575	6.14–6.35	s	m–w	CO$_2$– str
EDTA complexes	1650–1620	6.06–6.17	s	m–w	CO$_2$ str Cu(II), Zn(II)
	1610–1590	6.21–6.29	s	m–w	CO$_2$ str Cr(III), Co(III)
	500–400	20.00–25.00	s		M–N str

Metal–Acetylacetonato Compounds, Carboxylate Complexes and Complexes Involving the Carbonyl Group

Acetylacetone may coordinate to a metal atom through the oxygen atoms:

$$
\begin{array}{c}
CH_3 \\
\backslash \\
C{-}O \\
/ \qquad \diagdown \\
H{-}C \qquad M \\
\diagdown \qquad / \\
C{-}O \\
/ \\
CH_3
\end{array}
$$

In this (enol) type of complex, the band due to the C–O stretching vibration occurs at lower frequencies (coupling with C $\cdots$ C occurs), usually 1605–1560 cm^{-1} (6.23–6.41 μm), than that due to the free acetylacetone C=O stretching vibration which occurs at 1640 cm^{-1} (6.10 μm). A second strong band is observed near 1380 cm^{-1} (7.25 μm). The band due to the C–H stretching vibration tends to higher frequencies than might be expected because of the new benzene-type environment in which it is found. The bands due to C $\cdots$ C stretching vibrations for complexes are found at about 1540 cm^{-1} (6.49 μm) and 1290 cm^{-1} (7.75 μm). Acetylacetonates have two bands at 600–490 cm^{-1} (16.67–20.41 μm) and may also absorb near 430 cm^{-1} (23.26 μm) and at 390–290 cm^{-1} (25.64–34.48 μm).

The differences between the enol type acetylacetonate complexes, i.e. oxygen bound to metal atom, and the keto type, i.e. carbon bound, are as follows: (1) the γ-carbon hydrogen stretching frequency is lower by about 150–100 cm^{-1} in the keto form; (2) bands due to the asymmetric and symmetric C=O stretching vibrations appear at 1700–1650 cm^{-1}

(5.88–6.06 μm) and 1650–1610 (6.06–6.21 μm) respectively in the keto form, both bands being very strong, rather than the one at 1605–1560 cm^{-1} (6.23–6.41 μm) in the enol form.

Acetylacetone may also form coordination complexes by bonding through its γ-carbon atom. In this type of complex, the carbonyl band position is the same as for free acetylacetone. The bands due to the carbon–carbon stretching vibration are in different positions for the two types of complex. Coordination through the carbon–carbon double bond may also occur in some complexes.

Carboxylate Complexes and other Complexes Involving Carbonyl Groups

The carboxylate ion may bond to a metal atom through oxygen as a unidentate or bidentate ligand or act as a bridge between two metal atoms. For free acetate ion, the bands due to stretching vibrations occur near 1560 cm^{-1} (6.41 μm) and 1415 cm^{-1} (7.07 μm). In the Raman spectrum, strong-to-medium intensity bands are observed at 1470–1400 cm^{-1} (6.80–7.14 μm) and near 950 cm^{-1} (10.53 μm) and a medium-to-weak band near 1370 cm^{-1} (7.30 μm). In the case of covalent derivatives, a distinctive feature is the appearance of bands due to asymmetric and symmetric C–O vibrations and due to M–O stretching vibrations. In the unidentate situation the C=O stretching vibration occurs at higher frequencies than either of the two CO_2^- vibrations whereas the C–O stretching frequency is lower. As a bidentate ligand, the two bands due to the CO stretching vibration are closer together than for the free ion. When acting as a bridging ligand, these bands are near the positions of those for the free ion.

Table 22.19 Nitro- and nitrito- complexes

Functional Groups	Region		Intensity		Comments
	cm^{-1}	μm	IR	Raman	
M–ONO, nitrito-	1485–1400	6.73–7.14	s	m–s	N=O str
	1110–1050	9.01–9.52	s	m	N–O str
	850–820	11.77–12.20	m		def
	360–340	27.78–29.41			M–O str, Cr(III), Rh(III), Ir(III)
M–NO$_2$, nitro- terminal	1470–1370	6.80–7.30	s	m	Terminal NO$_2$ asym. str
	1340–1315	7.46–7.60	s	v	Terminal NO$_2$ sym. str
	~620	~16.13	m–s	m	Wagging NO$_2$
	455–300	21.98–33.33			M–N str (square planar complexes 380–300 cm^{-1})
	305–245	32.79–40.82	v	v	NO$_2$ rocking vib
M–(NO$_2$)–M bridging	1520–1470	6.58–6.80	s	m–s	Bridging
	~1200	~8.33	s	m–s	Bridging

Table 22.20   Thiocyanato-, isothiocyanato etc complexes

Functional Groups	Region		Intensity		Comments
	cm^{-1}	μm	IR	Raman	
M–NCS	2200–2045	4.55–4.89	s	m–s	Often broad CN str, Usually below 2050 cm^{-1}
	860–780	11.63–12.82	w	s	C–S str
	490–450	20.41–22.22	w–m		Sharp, NCS def vib
M–SCN	2185–2060	4.76–4.85	s–m	m–s	Usually sharp CN str and above 2100 cm^{-1}
	730–690	13.70–14.49	w	s	C–S str
	440–400	22.72–25.00	w		Usually several bands SCN def
M–NCSe	2145–2020	4.66–4.95	s	m–s	Usually below 2080 cm^{-1}
	700–620	14.29–16.13	w–m	m–s	
	440–410	22.73–24.39	m–w		NCSe def vib
M–SeCN	2135–2005	4.68–4.99	s	m–s	Usually above 2080 cm^{-1}
	550–500	18.18–20.00	w–m		
	410–360	23.39–24.39	w–m		
X=Y=Z and X=Y stretches					
Pt–NCO	2250–2190	4.44–4.57	m	m–s	
Pt–NCS	2200–2090	4.55–4.79	m	m–s	May be br.
Pt–SCN	2110–2080	4.74–4.81	m–s	m	May be br.
Pt–SeCN	2100–2085	4.76–4.80	m–s	m	Tri-phenylphosphine complexes may also have a band at ~2195 cm^{-1}
Pt–CN	2150–2135	4.65–4.68	m–w	m–s	
Bridging compounds M–SCN–M	2185–2150	4.58–4.65	m–s	m–s	
Pd–SCN bridging	2120–2100	4.72–4.76	m–s	m–s	Pd–SCN. Terminal 2185–2040 cm^{-1}
Pd–SeCN bridging	above 2100	below 4.76	m	m–s	Pd–SeCN. Terminal 2080–2040 cm^{-1}
M–SeCN–M M=Pt, Pd	~2140	~4.73	m–s	m–s	
Alkyl compounds					
Sn–SCN–Sn	2100–2060	4.76–4.85	m–s	m–s	
Pb–SCN–Pb	~2090	~4.78	m–s	m–s	
Au–SCN–Au	~2165	~4.62	m–s	m–s	
Sn⟍NCS⟋Sn	~1960	~5.10	m–s	m–s	
Al⟍SCN⟋Al	~2075	~4.82	m–s	m–s	
Ga⟍SCN⟋Ga	~2105	~4.75	m–s	m–s	
In⟍SCN⟋In	~2130	~4.69	m–s	m–s	

Table 22.20 (*continued*)

Functional Groups	Region		Intensity		Comments
	cm^{-1}	μm	IR	Raman	
Zn—SCN—Zn	2190–2130	4.57–4.69	m–s	m–s	
Cd—SCN—Cd	~2190	~4.57	m–s	m–s	
	~2140	~4.67	m–s	m–s	

Table 22.21   Isocyanato and fulminato complexes

Functional Groups	Region		Intensity		Comments
	cm^{-1}	μm	IR	Raman	
Isocyanato complexes M–NCO	2300–2180	4.35–4.89	vs	w	asym NCO str
	1505–1195	6.64–8.37	w–m	s	sym NCO str
	715–580	13.99–17.24	m	w	NCO def vib
Fulminato complexes M–CNO (M=Tl, Pb, Hg)	2135–2060	4.68–4.85	s–m	w	asym CNO str
	1230–1100	8.13–9.09	w–m	s	sym CNO str
	495–430	20.20–23.26			CNO def vib

In general, for unidentate complexes the coordinated carboxylate groups have a strong band due to the C=O stretching vibration in the range 1650–1590 cm^{-1} (6.06–6.29 μm). The position of this band is dependent on the metal atom, the frequency increasing as the metal–oxygen bond becomes more covalent. For example, the C=O group absorbs at 1675–1620 cm^{-1} (5.97–6.17 μm) for covalent derivatives such as Al(III), Co(III) and Cr(III) ions whereas the range is 1630–1575 cm^{-1} (6.14–6.35 μm) for Cu(II) and Zn(II) (see also data for carboxylic acid salts, page 129).

In the free oxalate ion, the CO bonds are equivalent. On coordination as a bidentate ligand, two CO bonds are strengthened and two weakened. The CO bonds change from C⋯O to C=O and C–O. Hence, as above, bands due to C=O stretching vibrations appear at higher frequency than the band due to the CO bond in the free oxalate ion and the band due to the C–O vibration occurs at lower frequency.

Oxalato complexes have a number of bands in the region 590–290 cm^{-1} (16.95–34.48 μm) due to M–O stretching and C–O–C deformation vibrations.

The band due to the C=O stretching vibration of bis (salicylaldehydato) complexes of divalent metals occurs in the region 1685–1575 cm^{-1} (5.93–6.35 μm) and its position is related to the stability constant of the complex, the greater the stability constant, the lower the frequency. In other words, the C=O stretching frequency decreases as the M–O bond becomes stronger and the M–O stretching frequency increases.

Un-ionised free ethylenediamine *N,N,N',N'*-tetraacetic acid, EDTA, absorbs strongly in the region 1750–1700 cm^{-1} (5.71–5.88 μm) due to the stretching vibration of the CO$_2$ group, whereas for coordinated EDTA this absorption occurs in the region 1650–1590 cm^{-1} (6.06–6.29 μm). In complexes where two nitrogen atoms and three carboxyl groups are coordinated, an additional absorption near 1750 cm^{-1} (5.71 μm) is observed due to the fourth free carboxyl group.

Divalent metal 3,4-dihydroxy-3-cyclobutene-1,2-diones have a very strong absorption in the region 1700–1400 cm^{-1} (5.88–7.14 μm). Metal complexes of this ligand, in which all the oxygen atoms are coordinated, do not have bands above 1600 cm^{-1} (below 6.25 μm), whereas diketo metal complexes of this ligand, which do have uncoordinated carbonyl groups, have bands above 1600 cm^{-1} (below 6.25 μm). For these complexes, a strong band is observed at 1600–1500 cm^{-1} (6.25–6.67 μm) due to C=C or C$\equiv$O stretching vibrations.

Complexes involving acid halides with Friedel–Crafts catalysts such as AlCl$_3$, BF$_3$, SnCl$_4$ etc. (e.g. CH$_3$COCl·AlCl$_3$) have the band due to their C=O stretching vibration at much lower frequencies than the free acid halide, reduced by about 170 cm^{-1}. The intensity is also reduced. A strong band is observed in the region 2305–2200 cm^{-1} (4.34–4.55 μm) due to the –$^+$C$\equiv$O stretching vibration. In some cases, more than one band due to stretching is observed in this region. Complexes involving ketones have their C=O stretching frequency reduced by about 125 cm^{-1}. The stretching vibration frequency of a CH$_3$ group attached directly to the carbonyl group is also reduced by about 50–60 cm^{-1} compared with the free ligand.

Nitro- (–NO$_2$) and Nitrito- (–ONO) Complexes[1,10,11]

Linkage isomerism is possible in the case of metal complexes containing the unit NO$_2$. Coordination to the metal atom may occur through the nitrogen atom, resulting in a nitro- complex, or through an oxygen atom, resulting in a nitrito- complex.

Nitro- complexes exhibit bands due to asymmetric and symmetric –NO$_2$ stretching vibrations and, in addition, one due to a NO$_2$ deformation vibration. The nitrito- complexes exhibit bands due to asymmetric and symmetric –ONO stretching vibrations which are well separated and occur at 1485–1400 cm^{-1} (6.73–7.14 μm) and 1110–1050 cm^{-1} (9.01–9.52 μm) respectively, see Table 22.19.

Nitro- groups in metal coordination complexes may exist as bridging or as end groups. Terminal nitro- groups absorb at 1470–1370 cm^{-1} (6.80–7.30 μm) and 1340–1315 cm^{-1} (7.46–7.61 μm) due to the asymmetric and symmetric stretching vibrations respectively of the NO$_2$ group. Nitrito- complexes do not have a band near 620 cm^{-1} (16.13 μm) which is present for all nitro- complexes. Nitro- groups acting as bridging units between two metal atoms absorb at 1485–1470 cm^{-1} (6.73–6.80 μm) and at about 1200 cm^{-1} (8.33 μm), these bands being broader than those for terminal nitro- groups.

Thiocyanato- (–SCN) and Isothiocyonato- (–NCS) Complexes[1,3,10–13]

The thiocyanate ion may act as an ambidentate ligand, i.e. bonding may occur either through the nitrogen or the sulphur atom. The bonding mode may easily be distinguished by examining the band due to the C–S stretching vibration which occurs at 730–690 cm^{-1} (13.70–14.49 μm) when the bonding occurs

through the sulphur atom and at 860–780 cm^{-1} (11.63–12.82 μm) when it is through the nitrogen atom.

The C$\equiv$N stretching vibration of thiocyanato- complexes (sulphur- bound, i.e. M–SCN) gives rise to a sharp band at about 2100 cm^{-1} (4.76 μm) (N.B. alkyl compounds: Al–SCN ∼2095 cm^{-1} (4.77 μm) and Ga–SCN 2095–2060 cm^{-1} (4.77–4.85 μm)), whereas for isothiocyanato- complexes (i.e. nitrogen bound), the resulting band is often broad and occurs near and below 2050 cm^{-1} (above 4.88 μm). In addition, deformation vibrations give several weak bands in the region 440–400 cm^{-1} (22.73–25.00 μm) for thiocyanato- complexes, which appears at 490–450 cm^{-1} (20.41–22.22 μm) for isothiocyanato- complexes (a single sharp band being observed).

M–SCN–M bridges absorb well above 2100 cm^{-1} (below 4.76 μm). Thiocyanates acting as bridging groups in platinum and palladium complexes absorb in the region 2185–2150 cm^{-1} (4.58–4.65 μm) (see Table 22.20).

The SeCN group also coordinates to metals through the nitrogen or selenium atoms, as well as forming bridges. For M–NCSe, the band due to the CN stretching vibration occurs below 2080 cm^{-1} (4.81 μm) whereas for M–SeCN it is higher.

Isocyanates, M–NCO

The band due to the asymmetric NCO stretching vibration occurs at 2300–2180 cm^{-1} (4.35–4.89 μm), that due to the symmetric stretching vibration occurs at 1505–1195 cm^{-1} (6.64–8.37 μm) and that due to the deformation vibration at 715–580 cm^{-1} (13.99–17.24 μm) (see Table 22.21). A band due to the M–N stretching vibration is also expected.

Nitrosyl Complexes[10–12]

The free nitrosonium ion NO$^+$ absorbs near 2370–2230 cm^{-1} (4.22–4.48 μm). In nitrosyl complexes, the MNO moiety may be linear or bent. The NO stretching vibration occurs in the range 1945–1500 cm^{-1} (5.14–6.67 μm), the band for the bent form occurring at lower frequencies than the linear form.

Nitrosyl complexes are also expected to exhibit bands for both M–N stretching and MNO deformation vibrations. However, since coupling of these vibrations often occurs, the bands are not always observed. The band due to the MN stretching vibration is normally found in the region 650–520 cm^{-1} (15.38–19.23 μm) and that due to the MNO deformation vibration in the region 660–300 cm^{-1} (15.15–33.33 μm).

Table 22.22 Nitrosyl complexes: N–O stretching vibration bands

Functional Groups	Region cm^{-1}	Region µm	Comments
Transition metal M–NO	1945–1700	5.14–5.88	Terminal, linear nitrosyls
	1700–1500	5.88–6.67	Terminal, bent
	1550–1450	6.45–6.90	Bridging nitrosyls
Fe(NO)(CN)$_5{}^{2-}$	~1945	~5.14	
Mn(NO)(CN)$_5{}^{3-}$	~1730	~5.78	
Cr(NO)(CN)$_5{}^{4-}$	~1515	~6.60	
Cr(NO)$_4$	~1720	~5.81	M–N ~650 cm^{-1} M–O ~495 cm^{-1}
Co(NO)$_3$	~1860	~5.38	Linear, M–N ~610 cm^{-1}
	~1795	~5.57	Bent, M–O ~565 cm^{-1}
Co(CO)$_3$NO	~1820	~5.50	
CoL$_2$Cl$_2$(NO)	~1750	~5.71	Linear
	~1650	~6.06	Bent
NiCl$_2$(NO)$_2$	~1870	~5.35	Linear, M–N ~525 cm^{-1}
	~1840	~5.44	Bent, M–O ~655 cm^{-1}
Mn(CO)$_4$NO	~1780	~5.62	

Table 22.23 Azides, dinitrogen and dioxygen complexes etc

Functional Groups	Region cm^{-1}	Region µm	Intensity IR	Intensity Raman	Comments
Azides	2195–2030	4.56–4.93	s	m–s	asym N$_3$ str
	1375–1175	7.27–8.51	w–m	s	sym N$_3$ str
	680–410	14.71–24.39	w		N$_3$ def vib
Superoxo complexes	1200–1100	8.33–9.09			
Peroxo complexes	920–750	10.87–13.33			
N$_2$ Complexes	2220–1850	4.50–5.41	v		
N$_2$ σ-Bonded	2150–1920	4.65–5.21	s		
N–Si	1235–570	8.01–17.54			N–Si str
N–Sn	560–250	17.86–40.00			N–Sn str
N–Ti	565–530	17.70–18.87			N–Ti str
Zn^{II}–N	440–220	22.73–45.45			N–Zn str
Pd^{II}–N	500–260	20.00–38.46			N–Pt str
Pt^{II}–N	605–295	16.53–33.90			N–Pt str
Pt^{IV}–N	530–390	18.87–25.64			N–Pt str
Rh–N	525–360	19.05–27.78			N–Rh str
S–N	1070–830	9.35–12.05			N–S str
N–N	1115–870	8.97–11.49	w	s–m	N–N str
N=N	1630–1380	6.13–7.25	w	s–m	N=N str
N≡N	2335–2200	4.28–45.45	w	s–m	N≡N str

Azides, M–N$_3$, Dinitrogen and Dioxygen Complexes and Nitrogen Bonds

In general, for azides the band due to the asymmetric N$_3$ stretching vibration is strong and occurs in the region 2195–2030 cm^{-1} (4.56–4.93 µm), while that due to the symmetric vibration is much weaker and occurs in the region 1375–1175 cm^{-1} (7.27–8.51 µm) and the band due to the deformation vibration is also weak and occurs at 680–410 cm^{-1} (14.71–24.39 µm). The frequency separation of the asymmetric and symmetric bands for alkyl–metal azides decreases as the electron density of the azide ligand is reduced by either changes in the electronegativity of the other ligands bonded to the metal atom or through the formation of an α-nitrogen bridging bond.

Dioxygen adducts of metal complexes may absorb at 1200–1100 cm^{-1} (8.33–9.09 µm) (superoxo-) or at 920–750 cm^{-1} (10.87–13.33 µm) (peroxo-) due to the O$_2$ stretching vibration.

Nitrido complexes of transition metals absorb at 1200–950 cm^{-1} (8.33–10.53 µm) due to the M≡N stretching vibration.

Hydrides[1,3,4,10–13]

The frequency ranges for O–H, N–H and F–H stretching vibrations are wide due to hydrogen bonding which results in bands appearing at lower frequencies than would otherwise be the case. The stretching vibrations of alkyl hydrides of elements in Groups IVb, Vb and VIb give rise to medium to strong bands. The bands due to deformation vibrations are difficult to identify due to the mixing of vibrational modes. In general, there is little difference between the M–H stretching frequencies of heterocyclic hydrides and the corresponding non–cyclic compound e.g. (CH$_2$)$_2$P–H and (CH$_3$)$_2$P–H.

Bands due to terminal metal hydride stretching vibrations occur in the region 2300–1675 cm^{-1} (4.35–5.97 µm). The band is of low-to-medium intensity, this being dependent on the polarity of the M–H bond. Assignment due to frequency alone can lead to errors since bands associated with other groups e.g. CO, CN etc may occur in this region. Cyclopentadienyl- and carbonyl cyclopentadienyl-stabilised metal hydrides absorb in the region 2055–1735 cm^{-1} (4.87–5.76 µm) due to the M–H stretching vibration. *Trans*-dihydrides generally have low M–H stretching vibration frequencies at 1750–1615 cm^{-1} (5.71–6.21 µm).

Table 22.24   Hydride A–H stretching vibration bands (all bands of medium to strong intensity)

Functional Groups	Region		Intensity	Comments
	cm^{-1}	μm		
O–H	3800–3000	2.63–3.33	m–s	
N–H	3500–3000	2.86–3.33	m	
C–H	3050–2850	3.28–3.51	m–s	
Si–H	2250–2100	4.44–4.76	m–s	
Ge–H	2160–1990	4.63–5.03	m–s	Strong band ~720 cm^{-1} def vib
S–H	2580–2450	3.88–4.08	m–s	
Se–H	2400–2200	4.17–4.55	m–s	
B–H (terminal)	2565–2440	3.90–4.10	m–s	
B–H (bridging)	2100–1600	4.76–6.25	m	Broad
P–H	2450–2200	4.08–4.55	m–s	
As–H	2300–2070	4.35–4.83	m–s	As–D ~1530 cm^{-1}
Metal–H	2270–1700	4.41–5.88	m	Sharp, M–H str
	800–600	12.50–16.67	m	M–H def vib
Mn–H	1845–1780	5.42–5.62	m	
Al–H	1910–1675	5.24–5.97	m–s	Usually br. strong band. Bridging hydrogens may give band as low as 1550 cm^{-1}
Ga–H	1855–1820	5.33–5.50		Bending vib ~700 cm^{-1}
Sn–H	1910–1790	5.24–5.59	m	Usually broad, strong broad band ~570 cm^{-1} def vib
Fe–H	1900–1725	5.26–5.80		Dihydrides usually lower end of range
Ni–H	1985–1800	5.08–5.56		
Co–H	2050–1755	4.88–5.70		
M–H (M=Pt,Ir,Ru,Os,Re)	2200–1890	4.55–5.29	s	Strong band. Highest freq. for Pt, lowest for Re; H trans to a Halogen, freq. ~100 cm^{-1} for higher than when *trans* to a phosphine
Ru–H	2020–1750	4.95–5.71	s	Strong band. Dihydrides as low as 1615 cm^{-1}
Os–H	2105–1845	4.75–5.42	s	Dihydrides as low as 1720 cm^{-1}
Rh–H	2140–1880	4.67–5.35		
RhH(CO)	~2005	~4.99		
Ir–H	2245–2000	4.45–5.00	s	Dihydrides as low as 1740 cm^{-1}
Ir–H *trans* to phosphorus, arsenic or carbonyl	2100–2000	4.67–5.00	s	
Ir–H *trans* to halogen	2240–2195	4.46–4.56	s	
Pt–H	2265–2005	4.42–4.99	s	Pt(II)–H str, Dihydrides as low as 1670 cm^{-1}
	870–810	11.49–12.35		Pt–H def vib
Pd–H	2025–1990	4.94–5.03	s	Pd(II)–H str
	740–710	13.51–14.08		Pd–H def vib
U–H	~2200	~4.55		
Re–H	2070–1760	4.83–5.68	s	

In transition metal complexes, correlations between the metal–hydrogen stretching vibration and the *trans*-effect have been observed. The band due to the M–H deformation vibration occurs in the region 870–600 cm^{-1} (11.49–16.67 μm). In complexes having bridging hydrogens, the M–H–M stretching frequency is in the range 1550–1000 cm^{-1} (6.45–10.00 μm). These bands may be very broad and weak.

In general, the M–H stretching band of dihydrides occurs at lower frequencies than that of the equivalent monohydride.

Table 22.25 Dihydride M–H stretching vibration bands

Functional Groups	Region cm^{-1}	µm	Comments
LM(CO)H$_2$ compounds, L=phosphine, arsine			
M=Fe	1875–1810	5.33–5.41	Fe–H str. *Trans*-dihydrides may be as low as 1725 cm^{-1}
M=Ir	2245–2005	4.45–4.99	Ir–H str. *Trans*-dihydrides may be as low as 1740 cm^{-1}
M=Pt	2255–2005	4.43–4.99	Pt–H str. *Trans*-dihydrides may be as low as 1670 cm^{-1}
M=Rh	2140–1960	4.67–5.10	Rh–H str
M=Ru	2020–1750	4.95–5.71	Ru–H str. *Trans*-dihydrides may be as low as 1615 cm^{-1}
M=Os	2105–1940	4.75–5.15	Os–H str. *Trans*-dihydrides may be as low as 1720 cm^{-1}

Metal Oxides and Sulphides[1–3,5,7,13]

Many simple metal oxides do not absorb in the region 4000–650 cm^{-1} (2.50–15.38 µm). However, oxides with more than one oxygen atom bound to a single metal atom usually absorb in the region 1020–970 cm^{-1} (9.80–10.31 µm) and, in general, metal oxides containing the group M=O have a strong absorption at 1100–825 cm^{-1} (9.09–12.12 µm). In some dioxo compounds, this band may be as low as 750 cm^{-1} (13.33 µm).

Different polymorphic forms can be distinguished in the region 700–300 cm^{-1} (14.29–33.33 µm). Cubic crystalline forms of rare earth oxides have a characteristic band at 570–530 cm^{-1} (17.54–18.87 µm).

The band due to the Ti–O stretching vibration has been found to vary in the range 1000–400 cm^{-1} (10.00–25.00 µm). Ti–O–Ti

Table 22.26 Approximate stretching vibration frequencies for AX$_4$

Group	Region cm^{-1}	µm
Ti^{IV}–O	625–310	16.00–32.26
O–O	1000–770	10.00–12.99
Al–O	750–490	13.33–20.41
Ti=O	1090–695	9.17–14.39
V=O	1035–890	9.66–11.24
Sn–O	780–300	12.82–33.33
Tc–O	~245	~40.82
Ge–O	1000–900	10.00–11.11
Ge–O–Ge	900–700	11.11–14.29
Sn–O	780–580	12.82–17.74
Pb–O	~625	~16.00

Table 22.27 Carbon clusters

Functional Groups	Region (cm^{-1})	Comment
C$_3$	~2042	
	~63	Bending vib
C$_4$	~1543	
C$_5$	~2164	
	~1447	
C$_6$	~1952	
	~1197	
C$_7$	~2138	
C$_8$	~1998	See ref. 38
C$_9$	~2128	
C$_{13}$	~1809	
C$_{60}$	~1428	Cyclic structure (on KBr)
	~1183	
	~577	
	~527	
C$_{70}$	~1460	Cyclic structure (on KBr)
	~1430	
	~1414	
	~1134	
	~795	
	~674	
	~642	
	~578	
	~565	
	~535	
	~458	

absorbs in the range 1000–700 cm^{-1} (10.00–14.29 µm) and Ti–O–Si at 950–900 cm^{-1} (10.53–11.11 µm). The stretching frequencies of silicate, Si–O, borate, B–O, metaphosphate, P–O, and germanate, Ge–O, bonds are 1100–900 cm^{-1} (9.09–11.11 µm), 1380–1310 cm^{-1} (7.25–7.63 µm),

Chart 22.5 Infrared – band positions of metal oxides and sulphides

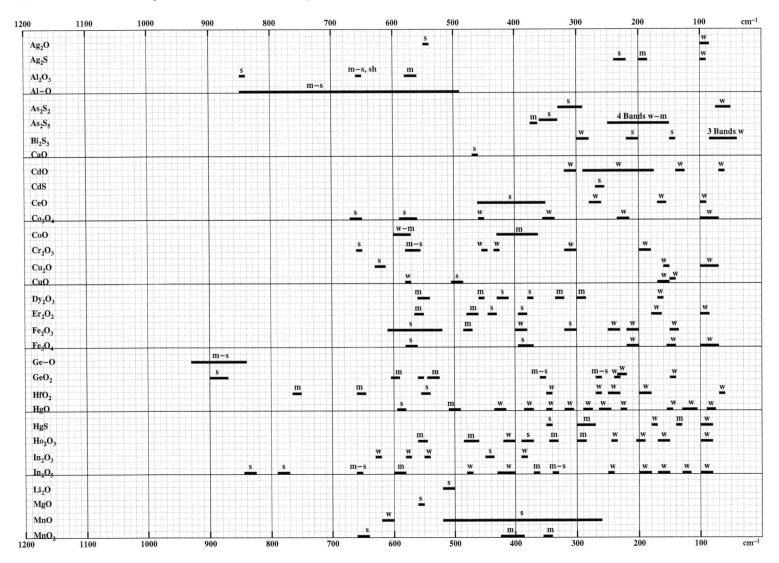

1300–1140 cm^{-1} (7.69–8.77 μm) and 930–840 cm^{-1} (10.75–11.90 μm) respectively. In general, the band due to the Pt–O stretching vibration occurs near 390 cm^{-1} (25.64 μm) and that for CoO in the region 430–360 cm^{-1} (23.25–27.77 μm). BeO, BaO$_2$, CrO$_3$, PbO$_2$, RuO$_2$ and Ti$_2$O$_3$ have no strong bands in the region 600–250 cm^{-1}

(16.67–40.00 μm). TiO$_2$ has a band of strong intensity at 700–660 cm^{-1} (14.29–15.15 μm), a band of medium-to-strong intensity at 360–320 cm^{-1} (27.28–31.25 μm) and weak bands at 185–170 cm^{-1} (54.05–58.82 μm) and 100–80 cm^{-1} (100.00–125.00 μm). Sb$_2$O$_3$ absorbs strongly at 770–740 cm^{-1} (12.99–13.51 μm) and 385–355 cm^{-1} (25.97–28.17 μm) and has a

Chart 22.5   (*continued*)

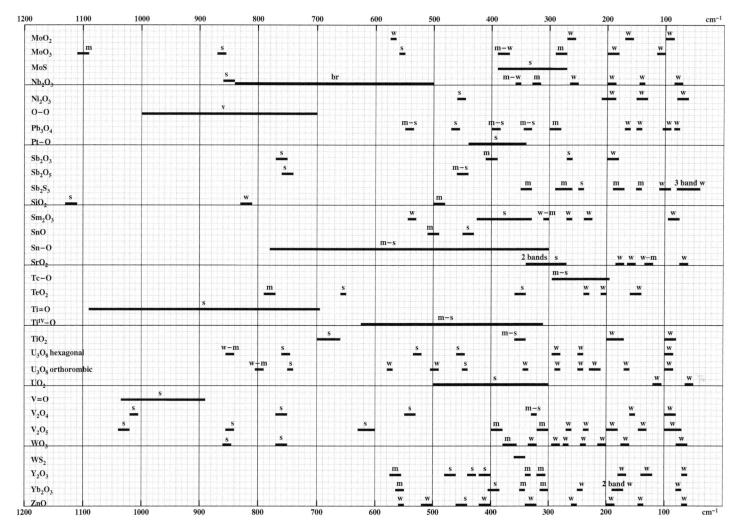

medium-intensity band at 415–395 cm^{-1} (24.10–25.32 µm) and a weak band at 200–180 cm^{-1} (50.00–55.56 µm).

Metal alkoxides have a band near 1000 cm^{-1} (10.00 µm) due to the C–O stretching vibration and another due to the M–O stretching vibration in the region 600–300 cm^{-1} (16.67–33.33 µm).

In general, the absorptions of metal sulphides occur below 400 cm^{-1} (25.00 µm).

Glasses

The fundamental vibrational frequencies of orthosilicates, SiO_4^{4-}, are ~955 cm^{-1} (~10.47 µm), ~820 cm^{-1} (~12.20 µm), ~525 cm^{-1} (19.05 µm) and ~355 cm^{-1} (~28.17 µm). In Raman spectra, the structural units of silicate glasses may be differentiated by using the Si–O stretching frequencies: silicates with four terminal oxygen atoms absorb near 850 cm^{-1}

Chart 22.5 (*continued*)

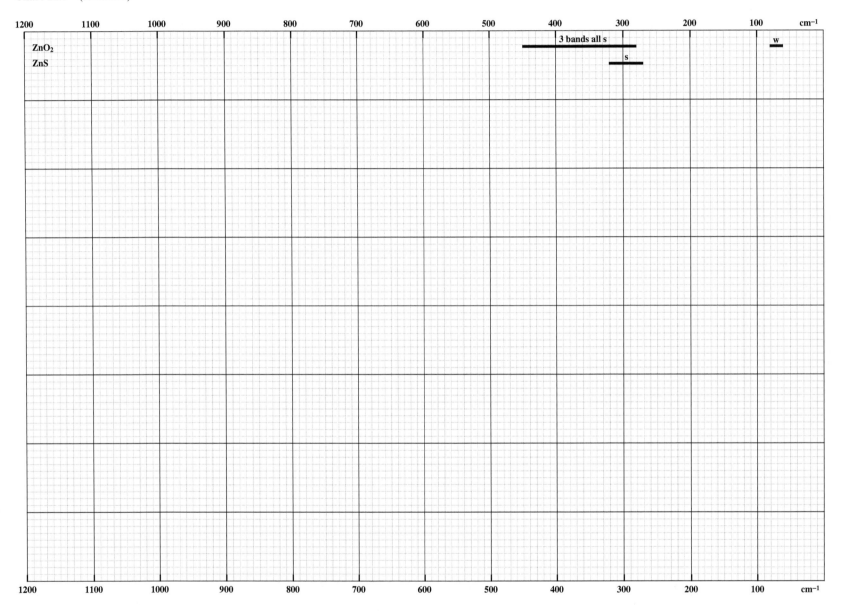

($\sim$11.76 µm), those with three terminal oxygen atoms absorb near 900 cm^{-1} ($\sim$11.11 µm), those with two terminal oxygen atoms absorb at 1000–900 cm^{-1} (10.00–11.11 µm) and those with only one terminal oxygen atom absorb at 1100–1050 cm^{-1} (11.11–9.52 µm). The Raman spectra of aluminosilicates exhibit an absorption due to the Al–O stretching vibration at 750–650 cm^{-1} (13.33–15.38 µm). A review of the use of Fourier transform infrared in the analysis of inorganic and organic surface coatings on insulating substrates has been made.[33]

Carbon Clusters

Uncharged small carbon clusters have been found to have linear structures.[19-21] Carbon clusters absorb in the infrared region. Some of the bands observed are given (for small clusters being mainly in an argon matrix) Table 22.27. The majority of bands given in the table for small carbon clusters are due to stretching vibrations. Buckminsterfullerene,[22-27,34] C_{60}, has a carbon cyclic structure that resembles the surface pattern of a modern football, i.e. composed of 20 hexagonal faces and 12 pentagonal faces. Four bands are observed in the infrared spectrum[24-26] of C_{60} near 1428, 1183, 577 and 527 cm^{-1} and ten bands in its Raman spectrum[23] near 1575, 1470, 1428, 1250, 1099, 774, 710, 496, 437 and 273 cm^{-1}. In general, the bands above 1000 cm^{-1} are mainly due to the tangential motions of the carbon atoms, i.e. along the C–C bonds, and those below 800 cm^{-1} are due to the radial motions of the carbon atoms, in effect, deformations, an exception to this being the band near 496 cm^{-1} which has been assigned to tangential motion. When the symmetry of the C_{60} is lowered either by adsorption or doping, additional bands are observed in the spectrum.

The Raman spectrum of diamond exhibits a strong, sharp band near 1330 cm^{-1} ($\sim$7.52 µm),[28] the position of which has been studied at high pressures.[29,30] In the infrared, thick crystals used for diamond anvil cells, depending on the type, absorb near 2000 cm^{-1} ($\sim$5.00 µm) and 1200 cm^{-1} ($\sim$8.33 µm) or only near 2000 cm^{-1} ($\sim$5.00 µm).[30,31] These absorption bands have been attributed to defects and multiphonon effects respectively.

No infrared bands are observed for graphite. However, graphite does have a band near 1575 cm^{-1} ($\sim$6.35 µm) in its Raman spectrum.[32] Carbon, as graphite (amorphous, crystalline etc.), may be characterised using Raman spectroscopy by examining the relative intensity of bands near 1575 cm^{-1} ($\sim$6.35 µm) and 1350 cm^{-1} ($\sim$7.41 µm).

References

1. K. Nakamoto, *Infrared and Raman Spectra of Inorganic and Coordination Compounds*, 5th edn, Wiley, New York, 1997.
2. I. A. Cadsen, *Infrared Spectra of Minerals and Related Inorganic Compounds*, Butterworth, London, 1975.
3. S. D. Ross, *Inorganic Infrared and Raman Spectra*, McGraw- Hill, London, 1972.
4. R. A. Nyquist and R. O. Kagel, *Infrared Spectra of Inorganic Compounds*, Academic Press, New York, 1971.
5. V. C. Farmer (ed.), *The Infrared Spectra of Minerals*, Mineralogical Society, London, 1974.
6. N. B. Colthurp, L. H. Daly, and S. E. Wiberley, *Introduction to Infrared and Raman Spectroscopy*, Academic Press, Boston, 1990.
7. A. A. Davydov, *IR Spectroscopy of Adsorbed Species on the Surface of Transition Metal Oxides*, C. H. Rochester (Ed.), Wiley, New York, 1984.
8. A. Johnson, and H. J. Taube, *Indian Chem. Soc.*, 1989, **66**, 503.
9. A. Rulmont *et al.*, *Eur. J. Solid State Inorg. Chem.*, 1991, **28**, 207.
10. 'Vibrational Spectra of Some Coordinated Ligands', *Spectrosc. Prop. Inorg. Organomet. Compds.*, Vols. 1 and on, 1968 to date.
11. 'Vibrational Spectra of Transition Element Compounds', *Spectrosc. Prop. Inorg. Organomet. Compds.*, Vols. 1 and on, 1968 to date.
12. 'Inorganic and Organometallic Compounds', *Spectrosc. Prop. Inorg. Organomet. Compds.*, Vols. 1 and on, 1968 to date.
13. 'Characteristic Vibrations of Compounds of Main Group Elements I to VIII', *Spectrosc. Prop. Inorg. Organomet. Compds.*, Vols. 1 and on, 1968 to date.
14. D. A. Thorton, *Coord. Chem. Rev.*, 1990, **104**, 173.
15. I. Eliezer and A. Reyer, *Coord. Chem. Rev.*, 1972, **9**, 189.
16. A. S. Braterman, *Metal Carbonyl Spectra*, Academic Press, New York, 1975.
17. R. L. Davidovich and V. I. Kostin, *Atlas of Long Wavelength Infrared Absorption Spectra of Complex Group III–V Metal and Uranyl Fluorides*, Nauka, Moscow, 1977.
18. R. L. Davidovich *et al.*, *Atlas of Infrared Absorption Spectra of and X-Ray Measurement Data for Complex Group IV and V Metal Fluorides*, Nauka, Moscow, 1972.
19. W. Weltner, Jr. and R. J. van Zee, *Chem. Rev.*, 1989, 89, 1713.
20. M. Vala *et al.*, *J. Chem. Phys.*, 1989, **90**, 595.
21. T. F. Giesen *et al.*, *Science*, 1994, **265**, 756.
22. H. W. Kroto *et al.*, *The Fullerenes*, Pergamon, Oxford 1993.
23. T. J. Dennis *et al.*, *Spectrochim. Acta.*, 1991, **47A**, 1289.
24. D. S. Bethune *et al.*, *Chem. Phys. Lett.*, 1991, **179**, 181.
25. C. I. Frum *et al.*, *Chem.. Phys. Lett.*, 1991, **176**, 504.
26. J. P. Hare *et al.*, *J. Chem. Soc. Chem. Commun.*, **1991**, 412.
27. R. E. Stanton and M. D. Newton, *J. Phys. Chem.*, **1988**, 2141
28. D. S. Knight and W. B. J. Materials Research, 1989, **4**, 385.
29. J. R. Ferraro, *Vibrational Spectroscopy at High Pressures*, Academic Press, New York, 1984.
30. E. R. Lippincott *et al.*, *Anal. Chem.*, 1961, **33**, 137.
31. A. Tardieu *et al.*, *J. Appl. Phys.*, 1990, **68**, 3243.
32. F. Tuinstra and J. L. Koenig, *J. Chem. Phys.*, 1970, **53**, 1126.
33. R. L. De Rosa *et al.*, *Glass Res.*, 1999, **9(1)**, 7.
34. R. Moret *et al.*, *Eur. Phys. J. B*, 2000, **15(2)**, 253.
35. F. Quiles and A. Burneau, *Vib. Spectrosc.*, 2000, **23(2)**, 231.
36. J. P. Devlin *et al.*, *J. Phys. Chem. A*, 2000, **104(10)**, 1974.
37. L. H. Johnson *et al.*, *Geochim. Cosmochim. Acta*, 2000, **64(4)**, 717.
38. S. L. Wang *et al.*, *J. Chem. Phys.*, 2000, **112(3)**, 1457.

23 Biological Molecules – Macromolecules

Introduction

The purpose of this chapter is to help those interested in the characterisation/identification of biological molecules/samples. The intention is not to deal with infrared or Raman techniques, nor to deal with sampling methods for the two techniques. There are several good books dealing with these aspects.

Both infrared and Raman spectroscopic techniques have proved to be valuable tools in the study of biological molecules. Many of the comments made in the introduction to Chapter 21 dealing with polymers, regarding the relative merits, advantages and disadvantages of the two techniques, are also true for biological samples. Both of these techniques can help in the understanding of the relationship between the molecular structure and the function of a biological substance. As mentioned in the introduction to Chapter 21, in general, the infrared and Raman spectra of biological macromolecules are broad and so it is often difficult to identify the origins of some bands. This is particularly true of large naturally occurring substances such as proteins, carbohydrates, nucleic acids, cell membranes and tissues. Nonetheless, many advances have been made in the application of both techniques to biological samples.

It should be borne in mind that the physical state of a biological material may affect the spectral features observed. For example, in the solid phase, differences in some of the features may be observed for different polymorphs. Differences will also be observed for crystalline and amorphous substances and also the presence or absence of hydrogen bonding will affect the spectrum obtained.

Sample Preparation

Water is a commonly used solvent for biological systems. Hence for infrared, sodium chloride and potassium bromide (as well as other water-soluble salts) cannot be used as window materials for cells. Silver chloride, calcium fluoride, and barium fluoride are more commonly used. Small pathlengths, of the order of 0.010 mm, are often used to reduce the intensity of the very strong infrared bands produced by water. In addition, the strong bands due to water may overlap sample bands of interest. This problem may be overcome by making use of heavy water, D_2O. The bands due to D_2O occur at lower wavenumbers than those of ordinary water (see Chapters 1 and 22). The importance of D_2O to reveal bands overlapped by bands due to water cannot be over emphasised. One advantage of using small pathlengths is that samples in nano- and microgramme quantities may be examined.

In some cases, the use of isotopic substitution may be helpful. Two problems often encountered in the assignment of infrared and Raman bands for a biological molecule are those of ambiguity or the bands of interest being overlapped by other bands. To resolve the problem, isotopic substitution may be used. For example, the deuterated equivalent molecule will have its hydrogen vibrations shifted to lower frequency. This frequency shift may easily be estimated using reduced masses. The presence of a group may be confirmed (or otherwise) or the band of interest is no longer overlapped. The use of other isotopes, not just those of hydrogen, may be helpful. Hence, isotopic considerations may help to reduce ambiguity or simplify spectra.

Raman spectroscopy has a definite advantage for biological systems in that absorption bands due to water do not present a problem. However, fluorescence may present a major problem for some samples when examined by Raman techniques and photochemical interactions may need to be borne in mind. The use of horizontal attenuated total reflectance techniques has become more popular for the study of aqueous solutions of biological samples.

Carbohydrates

There are two important limitations to the spectral identification of carbohydrates which should be borne in mind and those are that the differences between the spectra of consecutive members can become very small after the first five or so members and that the spectra of D and L enantiomorphs, if they occur, are identical.

Carbohydrates[19] have broad, medium-strong intensity bands in the region $3520-3100 \, \text{cm}^{-1}$ $(2.84-3.22 \, \mu\text{m})$ due to O–H stretching vibrations. The OH deformation vibrations result in a band in the region $1080-1030 \, \text{cm}^{-1}$ $(9.26-9.71 \, \mu\text{m})$. Strong absorptions are also observed in the region $1290-1030 \, \text{cm}^{-1}$ $(7.75-9.71 \, \mu\text{m})$ due to C–O stretching vibrations. Medium-to-weak intensity bands are observed in the region $960-730 \, \text{cm}^{-1}$ $(10.42-13.70 \, \mu\text{m})$ which may be used to distinguish between α and β anomers of pyranose compounds. The weak bands observed near $905 \, \text{cm}^{-1}$ $(11.05 \, \mu\text{m})$ and $780 \, \text{cm}^{-1}$ $(\sim 12.82 \, \mu\text{m})$ for disaccharides appear near $930 \, \text{cm}^{-1}$ $(\sim 10.75 \, \mu\text{m})$ and $760 \, \text{cm}^{-1}$ $(\sim 13.16 \, \mu\text{m})$ for polysaccharides.

Free hydroxyl groups absorb in the region $3730-3520 \, \text{cm}^{-1}$ $(2.68-2.84 \, \mu\text{m})$. If the carbohydrate is soluble in solvents to which hydrogen bonding is not possible, for example carbon tetrachloride, then in the spectra of dilute solutions, bands for bonded hydroxyl groups may still be observed if intramolecular hydrogen bonding is possible. Depending on the possible structural arrangements (conformations) of the carbohydrate, bands for both free and bonded hydroxyl groups may still be observed. In other words, infrared spectroscopy provides a simple means of distinguishing not only between intermolecular and intramolecular hydrogen bonding, but also chelation (which involves strong intramolecular bonding). The main factors that may affect the position of bands associated with the OH and NH groups are temperature and concentration. Dilution effects may also sometimes be observed. In a nonpolar solvent, the position, intensity and shape of a band associated with an intramolecular or chelated hydroxyl stretching vibration is unaffected on dilution whereas for intermolecular OH stretching vibrations the opposite is true. The bands associated with intramolecular and chelated hydroxyl stretching vibrations are generally sharp. For carbohydrates, other bands that may be affected by hydrogen bonding are those due to C=O and C–O–C stretching vibrations.

The chemical modification of a sample may sometimes assist in its identification. For example, deuterium may be substituted for the hydrogen atom of hydroxyl groups, either partial or complete deuteration being used. Acetylated glycosides have a band due to the acetyl C=O stretching vibration at $1775-1735 \, \text{cm}^{-1}$ $(5.63-5.76 \, \mu\text{m})$ which is not observed for the non-acetylated compounds. In addition, the non-acetylated glycosides exhibit a band at $3340-3275 \, \text{cm}^{-1}$ $(2.99-3.05 \, \mu\text{m})$ due to the OH stretching vibration which is not observed in the acetylated form.

The spectral examination of aqueous solutions is important since water is the environment of the natural system and is relatively easy to use. The problem is that water absorbs strongly over a wide region when infrared spectroscopy is used. However, in the case of Raman spectra, only one relatively weak band is observed over the region $2000-200 \, \text{cm}^{-1}$ $(5.00-50.00 \, \mu\text{m})$ and that is near $1640 \, \text{cm}^{-1}$ $(\sim 6.10 \, \mu\text{m})$.

To assist in the interpretation of carbohydrate infrared spectra, one approach is to make use of model compounds which are of a simple nature but have similarities with the structure of the carbohydrate being examined. For example, tetrahydropyran is such a model compound since it contains the pyranose ring which is found in sugars. A knowledge of its spectrum can assist in the study of, and interpretation of, carbohydrate spectra.

Cellulose and its Derivatives

Cotton absorbs at about $3355 \, \text{cm}^{-1}$ $(\sim 2.98 \, \mu\text{m})$ due to the O–H stretching vibration. After chemical modification, this band appears at higher frequencies – for example, for methyl cellulose, it is near $3400 \, \text{cm}^{-1}$ $(\sim 2.94 \, \mu\text{m})$, for ethyl cellulose near $3425 \, \text{cm}^{-1}$ $(\sim 2.92 \, \mu\text{m})$, for cellulose acetate, in the region $3510-3490 \, \text{cm}^{-1}$ $(2.85-2.87 \, \mu\text{m})$, for carboxymethyl cellulose near $3380 \, \text{cm}^{-1}$ $(\sim 2.95 \, \mu\text{m})$ and for regenerated cellulose from the acetate, the band appears near $3400 \, \text{cm}^{-1}$ $(\sim 2.94 \, \mu\text{m})$.

The positions and intensities of the OH stretching vibration bands vary for the different polymorphic forms of cellulose. For example, one form has two strong bands near $3480 \, \text{cm}^{-1}$ $(\sim 2.87 \, \mu\text{m})$ and $3345 \, \text{cm}^{-1}$ $(\sim 2.99 \, \mu\text{m})$, whereas another form has its dominant OH band near $3350 \, \text{cm}^{-1}$ $(\sim 2.99 \, \mu\text{m})$. These bands all exhibit dichroism. Intensity differences are also observed for different forms near $1430 \, \text{cm}^{-1}$ $(\sim 6.99 \, \mu\text{m})$ and $1110 \, \text{cm}^{-1}$ $(\sim 9.01 \, \mu\text{m})$.

Amino Acids

Amino acids[4-9] are amine derivatives of carboxylic acids and may, in fact, contain a number of amino and carboxylic acid groups. In the simplest case, with one acid and one amino group, the amino group may occupy α or β or γ, etc., positions. Amino acids, polypeptides and proteins are related compounds

Table 23.1 Characteristic bands observed for the pyranose ring

Vibration	Frequency cm^{-1}	Wavelength μm
Asym ring	930–900	10.75–11.11
Sym ring breathing	785–755	12.74–13.25
Anomeric C–H def	855–835	11.70–11.78
Anomeric C–H def	900–880	11.11–11.36
Equatorial CH def other than anomeric C–H def	870–865	11.49–11.56
Terminal methyl def	975–960	10.26–10.42

Table 23.2 Carbohydrates

Functional Groups	Region		Intensity		Comments
	cm^{-1}	μm	IR	Raman	
Carbohydrates	~3350	~2.99	s	w	OH str, br.
	~2900	~3.45	m−w	m	CH str
	1460–1200	6.85–8.33	m	m−w	CH and OH def vib. Numerous bands.
	1160–1000	8.62–10.00	s	m−w	C–O str
	960–730	10.42–13.70	m−w	m−w	CH def
Pyranose compounds	1200–1030	8.33–9.70	s	m−w	C–O str
Tetrahydropyranose compounds	~875	~11.43	m−s	m−s	asym ring str
	~815	~12.27	m	m−s	sym ring str
α-Pyranose compounds	985–955	10.15–10.47	m−s	m−s	sym ring vib
	975–960	10.26–10.42	m	m−w	Terminal CH$_2$ def vib
	935–905	10.70–11.05	m−w	m−s	asym ring vib
	890–870	11.24–11.49	w	m	Equatorial CH def vib (non-glycosidic)
	855–835	11.70–11.98	m−w	m	C–H def vib characteristic of α
	785–755	12.74–13.25	m−w	m−s	Ring vib
β-Pyranose compounds	985–955	10.15–10.47	m−s	m−s	sym ring vib
	975–960	10.26–10.42	m−s	m−s	sym ring vib
	935–905	10.67–11.05	m−w	m−s	asym ring vib
	900–880	11.11–11.36	m−w	m−w	C–H def vib characteristic of β
	890–870	11.24–11.49	w	m−w	Equatorial CH def vib (non-glycosidic)
	785–760	12.74–13.16	m−w	m−s	Ring vib

Table 23.3 Cellulose and its derivatives

Functional Groups	Region		Intensity		Comments
	cm^{-1}	μm	IR	Raman	
Cellulose	3575–3125	2.80–3.20	m	w	br. OH str
	1750–1725	5.71–5.80	s	w–m	C=O str, after oxidation
	1635–1600	6.12–6.25	m	m	OH def vib
	1480–1435	6.76–6.95	w	m	CH$_2$def vib. Intensity affected by degree of crystallinity
	~1375	~7.27	w	m−w	CH def vib
	~1340	~7.46	w	m−w	OH def vib
	1320–1030	7.58–9.71	w	m−w	Numerous bands
	~830	~12.05	w	w	CH$_2$ def vib

Table 23.4 Amino acid −NH+ and N−H vibrations

Functional Groups	Region cm⁻¹	Region μm	Intensity IR	Intensity Raman	Comments
Free amino acids (NH₃⁺)···COO⁻ and amino acid hydrohalides X⁻(NH₃⁺)···COOH (X = halogen)	3200–3000	3.13–3.31	m	m–w	asym −NH₃⁺ str
	2760–2530	3.62–3.95	m	m–w	br } −NH₃⁺ sym str
	2140–2050	4.67–4.88	w–m	m–w	
	1660–1590	6.03–6.29	w	w	asym −NH₃⁺ def vib
	1550–1485	6.46–6.74	w–m	w	sym NH₃⁺ def vib
	1295–1090	7.72–9.18	w	m–w	NH₃⁺ rocking vib
Deuterated amino acids	1190–1150	8.40–8.70	w	m–w	ND₃⁺ def vib
	~800	~12.50	w	w	ND₃⁺ rocking vib
Amino acid salts (NH₂)···COO⁻M⁺ (M = metal atom, e.g. Na)	3400–3200	2.94–3.13	m	m	Two bands, −NH₂ str

Table 23.5 Amino acid carboxyl group vibrations

Functional Groups	Region cm⁻¹	Region μm	Intensity IR	Intensity Raman	Comments
Free amino acids and amino acid salts	1600–1560	6.25–6.41	s	m–w	asym CO₂⁻ str
	1425–1390	7.02–7.19	w	m–w	sym CO₂⁻ str
Amino acid hydrohalides and dicarboxylic amino acids	1755–1700	5.70–5.88	s	m–w	C=O str of −COOH, α-amino acids absorb at 1755–1730 cm⁻¹, other amino acids absorb at 1730–1700 cm⁻¹

Table 23.6 Amino acids: other bands

Functional Groups	Region cm⁻¹	Region μm	Intensity IR	Intensity Raman	Comments
Free amino acids	3000–2850	3.33–3.51	m–s	m–s	CH str
	2650–2500	3.77–4.00	m	m	
	2120–2010	4.72–4.98	m	m	
	1340–1315	7.46–7.61	m	m–w	CH def vib
	560–500	17.86–20.00	s		
Amino acid hydrohalides	3000–2500	3.33–4.00	w	m–w	Series of broad bands
	~1300	~7.69	m	m	
	1230–1215	8.13–8.23	s	w	C−O str

Table 23.7 Amido acids

Functional Groups	Region		Intensity		Comments
	cm^{-1}	μm	IR	Raman	
Amido acids	3390–3260	2.95–3.07	m	w–m	N–H str
	2640–2360	3.79–4.24	w	m	Not always present
	1945–1835	5.14–5.45	w	m	Not always present
	1750–1695	5.71–5.90	s	m–s	Acid C=O str
	1565–1505	6.39–6.64	s	w	Amide II band
	1230–1215	8.13–8.23	s	w	C–O str
α-Amido acids	1620–1600	6.14–6.25	s	m–s	Amide C=O str
Other amido acids	1650–1620	6.06–6.14	s	m–s	Amide C=O str

and their infrared spectra reflect this to a certain extent.[5] Amino acids may be found in three forms:

(a) as a free acid,

,·COO$^-$

`·NH$_3$

where the dotted line represents any carbon backbone structure.

(b) as the salt, e.g. sodium, of the acid,

,·COO$^-$Na$^+$

`·NH$_2$

(c) as the amine hydrohalide,

,·COOH

`·NH$_3^+$X$^-$

Free Amino Acid –NH$_3^+$ Vibrations[7,8]

Free amino acids have –NH$_3^+$ stretching and deformation vibrations. In the solid phase, a broad absorption of medium intensity is observed in the region 3200–3000 cm^{-1} (3.13–3.33 μm) due to the asymmetric –NH$_3^+$ stretching vibration. Weak bands due to the symmetric stretching vibration of the NH$_3^+$ group are observed near 2600 cm^{-1} (~3.85 μm) and 2100 cm^{-1} (~4.76 μm). A fairly strong –NH$_3^+$ deformation band is observed at 1550–1485 cm^{-1} (6.46–6.74 μm) and a weaker band, which is not resolved for most amino acids, at 1660–1590 cm^{-1} (6.03–6.29 μm).

Free Amino Acid Carboxyl Bands[4–8]

Free amino acids also have carboxylate ion CO$_2^-$ stretching vibrations, a strong band occurring in the region 1600–1560 cm^{-1} (6.25–6.41 μm). Dicarboxylic acids have a strong band due to the C=O stretching vibration of the carboxyl group at 1755–1700 cm^{-1} (5.70–5.88 μm) and another strong band at 1230–1215 cm^{-1} (8.13–8.23 μm) due to the stretching vibration of the C–O bond. A band of medium intensity and of uncertain origin is usually observed near 1320 cm^{-1} (~7.68 μm). A strong band at 560–500 cm^{-1} (17.86–20.00 μm), which is due to the CO$_2^-$ or C–C–N group deformation vibrations, is observed for amino acids, except for cyclic amino acids. Skeletal deformation bands occur in the region 500–285 cm^{-1} (20.00–35.09 μm).

Amino Acid Hydrohalides

In addition to the –NH$_3^+$ stretching and deformation absorption bands, which are as given above, a series of weak, fairly broad bands is observed in the region 3000–2500 cm^{-1} (3.33–4.00 μm). In Raman spectra, the NH stretching vibration bands are lost under the much stronger CH stretching bands. Also, the band due to the C–O stretching vibration is the same as for free amino acids. The band due to the C=O stretching vibration of the carboxyl group is observed in the range 1755–1700 cm^{-1} (5.70–5.88 μm). In the Raman spectra of aqueous solution, the C=O band is usually at 1745–1725 cm^{-1} (5.73–5.80 μm).

Amino Acid Salts

For amino acid salts, two bands of medium intensity are observed at 3400–3200 cm^{-1} (2.94–3.13 μm) due to the asymmetric and symmetric stretching vibrations of the –NH$_2$ group. In Raman spectra, the symmetric NH$_2$

stretching band, $\sim$3305 cm^{-1} ($\sim$3.03 µm), is stronger than the asymmetric band, $\sim$3370 cm^{-1} ($\sim$2.97 µm). A strong band at 1600–1560 cm^{-1} (6.25–6.42 µm) due to the carboxylate ion is observed. In the Raman spectra of aqueous solutions, the C=O stretching band is usually difficult to observe, being too near the HOH deformation band near 1630 cm^{-1} ($\sim$6.13 µm). However, heavy water may be used to overcome this problem. In Raman spectra, the COO$^-$ asymmetric and symmetric stretching bands occur at 1600–1570 cm^{-1} (6.25–6.37 µm) and 1415–1400 cm^{-1} (7.07–7.13 µm) respectively.

Nucleic Acids

The full infrared spectra of nucleic acids[32] in aqueous media may be obtained by using water, H_2O, and deuterium oxide, D_2O. In this way, regions in which there are strong absorption due to H_2O may be covered by using D_2O. Specific regions contain information relating to particular groups within nucleic acids. In the spectral region 1780–1530 cm^{-1} (5.62–6.54 µm), the bands due to the in-plane double bond vibrations of bases occur. The absorptions in this region are sensitive to pairing and stacking effects. In the region 1550–1250 cm^{-1} (6.45–8.00 µm), base deformations coupled through the glycosidic linkage to the vibrations of saccharides are observed, the band positions being greatly influenced by the glycosidic torsion angle. In the region 1250–1000 cm^{-1} (8.00–10.00 µm), two strong absorptions are observed due to the asymmetric and symmetric stretching vibrations of the phosphate group, $\sim$1230 cm^{-1} ($\sim$8.13 µm) and $\sim$1090 cm^{-1} ($\sim$9.17 µm) and in addition there are other bands due to vibrations involving sugar components. The region 1000–700 cm^{-1} (10.00–14.29 µm) contains bands due to the vibrations of the phosphate-sugar backbone, ring puckering and out-of-plane base vibrations.

For a review of surface enhanced Raman spectroscopy of nucleic acid components see reference 36.

Amido Acids, $\diagdown$N–CO–$\cdots$COOH$\diagup$

In the solid phase, α-amido acids have a medium-intensity absorption at 3390–3260 cm^{-1} (2.95–3.07 µm) due to the stretching vibration of the N–H bond and strong bands at 1750–1695 cm^{-1} (5.71–5.90 µm), 1620–1600 cm^{-1} (6.14–6.25 µm), and 1565–1505 cm^{-1} (6.39–6.64 µm). The first two of these three bands are due to the carbonyl stretching vibration, the first being due to the acid and the other to the amide group. The third band is an amide II band. The band near 1610 cm^{-1} ($\sim$6.21 µm) is characteristic of α-amido acids, other amido acids having this absorption at 1650–1620 cm^{-1} (6.06–6.14 µm).

Proteins and Peptides

In general, proteins[3,9,29,30,37] and peptides[37] have broad, strong bands which, due to considerable overlap, are difficult to differentiate. The reason for this is the large number of different amino acids which form a complex protein.

In general, spectral changes are observed to accompany the denaturation of proteins. The second order structures of proteins may adopt a spiral form (α-helix), an extended chain (β-form) and a random coil arrangement. The position of the amide I, II and III bands (see below) are affected by the structural arrangement of the protein.

The spectra of all proteins exhibit absorption bands due to their characteristic amide group, CO·NH. Hence, the characteristic bands of the amide group of protein chains are similar to those of ordinary secondary amides. The bands of proteins are labelled in the same way as amide bands. This is in order to reflect the various contributions to the bands made by the vibrations. The strongest bands in the infrared spectra of proteins are the amide I and II bands. These bands are broad and, without deconvolution techniques, do not have enough definition to give useful structural information. Usually proteins have a strong, polarised band in their Raman spectra which occurs at 900–800 cm^{-1} (11.11–12.50 µm) due to the symmetrical CNC stretching vibration. This vibration results in a band of weak intensity in the infrared. Table 23.8 gives the approximate position of the important main bands observed for proteins and a summary of the contributions to the bands.

The most useful infrared bands for the characterisation of proteins in aqueous solution are the amide I and II bands. Raman studies tend to make use of the amide I and III bands. The amide II band is generally inactive or very weak in Raman. The amide I band occurs near 1655 cm^{-1} ($\sim$6.04 µm) and the precise position of this band is dependent on the nature of the hydrogen bonding between the CO and N–H groups. The nature of the hydrogen bonding is determined by the particular molecular arrangement adopted by the part of the protein responsible for the band. In general, proteins have a variety of domains, these having different conformations and, as a result, the amide I band is usually a complex composite which is composed of a number of overlapping bands resulting from the different types of structure that may be adopted – α-helixes, β-sheet structures and non-ordered structures, etc. (In fact, the positions of the amide I, II and III bands are sensitive to the torsional angles about the C^α–N and the C^α–C bonds. These two torsional angles have definite characteristic values which result in the α-helixes, β-sheet structures and non-ordered structures. Most proteins have a distribution of conformers and as a result certain bands are broad.) Curve fitting is used for the amide I band to determine the structural arrangement of the protein. The number of component bands, their positions and other parameters required may be obtained from the study of derivative spectra and the deconvolution of spectra. The relative

Table 23.8 Proteins

Functional Groups	Region cm^{-1}	Region μm	Intensity IR	Intensity Raman	Comments
Proteins	~3300	~3.03	m	w–m	N–H str
	~3100	~3.23	w		Overtone of amide II band
	~1655	~6.04	s	m–s	Amide I band. (~80% CO str, ~10% CN str, ~10% NH bending vib)

Region (cm^{-1})	Intensity IR	Intensity Raman	Comments
1675–1665	s	m–s	β-sheet structure
1670–1660	s, br	m–s	random chain
1655–1645	s	m–s	α-helix

Functional Groups	Region cm^{-1}	Region μm	Intensity IR	Intensity Raman	Comments
	~1565	~6.39	s	w	Amide II band. (~60% NH bending vib, 40% CN str)
	~1300	~7.69	w–m	v	Amide III band. (30% CN str, 30% NH bending vib, 10% CO str, 10% O=C–N bending vib, rest other vibs)

Region (cm^{-1})	Intensity IR	Intensity Raman	Comments
1300–1270	w	w	α-helix
1255–1240	w–m, br	m, br	random chain
1235–1225	w–m,	m–s	β-sheet structure

Functional Groups	Region cm^{-1}	Region μm	Intensity IR	Intensity Raman	Comments
	900–800	11.11–12.50	w	s, p	sym CNC str
	~725	~13.79	m	s–m	Amide V band. (N–H bending vib)
	~625	~16.00	s	m–s	Amide IV band.(40% O=C–N bending vib, rest other vibs)
	~600	~16.67			Amide VI band. (CO bending vib)
	~200	~50.00			Amide VII band. C–N torsional vib

amount of each structural arrangement of a domain is directly proportional to the area of its fitted component(s). These days, this curve-fitting task can be easily accomplished by making use of computer programs. Obviously, this approach can be used for other bands, such as the amide III band.

Unfortunately, for some proteins, the amide I for the α-helix may be hidden by strong absorptions by water, in which case the amide III band may be investigated. Changes may also be observed in the C–C stretching vibrations that occur in the region 1000–945 cm^{-1} (10.00–10.58 μm).

In order to maintain a given protein conformation, disulphide bonds are often present in the structure. The conformation about these bonds is related to the structure of the protein. Although, in general, in infrared spectra the S–S stretching vibration results in a weak absorption, in Raman spectra, the vibration leads to a strong band. The S–S stretching vibration occurs near 490 cm^{-1} (~20.41 μm). Different conformational arrangements

about the disulphide bond lead to the S–S stretching vibration bands occurring in different positions. The stretching vibration of the S–S bond is determined by the rotational conformation about the C–C and S–C bonds. The disulphide bond in the *gauche–gauche–gauche* arrangement absorbs near 510 cm^{-1} (~19.61 μm) and, for the *trans–gauche–gauche* and *trans–gauche–trans* arrangements, the band positions are near 525 cm^{-1} (~19.05 μm) and 540 cm^{-1} (~18.52 μm) respectively.

Valuable information may be obtained from protein spectra by varying parameters such as concentration, pH, ionic strength, etc.

The infrared spectra of optically active peptide enantiomers are identical (e.g. D, D and L, L isomers) but there may be differences in the spectra of isomers where the optical activity differs at different asymmetric carbons (e.g. D, L and L, D isomers). The use of polarised infrared radiation can be helpful in peptide studies – for example, to determine the orientation of

Table 23.9 Proteins and peptides

Functional Groups	Region		Intensity		Comments
	cm⁻¹	μm	IR	Raman	
Polyglycines, $NH_2(CH_2-CO-NH)_n-CH_2COOH$	~3300	~3.03	m–s	m–w	NH str intensity increases with molecular weight
	~3080	~3.25	m–s	m–w	NH str intensity increases with molecular weight
	~2925	~3.42	m–s	m	asym CH_2 str
	~2860	~3.50	m–s	m	sym CH_2 str
	~1680	~5.95	v	m	CO_2^- str (not always present)
	~1650	~6.06	s	m–s	C=O str
	~1630	~6.14	v	w–m	NH_3^+ def vib (not always present)
	~1575	~6.35	v	w–m	NH_3^+ def vib(not always present)
	~1515	~6.60	m–s	w–m	NH def vib
	~1400	~6.25	v	w–m	CO_2^+ str (not always present)
	~1015	~9.85	v	w–m	May be strong but often absent
	~700	~14.29	s	w	CH_2 rocking vib, NH def vib
Polypeptides α = folded chain β = extended chain	~3460	~2.89	m	w–m	Free NH str, may be doublet
	3330–3280	3.00–3.05	m	w–m	NH str, intramolecular hydrogen bonded
	1700–1680	5.88–5.95	s	s	Free C=O str, amide I band
	1660–1650	6.02–6.06	s	s	α form, C=O str, amide I band
	~1630	~6.14	s	s	β form, C=O str, amide I band
	1550–1540	6.45–6.49	s	w	α form, NH def vib, amide II band
	1525–1520	6.56–6.58	s	w	β form, NH def vib, amide II band. Amide II band of free form is above 1520 cm⁻¹
	1300–1270	7.69–7.87	w–m	w	Amide III α-helix
	1230–1235	8.13–8.10	w–m	s	Amide III β-helix

particular groups. Deuteration is also a useful tool in the study of proteins and polypeptides – for example, to determine interactions/overlap between group frequencies.

The phrase 'not always seen' in Table 23.9 indicates bands only observed in low molecular species (i.e. three to four amino acids). In larger molecules, these absorptions may either appear as shoulders on neighbouring bands or completely disappear if the Zwitter ion does not exist (or dimerisation of carboxyl groups takes place).

As mentioned above, the use of Raman spectroscopy has a distinct advantage in that aqueous solutions may easily be examined.

Respiratory proteins involve either iron or copper atoms and either transport oxygen or are involved in its conversion to water and energy. Haemoglobin[28,31] and myoglobin[29,30] and their complexes with various ligands, including oxygen and carbon monoxide, have been studied extensively. Free carbon monoxide gas absorbs near 2145 cm⁻¹ (~4.66 μm), whereas in haemoglobin, the absorption may be in the range 2000–1900 cm⁻¹ (5.00–5.26 μm). For ordinary copper complexes, the CO absorption is found at 2090–2030 cm⁻¹ (4.78–4.93 μm).

Lipids

Lipids are insoluble organic substances found in biological tissues[10,11,16] – they are fats found in biological membranes[2,7,22] Lipid molecules consist of polar heads and hydrophobic tails which usually consist of a very long hydrocarbon chain. For membrane lipids, the infrared spectrum may be split into regions which originate from the molecular vibrations of different parts of the lipid molecule.[33] These origins are the hydrocarbon tail, the interface region and the head group. The approximate positions of the main important lipid bands are given in Table 23.10. Bands originating from the acyl chain, for example, those due to CH_3 and CH_2 asymmetric and symmetric stretching

Table 23.10 Lipids

Functional Groups	Region		Intensity		Comments
	cm^{-1}	μm	IR	Raman	
Lipids	3030–3020	3.30–3.31	s–m	w–m	CH$_3$ asym str, (CH$_3$)$_3$N$^+$
	~3010	~3.32	s–m	m	=C–H str
	~2955	~3.38	s–m	m	CH$_3$ asym str
	~2930	~3.41	s–m	m	CH$_2$ asym str
	~2880	~3.47	s–m	m	CH$_3$ sym str
	~2850	~3.51	s–m	m	CH$_2$ sym str
	~1730	~5.78	s	w–m	C=O str
	1490–1470	6.71–6.80	m–s	m–w	CH$_3$ asym bending, (CH$_3$)$_3$N$^+$
	~1475	~6.78	m	m–w	CH$_2$ scissoring vib
	~1470	~6.80	m	m–w	Two bands CH$_2$ scissoring vib
	~1460	~6.85	m	m–w	CH$_2$ scissoring vib
	~1460	~6.85	m	m–w	CH$_3$ asym bending vib
	1405–1395	7.12–7.17	m–s	m	CH$_3$ sym bending vib, (CH$_3$)$_3$N$^+$
	~1380	~7.25	m–s	m–w	CH$_3$ sym bending vib
	1400–1200	7.14–8.33	m–w	m–w	CH$_2$ wagging vib
	~1230	~8.13	s	s	PO$_2^-$ asym str
	~1170	~8.55	s	w	CO–O–C asym str
	~1085	~9.22	s	s	PO$_2^-$ sym str
	~1070	~9.35	s	w–m	CO–O–C sym str
	~1045	~9.57	s	m–w	C–O–P str
	~970	~10.31	w	s–m	CN asym str, (CH$_3$)$_3$N$^+$
	~820	~12.20	s	w–m	P–O asym str
	~730	~13.71	m–w	m	CH$_2$ rocking vib
	~720	~13.89	m–w	m	CH$_2$ rocking vib
	~715	~13.99	m–w	m	CH$_2$ rocking vib
Other useful bands for lipids					
C–CD$_3$	~2210	~4.52	s–m	m	CD$_3$ asym str
	~2170	~4.61	s–m	m	CD$_3$ sym str
N–(CD$_3$)$_3^+$	980–950	10.20–10.53	s–m	m	CD$_3$ sym def vib
＼	2200–2190	4.55–4.57	s–m	m	CD$_2$ asym str
⟩(CD$_2$)$_n$					
／					
	2095–2085	4.77–4.80	s–m	m	CD$_2$ sym str
	~1095	~9.13	s	w	CD$_2$ def vib
	~1090	~9.17	s	w	CD$_2$ def vib
	~1085	~9.22	s	w	CD$_2$ def vib
	~515	~19.42	m–w	m	CD$_2$ rocking vib, separation of this band and one below depends on packing
	~520	~19.23	m–w	m	CD$_2$ rocking vib, separation of this band and one above depends on packing
ROH (Free)	3650–3590	2.74–2.79	m–s	m–w	O–H str, non-hydrogen-bonded
ROH	3400–3200	2.94–3.13	m	m–w	O–H str, br, hydrogen-bonded
	1400–1200	7.14–8.33	m	m	OH def vib
–CH$_2$OH	~1050	~9.52	s	m	C–O str
–CHROH	~1100	~9.09	s	m	C–O str
–CR$_2$OH	~1150	~8.70	s	m	C–O str

Table 23.10 (*continued*)

Functional Groups	Region cm^{-1}	Region μm	Intensity IR	Intensity Raman	Comments
$-CH_2-COOR$	1750–1720	5.71–5.81	s	m–w	C=O str. (^{13}C=O str absorbs at 1725–1700 cm^{-1})
	1425–1410	7.02–7.09	m	m	CH$_2$ def vib, band affected by conformational changes
RCOOH (Free)	3560–3500	2.81–2.86	m–s	m–w	O–H str, non-hydrogen-bonded
ROOH	2700–2500	3.70–4.00	m–s	m–w	O–H str, hydrogen-bonded
	1320–1210	7.58–8.26	m	m	OH bending vib
RCOO$^-$	1610–1550	6.21–6.45	s	m–w	COO$^-$ asym str
	1420–1300	7.04–7.69	s	m–w	COO$^-$ sym str
Cyclohexyl group	~1445	~6.92	m	m	CH$_2$ def vib
Cis–C=C–	~3010	~3.32	m	m	CH asym str
	1680–1600	5.95–6.25	m–w	s	C=C str
RNH$_2$	3500–3000	2.86–3.33	w–m	w	NH$_2$ asym and sym strs, position, intensity and shape affected by hydrogen bonding
	1650–1590	6.06–6.29	m–s	w–m	NH$_2$ bending vib
	1220–1020	8.20–9.80	w	w–m	C–N str
R$_2$NH	3500–3300	2.86–3.03	w–m	w	NH str
	1650–1520	6.06–6.58	m–s	w–m	N–H bending vib
RNH$_3$$^+$	~3200	~3.13	w–m	w	NH$_3$$^+$ asym str
	~3020	~3.31	w–m	w	NH$_3$$^+$ sym str
	1620–1570	6.17–6.37	m–s	w	NH$_3$$^+$ asym bending vib
	~1520	~6.58	m–s	w	NH$_3$$^+$ sym bending vib
$-N(CH_3)_3$$^+$	3030–3020	3.30–3.31	w–m	w	NCH$_3$ str
	1490–1470	6.71–6.80	m–s	m–w	NCH$_3$ asym def vib, affected by symmetry of group
	1405–1395	7.12–7.17	w	w	NCH$_3$ sym def vib
	970–950	10.31–10.53	m	w	NCH$_3$ asym def vib
$-NH(CH_3)_2$$^+$	1510–1470	6.62–6.80	m	w	NCH$_3$ asym def vib
RCONHR	~3300	~3.03	m–w	w	NH str
	~3100	~3.23	m–w	w	NH str
	~1650	~6.06	s	w	Amide I band, affected by hydrogen bonding
	~1550	~6.45	s	w	Amide II band
ROP(OR)O$_2$$^-$	1260–1200	7.94–8.33	s–m	s	PO$_2$$^-$ asym str
	1110–1085	9.01–9.22	s–m	s	PO$_2$$^-$ sym str

vibrations, CH$_2$ bending and rocking vibrations, the headgroup, that is the PO$_2$$^-$ stretching vibration, and the interface region, that due to the C=O stretching vibration, may be used to obtain information relating to the conformation of the lipid. The deconvolution of the ester group C=O band is particularly useful – absorptions may be assigned to hydrogen-bonded and non-hydrogen-bonded C=O groups.

Lipid–water gels undergo phase transitions as the ratio of lipid to water is changed or as the temperature is altered.[1,2,11,24] The phase change is endothermic. Below the phase transition temperature, chains are mainly in a *trans* configuration. Above the transition temperature, a significant number of *gauche* conformers are present, this being considered as a melting of the hydrocarbon chains. The transport properties across a membrane are dependent on the phase of the lipid. The phase transitions of lipids may easily be studied by infrared or Raman spectroscopy. Characteristic changes in the spectrum are monitored with temperature of the sample or concentration. Changes to the CH and C–C stretching vibration bands are observed as the *trans–gauche* in the hydrocarbon chains is altered. For example, as the structural changes occur, the intensities of bands near 2930 cm^{-1} (~3.41 μm) and 2880 cm^{-1} (~3.47 μm) are observed to alter.

As mentioned, hydrated lipids can exist in one or more polymorphous forms and, depending on the environmental conditions they experience, they can undergo transformations between forms, i.e. undergo phase transitions.

In theory, phase transitions involving the melting of the lipid hydrocarbon chains can be followed using any infrared absorptions of the CH_2 group. However, the most commonly used are those bands due to the asymmetric and symmetric stretching vibrations, these occurring near $2930\,cm^{-1}$ ($\sim3.41\,\mu m$) and $2850\,cm^{-1}$ ($\sim3.51\,\mu m$) respectively.

All hydrocarbon chain-melting phase transitions are accompanied by discontinuous changes in both wavenumber of the bands (i.e. the positions of the maximum of the absorptions) and the band-widths involved. The ratio of the intensities of these and other bands may be followed with temperature in order to determine the transition point, and other parameters affecting phase transitions may also be used in a similar manner. During hydrocarbon chain-melting, the absorption maxima and bandwidths increase, indicating greater hydrocarbon chain disorder and the start of the change to the *gauche* form. In the *gauche* form, the band near $2850\,cm^{-1}$ ($\sim3.51\,\mu m$) is weakened due to vibrational decoupling.

The phase change is accompanied by a shift in the position of the maximum absorption of the symmetric stretching vibration of 1.5 to $2.5\,cm^{-1}$, the magnitude of the change being dependent on the chemical structure of the lipid, the length of the hydrocarbon chain, the nature of the polar group and the nature of the phase transition. Unfortunately, the band due to the asymmetric stretching vibration may be overlapped by contributions due to methyl groups and, depending on the nature of the lipid phase, can be affected by a Fermi resonance interaction with the first overtone of the CH_2 scissor vibration. Hence, if the methylene groups of the lipid hydrocarbon chain form the vast majority of the methylene groups in the lipid, then it is preferable to use the band due to the symmetric stretching vibration, which is relatively free of interactions, to follow phase transitions.

The scissoring and rocking vibrations near $1460\,cm^{-1}$ ($\sim6.85\,\mu m$) and $725\,cm^{-1}$ ($\sim13.79\,\mu m$) can also be used to monitor chain-melting phase changes. These bands are sharp when the lipid is in the *trans* configuration and become broad as melting proceeds, the overall intensities[25,26] of these bands also decreasing with melting. The contours of these bands are sensitive to lateral packing interactions.

The infrared absorptions of hydrated lipids in a crystalline or semi-crystalline phase consist of sharp bands. Obviously, for crystalline or semi-crystalline lipids, close interactions between molecules are important and affect the spectra observed. The infrared spectra of crystalline or semi-crystalline lipids are sensitive to structural changes, the CH_2 and $C{=}O$ bands being affected.[27] These changes may not always be huge but they are distinct and easily observed. Spectroscopic changes for transformations from lamellar to non-lamellar forms also occur, but they are sometimes not easily observed as they are overlapped by other stronger features. The CH_2 wagging vibration bands in the region $1400{-}1300\,cm^{-1}$ ($7.14{-}7.69\,\mu m$) may be used to provide estimates of the concentrations of the various non-planar concentrations.

In aqueous media, strong absorptions due to water can prevent the observation of bands, hence the use of deuterium oxide, D_2O, can be helpful. The shift in the solvent absorptions allows the observation of overlapped sample bands. For example, the strong band near $1645\,cm^{-1}$ ($\sim6.08\,\mu m$) is moved to $1215\,cm^{-1}$ ($\sim8.23\,\mu m$). The absorptions of D_2O have lower wavenumbers than those of H_2O and may now overlap other sample bands. The use of D_2O will also result in the loss of absorptions due to H/D exchange of groups. This could, for example, result in the loss of the amide II band near $1550\,cm^{-1}$ ($\sim6.45\,\mu m$) and a new band appearing near $1465\,cm^{-1}$ ($\sim6.78\,\mu m$).

As mentioned earlier, both transmission and attenuated total reflectance, ATR, techniques may be used to examine lipids. It should be borne in mind that lipids may exhibit dichroic behaviour.[23,24]

Bacteria

The spectral examination of bacteria grown on a culture medium may be accomplished as outlined below. The bacteria are harvested using a spatula and then dispersed in water which is then transferred to an infrared-transparent plate, for example, one of ZnSe, and allowed to dry. A thin transparent film of bacteria is left on the plate.

The vibrational spectra of bacteria exhibit characteristic bands in the infrared.[12,18] Absorptions are observed near $3300\,cm^{-1}$ ($\sim3.03\,\mu m$) due to the NH stretching vibration, $1650\,cm^{-1}$ ($\sim6.06\,\mu m$) and $1550\,cm^{-1}$ ($\sim6.45\,\mu m$), these last two being due to the amide I and the amide II bands of secondary polyamides, that is, due to the protein portion. In addition, a broad band with two or three peaks is observed in the region $1150{-}1050\,cm^{-1}$ ($8.70{-}9.52\,\mu m$) due to the polysaccharide component and a band near $1260{-}1220\,cm^{-1}$ ($7.94{-}8.20\,\mu m$) may be assigned to the presence of the PO_2^- group.

It is possible to categorise bacteria by FT-IR by making use of derivative spectra[12] of the original absorption spectrum. The first to fourth derivatives are normally used. By making use of the derivatives, it is possible to resolve subtle differences hidden by overlapping bands. A weak, sharp spectral feature appears more prominent, the higher the derivative examined – large numbers of peaks are observed for the higher derivative spectra. Since even a weak spectral feature overlapping the spectra of bacteria can become more conspicuous, it is important to eliminate any contributions by carbon dioxide and water to a spectrum. Computer software programs are available for these functions.

Table 23.11 contains the infrared spectral bands of common groups to be found in many bacteria. It should be borne in mind that, in general, the infrared spectra of bacteria consist of broad overlapping bands.

Table 23.11 Bands of common functional groups found in the spectra of bacteria

Functional Groups	Region		Intensity		Comments
	cm^{-1}	μm	IR	Raman	
CH$_2$	~2920	~3.42	m–s	m	asym CH$_2$ str
	~2850	~3.51	m–s	m	sym CH$_2$ str
	~1465	~6.83	m–s	m	CH$_2$ scissoring vib, also weak shoulder at ~1480 cm^{-1}
	~1380	~7.25	m–s	m	CH$_2$ sym bending vib
	~720	~13.89	m–w	w	CH$_2$ rocking vib
⟩C—O (ester)	~1730	~5.78	s	m–w	C=O str, usually 2 or 3 components
	~1170	~8.55	s	m–w	C–O str, shoulder ~1180 cm^{-1}
C=C–H	~3015	~3.32	w	m–w	C=C–H str
α,β-Unsaturated ester	~1650	~6.06	m–s	s	C=C str
⟩C—O (ester)	~1650	~6.06	s	m–w	Amide I band, usually 2 or 3 components
	~1550	~6.45	s	m–w	Amide II band, usually 2 or 3 components
Carboxylic acid, –COOH	~1730	~5.78	s	m–w	C=O str
Carboxylate group, COO$^-$	1630–1605	6.13–6.23	s	w	COO$^-$ asym str
	~1410	~7.09	s	m–s	COO$^-$ sym str
PO$_2^-$	1260–1220	7.94–8.20	m–s	s	asym PO$_2^-$ str
	~1110	~9.01	m–s	s	sym PO$_2^-$ str
Saccharide components	1155–1130	8.66–8.85	m	m	Ring str vib
	~1150	~8.70	s	m–w	C–O and C–C str, number of components
	1120–1020	8.93–9.80	s	m–w	C–O–C str, number of components
	1115–1005	8.97–9.95	m	m	Ring str vib
	1080–1060	9.26–9.43	m	m	Ring str vib
	1080–1000	9.26–10.00	s	m–w	C–OH str, number of components
	1055–1015	9.48–9.85	m	m	Ring str vib
Sulpholipids, C–O–S	~1240	~8.06	s	m–w	C–O–S str
	830–820	12.05–12.20	m	m–w	C–O–S str
Acetyl groups, C–O–CH$_3$	~1235	~8.10	s	m–w	C–O–C str

Food, Cells and Tissues

Infrared and Raman instrumental advances, microspectroscopic techniques and fibre optics and new sampling methods have made possible many biological and medical applications. Correction for background and interference is automatically performed by most modern instruments. The use of statistical techniques and of derivative spectra for the examination of subtle differences in cases where bands overlap have been very useful. The direct examination of cells and tissues by infrared[8–11] can provide useful information on cellular composition, packing of cellular components, cell structure, metabolic processes and disease.[14,15] Near infrared and Fourier Transform techniques may be applied to the study of food.[35]

Proteins are the most abundant species in cells[21] and tissues and hence their absorptions dominate the spectra of cells and tissues (see the section above on proteins). The spectra of proteins vary with the second order structures of proteins, that is, spiral form (α-helix), an extended chain (β-form) and a random coil arrangement, the protein's state of hydration and the ionic strength of the solvent. The spectra of metabolic and structural proteins found in cells have similar features. The only proteins to exhibit distinctly different features are found in connective tissue, such as collagen.

The spectra of proteins found in cells have a strong amide I band near 1650 cm^{-1} (~6.06 μm). This band is affected by the environment of the peptide linkage and the protein's secondary amine. The amide II band occurs near 1530 cm^{-1} (~6.54 μm) and the amide III band occurs near 1245 cm^{-1}

(~8.03 μm). Other bands are found near 1450 cm^{-1} (~12.45 μm), 1390 cm^{-1} (~7.19 μm) and 1310 cm^{-1} (~7.63 μm).

The infrared spectra of DNA and RNA also depend on the state of hydration of the nucleic acid and its secondary structure. Both DNA and RNA have bands due to the C=O and aromatic CC stretching vibrations in the region 1700–1580 cm^{-1} (5.88–6.33 μm). The ionised PO$_2^-$ and the ribose groups exhibit bands of medium intensity near 1095 cm^{-1} (~9.13 μm), 1085 cm^{-1} (~9.22 μm) and 1070 cm^{-1} (~9.35 μm). For RNA, the band near 1085 cm^{-1} (~9.22 μm) is stronger than the other two bands. For DNA, these three bands are almost of equal intensity. In addition, DNA has bands near 1245 cm^{-1} (~8.03 μm) and 965 cm^{-1} (~10.58 μm) due to the phosphodiester group. The spectra of DNA and RNA are easy to distinguish. The interpretation of the spectral features of DNA and RNA may be complicated by the dehydrating conditions experienced which significantly alter the spectra. For example, the spectrum of DNA precipitated from alcohol resembles that of RNA (DNA undergoes a phase change). Also it should be borne in mind that the spectrum of DNA in cells is not composed of a simple addition of the spectra of water and protein. This is to be expected since the DNA in a cell adopts a more complicated tertiary structure, since the DNA is in solution. For some cells, for example those not actively involved in division or not involved in certain immune function aspects, their spectra do not show any features due to DNA, or they exhibit only very small absorptions

The molecules forming the cell membrane bilayer are phospholipids (see section above dealing with lipids). The most pronounced feature of the infrared spectra of phospholipids is the band due the C=O stretching vibration of the ester group (part of the fatty acid or triglyceride or other polar head-group) near 1735 cm^{-1} (~5.76 μm). In the spectra of certain cells and tissues, some phospholipids exhibit a shoulder near 1740 cm^{-1} (~5.75 μm). An idea of the conformation and fluidity of phospholipids may be gained by examining the ratio of the intensities of the bands near 3000 cm^{-1} (~3.33 μm) and 2900 cm^{-1} (~3.45 μm).

Phosphorylated proteins, which may be present in cells, exhibit a strong band in their infrared spectra near 950 cm^{-1} (~10.35 μm).

The infrared spectrum of hydrated glycogen has strong absorptions due to the stretching vibrations of C–O and C–C and due to the C–O–H deformation vibrations near 1150 cm^{-1} (~8.70 μm), 1080 cm^{-1} (~9.26 μm) and 1030 cm^{-1} (~9.71 μm) respectively.

Differences have been observed in the infrared spectra of normal and abnormal tissues for cervical epithelium,[17,20] colon, prostate gland, etc. In the region 1100–950 cm^{-1} (9.09–10.53 μm), a loss of structure and a slight increase in intensity has been observed between normal and cancerous cells.

Infrared spectroscopy can be used to estimate the sucrose content of sugar cane juice,[13,19] the region 1250–800 cm^{-1} (8.00–12.50 μm) being monitored.

Absorptions near 1140 cm^{-1} (~8.77 μm), 1115 cm^{-1} (~8.97 μm), 1055 cm^{-1} (~9.48 μm), 995 cm^{-1} (~10.05 μm) and 930 cm^{-1} (~10.75 μm) are observed.

In a recent review the characterisation of wood pulp by Raman spectroscopy has been given.[34]

References

1. R. J. H. Clark and R. E. Hester, (eds), *Advances in Infrared and Raman Spectroscopy*, Wiley, New York, 1984.
2. R. Mendlelsohn *et al.*, *Biochemistry*, 1981, **20**, 6699.
3. W. K. Sutewiz *et al.*, *Biochemistry*, 1993, **32(2)**, 389.
4. R. J. Koegel *et al.*, *Ann. NY Acad. Sci.*, 1957, **69**, 94.
5. G. B. B. M. Sulherland, *Adv. Protein Chem.*, 1952, **7**, 291.
6. R. J. Koegel *et al.*, *J. Am. Chem. Soc.*, 1955, **77**, 5708.
7. M. Tsuboi *et al.*, *Spectrosc. Acta*, 1963, **19**, 271.
8. E. Steger *et al.*, *Spectrosc. Acta*, 1963, **19**, 293.
9. W. B. Fischer and H. H. Eysel, *Mol. Struct.*, 1997, **415(3)**, 249.
10. J. L. R. Armondo and F. M. Goni, *NATO ASI Ser. Ser. E*, 1997, **342**, 249.
11. J. Cast, in Dev. Oils, Fats, R. J. Hamilton (ed), *Blackie*, 1985, p. 224.
12. D. Naumann, *Mikrochim. Acta*, 1988, **1**, 373.
13. P. B. Lipescheski, *Semin. Food Anal.*, 1996, **1(2)**, 85.
14. H. Fabian *et al.*, *Biospectrosc.*, 1995, **1**, 37.
15. P. Lasch and D. Naumann, *Cell Mol. Biol.*, 1998, **44**, 189.
16. C. P. Schultz *et al.*, *Cell Mol. Biol.*, 1998, **44**, 203.
17. L. Chiriboga *et al.*, *Biospectrosc.*, 1998, **3**, 47.
18. D. Helm *et al.*, *J. Gen. Microbiol.*, 1991, **137**, 69.
19. F. Cadet *et al.*, *Appl. Spectrosc.*, 1991, **45**, 166.
20. L. Chiriboga *et al.*, *Cell Mol. Biol.*, 1998, **44**, 219.
21. E. Benedetti *et al.*, *Appl. Spectrosc.*, 1997, **51**, 792.
22. D. R. Scheuing (ed.), *Fourier Transform Infrared Spectroscopy in Colloid and Interface Sci.*, ACS Symp. Ser. 447, American Chemical Society, Washington, DC, 1991.
23. A. Holmgren *et al.*, *J. Phys. Chem.*, 1991, **91**, 5298.
24. A. Nilsson *et al.*, *Chem. Phys. Lipids*, 1994, **71**, 119.
25. D. J. Moore *et al.*, *Biochemistry*, 1993, **32**, 6281.
26. N. C. Chia *et al.*, *J. Am. Chem. Soc.*, 1993, **115**, 10250.
27. R. N. A. H. Lewis and R. N. McElhaney, *Biophys. J.*, 1992, **61**, 63.
28. A. Dong and W. S. Caughey, *Methods Enzymol.*, 1994, **232**, 139.
29. J. A. Larrabee and S. Choi, *Methods Enzymol.*, 1993, **226**, 289.
30. K. Nakamoto and R. S. Czernuszewicz, *Methods Enzymol.*, 1993, **226**, 259.
31. M. F. Perutz *et al.*, *Acc. Chem. Res.*, 1987, **20**, 310.
32. E. G. Brame, Jr (ed.), *Applied Spectroscopy Reviews*, Marcel Dekker, New York, 1969.
33. A. A. Ismail *et al.*, in *Spectral Properties of Lipids*, R. J. Hamilton and J. Cast, eds, Sheffield Academic Press, 1999, p. 235.
34. U. P. Agarwal, *Advances in Lignoallul Characterisation* D. S. Argyropoulos, ed., 1999, p. 201.
35. A. Goanker (ed.), *Characterisation of Food*, 1995, Elsevier.
36. B. Giese and D. McNaughton, *Spectrosc. Bio. Mol. New Dir., Eur. Conf. 8th*, 1999, 255.
37. S. Krimm, *ACS Symp. Ser.*, 2000, **750**, 38.

Appendix Further Reading

1. A. P. Arzamaster and D. S. Yashkina, *UV and IR Spectra of Drugs*, No. 1, Steroids. Meditsina, USSR, 1975

2. G. A. Atkinson, *Time Resolved Vibrational Spectroscopy*, Academic Press, New York, 1983.

3. L. J. Bellamy, *The Infrared Spectra of Complex Molecules*, 3rd edition, Chapman Hall, London, 1975.

4. L. J. Bellamy, *Advances in Infrared Group Frequencies*, 2nd edition, Chapman Hall, London, 1980.

5. J. Bellanato and A. Hidalgo, *Infrared Analysis of Essential Oils*, Sadtler Research Laboratories, 1971.

6. F. F. Bentley, D. L. Smithson, and A. L. Rozek, *Infrared Spectra and Characteristic Frequencies* ~ 700–$300\,cm^{-1}$, Interscience, New York, 1968.

7. D. I. Bower and W. F. Maddams, *The Vibrational Spectroscopy of Polymers*, Cambridge Univ. Press, Cambridge, 1992.

8. D. A. Burnes and E. W. Ciurczak, *Handbook of Near Infrared Analysis*, in *Pract. Spectrosc.*, 1992, p. 13.

9. R. J. H. Clark and R. E. Hester (eds), *Advances in Infrared and Raman Spectroscopy*, Wiley, New York, 1984.

10. Chicago Soc. For Paint Tech., *An Infrared Spectroscopy Atlas for Coatings Industry*, Fed. of Soc. Paint Tech., 1315 Walnut St., Philadelphia, PA, 1980.

11. R. J. H. Clark and R. E. Hester (eds), *Biomolecular Spectroscopy*, Vol. 20, Part A, Wiley, New York, 1993.

12. N. B. Colthurp, L. H. Daly, and S. E. Wiberley, *Introduction to Infrared and Raman Spectroscopy*, Academic Press, Boston, 1990.

13. R. A. Cromcobe, M. L. Olson and S. L. Hill, *Computerised Quantitative Infrared Analysis*, ASTM STP 934, G. L. McGlure (ed.), American Soc. For Testing Materials, Philadelphia, PA, 1987, pp 95–130.

14. F. R. Dollish, W. G. Fateley, and F. F. Bentley, *Characreristic Raman Frequencies of Organic Compounds*, Wiley, New York, 1974.

15. J. R. Durig (ed.), *Spectra and Structure*, Elsevier, Amsterdam 1982 to date.

16. V. C. Farmer (ed.), *The Infrared Spectra of Minerals*, Mineral Society, London, 1974.

17. V. C. Farmer, *Infrared of Anhydrous Mineral Oxides*, Infrared Spectra Miner., 1974, p. 183.

18. J. R. Ferraro and L. J. Basile (eds), *FT-IR Spectroscopy*, Academic Press, Boston, 1985.

19. J. R. Ferraro and K. Krishnan (eds), *Practical FTIR, Industrial and Laboratory Chemical Analysis*, Academic Press, Boston, 1990.

20. W. O. George, H. A. Willis (eds), *Computer Methods in UV–Visible and Infrared Spectroscopy*, Royal Society, 1990.

21. P. R. Griffiths and J. A. de Haseth, *FT-IR Spectroscopy*, Wiley, New York, 1986.

22. J. Haslam *et al.*, *Identification and Analysis of Plastics*, Iliffe, London, 1972.

23. J. R. Heath and Saykally, *The Structures and Vibrational Dynamics of Small Carbon Clusters*, P. J. Reynolds (ed.), North-Holland, Amsterdam, 1993, pp 7–21.

24. C. J. Heinniker, *Infrared Analysis of Industrial Polymers*, Academic Press, New York, 1967.

25. R. E. Hester and R. J. H. Clark (eds), *Advances in Infrared and Raman Spectroscopy*, Heyden, London, 1981.

26. R. E. Hester, Infrared spectra of molten salts, *Adv. Molten Salt Chem.*, 1971, **1**, 1.

27. D. O. Hummel, *Infrared Analysis of Polymers, Resins and Additives: An Atlas*, Wiley, New York, 1972.

28. D. E. Irish, Infrared spectra of fused salts, *Ionic Interactions*, 1971, **2**, 187.

29. M. Iwamoto and S. Kawano (eds), *The Proceedings of the Second International Near I. R. Conference*, Tsukuba, Japan, Korin Pub., Tokyo, 1990.

30. C. Karr (ed.), *Infrared and Raman Spectroscopy of Lunar and Terrestial Minerals*, Academic Press, New York, 1975.

31. J. E. Katon and A. J. Sommer, Infrared microspectroscopy, *Anal. Chem.*, 1992, **64**(19) 931.

32. S. Kawano, in *Characterisation of Food*, A. G. Goankar (ed.), Elsevier, Amsterdam, 1995.

33. K. P. Kirkbride, The application of infrared microspectroscopy to the analysis of single fibres in, *Forensic Exam. Fibres*, Vol. 181, J. Robertson (ed.), Harvard, 1992.

34. O. Kirret and L. Lahe, *Atlas of Infrared Spectra of Synthetic and Natural Fibres*, Valgus, Talliun, 1988.

35. J. L. Koenig, Spectra of polymers, *Amer. Chem. Soc., Dept.* 31, 1155 Sixteenth St., N.W. Washington, D.C. 20036.

36. V. A. Koptyug (ed.), *Atlas of Spectra of Organic Compounds in the Infrared, UV and Visible Regions*, Nos 1–32, Novosibirsk, 1987.

37. H. W. Kroto *et al.*, *The Fullerenes*, Pergamon, Oxford, 1993.

38. D. Lin-Vien, N. B. Colthurp, W. G. Fateley, and J. G. Grasselli, *The Handbook of IR and Raman Characteristic Frequencies of Organic Molecules*, Academic Press, Boston, 1991.

39. J. C. Merlin, Infrared and Raman spectroscopy: novel techniques for studying the molecular structure of flavonoids, *Bull. Liaison – Group Polyphenols*, 1990, **15**, 219.

40. R. G. Messerschmidt and M. A. Harthcock (eds), *Infrared Microscopy*, Marcel Dekker, New York, 1988.

41. I. Murray and I. A. Cowe (eds), *Advances in Near Infrared Spectroscopy*, International Conference, VCH, Germany, 1992.

42. K. Nakamoto, *Infrared and Raman Spectra of Inorganic and Coordination Compounds*, 5th edition, Wiley, New York, 1997.

43. K. N. Nakanishi, P. H. Solomon, and N. Furutachi (eds), *Characterisation Exercises in Infrared Absorption Spectroscopy*, 22nd edition, Naukodo, Tokyo, 1987.

44. N. Neuroth, Infrared spectroscopy of glass, *Fachausschussber. Dtsch. Glasstech. Ges.*, 1974, **70**, 141.

45. T. Ogawa, *Handbook for Polymer Analysis*, Japan Soc. For Anal. Chem., Asakura Shoten, Tokyo, 1985.

46. B. G. Osborne and T. Fearn, *Food Analysis*, Longman, 1986.

47. P. C. Painter *et al.*, *The Theory of Vibrational Spectroscopy and its Application to Polymeric Materials*, Wiley, 1982.

48. F. S. Parker, *Application of Infrared, Raman and Resonance Raman Spectroscopy in Biochemistry*, Plenum, New York, 1983.

49. W. B. Pearson and G. Zerbi (eds), *Vibrational Intensities in Infrared and Raman Spectroscopy*, Elsevier, Amsterdam, 1982.

50. O. R. Sammul *et al.*, Pharmaceutical compounds, in *Infrared and Ultraviolet Spectra of Some Compounds of Pharmaceutical Interest*, revised edition, Association of Official Analytical Chemists, Washington, DC, 1972, pp. 102–175

51. D. R. Scheuing (ed.), *Fourier Transform Infrared Spectroscopy in Colloid and Interface Sci., ACS Symp. Ser. 447*, American Chem. Soc., Washington, DC, 1991.

52. H. W. Siesler and K. Holland-Moritz, *Infrared and Raman Spectroscopy of Polymers*, Marcel Dekker, New York, 1980.

53. N. A. Shimanko and M. V. Shishkina, *Infrared and UV Absorption Spectra of Aromatic Esters*, Nauka, Moscow, 1987.

54. R. M. Silverstein, *Spectrometric Identification of Organic Compounds*, Wiley, New York, 1991.

55. R. M. Silverstein and R. Milton (eds), *Spectrometric Identification of Organic Compounds*, Wiley, New York, 1981.

56. C. G. Smith *et al.*, Infrared analysis of polymers, *J. Anal. Chem.*, 1991, **63**, 11R.

57. T. A. Skotheim *et al.*, *Handbook of Conducting Polymers*, Marcel Dekker, New York, 1998.

58. H. A. Szymanski, *Interpreted Infrared Spectra*, Plenum, New York, 1971.

59. H. A. Szymanski and R. E. Erickson, *Infrared Band Hand Book*, Vols I and II, Plenum, New York, 1970.

60. H. A. Szymanski, *Infrared Band Hand Book*, Vols I–III Plenum, New York, 1964, 1966, 1967, and *Correlation of Infrared and Raman Spectra of Organic Compounds*, Hertillon, 1969.

61. R. S. Tipson, *Infrared Spectroscopy of Carbohydrates, U. S. Dept. Commerce Nat. Bur. Stand. Monogr.*, **110**, 1968.

62. V. N. Vatulev, S. V. Lapti, and Y. Y. Kercha, *Infrared Spectra and Structure of Polyurethanes*, Naukova Dunka, Kiev, 1987.

63. P. Williams and K. Norris (eds), *Near Infrared Technology in the Agricultural and Food Industries*, American Assoc. of Cereal Chemists, 1987.

64. J. T. Yates and T. E. Madely, *Vibrational Spectroscopy of Molecules on Surfaces*, Plenum, New York, 1987.

65. R. Zbinden, *Infrared Spectra of Polymers*, Academic Press, New York, 1964.

66. G. Zerbi, in *Conjugated Polymers*, J. L. Bredas and R. Silbey (eds), Kluwer, Amsterdam, 1991, p. 435.

67. J. E. D. Davies, *Vibrational Studies of Host–Guest Compounds (Inclusion Compounds)*, ARI, 1998, **51(2)**, 120.

68. M-I. Baraton, Surface spectroscopy of nanosized particles, *Handbook Nanostruct. Matter. Nanotechnol.*, 2000, **2**, 89.

Index

Page numbers in italics refer to detailed treatment of the item listed.